METHODS IN MOLECULAR BIOLOGY™

Lipidomics

Volume 2: Methods and Protocols

Edited by

Donald Armstrong

University at Buffalo, Buffalo, NY, USA
University of Florida, Gainesville, FL, USA

Humana Press

Editor
Donald Armstrong
University of Buffalo
Buffalo, NY, USA
and
University of Florida
Gainesville, FL, USA

The cover image are overlapping snapshots of a single peroxidized lipid, taken from computer stimulations. Peroxidation mofidies the local conformational preferences of acyl chains and increases their mobility, with implications for structural and dynamic properties of the membrane.

ISSN 1064-3745 e-ISSN 1940-6029
ISBN 978-1-60761-324-4 e-ISBN 978-1-60761-325-1
DOI 10.1007/978-1-60761-325-1
Springer New York Dordrecht Heidelberg London

Library of Congress Control Number: 2009927725

Cover illustration: Snapshots of a single PLPC and a peroxidized analogue (13-tc), taken at 5 ns intervals. Molecules are oriented along the membrane and superimposition was done on the phosphorus and oxygen atoms. Images were obtained from Chapter 18, Vol. 1.

Cover design: Karen Schulz

Printed on acid-free paper

Humana Press is part of Springer Science+Business Media (www.springer.com)

Preface

Lipidomics is a sub-discipline of metabolomics and is defined as the large-scale study of non-water-soluble metabolites (lipids and lipidome) that utilize system-level analysis to characterize lipids and their interacting moieties (*1*). A literature search at the end of 2008 showed there were 200 articles published on lipidomics encompassing glycerophospholipids, sphingolipids, polyunsaturated fatty acids, glycolipids, sterol lipids, and proteolipids. It has been predicted that the combination of these lipid classes totals between 1,000 and 2,000 molecules. Lipids can also act as second messengers, or mitogens, and participate in profiling and signaling via specialized microdomains that have large amounts of lipids. When lipids are disturbed, their metabolites probably contribute to disease. In prostate cancer, for example, cyclooxygenase and lipooxygenase are upregulated reducing angiogenesis and tumor growth.

Early separation and identification of lipids started with TLC, and as technology advanced, it progressed to the use of GC and HPLC. Technical improvements to HPLC include reversed-phase methods, ESI, evaporative light scattering, electrochemical detection, APCI, suppressed conductivity and multi-dimensional electrophoresis. Other technologies coupled to chromatographic methods, such as MS/MS-MSn/MALDI/TOF, NMR, and MRM, provide a powerful approach to the global analysis of complex lipid mixtures, understanding structural changes through biophysical approaches and the effects of lipids on physiology, i.e. atherosclerosis. This has given us a clearer understanding of human and animal pathology, i.e. diabetes, cancer, neurodegeneration, and infectious disease. A new approach to measure oxidized lipids, referred to as "oxidative lipidomics," has recently been described which provides methodology for separation and identification of these highly reactive lipids, especially in mitochondria. Many novel techniques are described in these volumes, including an imaging lipidomics approach. For another lipidomic approach of a lipid-derived radical technique, the reader is referred to Iwabashi, H., 2008. *Advanced Protocols in Oxidative Stress I*, volume 477, Chapter 6, Humana Press. In that same volume, a lipidomics technique for sphingolipids is also described, i.e. Wilder, AJ and Cowart, LA, Chapter 28.

The present volumes have taken a "shotgun" approach and are divided into seven parts in order to include as many different varieties of technology as possible. Chapters by international experts present a wide variety of reviewed as well as unpublished data on isolation techniques, structural analysis, lipid rafts, lipid trafficking and profiling, biomarkers, lipid peroxidation, biostatistics applied to lipids, software tools, and bioinformatics. These studies range from simple systems, such as in yeast, to complex biological models.

The ever increasing utilization of lipidomics will lead to more powerful technology, improved diagnostic–prognostic capabilities for medical disorders, and for the identification of new classes of lipids.

I thank my son, Dennis Armstrong, and my grandson, David Armstrong of On-Staff Technology, Inc., for assistance with technical support, information technology, and multi-media services.

Buffalo, NY
Gainesville, FL

Donald Armstrong

Reference

1. Wenk, M.R. 2005. The emerging field of lipidomics. Nature Reviews 4: 595–610.

Contents

Contributors

ANDREW A. AMOSCATO • *Department of Pathology, University of Pittsburgh, Pittsburgh, PA, USA*
ELSA ARIAS-BARRAU • *Department of Biochemistry, University of Nebraska, Lincoln, NE, USA*
RUTH BIRNER-GRUENBERGER • *ZMF Center for Medical Research, Medical University of Graz, Graz, Austria*
PAUL N. BLACK • *Department of Biochemistry, University of Nebraska, Lincoln, NE, USA*
KRISHNA B. CHAITHANYA • *Department of Biochemistry, Institute of Biomedicine, University of Helsinki, Helsinki, Finland*
VINEET CHOUDHARY • *Department of Medicine, Division of Biochemistry, University of Fribourg, Fribourg, Switzerland*
CHRISTINE A. CURCIO • *Department of Ophthalmology, Callahan Eye Foundation Hospital, University of Alabama School of Medicine, Birmingham, AL, USA*
JOSEF CVAČKA • *Institute of Organic Chemistry and Biochemistry, Academy of Sciences of the Czech Republic, Prague, Czech Republic*
FABIO LUIZ D'ALEXANDRI • *Department of Medical Biochemistry and Biophysics, Division of Physiological Chemistry II, Karolinska Institutet, Stockholm, Sweden Department of Parasitology, Department of Biochemical Sciences, University of San Paulo, San Paulo, Brazil*
CONCETTA C. DIRUSSO • *Department of Biochemistry and Nutrition and Health Sciences, University of Nebraska, Lincoln, NE, USA*
MICHAEL W. EPPERLY • *Department of Radiation Oncology, University of Pittsburgh, Pittsburgh, PA, USA*
AUDREY J. GASKINS • *Division of Epidemiology, Statistics and Prevention Research, Eunice Kennedy Schriver National Institute of Child Health and Human Development, NIH, Rockville, MD, USA*
SUSUMU GOTO • *Bioinformatics Center, Institute for Chemical Research, Kyoto University, Kyoto, Japan*
JOEL S. GREENBERGER • *Department of Radiation Oncology, University of Pittsburgh, Pittsburgh, PA, USA*
PERTTU HAIMI • *Department of Biochemistry, Institute of Biomedicine, University of Helsinki, Helsinki, Finland*
MARTIN HERMANSSON • *Department of Biochemistry, Institute of Biomedicine, University of Helsinki, Helsinki, Finland*
FRANCOISE HULLIN-MATSUDA • *INSERM-RIKEN Lipidomics Unit, Lipid Biology Laboratory, RIKEN Advanced Science Institute, Saitama, Japan*
REIKO ISHITSUKA • *VCAD System Research Program, Bio-Research Infrastructure Construction Team, RIKEN, Saitama, Japan*
IRIS JESSE • *Institute of Biochemistry, Graz University, Graz, Austria*

SI JIN • *Department of Pharmacology & Toxicology, Medical College of Virginia Campus, Virginia Commonwealth University, Richmond, VA, USA*

VALERIAN E. KAGAN • *Center for Free Radical and Antioxidant Health and Department of Environmental and Occupational Health, University of Pittsburgh, Pittsburgh, PA, USA*

VILLE KAINU • *Department of Biochemistry, Institute of Biomedicine, University of Helsinki, Helsinki, Finland*

PATTANA KAKUMYAN • *Department of Biological Chemistry, Faculty of Agriculture and Department of Applied Molecular Bioscience, Graduate School of Medicine, Yamaguchi University, Yamaguchi, Japan*

HEIKKI KALLIO • *Department of Biochemistry and Food Chemistry, University of Turku, Turku, Finland*

VALENTINA I. KAPRASLOVA • *Center for Free Radical and Antioxidant Health, and Department of Environmental & Occupational Health, University of Pittsburgh, Pittsburgh, PA, USA*

YOJI KATO • *Laboratory of Food and Biodynamics, Nagoya University Graduate School of Bioagricultural Sciences, Chikusa, Japan*

ALEJANDRO MIGUEL KATZIN • *Department of Parasitology, Institute of Sciences, University of San Paulo, San Paulo, Brazil*

SERGEY A. KHOROBRYKH • *Science Research Center, Yamaguchi University, Yamaguchi, Japan; Institute of Basic Biological Problems, Russian Academy of Sciences, Moscow, Russia*

EMILIA A. KIMURA • *Department of Parasitology, Institute of Biomedical Sciences, University of San Paulo, San Paulo, Brazil*

CHRISTIAN KLUKAS • *Leibniz Institute of Plant Genetics and Crop Plant Research (IPK), Gatersleben, Germany*

TOSHIHIDE KOBAYASHI • *Lipid Biology Laboratory/INSERN-RIKEN Lipidomics Unit, RIKEN, Saitama, Japan; INSERM U870/INRA U1235/INSA, University of Lyon, Lyon, France*

EDITA KOFROŇOVÀ • *Institute of Organic Chemistry and Biochemistry, Academy of Sciences of the Czech Republic, Prague, Czech Republic*

ARNIS KUKSIS • *Banting and Best Department of Medical Research, University of Toronto, Toronto, ON, Canada*

THOMAS LANGMANN • *Institute of Human Genetics, University of Regensburg, Regensburg, Germany*

PIN-LAN LI • *Department of Pharmacology and Toxicology, Medical College of Virginia Campus, Virginia Commonwealth University, Richmond, VA, USA*

GERHARD LIEBISCH • *Institute of Clinical Chemistry and Laboratory Medicine, University of Regensburg, Regensburg, Germany*

XUEBO LIU • *Laboratory of Food and Biodynamics, Nagoya University Graduate School of Bioagricultural Sciences, Chikusa, Japan*

JUN'ICHI MANO • *Institute of Basic Biological Problems, Russian Academy of Science, Moscow, Russia*

KENJI MATSUI • *Department of Biological Chemistry, Faculty of Agriculture and Department of Applied Molecular Bioscience, Graduate School of Medicine, Yamaguchi University, Yamaguchi, Japan*
RICHARD MAUERER • *Synlab Medical Care Service, Medical Care Center, Weiden, Germany*
LUCA MONTICELLI • *Department of Applied Physics, Helsinki University of Technology, Espoo, Finland; Department of Physics, Tampere University of Technology, Tampere, Finland*
MATEJ OREŠIČ • *VTT Technical Research Centre of Finland, Espoo, Finland*
TOSHIHIKO OSAWA • *Laboratory of Food and Biodynamics, Nagoya University Graduate School of Bioagricultural Sciences, Chikusa, Japan*
MARTIN RUDOLF • *University Eye Hospital, Lubeck, Universitatsklinikum, Schleswig-Holstein, Lubeck, Germany*
MAXIMILIAN SCHICHER • *Institute of Biochemistry, University of Graz, Graz, Austria*
ENRIQUE SCHISTERMAN • *Division of Epidemiology, Statistics and Preventive Research, Eunice Kennedy Schriver National Institute of Child Health and Human Development, NIH, Bethesda, MD, USA*
GERD SCHMITZ • *Institute for Clinical Chemistry and Laboratory Medicine, University Hospital, Regensburg, Regensburg, Germany*
ROGER SCHNEITER • *Department of Medicine, Division of Biochemistry, University of Fribourg, Fribourg, Switzerland*
FALK SCHREIBER • *Leibniz Institute of Plant Genetics and Crop Plant Research (IPK), Gatersleben and Institute of Computer Science, Martin-Luther-University, Halle-Wittenberg, Germany*
PENTTI SOMERHARJU • *Department of Biochemistry, Institute of Biomedicine, University of Helsinki, Helsinki, Finland*
KOICHI SUGIMOTO • *Department of Biological Chemistry, Faculty of Agriculture and Applied Molecular Bioscience, Graduate School of Medicine, Yamaguchi University, Yamaguchi, Japan; Center for Ecological Research, Kyoto University, Kyoto, Japan*
JUKKA-PEKKA SUOMELA • *Department of Biochemistry and Food Chemistry, University of Turku, Turku, Finland*
MIWA TAKAHASHI • *VCAD System Research Program, BioResearch Infrastructure Construction Team, RIKEN, Saitama, Japan*
MARKO TARVAINEN • *Department of Biochemistry and Food Chemistry, University of Turku, Turku, Finland*
JELENA TELENIUS • *Department of Applied Physics, Helsinki University of Technology, Espoo, Finland*
RENATA TONHOSOLO • *Department of Parasitology, Institute of Biomedical Sciences, University of San Paulo, San Paulo, Brazil*
VLADIMIR A. TYURIN • *Center for Free Radical and Antioxidant Health and Department of Environmental & Occupational Health, University of Pittsburgh, Pittsburgh, PA, USA*
YULIA Y. TYURINA • *Center for Free Radical and Antioxidant Health and Department of Environmental & Occupational Health, University of Pittsburgh, Pittsburgh, PA, USA*

ILPO VATTULAINEN • *Department of Applied Physics, Helsinki University of Technology, Espoo, Finland; Department of Physics, Tampere University of Technology, Tampere, Finland MEMPHYS – Center for Biomembrane Physics, Physics Department, University of Southern Denmark, Odense, Denmark*

YANA WALCZAK • *Institute of Human Genetics, University of Regensburg, Regensburg, Germany*

LAN WANG • *Department of Ophthalmology, Callahan Eye Foundation Hospital, University of Alabama School of Medicine, Birmingham, AL, USA*

CRAIG E. WHEELOCK • *Department of Medical Biochemistry and Biophysics, Division of Physical Chemistry II, Karolinska Institutet, Stockholm, Sweden Bioinformatics Center, Institute for Chemical Research, Kyoto University, Kyoto, Japan*

INGRID WISWEDEL • *Department of Pathological Biochemistry, Medical Faculty of the Otto-von-Guericke-University, Magdeburg, Germany*

NARUOMI YAMADA • *Laboratory of Food and Biodynamics, Nagoya University Graduate School of Bioagricultural Sciences, Chikusa, Japan*

LAXMAN YETUKURI • *VTT Technical Research Centre of Finland, Espoo, Finland*

FAN YI • *Department of Pharmacology & Toxicology, Medical College of Virginia Campus, Virginia Commonwealth University, Richmond, VA, USA*

Part I

Oxidized Lipids

Chapter 1

F_2-Isoprostanes: Sensitive Biomarkers of Oxidative Stress In Vitro and In Vivo: A Gas Chromatography-Mass Spectrometric Approach

Ingrid Wiswedel

Summary

A gas chromatography-mass spectrometric method was developed that allowed the accurate, highly sensitive and specific quantification of F_2-isoprostanes (F_2-IsoPs) in different tissues and body fluids. Measurement of F_2-IsoPs in isolated rat brain mitochondria, HaCaT keratinocytes, human plasma, and microdialysates of human skin has established the occurrence of oxidative stress in a variety of model systems and disease states. F_2-IsoPs correlated with other markers of lipid peroxidation (e.g., TBARS, HETEs) in experimental models of oxidative stress. F_2-IsoPs were elevated about 100-fold after iron/ascorbate-induced oxidative stress and 2- to 4-fold after pentylenetetrazol (PTZ)-induced seizures, in hemodialysis patients with end stage renal disease, in psoriasis patients, in HaCaT keratinocytes, and in microdialysates of human skin following UVB irradiation.

Both human and experimental studies have indicated associations of F_2-IsoPs and inflammatory conditions. Anti-inflammatory drugs such as diclofenac did not only suppress the prostaglandin but also the F_2-IsoP pathway.

Microdialysis allows the "near-in vivo" measurement of prostanoid mediators, released in the interstitial space of the dermis under inflammatory conditions.

Key words: F_2-isoprostanes-GC-NICI-MS-inflammation

1. Introduction

F_2-isoprostanes (F_2-IsoPs), discovered in 1990 by Morrow and Roberts *(1)*, are prostaglandin (PG)-like compounds produced nonenzymatically by free radical-catalyzed peroxidation of arachidonic acid. They are chemically stable molecules, formed primarily in situ esterified to phospholipids and released by phospholipase(s), whereas PGs are generated only from free arachidonic acid *(2)*.

Donald Armstrong (ed.), *Lipidomics*, Methods in Molecular Biology, vol. 580,
DOI 10.1007/978-1-60761-325-1_1, © Humana Press, a part of Springer Science+Business Media, LLC 2009

F_2-IsoPs contain side chains that are oriented predominantly *cis* to the prostane ring, whereas PGs possess exclusively *trans* side chains *(1)*. F_2-IsoPs can be detected in all human tissues and body fluids and "physiological" levels have been defined *(3)*. F_2-IsoPs exert biological actions such as 8-iso-PGF$_{2\alpha}$ (first F_2-IsoP discovered from biological sources) known as potent renal and pulmonary vasoconstrictor and modulator of platelet aggregation *(4, 5)*.

F_2-IsoPs are elevated in animal models of oxidant injury and in human diseases associated with oxidative stress. Antioxidants, cessation of smoking, and weight loss decrease F_2-IsoP levels *(6, 7)*. F_2-IsoPs are formed via hydrogen abstraction from arachidonic acid resulting in four F_2-IsoP groups with cyclopentane ring structure denoted as 5-, 8-, 12-, and 15-series regioisomers depending on the carbon atom where the side chain hydroxyl group is located *(8, 9)*. Series of 8- and 12-regioisomers readily undergo further oxidation, whereas 5- and 15-series F_2-IsoPs accumulate at higher concentrations in tissues and fluids *(10)*. Therefore, 15-series F_2-IsoPs were mostly analyzed, but recently 5-series regioisomers too *(11)*.

F_2-IsoPs are measured either free or after base-hydrolysis of esterified F_2-IsoPs from lipids. They can be analyzed highly specifically, precisely, and very sensitively in the low picogram range using gas chromatography/negative ion chemical ionization mass spectrometric approach GC-NICI-MS) with tetradeuterated internal standards *(12)*. This method was named as the "gold standard" to assess in vivo oxidant stress status *(13)*. The disadvantage of the method is that it is labor intensive (extraction and derivatization steps are included). But other methods such as immunological approaches and LC-MS/MS are less sensitive, precise, and accurate *(10)*.

Our laboratory uses GC-NICI-MS since more than 10 years *(14–18)*. The principles of this method are described in this chapter, and from the large number of studies we carried out some selected examples are also presented showing that F_2-IsoPs are not only reliable biomarkers of oxidative stress but also mediators of inflammation.

2. Materials

2.1. Equipment

1. GC-MS system SSQ 710 Finnigan MAT, Bremen, Germany (1994–2005) including a single stage quadrupol mass spectrometer, a Varian 3400 gas chromatograph, a KAS III injector from Gerstel GmbH, Mühlheim/Ruhr, Germany, a CTC autosampler from Finnigan MAT, a DEC station with Finnigan software ICIS 8.0/Ultrix 4.3, and a laser printer.

2. GC-MS system DSQ/Trace GC Ultra, Thermo Fisher Scientific, Dreieich, Germany (2005–2008) including PTV injector TRI plus autosampler, computer with Xcalibur software and printer Deskjet 990 Cxi HP, Böblingen, Germany.
3. A DB5-MS column (30 or 50 m × 0.25 mm ID; 0.25 μm film thickness; J & W Scientific, Folsom, California).
4. 1.5 ml Glas autosampler vials with caps and glas inserts, VWR Dresden, Germany.
5. C_{18} and NH_2 cartridges Vac 3cc 500 mg for solid phase extraction (Waters Corporation, Milford, Ireland).

2.2. Regents and Supplies

1. $PGF_{2\alpha}$-d_4 and PGE_2-d_4 as internal standards and 8-iso-$PGF_{2\alpha}$, 9α,11α-$PGF_{2\alpha}$, and PGE_2 as standards (Cayman Chemicals Co., Ann Arbor, Michigan).
2. HCL and KOH, VWR Dresden, Germany.
3. Ethylacetate, VWR Dresden, Germany.
4. Hexane, VWR Dresden, Germany.
5. Acetonitrile, VWR Dresden, Germany.
6. Methanol, VWR Dresden, Germany.
7. Acetic acid, Merck, Darmstadt, Germany.
8. *N*,*N*-di-isopropylethylamine, Sigma, Steinheim, Germany.
9. Pentafluorobenzylbromide (PFB), Aldrich, Steinheim Germany.
10. *Bis*-(trimethylsilyl)trifluoroacetamide (BSTFA), Sigma, Steinheim, Germany.
11. Isooctane, Merck, Darmstadt, Germany.

3. Methods

3.1. Sample Processing for the Quantification of F_2-IsoPs

3.1.1. Isolated Rat Brain Mitochondria and Brain Homogenates

For quantification of F_2-IsoPs, the method of Nourooz-Zadeh et al. *(19)* was modified and improved *(14)*. Two hundred fifty microliter of mitochondrial suspension (1 mg protein/ml) or brain homogenate (10% in phosphate-buffered saline) was incubated with 250 μl of 4 M KOH at 40°C for 30 min to cleave esterified lipids. (For the analysis of free F_2-IsoPs, alkaline hydrolysis was omitted.) The pH was adjusted to about 2 using about 250 μl of 4 M HCL, and $PGF_{2\alpha}$-d_4 was added as internal standard (*see also* **Note 1**). Subsequently, five volumes of ethylacetate were added and samples were vortex-mixed for 30 s. The samples were centrifuged at 2,000 × *g* for 5 min for proper phase separation and the organic (upper) layer was transferred into a new glass tube.

The total lipid extract then was applied onto an NH_2 cartridge prewashed with hexane (5 ml). The columns were subsequently washed with 10 ml of hexane/ethyl acetate (30/70, v/v), acetonitrile/water (90/10, v/v), and acetonitrile. F_2-IsoPs were eluted washing the column with 5 ml of ethylacetate/methanol/acetic acid (10/85/5, v/v/v).

Final lipid extracts following NH_2 chromatography step were dried under a stream of argon at 45°C. 40 μl PFB (10% in acetonitrile) and 20 μl of *N,N*-di-isopropylethylamine (10% in acetonitrile) were added to the residues and the samples were kept at 45°C for 30 min.

50 μl of BSTFA and 5 μl of *N,N*-di-isopropylethylamine were added to the dried sample. The samples were kept at 45°C for 45 min, the solvents were removed and the samples were dissolved in 40 μl isooctane (containing 0.1% BSTFA).

3.1.2. Plasma or Serum

For the determination of F_2-IsoPs in plasma or serum samples, some modifications were included *(16)*: After alkaline splitting, addition of internal standard and adjustment of pH, the samples (at least 0.2 ml of plasma or serum) were centrifuged at 5,000 × *g* for 15 min and the supernatant was applied onto C_{18} cartridges and prewashed with 5 ml of methanol and 5 ml of water. The cartridge was then washed with 10 ml HCL (0.1 M) and 10 ml of acetonitrile/water (15/85, v/v). F_2-IsoPs were eluted from the column with 5 ml of *n*-hexane/ethylacetate/2-propanol (30/65/5, v/v/v). The prostanoid extract was applied then onto NH_2 cartridges as described earlier, eluted, and evaporated to dryness under a stream of argon. Derivatization steps were performed as described under **Subheading 3.1.1.**

3.1.3. HaCaT Keratinocytes and Microdialysates of Human Skin

To hydrolyze esterified F_2-IsoPs, HaCaT keratinocyte samples (150 μl; about 2.0 mg protein/ml) were treated with 75 μl KOH (1 M) at 40°C for 30 min. Thereafter, 5 ng deuterated internal standard was added and the pH of 2 was adjusted by addition of HCL. The samples were applied onto a C_{18} cartridge and treated as described in **Subheading 3.1.2**. The prostanoid extract was evaporated to dryness under a stream of argon at 45°C and derivatized with PFB and BSTFA as described under **Subheading 3.1.1.**

In microdialysate samples, only nonesterified F_2-IsoPs were observed and the alkaline splitting was omitted. The levels of F_2-IsoPs in microdialysates were very small, and under control conditions, they were near the limit of detection (*see also* **Note 2**). Therefore, solid-phase extraction was also omitted. Further details of the method are described in refs. *(18, 19, 26)*.

3.2. GC-MS Analysis

F_2-IsoPs were gas chromatographically separated using a DB5-MS column with the following temperature program: initial temperature of 175°C for 2 min; with a rate of 30°C/min to the final temperature of 270°C maintained for 27 min; final time: 32.2 min. Quantitative analysis was performed with an SSQ 710 Finnigan MAT machine and since 3 years, a DSQ/Trace GC Ultra using NICI at first with methane and for better sensitivity since several years with ammonia as reagent gas. Selected ion monitoring of the carboxylate anion $[M-181]^-$ at *m/z* 569 and 573 for F_2-IsoPs and deuterated internal standard ($PGF_{2\alpha}d_4$) was used. For the simultaneous analysis of PGE_2 and PGE_2-d_4 in microdialysate samples, the carboxylate anion at *m/z* 495 and 499 was recorded.

3.3. Sample Storage and Standardization

All samples, which have been prepared and measured, were flash-freezed in liquid nitrogen and stored at –80°C (*see also* **Note 3**). Butylated hydroxytoluene was added as antioxidant (100 μM) and the acceptable storage period was not more than 6 months.

In all experiments concerning F_2-isoPs, 9α,11α-$PGF_{2\alpha}$-d_4 was used as internal standard. Studies with 8-iso-$PGF_{2\alpha}$-d_4 as internal standard were not successful because this commercially available compound was not free of 8-iso$PGF_{2\alpha}$ (*see also* **Note 4**). The internal standard was quantified using a five-point calibration curve. Each sample contained 0.5 ng of 9α,11α-$PGF_{2\alpha}$ or 0.5 ng 8-iso-$PGF_{2\alpha}$ and either 0.05, 0.25, 0.5, 1, or 2.5 ng of $PGF_{2\alpha}d_4$. Each day the sensitivity of the GC-MS was checked by injecting the internal standard and a standard mixture consisting of $PGF_{2\alpha}$, 8-iso-$PGF_{2\alpha}$, and recently 5-iso-$PGF_{2\alpha}$(VI) too.

3.4. Results

3.4.1. F_2-IsoPs Analysis: Representative Chromatograms and Calibration Curves

Figure 1 shows a representative chromatogram for the GC-MS analysis of F_2-IsoPs in plasma samples with NICI. Deuterated $PGF_{2\alpha}$ (9α,11α-$PGF_{2\alpha}$) was used as internal standard (upper chromatogram; panel a). The *m/z* 569 ion current chromatogram in panel b shows the separation of two major peaks in plasma samples, the first peak coeluating with authentic 8-iso-$PGF_{2\alpha}$ and the second peak coeluating with authentic 9α,11α-$PGF_{2\alpha}$. The first peak, named "8-iso-$PGF_{2\alpha}$" or "peak I", is expected to contain exclusively F_2-IsoP isomers and is therefore most appropriate for the quantification of F_2-IsoPs *(15, 20)*. This is suggested by its larger broadness compared with the second peak (9α,11α-$PGF_{2\alpha}$), despite shorter retention times. Recently, 5-iso-$PGF_{2\alpha}$ (VI) was detected as a further constituent of peak I, but in spite of a longer column (50 m), both isomers were only successfully separated from each other in microdialysate samples of human skin *(11)*. Peak II may contain both F_2-IsoP isomers and cyclooxygenase-derived $PGF_{2\alpha}$. The occurrence of enzymatically generated prostanoids in peak II is also derived from the observations that in plasma samples, the shares of free prostanoids of peak II are higher (20–25%) than those of peak I (10–20%) *(15)*. In HaCaT keratinocytes *(18)*, the isomers of peak I completely

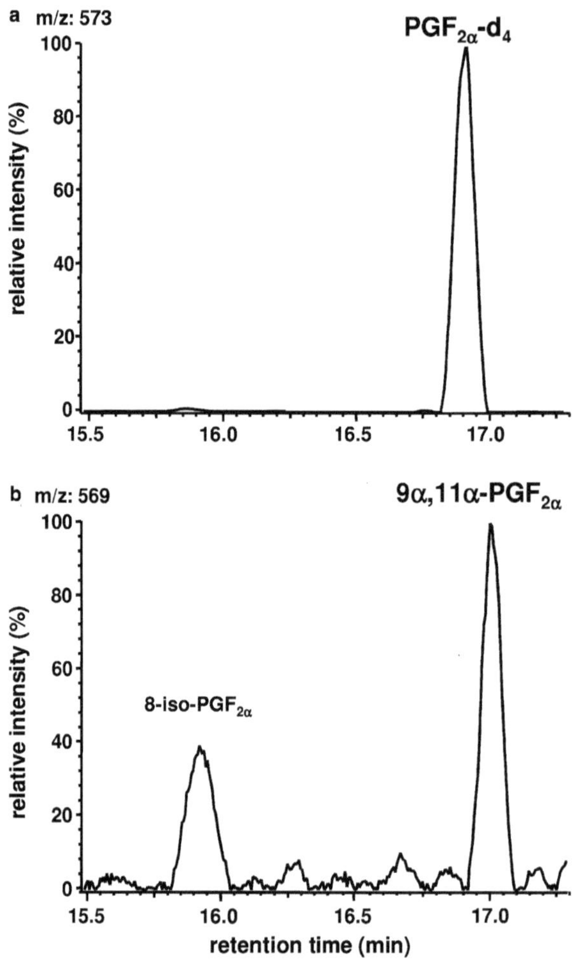

Fig. 1. Gas chromatography-mass spectrometric separation (in the negative ion chemical ionization mode) of F_2-isoprostanes of human plasma. $PGF_{2\alpha}$-d_4 was used as internal standard (panel **a**). Panel **b** shows a representative example for the separation of F_2-isoprostanes from the plasma of a healthy control person. The first peak coeluated with authentic 8-iso-$PGF_{2\alpha}$ and the second peak with authentic 9α,11α-$PGF_{2\alpha}$. According to Morrow et al. ***(20)***, only the first peak is expected to contain exclusively F_2-IsoP isomers, whereas the second peak may contain both, F_2-isoprostane isomers and cyclooxygenase-derived $PGF_{2\alpha}$.

occur in the esterified form and were not released and detected in the cell-free supernatant. The isomers of peak II predominantly occur in the free or nonesterified form (77%), and a considerable part was released in the supernatant *(21)*.

In **Fig. 2**, typical standard calibration curves for the main isomers of the peaks I and II: 8-iso-$PGF_{2\alpha}$ and 9α,11α-$PGF_{2\alpha}$ are presented. Standard curves were generated for each isomer by adding different amounts of each F_2-IsoP/PG (0.1–1 ng) to 5 ng of the internal (deuterated) standard in presence of HaCaT keratinocytes (150 μl; as in **Subheading 3.1.3**), followed by complete

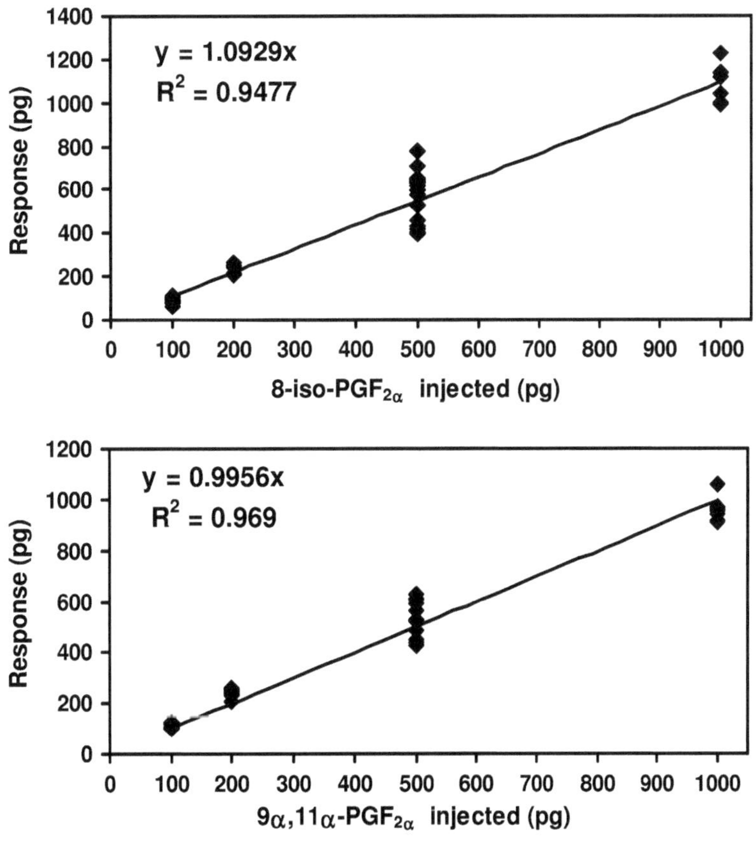

Fig. 2. Standard calibration curves for 8-iso-$PGF_{2\alpha}$ and 9α,11α-$PGF_{2\alpha}$. Plotted is the amount in picograms of 8-iso-$PGF_{2\alpha}$ and of 9α,11α-$PGF_{2\alpha}$ versus the response ratio of the prostanoid to the $PGF_{2\alpha}$-d_4 internal standard. Standard curves were generated for each isomer by adding of 0.1–1 ng of each prostanoid to 5 ng internal standard in presence of 150 μl HaCaT keratinocytes (2.5 mg protein/ml corresponding to 15 × 10^6 cells/ml) followed by complete sample preparation procedure. The measured concentrations were corrected by the endogenous isoprostane /prostaglandin levels of keratinocytes (n = 11). The standard curves for both eicosanoid isomers were linear over the range of interest and had correlation coefficients of 0.95 and 0.97, respectively.

sample preparation procedure. The measured concentrations were corrected by the endogeneous prostanoid levels of keratinocytes. Standard curves for both eicosanoid isomers were linear over the range of interest and had correlation coefficients of 0.95 and 0.97, respectively. The response factors were 1.022 for 8-iso-$PGF_{2\alpha}$ and 1.161 for 9α,11α-$PGF_{2\alpha}$.

3.4.2. F_2-IsoP Analysis in Rat Brain Mitochondria After Induction of Oxidative Stress

Figure 3 shows the kinetics of appearance of F_2-isoprostanes (the peak which coeluates with authentic 8-iso-$PGF_{2\alpha}$), TBARS, and monohydroxyeicosatetraenoic acids (HETEs) and the disappearance of α-tocopherol in isolated rat brain mitochondria after iron/ascorbate-induced peroxidation. In this particular model system, the content of total HETE isomers and TBARS was

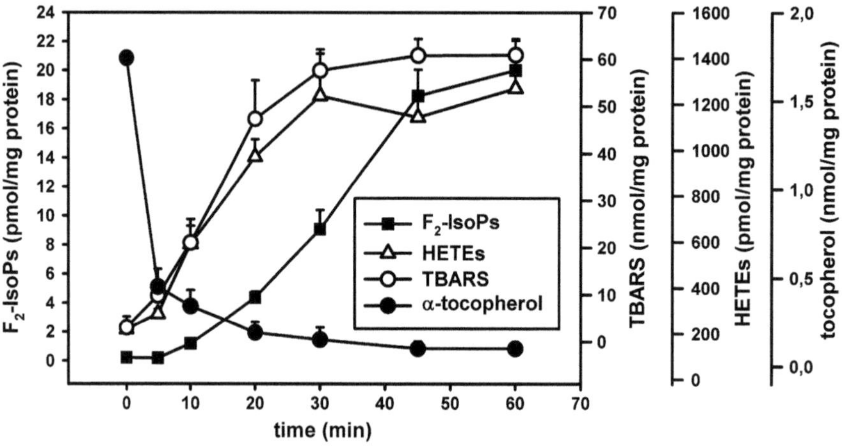

Fig. 3. Time course of formation of F_2-isoprostanes (8-iso-PGF$_{2\alpha}$), TBARS, and mono-hydroxyeicosatetraenoic acids (HETEs; sum of 5-, 8-, 12-, and 15-HETEs) and decline of α-tocopherol during the time course of iron/ascorbate-induced peroxidation of rat brain mitochondria. Values are means ± SEM of four different mitochondrial preparations. Baseline levels of F_2-IsoPs were 0.21 ± 0.10 pmol/mg protein and that of HETEs were 220 ± 40 pmol/mg protein ***(14)***.

elevated until sevenfold compared with baseline levels, whereas that of F_2-isoprostanes as minor compounds were elevated about 100-fold. F_2-IsoPs occurred in mitochondrial preparations predominantly in the esterified form and following induced oxidative stress (iron plus ascorbate), only esterified F_2-isoprostanes were elevated *(14)*.

In further experiments, we studied the role of oxidative stress in PTZ-induced seizures as an established animal model for epilepsy *(22)*. PTZ is a chemical convulsant and acute seizures are induced by a single dose and kindling by repeated application of PTZ over 4 weeks. Concentrations of F_2-isoprostanes were significantly enhanced in isolated rat brain mitochondria, in particular of kindled animals, but of acutely convulsed rats, too (**Fig. 4**). This was in parallel with the consumption of α-tocopherol and the expression of MnSOD *(22)*, measured under similar conditions and was more expressed in isolated mitochondria than in the whole brain tissue (not shown).

3.4.3. Role of F_2-IsoPs as Prognostic Indicators of Oxidative Stress and Mediators of Inflammation in Chronic Renal Failure

Figure 5 shows the increase of F_2-isoprostanes in serum samples of hemodialysis patients with end stage renal disease. F_2-IsoP concentrations of patients before and after dialysis are fourfold higher than those of healthy control persons with normal renal function. There are no statistically significant differences in F_2-IsoP concentrations before and after dialysis.

There are several reports that F_2-isoprostanes are not only biomarkers of oxidant stress but also mediators of inflammation involving cyclooxygenases and/or cytokines and that they promote a possible link between oxidative stress and inflammation *(11, 16, 18, 23, 24)*. **Figure 6** exhibits that F_2-isoprostane levels in serum

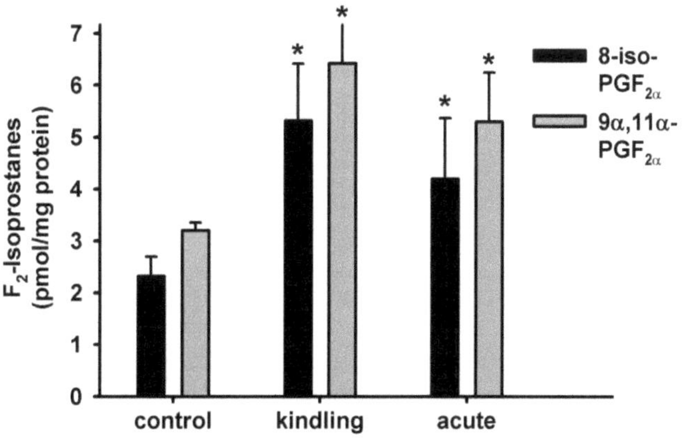

Fig. 4. Levels of F_2-isoprostanes (8-iso-PGF$_{2\alpha}$ and 9α,11α-PGF$_{2\alpha}$) in mitochondrial preparations of control, kindled, and acutely convulsing rats. The values are means ± SEM of n = 3–4 mitochondrial preparations. $^*p < 0.05$; values are significantly different from control. Behavioral investigations are described in *(22)*.

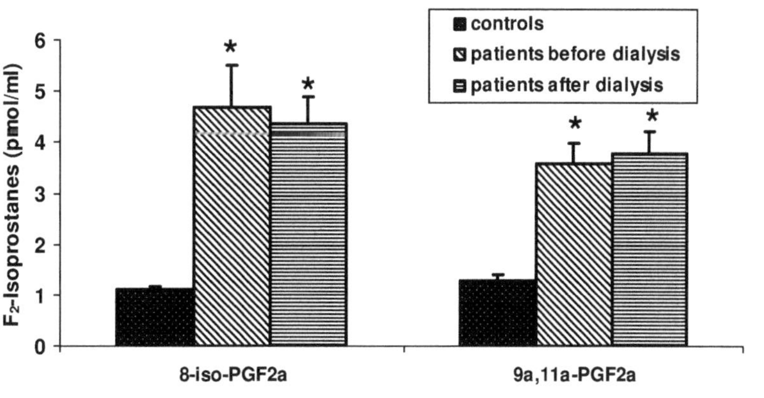

Fig. 5. F_2-Isoprostane concentrations (8-iso-PGF$_{2\alpha}$ and 9α,11α-PGF$_{2\alpha}$) in serum samples of patients before and after hydrolysis (n = 19) and of healthy control persons (n = 65). Mean values ± SEM; two-sample t test; $p < 0.001$ versus control. The differences before versus after hydrolysis were not significant (t test for paired data).

samples of hemodialysis patients correlate to the concentrations of the C-reactive protein, which is significantly enhanced in comparison to healthy controls. The causal relationship remains open, but inflammation may be responsible for free radical production, for example, via activation of NADPH oxidase by myeloperoxidase and/or interleukins *(25)*.

3.4.4. Increase of F_2-IsoP Concentrations in Patients with Inflammatory Skin Diseases and Following UVB Irradiation of Human Skin Using Microdialysis Technique

Elevations of F_2-isoprostanes in human body fluids and tissues have been found in a diverse array of human disorders, here shown for plasma samples of patients with skin diseases as psoriasis (**Fig.** 7).

Since several years, it is known that UVB irradiation does not only induce oxidative stress but also inflammatory reactions. We studied the influence of UVB irradiation on the generation and

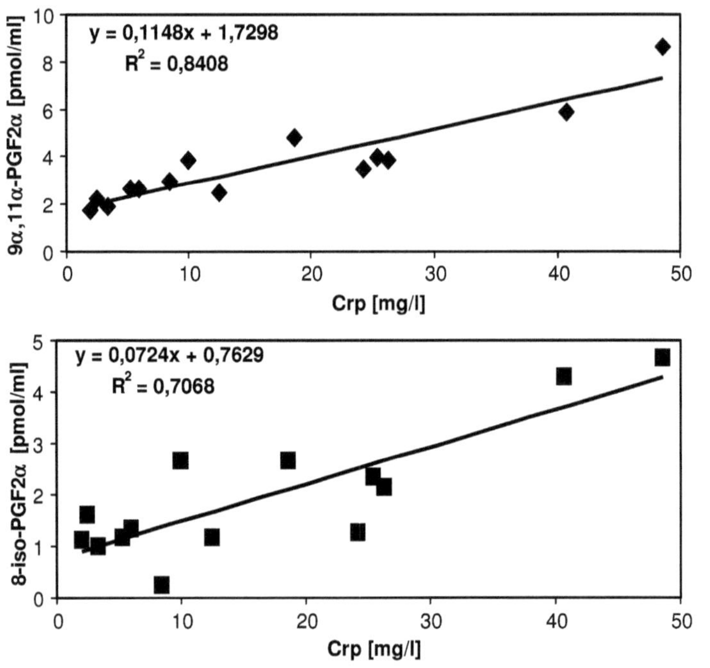

Fig. 6. Correlations between levels of F_2-isoprostanes [8-iso-PGF$_{2\alpha}$ (*lower part*) and 9α,11α-PGF$_{2\alpha}$ (*upper part*)] and C-reactive protein; $n = 14$ patients. Correlations have been investigated with linear regression analysis. This figure was taken from Wiswedel et al. [(Biomarker Insights 2008:3; 419–428), Fig. 3].

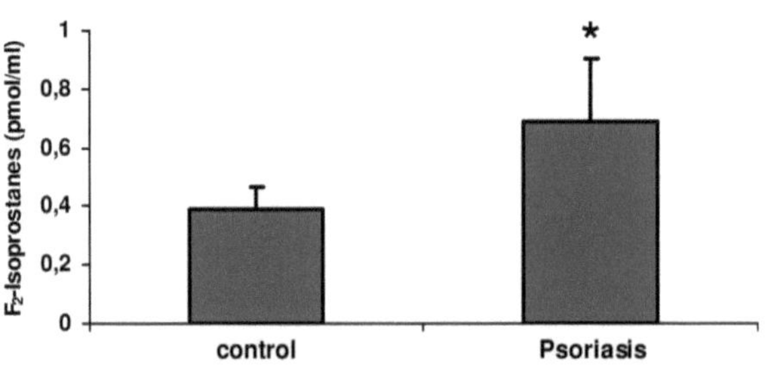

Fig. 7. Increased F_2-isoprostane concentrations (8-iso-PGF$_{2\alpha}$) in plasma samples of psoriasis patients. Values are means ± SD of $n = 5$, t test; $p < 0.02$.

release of F_2-IsoPs and PGs at first in HaCaT keratinocytes as well established in vitro model system *(18, 26)* and second in human skin using microdialysis *(11, 17, 18, 21)* technique. Low or moderate UVB doses enhanced the levels of F_2-IsoPs in a near dose-dependent fashion, until 50 mJ/cm^2 (**Fig. 8**). Higher UVB doses did not lead to further increase, but a serious damage of HaCaTs was observed *(26)*. As an adaptive response to the enhanced oxidative stress, protein levels of MnSOD increased about threefold until 50 mJ/cm^2 and decreased at higher doses *(26)*. This means that keratinocytes seem sufficiently protected at low UVB doses, whereas higher doses lead to irreversible cell damage.

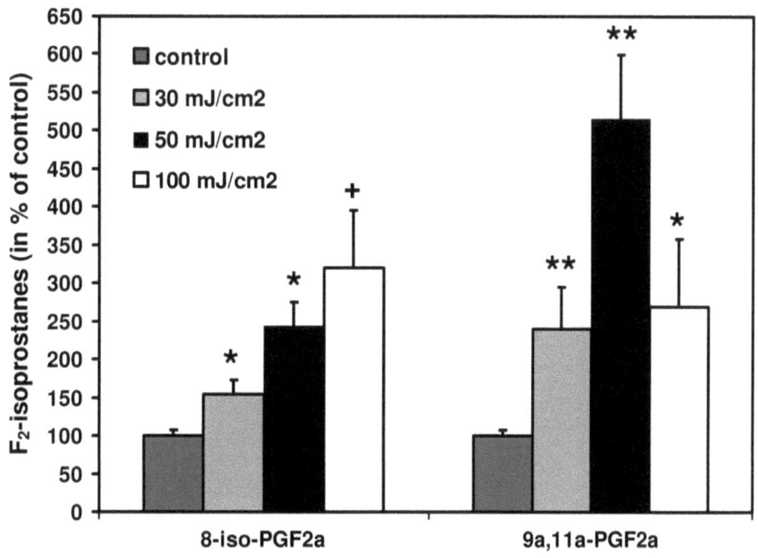

Fig. 8. Dose-dependent enhancement of F_2-isoprostanes in HaCaT keratinocytes following UVB irradiation. Values are means ± SEM. Endogenous concentrations (controls without UVB irradiation) were 85.1 ± 7.1 pg/mg protein for 8-iso-PGF$_{2\alpha}$ ($n = 28$) and 177.3 ± 14.2 pg/mg protein for 9α,11α-PGF$_{2\alpha}$ ($n = 28$). 30 mJ/cm^2 ($n = 20$) and 50 and 100 mJ/cm^2 ($n = 6$). Significance between different UVB doses (each versus the corresponding lower dose) was calculated by Student's *t* test. $^*p < 0.05$ and $^{**}p < 0.01$. In the case of 8-iso-PGF$_{2\alpha}$, the difference between 100 and 50 mJ/cm^2 was not significant ($p = 0.069$), but those between 100 and 30 mJ/cm^2 was significant (+; $p = 0.026$). [1] This figure was taken from Wiswedel et al. [***(26)***, Fig. 1)].

Microdialysis is a minimal-invasive technique that allows the continuous near-in vivo measurement of F_2-IsoPs and PGs in the interstitial space of the skin. In principle, a semipermeable membrane is inserted in the upper dermis and due to the concentration gradient between the interstitial space of the dermis and the perfusate, substances diffuse through the pores of the membrane and can be analyzed in the dialysate *(18, 27)*. The recovery of prostanoids determined by retroanalysis was higher than 80% using a flow rate of 0.5 or 1.0 µl/min (Quist and Wiswedel, unpublished results). Concentrations of F_2-IsoPs and PGE_2 in microdialysate samples were about three- to fourfold enhanced in 24 h following UVB treatment compared with healthy untreated skin (**Fig. 9**). Diclofenac (DCLF), topically applied as cream or gel, prevented the generation of erythema and reduced the levels of prostanoids remarkably *(18)*. As observed in HaCaT keratinocytes *(18)*, DCLF did not only influence the PG, but additionally the isoprostane pathway, which may be, at least partially, due to radical scavenging or antioxidant side effects of DCLF or cyclooxygenase-mediated F_2-IsoP generation. The additional observations that F_2-isoprostane concentrations do correlate with

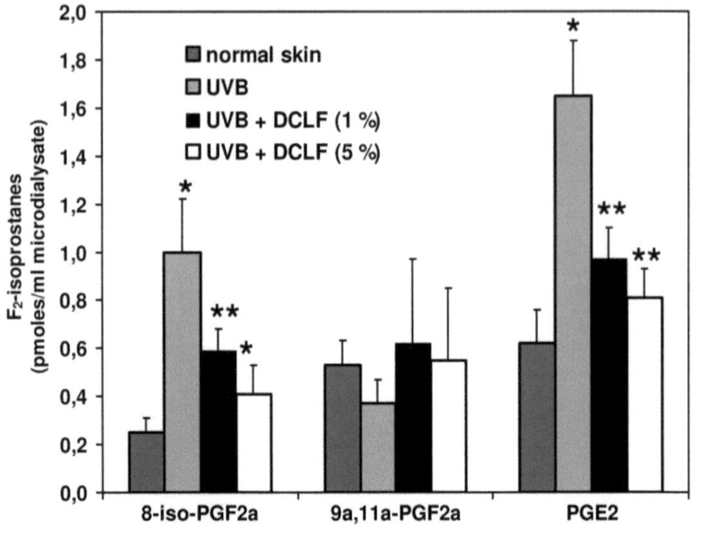

Fig. 9. Levels of F_2-isoprostanes/prostaglandins in microdialysis samples of human skin – influence of UVB irradiation and treatment with diclofenac (*DCLF*). Values are means ± SEM of $n = 6$ healthy volunteers. UVB irradiation (about 300 mJ/cm^2) caused the formation of erythema of stages 3–5, which were completely suppressed by DCLF immediately applied following UVB irradiation. Samples were taken 24 h later, after an additional hour of "baseline" microdialysis for equilibration. Microdialysate samples of about 2 h have to be combined for triplicate analyses. Differences in the levels of prostanoids are significant for 8-iso-PGF$_{2\alpha}$ and PGE$_2$ according to Student's *t* test: $^*p < 0.05$ versus nonirradiated control and $^{**}p < 0.05$ versus UVB without DCLF.

cytokines, measured simultaneously with prostanoids in microdialysate samples *(Quist et al, personal communication)* and with C-reactive protein, measured in plasma samples of hemodialysis patients with end stage renal disease (**Fig. 6**) suggest a role of F_2-IsoPs not only as biomarkers of oxidative stress but also as mediators of inflammation.

4. Notes

1. A pH of 2–3 has to be adjusted before extraction of F_2-isoprostanes.
2. The limit of detection in our experience was 2.5 pg.
3. It is very important that samples have to be thawed not earlier than immediately before analysis.
4. The commercially available deuterated internal standard may contain small amounts of the nondeuterated compound. This must be checked and taken into consideration.

Acknowledgments

The work was supported by grants of the Federal state of Saxony-Anhalt and by COST action B 35. I want to thank Drs. A. Gardemann and W. Augustin for continuous interest and discussions, Drs. H. Sies, H. Gollnick, W. Siems, T. Schewe and C. Rauca for cooperation, Daniela Peter for excellent work in MS analyses, and Heidemarie Faber and Elke Wölfel for skilful experimental assistance.

References

1. Morrow JD, Hill KE, Burk RF, Nammour TM, Badr KF, Roberts LJ. (1990) A series of prostaglandin F2-like compounds are produced in vivo by a non-cyclooxygenase, free radical-catalyzed mechanism. Proc Natl Acad Sci (USA) **23**:9383–9387.
2. Morrow JD, Awad JA, Boss HJ, Blair IA, Roberts LJ. (1992) Non-cyclooxygenase-derived prostanoids (F_2-isoprostanes) are formed in situ in phospholipids. Proc Natl Acad Sci (U S A) **89**:10721–10725.
3. Fam SS, Morrow JD. (2003) The isoprostanes: unique products of arachidonic acid oxidation-a review. Curr Med Chem **10**:1723–1740.
4. Lynch SM, Morrow JD, Roberts LJ, Frei B. (1994) Formation of non-cyclooxygenase-derived prostanoids (F2-isoprostanes) in plasma and low-density lipoprotein exposed to oxidative stress in vitro. J Clin Invest **93**:998–1004.
5. Montuschi P, Barnes PJ, Roberts LJ. (2004) Isoprostanes: markers and mediators of oxidative stress. FASEB J **18**:1791–1800.
6. Davi G, Ciabattoni G, Consoli A, Mezzetti A, Falco A, Santarone S, Pennese E, Vitacolonna E, Bucciarelli T, Constantini F, Capani F, Patrono C. (1999) In vivo formation of 8-iso-prostaglandin f2alpha and platelet activation in diabetes mellitus: effects of improved metabolic control and vitamin E supplementation. Circulation **99**:224–229.
7. Roberts LJ, Oates JA, Linton MF, Fazio S, Meador BP, Gross MD, Shyr Y, Morrow JD. (2007) The relationship between dose of vitamin E and suppression of oxidative stress in humans. Free Radic Biol Med **43**:1388–1393.
8. Roberts LJ, Morrow JD. (2000) Measurement of F_2-isoprostanes as an index of oxidative stress *in vivo*. Free Radic Biol Med **28**:505–513.
9. Taber DF, Morrow JD, Roberts LJ. (1997) A nomenclature system for the isoprostanes. Prostaglandins **53**:63–67.
10. Milne GL, Huiyong Y, Morrow JD. (2008) Human biochemistry of the isoprostane pathway. J Biol Chem **283**:15533–15537.
11. Quist SR, Wiswedel I, Quist J, Gollnick HP. (2008) Anti-inflammatory effects of sea-silt and sea-salt containing external formulations in human skin in vivo using cutaneous microdialysis. J Cosmet Dermatol, in press.
12. Milne GL, Sanchez SZ, Musiek ES, Morrow JD. (2007) Quantification of F2-isoprostanes as a biomarker of oxidative stress. Nature Protocols 221–226.
13. Musiek ES, Yin H, Milne GL, Morrow JD. (2005) Recent advances in the biochemistry and clinical relevance of the isoprostane pathway. Lipids **40**:987–994.
14. Wiswedel I, Hirsch D, Nourooz-Zadeh J, Flechsig A, Lueck-Lambrecht A, Augustin, W. (2002) Analysis of monohydroxyeicosatetraenoic acids and F_2-isoprostanes as markers of lipid peroxidation in rat brain mitochondria. Free Radic. Res **36**:1–11.
15. Wiswedel I, Hirsch D, Kropf S, Gruening M, Pfister E, Schewe T, Sies H. (2004) Flavanol-rich cocoa drink lowers plasma F_2-isoprostane concentrations in humans. Free Radic Biol Med **37**:411–421.
16. Wiswedel I, Hirsch D, Carluccio F, Hampl H, Siems W. (2005) F_2-Isoprostanes as biomarkers of lipid peroxidation in patients with chronic renal failure. Biofactors **24**:201–208.
17. Grundmann JU, Wiswedel I, Hirsch D, Gollnick HPM. (2004) Detection of mono-hydroxyeicosatetraenoic acids and F_2-isoprostanes in microdialysis samples of human UV-irradiated skin by gas chromatography-mass spectrometry. Skin Pharmacol. Physiol **17**:37–41.
18. Wiswedel I, Grundmann JU, Boschmann M, Krautheim A, Böckelmann R, Peter DS, Holzapfel I, Götz S, Müller-Goymann C, Bonnekoh B, Gollnick HP. (2007) Effects of UVB

irradiation and diclofenac on F_2-isoprostane/prostaglandin concentrations in keratinocytes and microdialysates of human skin. J Invest Dermatol **127**:1794–1797.

19. Nourooz-Zadeh J, Gopaul NK, Barrow S, Mallet AI, Anggard EE. (1995) Analysis of F2-isoprostanes as indicators of non-enzymatic in vivo lipid peroxidation by gas chromatography: development of a solid-phase extraction procedure. J Chromatogr **667**:199–208.
20. Morrow JD, Zackert WE, van der Ende DS, Reich EE, Terry ES, Cox B, Sanchez SC, Montine TJ, Roberts LJ. (2002) Qantification of isoprostanes as indicators of oxidant stress in vivo. In: Cadenas E and Packer L, ed. Handbook of Antioxidants, 2nd ed., Marcel Dekker, Inc., New York, Basel, 57–74.
21. Quist SR, Simmel F, Wiswedel I, Neubert R, Gollnick H. (2006) Influence of green and black tea epigallocatechin-3-gallate and theaflavin on prostanoid synthesis in vitro and in vivo using microdialysis. Proceedings of the 13th Biennial Congress of the International Society of Free Radical Research – SFRR. Medimond; Davos Switzerland, 293–297.
22. Rauca C, Wiswedel I, Zerbe R, Keilhoff G, Krug M. (2004) The role of superoxide dismutase and α-tocopherol in the development of seizures and kindling induced by pentylenetetrazol – influence of the radical scavenger α-phenyl-*N-tert*-butylnitrone. Brain Res **1009**: 203–212.
23. Basu S. (2004) Isoprostanes: novel bioactive products of lipid peroxidation. Free Radic Res **38**:105–122.
24. Praticò D, Rokach J, Lawson J, FitzGerald GA. (2004) F2-Isoprostanes as indices of lipid peroxidation in inflammatory diseases. Chem Phys Lipids **128**:165–171.
25. Morena M, Delbosc S, Dupuy AM, Canaud B, Cristol JP. (2005) Overproduction of reactive oxygen species in end-stage renal disease patients: a potential component of hemodialysis-associated inflammation. Hemodial Int **9**:37–46.
26. Wiswedel I, Keilhoff G, Dörner L, Navarro A, Böckelmann R, Bonnekoh B, Gardemann A, Gollnick H. (2007) UVB irradiation-induced impairment of keratinocytes and adaptive responses to oxidative stress. Free Radic Res **41**:1017–1027.
27. Gottlob A, Abels C, Landthaler M, Szeimies RM. (2002) Cutaneous microdialysis. Use in dermatology. Hautarzt 2002;**53**;174–178.

Chapter 2

Volatile Oxylipins and Related Compounds Formed Under Stress in Plants

Kenji Matsui, Koichi Sugimoto, Pattana Kakumyan, Sergey A. Khorobrykh, and Jun'ichi Mano

Summary

Plants form volatile oxylipins and related compounds under stress. Some of them are important flavor chemicals and give big impact on the flavor quality of food made from plant materials. They are also involved in defense responses of plants against pathogens and herbivores. Furthermore, in some instances, they cause harmful effects on plants themselves. Because of these significances of volatile oxylipins and related compounds, demands to perform comprehensive analyses of these compounds are increasing. In this chapter, we describe the simple but efficient procedures to reveal profiles of volatile oxylipins and related compounds by using HPLC and GC-MS. They are simple and can be performed in biochemical laboratories equipped with common facilities.

Key words: Volatile oxylipins, Reactive aldehydes, Plant stress responses, HPLC, GC-MS

1. Introduction

Oxylipin is a group name of compounds derived from fatty acids usually through at least one step of oxidation reaction. In mammals, C20 fatty acids such as arachidonic acid are the precursor of oxylipins. Prostaglandins and leukotrienes are the ones of most important mammalian oxylipins because of their diverse and prominent physiological functions. Therefore, various excellent protocols for analyses of mammalian oxylipins are available. Plants also form diverse arrays of oxylipins but usually from C18 and C16 fatty acids. Ones of most important plant oxylipins are jasmonic acid, its methyl ester, and its amino acid conjugates, which are involved in plant defense, growth, and development *(1)*. Plants

Donald Armstrong (ed.), *Lipidomics,* Methods in Molecular Biology, vol. 580,
DOI 10.1007/978-1-60761-325-1_2, © Humana Press, a part of Springer Science+Business Media, LLC 2009

have a unique oxylipin pathway forming short-chain, volatile oxylipins *(2)*. Fatty acid hydroperoxide lyase (HPL) that cleaves 13- and/or 9-hydroperoxides of linolenic or linoleic acid is the key branching enzyme in the pathway. The first products formed from 13-hydroperoxides are C6-aldehydes, whereas C9-aldehydes are formed from 9-hydroperoxides. These aldehydes could be further reduced by alcohol dehydrogenase to form the corresponding short-chain alcohols, and in some cases, acetyl transferase converts them to their acetates.

These volatile oxylipins have unique, green leaf-like flavor properties; thus, they are important flavor compounds in food materials of plant origin. With soybean, *n*-hexanal is a cause of unpleasant, beany flavor, whereas in tomato fruits, (*Z*)-3-hexenal and (*Z*)-3-hexen-1-ol are important for the freshness of the fruits. In intact and healthy plant tissues, the amounts of volatile oxylipins are usually low; however, once the tissues suffer stresses, such as mechanical wounding, pathogen infection, herbivore attack, high temperature, or draught, they are immediately formed *(2)*. From this, it has been assumed that volatile oxylipins are involved in stress responses of plants. For example, volatile oxylipins are involved in defense responses of Arabidopsis against necrotrophic fungal pathogens, such as *Botrytis cinerea (3)*. Also, their involvement in the tritrophic system consisting of plants, herbivores, and parasites to recruit parasitic wasps has been reported *(4)*. Volatile oxylipins can also be airborne signal compounds to prime neighboring maize plants in order to defend efficiently against forthcoming herbivore attack *(5)*.

Other than volatile oxylipins formed through HPL pathway, various reactive aldehydes are formed in plants especially when plants suffer stresses. Oxidation of polyunsaturated fatty acids with radicals and subsequent addition of dioxygen leads to the formation of peroxyl radicals (LOO•). LOO• may be reduced by other organic molecules to form a peroxide (LOOH). For example, lipoxygenase-independent formations of 12- and 16-hydroperoxides of linolenic acid are observed in leaves under oxidative stress *(6)*. Alternatively, LOO• radicals are converted to monocyclic peroxides and bicyclic endoperoxides *(7, 8)*. These peroxides decompose by nonenzymatic mechanisms to form a variety of molecules comprising carbonyl moiety. Typical aldehyde-producing reactions are α- and β-scission of LOOH. Bicyclic endoperoxides decompose to form malondialdehyde (MDA) *(8)* or highly reactive γ-ketoaldehydes *(9)*. Note that chemical property and biological activity of each carbonyl compound differ greatly, and the composition of them in plant cells changes depending on the physiological and stress status of the plant. Aldehyde profiles obtained by HPLC analysis, therefore, can provide more detailed information than does the conventional thiobarbituric acid assay, which detects aldehydes collectively.

In this chapter, three protocols to analyze volatile oxylipins and related compounds are described. First one is on HPLC analyses of short-chain aldehydes after derivatization with 2,4-dinitrophenyl-hydrazine (DNPH). This protocol can determine the amounts of even very short-chain (and therefore water-soluble) aldehydes in an accurate way. Although many noncarbonyl oxylipins escape from this, the DNP derivatization has an advantage to increase the sensitivity due to a strong light absorbance of the derivatives. In addition, focusing on carbonyl compounds has physiological significance because they generally have high biological activities. Second protocol uses solid phase microextraction (SPME) fiber to collect volatile compounds. Subsequent direct injection of the fiber to GC-MS gives detailed profiles of volatile compounds. SPME system is developed as a convenient and an efficient system to extract volatile organic chemicals on a fiber coated with absorbing matrix *(10)*. For volatile oxylipins, SPME fiber coated with 50/30 μm DVB/Carboxen/PDMS is suitable (**Note 1**). If enough amounts (>10 g) of plant tissues are available, the closed loop stripping (CLS) system with a charcoal trap is suitable to analyze highly volatile compounds *(11)*. Vaporization of the volatiles is facilitated by the airflow from a minipump, then the volatiles are splitted onto a charcoal trap. The trapped compounds are desorbed with dichloromethane for subsequent GC-MS analyses. Instead of the charcoal filter, other absorbing matrix, such as Porapak Q or Tenax, can be used. However, again, one must notice that the recovery of each compound significantly differs with each matrix *(12)*.

2. Materials

2.1. Equipment

1. Wakosil DNPH-II column (4.6 × 150 mm) for HPLC (Wako Pure Chemical Industries, Ltd., Osaka, Japan). Equivalent ODS column should work; however, resolution of some aldehydes might be worse.
2. Common HPLC system with an absorption detector (Such as Shimadzu LC-10 system equipped with a photodiode array detector, SPD-M10A).
3. Common GC-MS apparatus (such as Shimadzu GC-MS QP-5050). If MS detector is not available, an FID detector can work. However, the retention time of each standard compound must be carefully determined for assignment of the peak).
4. SPME fiber assembly (50/30 mm DVB/Carboxen/PDMS, Supelco, Bellefonte, PA) with a manual holder (also available from Supelco). Bake the fiber before use in the injection port of GC (set at 230°C) at least for 30 min just before use.

5. A glass vial (22 ml) with butyl stopper and crimp top seal (open top). Bake the vial at 180°C for 3 h before use.
6. ORBO-32 mini charcoal tubes (Supelco).
7. Glass apparatus for closed loop splitting system (**Fig. 1**).

2.2. Reagents

1. Acetonitrile, HPLC grade.
2. 2-Ethylhexanal as an internal standard (IS) for HPLC, 0.5 mM in HPLC grade acetonitrile.
3. Butylated hydroxytoluene as antioxidant.

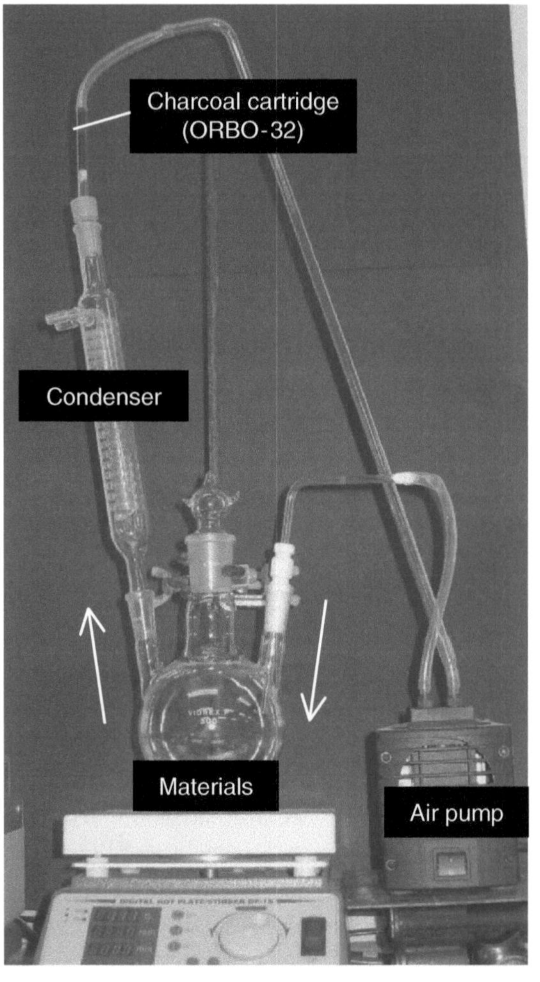

Fig. 1. Closed loop splitting (*CLS*) system. Plant materials are put into the three-way round-shaped flask (500 ml) equipped with a condenser to remove moisture. With the hot plate/stirrer, the materials can be warmed up and stirred to facilitate vaporization of volatile compounds. The airflow is introduced from the pump through the materials to the charcoal cartridge.

4. DNPH. DNPH must be purified by recrystallization because commercial DNPH contains water in order to avoid explosion and usually contaminated with spontaneously formed hydrazones. Dissolve DNPH as much as you can into hot acetonitrile (60°C), then gradually cool the solution to room temperature. Recover the crystal by simple filtration through filter paper. Handle the pure DNPH gently with avoiding any shock. Immediately dissolve the crystal to pure acetonitrile to be 20 mM.
5. Formic acid, HPLC grade.
6. Saturated NaCl solution.
7. $NaHCO_3$.
8. Deionized, purified water (equivalent to Milli-Q water).
9. HPLC Mobile Phases, Mobile phase A: Wakosil DNPH-II Eluent A (Wako Pure Chemical, Osaka, Japan), Mobile phase B: Wakosil DNPH-II Eluent B (Wako Pure Chemical), Mobile phase C: Acetonitrile. Mobile phase C is used just to facilitate the elution of highly hydrophobic compounds, and hence, it can be replaced by B, although a longer time is required for chromatography.
10. IS solution for headspace SPME-GC analyses. Take 80 mg of *n*-heptanal and mix it well with 280 mg of Tween 20. Add 100 ml of water, then, thoroughly suspend the contents with a sonicator. Fill up to 700 ml to make 1.0 mM *n*-heptanal solution in 0.04% Tween 20. The solution can be kept at –20°C, but resuspend the contents with sonication every time before use.
11. Saturated solution of $CaCl_2$. Put 500 g of $CaCl_2{\cdot}2H_2O$ in a glass bottle, then add 300 ml of distilled water, and mix them. A part of the crystal remains. Take the solution and pass it through a column packed with charcoal (3 cm id × 10 cm) in order to remove contaminating organic compounds. Since saturated $CaCl_2$ solution is very viscous, vacuum filtration is better.
12. Sodium phosphate buffer (50 mM, pH 6.3 containing 0.1-mM diethylenetriamine penta-acetic acid as a chelating agent). After preparing the buffer, remove organic chemicals with a charcoal column as described above for saturated $CaCl_2$ solution.
13. Freshly distilled dichloromethane.
14. IS solution for CLS system. *n*-Nonyl acetate dissolved in distilled dichloromethane to make final concentration of 2 μg/ml. Seal the container tightly in order to avoid evaporation of dichloromethane.

3. Methods

3.1. HPLC Analyses of Short-Chain Aldehydes After Derivatization with DNPH

1. Immerse leaf samples (0.20–0.30 g fresh weight) in acetonitrile (2.5 ml) containing 2-ethylhexanal (12.5 nmol) and butylated hydroxytoluene (0.005%) in a glass tube with a screw cap. Tighten the cap, and incubate them in a water -bath at 60°C for 30 min. (Methanol and ethanol should be avoided for extraction of the aldehydes from plant tissues because they contain significant amounts of carbonyl compounds that cannot be removed easily.
2. Collect the extract into another glass tube by decantation and add DNPH solution (62.5 μl; final concentration of 0.5 mM) and formic acid (48.4 μl; final concentration of 0.5 M; **Note 2**). Tighten the cap, mix well, and incubate the mixture at 25°C for 60 min.
3. Add 2.5 ml of saturated NaCl solution and 450-mg $NaHCO_3$ for neutralizing formic acid. Shake at intervals for 10 min. Small bubbles are formed, and cease when neutralization is accomplished.
4. After centrifugation to facilitate phase separation, collect the upper acetonitrile layer, and dry it up in vacuo.
5. Add 400 μl acetonitrile, vortex, and collect the solution.
6. To remove chlorophylls and other pigments, load the sample solution on a BondEluteC18 cartridge (sorbent mass 200 mg, Varian), which has been pre-washed with 2 ml acetonitrile, and collect the pass-through. Apply additional 350 μl acetonitrile and combine the eluted solution with the pass-through. This solution may be kept for a while at -20°C in a tightly capped vial. Avoid drying up the solution, else the hydrazones easily decompose spontaneously.
7. Subject the solution to the reversed phase HPLC system. The compounds can be separated with 100% A (0–5 min), a linear gradient from 100% A to 100% B (5–20 min), and subsequently, 100% B (20–25 min) with a flow rate of 1 ml/min. In order to clean the column, 100% C can be used. DNP-derivatives of *n*-alkanals have a characteristic absorption peak at around 360 nm, and those of 2-alkenals at around 380 nm (**Fig. 2**). The DNP-derivative of MDA, in contrast, has a peak at 307 nm because of its ring structure (*13*). For photometric detection of these DNP derivatives, a photodiode array detector is suitable. For single-wavelength detection, 340 nm is a compromise (**Fig. 2**). In order to assign each peak, corresponding standard aldehyde-DNP derivative must be prepared (**Note 3**).

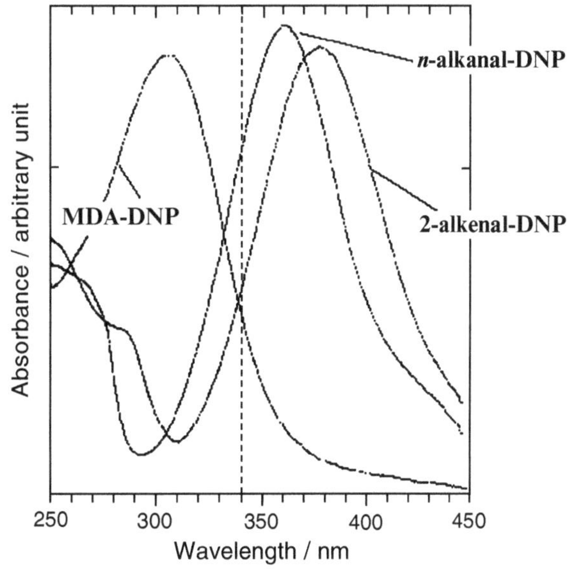

Fig. 2. Absorption spectra of DNP-derivatives of *n*-alkanal, 2-alkenal, and MDA.

3.2. Headspace SPME-GC-MS Analyses of Volatile Oxylipins

1. Plant leaves (0.5 g fresh weight) are homogenized with a Polytron mixer (Kinematica) with 5 ml of 50-mM sodium phosphate buffer (pH 6.3) containing 0.1-mM diethylenetriamine pentaacetic acid in the 22-ml-glass vial.
2. The vial is capped with the butyl stopper, and then incubated for 5 min at 25°C in order to facilitate enzymatic reactions to form volatile oxylipins. The IS (10 μl of 1.0-mM *n*-heptanal in 0.04% Tween 20) is added.
3. Afterward, 5 ml of saturated solution of $CaCl_2$ is added to inactivate the enzymes, and the stopper is tightly fastened with the crimp top seal. The homogenate can be stored at −80°C for a week. If volatiles in intact plant leaves are analyzed, homogenize the leaves with 5 ml of saturated solution of $CaCl_2$ to avoid any enzymatic reactions during and after homogenization. Thereafter, add the buffer and IS as described before.
4. The homogenate is warmed to 40°C, then insert the SPME fiber through the butyl stopper. The matrix is exposed to the headspace for 30 min at 40°C. The fiber is inserted into the insertion port (set at 230°C) of the GC-MS (Shimadzu QP-5050) equipped with 0.25 μm × 30 m Stabiliwax column (Restek, Bellefonte, PA). The column temperature is programmed as 40°C for 5 min to 180°C for 5 min at 10°C/min with a carrier gas (He) at 1 ml/min. The mass detector was operated in electron impact mode with ionization energy of 70 eV. The glass insert must be a special one for SPME

analysis (SPME Sleeve, available from Supelco), otherwise highly volatile compounds appear as broad peaks. Splitless injection with a sampling time of 2 min is used. The fiber is kept inserted into the injection port until the end of analysis in order to bake out any compounds from the matrix.

3.3. CLS System to Analyze Volatile Oxylipins and Related Compounds

1. Fresh mushrooms (50 g fresh weight) are diced and homogenized with 75 ml of 0.1 M sodium phosphate (pH 6.5) with a Polytron mixer (Kinematica).
2. Put the homogenate into to the round-shaped flask of close loop splitting system (**Fig. 1**). For highly volatile compounds such as C6-volatiles, the temperature of the hot stirrer is set at 30°C. For low volatile compounds, the temperature can be increased, but one should be careful not to make the charcoal wet (because of the high humidity of the airflow), and not to heat-denature labile volatile compounds. In order to facilitate vaporization of volatiles, the airflow can be introduced into the homogenate with bubbling. In some cases, addition of antifoam is essential; however, avoid the one consisted of silicon polymer because it results in contamination of Si compounds. Antifoam consisted of monoglycerides and lecithin works well. If one would like to avoid using antifoam, flow the air on the surface of homogenate with stirring the homogenate vigorously with a magnetic stirring bar.
3. The volatiles from the sample are split on a charcoal trap (ORBO-32 Mini) for 3 h with a flow rate of 4 L/min. In order to clean the pump, connect the outlet and inlet directly with the ORBO cartridge, and keep flowing for 1 h before use.
4. After splitting, take off the cartridge, and remove the retaining plugs and backup adsorbent and the additional retaining plugs. Slowly and carefully add 250 μl of dichloromethane containing 2 μg/ml of nonyl acetate as an IS into the outlet of the cartridge to elute volatiles. Rapid addition of the solvent results in explosion of the charcoal from the end. Repeat the elution once again with the same volume of the dichloromethane solution. Now, the eluted solution is ready for GC-MS analysis. It can be kept at –20°C for at least a week.
5. GC-MS analysis is performed as described for SPME-GC-MS analyses, but the insert glass must be replaced with the normal one for splitless injection. One microliter of the eluted solution usually gives satisfactory result.

3.4. Results

3.4.1. Peak Identification and Quantification of Aldehyde-DNP Derivatives

DNP-aldehydes can be identified by their retention time. For the plant sources that have not been analyzed before, however, it is better to confirm the identity of each peak with mass spectrometer. **Table 1** is a list of typical DNP-aldehydes with their retention

Table 1
Retention time and conversion coefficient (*k*) when detection is performed at 340 nm for DNP derivatives of aldehydes

Aldehyde or ketone	Retention time (min)	*k*
MDA	4.55	3.24
Formaldehyde	5.28	3.32
Acetaldehyde	6.96	0.49
4-Hydroxy-(*E*)-2-hexenal	7.45	0.61
Acetone	9.37	3.66
Acrolein	10.09	1.36
Propionaldehyde	11.22	0.56
Crotonaldehyde	13.78	2.89
Butylaldehyde	15.63	0.54
Phenylacetaldehyde	17.95	0.59
(*E*)-2-Pentenal	18.30	2.71
n-Pentanal	19.69	0.62
(*Z*)-3-Hexenal	20.30	0.52
(*E*)-2-Hexenal	22.19	2.36
n-Hexanal	23.37	0.46
2-Ethylhexanal (IS)	29.00	1.0
(*E*)-2-Nonenal	34.33	2.72

time. In order to determine the amount of an aldehyde from the peak area at 340 nm, the peak area ratio of the aldehyde and the IS (2-ethylhexanal) is determined first. The content of the aldehyde can be obtained by multiplying this ratio by the conversion factor *k* (**Table 1**). The *k* value of each aldehyde, empirically determined, reflects both the derivatization/extraction efficiency and the absorption coefficient of its DNP-derivative.

3.4.2. Peak Identification and Quantification of SPME-GC-MS

A representative chromatogram is shown in **Fig. 3**. For quantification of each compound, an aqueous solution of the corresponding authentic compound should be prepared with Tween 20 as described for IS solution (*n*-heptanal). Organic solvents must

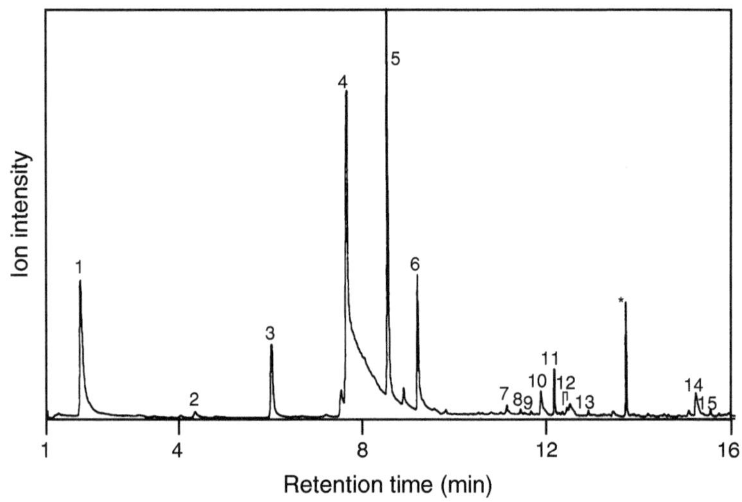

Fig. 3. A representative total ion chromatogram of volatile oxylipins formed in disrupted *Arabidopsis* leaves (ecotype Nossen) analyzed with SPME GC-MS. The labels represent following: 1, acetone; 2, 1-penten-3-one; 3, *n*-hexanal; 4, (*Z*)-3-hexenal; 5, *n*-heptanal (IS); 6, (*E*)-2-hexenal; 7, (*Z*)-2-penten-1-ol; 8, 6-methyl-hepten-2-one; 9, *n*-hexan-1-ol; 10, (*Z*)-3-hexen-1-ol; 11, butyl 1-isothiocyanate; 12, 2,4-hexadienal (two peaks of geometrical isomers); 13, 3-methylbutyl 1-isothiocyanate; 14, 5-ethyl-2(5H)-furanone; 15, b-cyclocitral; *, contamination from a plastic ware.

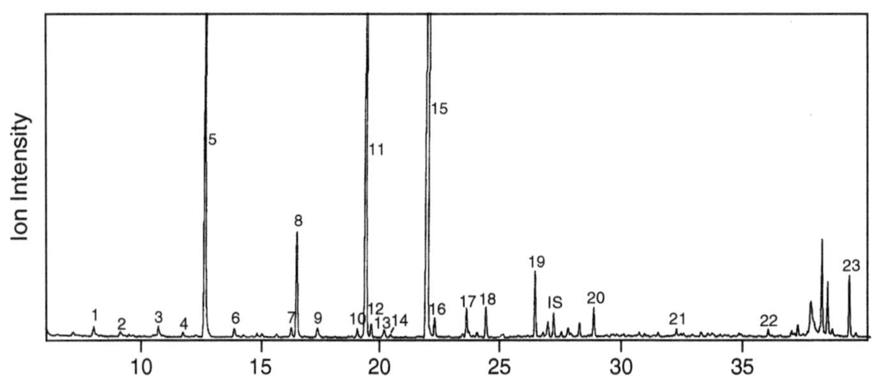

Fig. 4. A representative total ion chromatogram of volatiles formed in disrupted mushrooms (*Astraeus hygrometricus*) extracted with CLS system. The labels represent following: 1, 1-butanol; 2, pyridine; 3, 1-pentanol; 4, 2-*n*-pentylfuran; 5, 3-octanone; 6, 3-hydroxy-2-butanone; 7, unknown (M^+ = 111, $C_7H_{11}O$); 8, 6-methyl-5-hepten-2-one; 9, 4-hydroxy-4-methyl-2-pentanone; 10, *n*-nonanal; 11, 3-octanol; 12, tetradecane; 13, 2-cyclohexen-1-one; 14, 1-octen-3-one; 15, 1-octen-3-ol; 16, furfural; 17, 2-ethyl-1-hexanol; 18, benzaldehyde; 19, 1-octanol; 20, (*E*)-2-octen-1-ol; 21, 4-(5-methyl-2-furanyl)-2-butanone; 22, (*E,E*)-2,4-decadienal; 23, 4-(2-furanyl)-(*E*)-3-buten-2-one.

be avoided because they extensively affect adsorption kinetics of compounds on the matrix. With the aqueous solution, prepare serial dilution of it with the buffer and saturated $CaCl_2$, and calibration curves must be constructed by using peak area ratio between the compound and the IS.

3.4.3. Peak Identification and Quantification of CLS-GC-MS

A representative chromatogram is shown in **Fig. 4**. The compounds are identified with their GC retention times and mass spectra by comparing them against those obtained with authentic standards. The retention index of each compound is calculated using *n*-alkanes (C_7–C_{30}, Sigma) as external references. This procedure can also be used to analyze volatiles emitted from living plant parts without cutting even under their growing conditions. In that case, Tedlar bag is usually used to cover the plant organs. Because the air is circulated, humidity increases and CO_2/O_2 balance changes during extraction, which might cause artificial effect on the physiology of plants. In order to avoid this problem, the fresh air is flowed into the Tedler bag *(14)*, but a special care must be paid to avoid contaminations.

4. Notes

1. For selection of the matrix of SPME fiber, one must notice that the recovery of a compound differs significantly depending on the nature of the matrix. Detailed information on SPME is available on the web site of Sigma-Aldrich (http://www.sigmaaldrich.com).
2. Formic acid can be substituted with 0.5-M acetic acid, but the former results in a more efficient derivatization of MDA (*15*). Harder acid such as phosphoric acid and HCl will facilitate the isomerization of *g,d*-unsaturated carbonyl to *b,g*-unsaturated carbonyl [e.g., (*Z*)-3-hexenal to (*E*)-2-hexenal], and hence should be avoided.
3. In order to prepare standard aldehyde-DNP derivatives, add 1 mmol of aldehyde (some aldehydes are available only as their acetals; *see* below) or ketone and a few drops of formic acid to the DNPH solution (ca. 0.3 g of DNPH in 20-ml ethanol warmed to 50°C), and incubate it at 50°C with gentle stirring until the color changes. Keep the solution at ambient temperature overnight to form the crystals of the DNP-derivative, and collect them on a glass filter. When an aldehyde has to be prepared from its acetal, first dissolve the acetal at 1–10 mM in 0.1-M HCl, and incubate at 40°C for 1 h. Then add 0.035% DNPH (in 1-M HCl) to an equimolar amount, and incubate it at ambient temperature for 1 h. Extract the formed DNP-derivative with hexane, chill the hexane solution to –20°C to facilitate crystallization, and collect the crystals.

Acknowledgments

This research was supported by Grants-in-Aid for Scientific Research on Priority Area (Grant 19045021) and Scientific Research (C) (Grant 185801059) to K.M. from the Ministry of Education, Culture, Sports, Science, and Technology of Japan and by the JSPS-NRCT Scientific Cooperation Program under the Core University Program on Microbial Resources to K.M. and P.K.

References

1. Wasternack, C. (2007) Jasmonates: an update on biosynthesis, signal transduction and action in plant stress response, growth and development. *Ann. Bot. (Lond)* **100**, 681–697.
2. Matsui, K. (2006) Green leaf volatiles: hydroperoxide lyase pathway of oxylipin metabolism. *Curr. Opin. Plant Biol.* **9**, 274–280.
3. Kishimoto, K., Matsui, K., Ozawa, R., and Takabayashi, J. (2008) Direct fungicidal activities of C6-aldehydes are important constituents for defense responses in Arabidopsis against *Botrytis cinerea. Phytochemistry* **69**, 2127–2132.
4. Shiojiri, K., Kishimoto, K., Ozawa, R., Kugimiya, S., Urashimo, S., Arimura, G., Horiuchi, J., Nishioka, T., Matsui, K., and Takabayashi, J. (2006) Changing green leaf volatile biosynthesis in plants: an approach for improving plant resistance against both herbivores and pathogens. *Proc. Natl. Acad. Sci. USA.* **103**, 16672–1667
5. Engelberth, J., Seidl-Adams, I., Schultz, J.C., and Tumlinson, J.H. (2007) Insect elicitors and exposure to green leafy volatiles differentially upregulate major octadecanoids and transcripts of 12-oxo phytodienoic acid reductases in *Zea mays. Mol. Plant Microbe Interact.* **20**, 707–716.
6. Montillet, J.-L., Cacas, J.-L., Garnier, L., Montané, M.-H., Douki, T., Bessoule, J.-J., Polkowska-Kowalczyk, L., Maciejewska, U., Agnel, J.-P., Vial, A., and Triantaphylidès, C. (2004) The upstream oxylipin profile of *Arabidopsis thaliana*: a tool to scan for oxidative stresses. *Plant J.* **40**, 439–451.
7. Porter, N. A. (1984) Chemistry of lipid peroxidation. *Methods Enzymol.* **105**, 273–282.
8. Mueller, M. J. (2004) Archetype signals in plants: the phytoprostanes. *Curr. Opin. Plant Biol.* **7**, 441–448.
9. Bernoud-Hubac, N., Davies, S. S., Boutaud, O., Montine, T. J., and Roberts, L. J. II., (2001) Formation of highly reactive γ-ketaldehydes (neuroketals) as products of the neuroprostane pathway. *J. Biol. Chem.* **276**, 30964–30970.
10. Dietz, C., Sanz, J., and Cámara, C. (2006) Recent developments in solid-phase microextraction coatings and related techniques. *J. Chromatogr. A* **27**, 1103, 183–192.
11. Buttery, R.G. and Ling, L.C. (1996) Methods for isolating food and plant volatiles. *In* Biotechnology for improved foods and flavors (Takeoka, G.R., Teranishi, R., Williams, P.J., and Kobayashi, A., eds.). ACS Symposium Series 637, pp. 240–248.
12. Fäldt, J., Eriksson, M., Valterová, I., and Borg-Karlson, A.K. (2000) Comparison of headspace techniques for sampling volatile natural products in a dynamic system. *Z. Naturforsch [C].* **55**, 180–188.
13. Fenaille, F., Mottier, P., Turesky, R. J., Ali, S., and Guy, P. A. (2001) Comparison of analytical techniques to quantify malondialdehyde in milk powders. *J. Chromatogr. A* **921**, 237–245.
14. Kessler, A., Halitschke, R., Diezel, C., and Baldwin, I.T. (2006) Priming of plant defense responses in nature by airborne signaling between *Artemisia tridentata* and *Nicotiana attenuata. Oecologia.* **148**, 280–292.
15. Andreoli, R., Manini, P., Corradi, M., Mutti, A., and Niessen, W. M. A. (2003) Determination of patterns of biologically relevant aldehydes in exhaled breath condensate of healthy subjects by liquid chromatography/atmospheric chemical ionization tandem mass spectrometry. *Rapid Commun. Mass Spectrom.* **17**, 637–645.

Chapter 3

Quantification of Lysophosphatidylcholine Species by High-Throughput Electrospray Ionization Tandem Mass Spectrometry (ESI-MS/MS)

Gerhard Liebisch and Gerd Schmitz

Summary

Lysophosphatidylcholine (LPC) is a bioactive lipid implicated to play a functional role in various diseases including atherosclerosis, diabetes, cancer, and inflammation. Conventional methods are of limited value for a systematic evaluation of LPC species concentrations due to complicated, time-consuming procedures. Here we describe a methodology based on electrospray ionization tandem mass spectrometry (ESI-MS/MS) applicable for high-throughput LPC species quantification. This assay provides accuracy and precision sufficient for the analysis of large clinical studies as well as basic biochemical studies in a broad range of biological material including plasma, tissues, and cell culture material. This method may be combined with methods based on the same analytical setup for glycerophospholipid, sphingolipid, and sterol analysis to evaluate LPC species in relation to other lipid species profiles.

Key words: Tandem mass spectrometry, Electrospray ionization, Glycerophospholipid

1. Introduction

Lysophospholipids including lysophosphatidylcholine (LPC) are known as biological active lipids that elicit a wide range of biological functions including cell proliferation, cell migration, angiogenesis, and inflammation *(1–3)*. LPC is a relevant component of human plasma originating from lecithin-cholesterol acyltransferase, hepatic secretion, or action of phospholipase A2 (PLA2). Recently, lipoprotein-associated phospholipase A_2 activity, an enzyme that hydrolyzes oxidized phospholipids generating LPC, was associated with stroke risk *(1, 2)*. Plasma levels of LPC

Donald Armstrong (ed.), *Lipidomics*, Methods in Molecular Biology, vol. 580,
DOI 10.1007/978-1-60761-325-1_3, © Humana Press, a part of Springer Science+Business Media, LLC 2009

species were discussed as a biomarker in ovarian *(4)* and colorectal *(5)* cancer as well as sepsis *(6)*. Taken together, these data highlight the role LPC as bioactive lipid and indicate a potential use of LPC as diagnostic marker. Cellular LPC levels have been implicated in the phagocytosis of apoptotic cells both as ligand and as chemoattractant signal *(7)*. Another interesting function of LPC is its contribution to the therapeutic action of the antidiabetic drug metformin in hepatocytes *(8)*.

A variety of different techniques are used for LPC measurement. Conventional techniques like thin-layer chromatography and high performance liquid chromatography (HPLC) separations usually do not allow analysis of LPC species. To determine LPC species, a combination of thin-layer chromatography separation followed by gas chromatographic analysis or HPLC coupled to mass spectrometry (LC-MS) is needed. Apart from that, a direct analysis of LPC species is possible by direct flow injection electrospray ionization tandem mass spectrometry (ESI-MS/MS) of crude lipid extracts *(9)*. In contrast to other methodologies, this method is applicable for high-throughput due to its short run time and easy sample preparation. This LPC analysis was already applied for a number of different biological materials like plasma and different tissues like liver, lung, heart, gut, and cell culture material. Moreover, using this analytical setup, LPC profiles *(9)* may be combined with species level of phosphatidylcholine, sphingomyelin *(10)*, phosphatidylethanolamine, phosphatidylserine, ceramide *(11)*, and free cholesterol/cholesteryl ester *(12)*. Consequently, LPC species analyzed by the method described in this chapter may be integrated into a lipidomic profile using the same analytical setup.

2. Materials

2.1. Equipment

1. Triple quadrupole mass spectrometer, Quattro Ultima (Micromass, UK) equipped with an electrospray ion source.
2. Autosampler, HTS PAL autosampler (Zwingen, Switzerland).
3. Agilent 1100 binary pump (Waldbronn, Germany).
4. Glass centrifuge tubes (10 ml).
5. Glass autosampler vials (1.5 ml).
6. Glass vessels with Teflon screw cap (10 ml).
7. Glass Pasteur pipettes.
8. Electric pipette filler.
9. Alternative to 7 and 8: Pipetting robot.
10. Centrifuge with buckets and adapters for centrifuge tubes (10 ml).

11. Vacuum concentrator (Christ,Osterode, Germany).

2.2. Reagents

1. Methanol – HPLC grade.
2. Chloroform – HPLC grade (*see* **Note 1**).
3. Water – Millipore.
4. Ammonium acetate.

3. Methods

3.1. Sample Collection

Because LPC may be generated very fast by PLA_2 action from phosphatidylcholine, all sample materials have to be stored immediately on ice and frozen at –80°C until analysis.

3.2. Preparation and Storage of Lipid Standards and Calibrators

LPC standards were all 1-acyl-2-hydroxy-sn-glycero-3-phosphocholines purchased from Avati Polar Lipids (Alabaster, Alabama) with purity higher than 99%.

1. Dissolve LPC standards at a concentration of 1 mg/ml in chloroform in 10 ml screw cap glass vessels.
2. Store standards at –80°C.

ESI is highly matrix dependent. Therefore, for direct flow injection of crude lipid extracts, it is necessary to use calibrators based on the same matrix as the samples. For example, analysis of plasma samples requires calibration samples prepared by addition of naturally occurring LPC species to pooled plasma. Similarly, for the quantification of LPC from cell culture material, LPC standards are added to pooled cell culture material (*see* **Table 1** for species, calibration range, and material requirement).

1. Mix LPC standards (**Table 1**) to one calibrator solution in chloroform. Use 10 ml screw cap glass vessel.
2. Place a defined volume of the LPC calibrator into glass centrifuge tubes to get different calibrator levels.

Table 1
Calibrator concentrations

Species	Cultured cells	Human plasma
LPC 16:0	0–100 pmol	0–1000 pmol
LPC 18:0	0–100 pmol	0–1000 pmol
LPC 18:1	0–100 pmol	0–1000 pmol
LPC 22:0	0–100 pmol	0–1000 pmol
Sample used	100 µg protein	20 µl

Table 2
Internal standard concentrations

Material	Material used	LPC 13:0 and LPC 19:0 added
Cultured cells	100 μg protein	50 ng each
Human plasma	20 μl	1000 ng each

3. Add internal standards (*see* **Table 2**).
4. Remove the solvent in a vacuum concentrator.
5. Add sample pool and mix thoroughly.
6. Subjected to lipid extraction as described in 3.3 starting with step 3.

3.3. Lipid Extraction

To avoid the loss of lipids, extraction is performed in glassware. Lipid extraction is performed according to the protocol of Bligh and Dyer *(13)*:

1. Place internal standards into a glass centrifuge tube (*see* **Table 2**).
2. Remove solvent in a vacuum concentrator.
3. Add sample and fill to 800 μl with water.
4. Add 3 ml of a methanol/chloroform = 3/1 (v/v).
5. Vortex thoroughly and incubate 60 min at room temperature.
6. Add 1 ml chloroform.
7. Add 1ml water.
8. Vortex thoroughly.
9. Centrifuge at 1800 × *g* for 10 min.
10. Recover the lower chloroform phase manually using Pasteur pipette and electric pipette filler or automated by a pipetting robot (*see* **Note 2**).
11. Dispense chloroform phase into 1.5 ml autosampler vials.
12. Remove chloroform in a vacuum concentrator.
13. Dissolve samples in mobile phase (*see* **Subheading 3.4.1**).

3.4. Direct Flow Injection Analysis

3.4.1. Mobile Phase Preparation

1. Prepare a solution of 10 mM ammonium acetate in methanol. Dissolve 770.8 mg ammonium acetate in 1 L methanol by agitation.
2. Mix the methanolic 10 mM ammonium acetate solution with chloroform in a volume ratio of 3:1.

Table 3
Flow gradient

Time (min)	Flow (ml/min)
0.00	0.05
0.09	0.05
0.10	0.03
1.10	0.03
1.11	0.20
1.29	0.20
1.30	0.05

3.4.2. Sample Injection and Flow Gradient

1. Autosampler syringe: 100 μl.
2. Inject loop: 20 μl.
3. Injection volume: 20 μl.
4. Use inject ahead function, which means the next sample is drawn during the current run.
5. Wash solvent: Chloroform/methanol = 1/1 (v/v). Wash syringe three times and inject port once with 70 μl each.
6. Operate HPLC pump isocratic with a flow gradient (*see* **Table 3**, mobile phase *see* **Subheading 3.4.1**).
7. Run time: 1.3 min.

3.4.3. Mass Spectrometer Settings

1. Ionization mode: ESI positive.
2. Ionization voltage: 3500 V.
3. Source temperature: 300°C.
4. Cone voltage: 41 V.
5. Collision gas: Argon.
6. Collision gas pressure: 1.3×10^{-3} Torr.
7. Collision energy: 24 V.
8. MS/MS-mode: Precursor ion scan of m/z 184.1; for samples with low LPC level, use MRM transitions.
9. MS run time. 1.3 min.
10. Operate both Q1 and Q3 at unit resolution.

3.5. Data Analysis

3.5.1. Raw Data Analysis

Analysis of the raw data is performed by Neolynx, a tool of Masslynx (mass spectrometer software).

1. Combine spectra at half peak height (do not exceed the constant flow range from 0.1 to 1.1 min).

2. Smooth combined spectrum (not necessary using MRM mode).
3. Centroid spectrum (not necessary using MRM mode).
4. Pick peak intensities of LPC species.
5. Export peak intensities and sample information as comma separated value file.

3.5.2. Data Processing

Data processing is performed by Excel Macros.

1. Import peak intensities and sample information from comma separated value file into Excel.
2. Arrange data: Samples in separate rows and species in columns.
3. Correct isotope overlap using calculated isotope distribution (five isotope peaks including the monoisotopic) as described for phosphatidylcholine and sphingomyelin *(10)*. Peak intensities are corrected for the isotope overlap in a sequential algorithm starting from low mass species.
4. Calculate ratios to both internal standards LPC 13:0 and LPC 19:0.

3.5.3. Calibration and Quantification

Calibration and quantification are performed by Excel Macros.

1. Generate calibration lines for all species added to calibration samples.
2. Use calibration line slopes to quantitate LPC species. LPC species without calibration line are quantified with the closest related slope. For example, for LPC 14:0, LPC 15:0, LPC 16:1, and LPC 16:0, the slope of the LPC 16:0 calibration curve is used; for LPC 17:0, LPC 18:2, LPC 18:1, and LPC 18:0, the slope of the LPC 18:0 calibration curve is used; for LPC 20:4, LPC 20:3, LPC 22:6, and LPC 22:5, the slope of the LPC 22:0 calibration curve is used.

3.5.4. Quality Check

1. Compare ratio of internal standards LPC 13:0 to LPC 19:0, which are added in a constant ratio. Ratios should not deviate from the mean ratio by more than 5%. Otherwise the LPC species response of these samples may be disturbed by matrix interference or other analytical problems. A reanalysis is recommended for those samples.
2. For large clinical studies, it is recommended to include three quality control samples with different concentration levels into each batch. This allows monitoring of the performance from batch to batch. Quality control samples outside a defined range lead to reanalysis of the batch.

3.6. Results

The assay described was validated for plasma samples. Calibration lines of the spiked LPC species showed a linear response for both internal standards LPC 13:0 and LPC 19:0 in the spiked range.

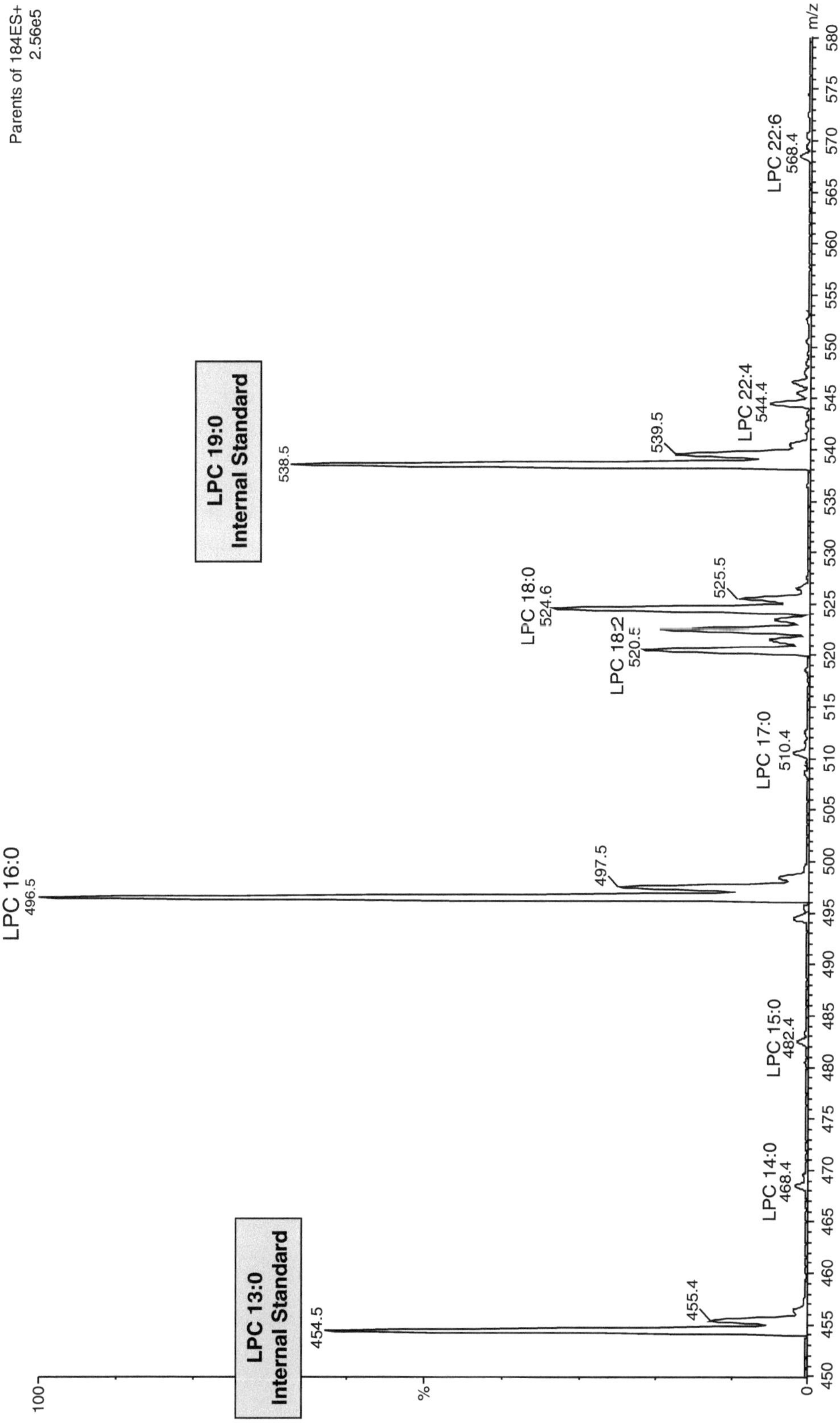

Fig. 1. Precursor ion scan of m/z 184 of human plasma.

Since ESI is highly affected by matrix components, we tested the influence of the lipid concentration and different plasma cholesterol concentrations on the LPC species response. Neither 50- to 400-fold dilution related to the extracted plasma volume nor different cholesterol level showed a significant influence on the calibration line slopes. The in-run precision for the total LPC concentration and the major species was ~3% coefficient of variation and 10–13% for the minor species. The day-to-day precision was significantly lower using LPC 13:0 compared with LPC 19:0 as internal standard with ~12% for the major and 25% for the minor species. The limit of detection was between 0.6 and 0.8 μM related to the LPC plasma concentration.

Analysis of crude plasma lipid extracts by a precursor-ion scan of *m/z* 184 showed the main species LPC 16:0, LPC 18:2, LPC 18:1, and LPC 18:0 and the minor species LPC 14:0, LPC 15:0, LPC 16:1, LPC 17:0, LPC 20:4, LPC 20:3, LPC 22:6, and LPC 22:5 (**Fig. 1**). The spectra contained no peaks that could not be assigned to distinct LPC species showing the specificity of the analysis. In summary, these data support the performance of this assay. Together with other assays *(10–12)* on the basis of ESI-MS/MS, the presented method may be a valuable tool for large clinical studies as well as basic biochemical research to further elucidate the function of lipid species.

4. Notes

1. Always use fresh chloroform because it may decompose to hydrochloric acid, phosgene, and chlorine.
2. Try to exclude contamination of the upper phase containing salt and other water-soluble material.

Acknowledgments

This work was supported in part by Deutsche Forschungsgemeinschaft (Li 923/2–1/2) and by the seventh framework program of the EU-funded “LipidomicNet” (proposal number 202272).

References

1. Matsumoto T, Kobayashi T, Kamata K. (2007) Role of lysophosphatidylcholine (LPC) in atherosclerosis. *Curr Med Chem.* **14**, 3209–3220.
2. Gorelick PB. (2008) Lipoprotein-associated phospholipase A2 and risk of stroke. *Am J Cardiol.* **101**, 34F–40F.

3. Meyer zu HD, Jakobs KH. (2007) Lysophospholipid receptors: signalling, pharmacology and regulation by lysophospholipid metabolism. *Biochim Biophys Acta.* **1768**, 923–940.
4. Okita M, Gaudette DC, Mills GB, Holub BJ. (1997) Elevated levels and altered fatty acid composition of plasma lysophosphatidylcholine(lysoPC) in ovarian cancer patients. *Int J Cancer.* **71**, 31–34.
5. Zhao Z, Xiao Y, Elson P et al (2007) Plasma lysophosphatidylcholine levels: potential biomarkers for colorectal cancer. *J Clin Oncol.* **25**, 2696–2701.
6. Drobnik W, Liebisch G, Audebert FX et al (2003) Plasma ceramide and lysophosphatidylcholine inversely correlate with mortality in sepsis patients. *J Lipid Res.* **44**, 754–761.
7. Mueller RB, Sheriff A, Gaipl US, Wesselborg S, Lauber K. (2007) Attraction of phagocytes by apoptotic cells is mediated by lysophosphatidylcholine. *Autoimmunity.* **40**, 342–344.
8. Wanninger J, Neumeier M, Weigert J et al (2008) Metformin reduces cellular lysophosphatidylcholine and thereby may lower apolipoprotein B secretion in primary human hepatocytes. *Biochim Biophys Acta.* **1781**, 321–325.
9. Liebisch G, Drobnik W, Lieser B, Schmitz G. (2002) High-throughput quantification of lysophosphatidylcholine by electrospray ionization tandem mass spectrometry. *Clin Chem.* **48**, 2217–2224.
10. Liebisch G, Lieser B, Rathenberg J, Drobnik W, Schmitz G. (2004) High-throughput quantification of phosphatidylcholine and sphingomyelin by electrospray ionization tandem mass spectrometry coupled with isotope correction algorithm. *Biochim Biophys Acta.* **1686**, 108–117.
11. Liebisch G, Drobnik W, Reil, M et. al (1999) Quantitative measurement of different ceramide species from crude cellular extracts by electrospray ionization tandem mass spectrometry (ESI-MS/MS). *J Lipid Res.* **40**, 1539–1546.
12. Liebisch G, Binder M, Schifferer R, Langmann T, Schulz B, Schmitz G. (2006) High throughput quantification of cholesterol and cholesteryl ester by electrospray ionization tandem mass spectrometry (ESI-MS/MS). *Biochim Biophys Acta.* **1761**, 121–128.
13. Bligh EG, Dyer WJ. (1959) A rapid method of total lipid extraction and purification. *Can J Biochem Physiol.* **37**, 911–917.

Chapter 4

Use of Lipidomics for Analyzing Glycerolipid and Cholesteryl Ester Oxidation by Gas Chromatography, HPLC, and On-line MS

Arnis Kuksis, Jukka-Pekka Suomela, Marko Tarvainen, and Heikki Kallio

Summary

Various analytical techniques have been adopted for the isolation and identification of the oxolipids and for determining their functionality. Gas chromatography in combination with mass spectrometry (MS) has been specifically utilized in analysis of isoprostanes and other low molecular weight oxolipids, although it requires derivatization of the solutes. In contrast, liquid chromatography (LC) in combination with on-line MS has proven to be well suited for analysis of intact oxolipids without (or minimal) derivatization. LC-MS has also been helpful for the identification of lipidomic changes resulting from covalent binding of lipid ester core aldehydes to amino lipids, amino acids, peptides, and proteins. This chapter reviews the use of the above techniques for lipidomic analysis of the autoxidation products of cholesteryl esters and glycerolipids as practiced in the authors' laboratories.

Key words: Mass spectrometry, Oxo-fatty acids, Oxo-glycerolipids, Oxo-cholesteryl esters, Sample preparation, Plasma, Cells and tissues, Urine, Atheroma, Amino acid and nucleic acid binding, Receptor binding

1. Introduction

Lipidomics has been defined as the characterization of the molecular species of lipids in biological samples *(1)*. For the purpose of this chapter, the term lipidomics is limited to the analysis of the oxidation products of glycerolipids and cholesteryl esters. The analysis of oxidized lipids has become increasingly popular in recent years due to the widespread recognition of the importance of lipid autoxidation in biology, medicine, and food industry. With few

Donald Armstrong (ed.), *Lipidomics*, Methods in Molecular Biology, vol. 580,
DOI 10.1007/978-1-60761-325-1_4, © Humana Press, a part of Springer Science+Business Media, LLC 2009

exceptions, autoxidation affects all lipid classes, although some lipids are more susceptible to it than others. The significance of oxidation arises from the specific physicochemical, chemical, and biological activities of the oxidized lipids, which are peculiar to each lipid class. In addition to the long established adverse metabolic effects of oxocholesterol and short-chain aldehydes, recent work has added the detrimental metabolic effects of the hydroperoxides, isoprostanes, and core aldehydes of the glycerolipids. Less specifically, analyses of oxolipids have provided an indication of a degenerating environment, food spoilage, ageing, and disease. Various analytical techniques have been employed in the isolation and identification of the oxolipids and in the investigation of their functionality. Tissue lipid oxidation products have been difficult to identify, until recently, when ultrasensitive detectors and improved gas chromatography (GC), liquid chromatography (LC), and mass spectrometry (MS) techniques have become available. GC and LC in combination with online electrospray ionization (ESI) and tandem mass spectrometry (MS/MS) have been especially well suited for this purpose and constitute major tools of lipidomics. However, certain techniques of proteomics such as gel electrophoresis, trypsin digestion of proteins, and LC of peptides along with MS/MS analysis are required for identification of oxolipid adducts of proteins.

Following a brief consideration of the specialized methodology, which guards against artifact formation during extraction and derivatization and LC/ESI-MS analysis, the chapter recounts in an orderly fashion the isolation and resolution of oxo-glycerolipids and oxo-cholesteryl esters followed by identification of the component oxo-fatty acids common to both oxo-glycerolipids and oxo-steryl esters. The chapter reviews the combination of chromatographic and MS techniques as practiced in the authors' laboratories. The main materials employed in these studies have been autoxidized plasma lipoproteins, dietary fats and oils, and lipid extracts of atherosclerotic tissues. In addition, extensive use has been made of synthetic preparations of oxolipids both as model products of lipid autoxidation and as reference standards for identification of oxolipids from natural sources.

2. Materials

2.1. Equipment

2.1.1. Gas Chromatographs

1. Hewlett-Packard (Palo Alto, CA) Model 5880 Gas Chromatograph equipped with a flexible quartz column (8 m × 0.30 mm ID) coated with a permanently bonded nonpolar SE-54 liquid phase (Hewlett-Packard) and a hydrogen

flame ionization detector. Introduce samples by means of a fused silica capillary needle via an air-cooled injector.

2. Hewlett-Packard Model 5890 Automatic Gas Chromatograph equipped with an automatic sampler Model 7673A and a flexible quartz column (15 m × 0.32 mm ID) coated with a bonded with a polar SP 2380 (Supelco, Mississauga, ON) equipped with an automatic injector and flame ionization detector.

2.1.2. Liquid Chromatographs

1. Hewlett-Packard Model 1050 Liquid Chromatograph equipped with a silica column (Spherisorb, 3 μm, 100 × 4.6 mm ID, Alltech, Guelph, ON) interfaced with a variable wavelength detector.
2. Hewlett-Packard Model 1050 Liquid Chromatograph equipped with a reversed phase Supelcosil LC-18 column (250 mm × 4.6 mm ID) and coupled to Varex ELSD II light scattering detector (Varex, MD).

2.1.3. Mass Spectrometers (MS and MS-MS)

1. Flow ESI-MS

   Hewlett-Packard Model 5989A single quadrupole mass spectrometer equipped with Model 59987A ESI interface using the flow injection mode.
2. Flow ESI/CID (collision-induced dissociation)-MS (pseudo MS/MS)

   Hewlett-Packard Model 5989A mass spectrometer equipped with Model 59987A ESI interface using the flow injection mode. For fragmentation studies, raise CapEx voltage from 120 V or 160 to 300 V. Use nitrogen gas both as nebulizing gas (40 psi) and as drying gas (60 psi, 270°C).
3. FAB-MS.

   Fast atom bombardment-mass spectrometry (FAB-MS) was performed using a VG-Analytical ZAB-SE instrument with a VG-11/250 data system (VG Analytical Ltd, Manchester, UK). Dissolve samples in $CHCl_3$, and use triethylamine as matrix for negative ion analysis. Effect bombardment with Xenon atoms at 8 kV anode potential and 1.2 mA anode current. Alternatively, use thioglycerol as the matrix.

2.1.4. Combined Instrumentation

1. Hewlett-Packard Model 5890 Series II Gas Chromatograph equipped with a flexible quartz column (8 m × 0.30 mm ID) coated with a permanently bonded nonpolar SE-54 liquid phase (Hewlett-Packard) coupled to a Hewlett-Packard Model 5989B single quadrupole mass spectrometer operated with an electron impact or chemical ionization with ammonia ionization source (GC-MS).

2. Hewlett-Packard Model 1084B Liquid Chromatograph equipped with a normal phase silica column (Spherisorb, 3 μm, 100 × 4.6 mm ID, Alltech) coupled to a Hewlett-Packard mass spectrometer equipped with a direct liquid inlet (DLI) interface. Admit about 1% of the LC column effluent to the mass spectrometer via DLI interface. (LC/DLI-MS).
3. Hewlett-Packard Model 1090 Liquid Chromatograph equipped with a normal phase silica column (Spherisorb, 3 μm, 100 × 4.6 mm ID, Alltech) coupled to a Hewlett-Packard Model 5985 Single quadrupole mass spectrometer interfaced with a thermospray ionization source (LC/TSI-MS).
4. Hewlett-Packard Model 1050 Liquid Chromatograph equipped with normal phase silica column (3 μm, 100 × 4.6 mm ID, Alltech) coupled to Hewlett-Packard Model 5880B Single quadrupole mass spectrometer interfaced with a nebulizer-assisted ESI interface (LC/ESI-MS). For CID/ESI-MS, raise CapEx to 300 V. (*See* **Note 1**).
5. An LC system consisting of a Hitachi (Tokyo) L-6200 Intelligent Pump with a Discovery HS C18 column (250 mm × 4.6 mm ID, Supelco, Bellefonte, PA) and running under a linear gradient of 20% IPA in MeOH to 80% IPA in MeOH in 20 min (0.85 ml/min) was interfaced (15% of effluent) with a Finnigan MAT TSQ 700 triple quadrupole mass spectrometer (Finnigan, San Jose, CA) equipped with a nebulizer-assisted ESI interface. Full scan MS spectra were collected in negative (*m/z* 600–1,200) and positive (*m/z* 450–1,100) ionization mode. The electrospray voltages were –4.5 and +4.5 kV, respectively.

2.2. Reagents and Supplies

2.2.1. Oxidizing and Reducing Agents

Osmium tetraoxide was obtained from British Drug Houses, Ltd. (Toronto, Canada); periodic acid was from Sigma Chemical Co. (St. Louis, MO); *tert*-butyl hydroperoxide (TBHP) as 70% solution in H_2O, sodium borohydride ($NaBH_4$), sodium cyanoborohydride ($NaCNBH_3$), and triphenylphosphine (TPP) were purchased from Sigma (St. Louis, MO). 3-Chloroperoxybenzoic acid was obtained from Aldrich Chemical Co. (Milwaukee, WI).

3-Morpholinosydnoimine (SIN-1) and *N*-bis(carboxymethylamino)-ethylglycinepentaacetic acid (DTPA) were from Avanti Polar Lipids, Inc. (Alabaster, AL).

2.2.2. Derivatizing Reagents

Methoxylamine HCl (MOX reagent), trimethylsilyl chloride, *N*,*O*-bis(trimethylsilyl)trifluoroacetamide (BSTFA), and 1-fluoro-2,4-dinitrobenzene were from Pierce Chemical Co. (Rockford, IL); 2,4-dinitrophenylhydrazine was from Aldrich Chemical Co. (Milwaukee, WI); *tert*-butyldimethylchlorosilane/imidazole reagent was obtained from Supelco (Bellefonte,

PA); diisopropylethylamine (DIPE) and 2,3,4,5,6-pentafluorobenzylbromide (PFB-Br) were purchased from Sigma-Aldrich (St. Louis, MO).

2.2.3. Enzymes

Phospholipase C (*Clostridium welchii and Bacillus cereus*) were from Sigma Chemical Co. (St. Louis, MO); pancreatic lipase (pancreatin) was from Nutritional Biochemicals, Inc. (Cleveland, OH); secretory phospholipases of group IIA, V, and X were gifts to Dr. Pruzanski from Drs. J. Browning (Biogen), W. Cho (University of Illinois), and G. Lambeau (Institute de Pharmacologie, Valbonne, France).

2.2.4. Lipid Standards

Synthetic 1-stearoyl-2-oleoyl-3-stearoyl-*sn*-glycerol (18:0/18:1/18:0), 1,2-dipalmitoyl-3-oleoyl-*sn*-glycerol (16:0/16:0/18:1), 1,2-dioleoyl-3-stearoyl-*sn*-glycerol (18:1/18:1/18:0), and 1,2-distearoyl-3-linoleyl-*sn*-glycerol (18:0/18:0/18:2) were available in the laboratory *(3)*. Reference monohydroperoxy, monohydroxy, and mono C_9 core aldehyde derivatives of the dioleoyl stearoyl and distearoyl-linoleoyl-glycerols were also available from the previous study *(3)*. Reference monohydroperoxides were prepared by photosensitized oxidation *(4, 5)*, while the dihydroperoxides were prepared by oxidation with TBHP *(3)*. The corresponding hydroxyl derivatives were obtained by reduction of the hydroperoxides with $NaBH_4$ *(6)* or TPP *(7)*.Monoepoxides were prepared by the method of Deffense *(8)*. Triacylglycerol (TAG) core aldehydes were prepared by ozonization and TPP reduction *(9)*. The oxo-TAG standards were purified by TLC using heptane-$(IP)_2O$-HOAc (60:40:4, by vol) and their structures were verified by reversed phase LC/ESI-MS *(10)*.

The palmitic, oleic, linoleic, and arachidonic acid esters of cholesterol were from Sigma Chemical Co. (St. Louis, MO). Cholesteryl 5-oxovalerate and cholesterol 9-oxononanoate were prepared from cholesteryl arachidonate and cholesteryl oleate, respectively, by osmium tetroxide oxidation and periodate cleavage *(11)*. The cholesteryl ester core aldehydes were purified by TLC on silica gel H using heptane/$(IP)_2O$/glacial HOAc 60:40:4 (by vol) as the developing solvent. The aldehydes were localized by spraying the plate with Schiff's reagent, which, however, destroys the sample.

The 1-palmitoyl-2-(9-oxo)nonanoyl GroPCho was prepared from egg yolk PtdCho as described for the cholesteryl ester core aldehydes. The core aldehydes derived from the egg yolk PtdCho were purified by TLC using $CHCl_3$/MeOH/H_2O 65:35:6 (by vol) as the developing solvent *(10)*. The fluorescent silica gel zones were scraped off the plate and extracted with $CHCl_3$/MeOH 2:1 (v/v). (*See* **Note 2**).

2.2.5. Solvents and Other Chemicals

Propionitrile (EtCN) was from Romil (Loughborough, England). Other chromatographic grade solvents and reagent grade chemicals were obtained from Caledon Laboratories (Georgetown, ON) or other local suppliers.

3. Methods

3.1. Isolation of Substrates

Since oxolipids are formed by mere exposure to air, a critical methodological issue is the use of techniques that prevent any significant artifact formation by oxidation during sample preparation, purification by chromatography, as well as sample analysis. Lipid peroxidation and peroxide loss on chromatographic media has been recognized as a major concern *(13, 14)* and has led to the elimination of the chromatographic step in some tandem mass spectrometric (MS/MS) analyses *(15)*. The complexity of the oxolipids, however, has prevented the complete elimination of the chromatographic step, although new mass spectrometric methods are constantly being developed *(16–18)*.

The adsorbents for TLC and column cartridges must be washed with chelating agents to remove such divalent ions as Cu^{2+} and Fe^{2+}. It is recommended to rinse all aqueous solvents with Chelex-100 prior to use *(13, 15)*. Phosphate-buffered saline (PBS, pH 7.4, 50 mM) may need to be stored over Chelex-100 at least 24 h to remove transition metal contaminants. Furthermore, all columns and filters are to be prerinsed with diethylenetriaminepentacetic acid (DTPA)-containing solvents (at neutral pH). The washing may be done in situ, at the normal flow rate of the column. It is also good practice to complete the extraction as quickly as possible and, if necessary, to store the extracts at as low a temperature as possible (e. g. –80°C).

In most instances, tolerable recoveries of oxolipids have been obtained using the $CHCl_3$/MeOH method *(19, 20)* of extraction under inert atmosphere (glove box) and in the presence of BHT. The synthetic antioxidant, S20478 (50 μM), was capable of inhibiting initiation and propagation of copper-mediated low-density lipoprotein (LDL) oxidation as determined by the time- and dose-dependent inhibition of the formation of conjugated dienes and thiobarbituric acid-reactive substances *(21)*.

3.1.1. Lipoproteins and Oxolipoproteins

3.1.1.1. Density Centrifugation; Anonymous *(22)*

1. Obtain whole blood by venipuncture from healthy volunteers, who have fasted 12 h.
2. Collect blood samples into tubes containing EDTA (1 mg/ml).
3. Separate plasma by low-speed centrifugation (1,500 × *g* for 20 min), and add BHT (20 μM).
4. Perform sequential ultracentrifugation as follows (based on the use of Beckman Preparative Ultracentrifuge, Model L2–65 or later equivalent).
5. Precool the 40.3 rotor to 10°C in the ultracentrifuge.
6. Label cellulose nitrate tubes with a water-proof marker and assemble in tube rack.
7. Assemble the centrifuge tube caps except for the fill-hole set screw.

8. Allow samples to warm to room temperature (~23°C) and pipette 5.0 ml of plasma into the centrifuge tubes.
9. Carefully layer 0.15-M saline solution of the sample within 5 mm of the top of the tube.
10. Cap the tubes tightly. Fill the tubes completely with saline using a syringe with 26 gauge needle and seal the tubes by inserting the fill-hole set screw.
11. Place the tubes in the rotor sockets and balance them precisely.
12. Place a very thin film of vacuum grease on the rotor gasket and seal the rotor.
13. Centrifuge samples for 18 h at 10°C at 105,000 × *g* (4,000 rpm).
14. Stop centrifuge after 18 h and gently remove the rotor.
15. Using the extraction tool, remove each tube slowly.
16. Insert the tube into the tube slicer until the bottom edge of the cap is 0.8 cm from the top of the slicer.
17. Remove the set screw from the fill-hole and cover the hole with a finger.
18. Slice the tube with a quick, smooth thrust of the blade.
19. Withdraw as much as possible of the top fraction through the fill-hole using a 2.5 ml disposable syringe, label fraction as LDL, and save for further oxidation and lipid extraction.
20. To isolate total HDL by ultracentrifugation, adjust plasma to a density of 1.063 g/ml with NaBr and centrifuge at 110,000 × *g* for 25 h.
21. Recover infranatant fraction and adjust to density of 1.21 g/ml and centrifuge at 110,000 × *g* for 48 h.
22. For subfractionation, isolate HDL_2 between the densities of 1.063 and 1.125 g/ml with centrifugation for 48 h *(23)*.
23. Isolate HDL_3 between the densities of 1.125 and 1.21 g/ml by centrifugation at 110,000 × *g* for 48 h *(23)*.

3.1.1.2. Ultracentrifugation with TFT 45.6 Rotor; Suomela et al. *(24)*

1. Obtain a chylomicron-rich fraction from plasma by overlaying 1.8 ml plasma (2 × 0.9 ml) with 1.6-ml NaCl solution (*d* = 1,006 g/l, including 11.4-g NaCl and 0.1-g disodium EDTA dihydrate) and ultracentrifuge with a TFT 45.6 rotor (Kontron Instruments, Italy) at 38,000 × *g* (18,000 rpm) and 16°C for 30 min.
2. Aspirate the top 1.1 ml to collect the chylomicron-rich fraction.
3. Overlay the infranate again with NaCl solution (0.9 ml) and ultracentrifuge the samples to separate the VLDL-rich fraction at 160,000 × *g* (37,000 rpm) and 16°C for 15 h. (*See* **Note 3**).

3.1.1.3. Oxolipoproteins (LDL and HDL); Kamido et al. *(11, 27)*

1. Dialyze LDL for 48 h at 4°C in the dark against vacuum-degassed 0.01-M phosphate buffer, pH 7.4, containing

0.16-M NaCl and 0.1-mg/ml chloramphenicol using Spectropor 20.4 diameter dialysis tubing.

2. For oxidation, dilute dialyzed LDL solution with the dialysis buffer to a final concentration of 1.5-mg LDL protein/ml.
3. Transfer into a dialysis bag and immerse in a threefold volume of dialysis buffer with 5-μM $CuSO_4$.
4. Keep system in the dark for 24 h at room temperature, while bubbling O_2 continuously through external buffer.
5. Take samples for chemical analysis of peroxidation products at 0–24 h.
6. Extract total lipids from LDL samples (250 μl of 1.5-mg LDL protein/ml) with 20 ml of $CHCl_3$/MeOH (2:1, v/v).

3.1.2. Tissue Homogenates

3.1.2.1. Atheroma; Kamido et al. *(27)*

1. Obtain atherosclerotic plaque from individual subjects (69 ± 2 years).
2. Following removal, place samples immediately in ice-cold PBS containing BHT (0.02 mg/ml) and EDTA (0.75 mg/ml).
3. Rinse to remove any loosely adherent blood clots and store at −80°C.
4. For isolation of F_2-isoprostanes, thaw samples at room temperature and cut into 5 mm^3 pieces.
5. Place minced plaque in 5.5 ml of IPA containing BHT (0.044 g/L) and homogenize using Ultra-Turrax T8 blade homo-genizer (IKA Labortechnik, Germany) for 5 min at the highest speed.
6. Incubate homogenized sample at room temperature for 30 min.
7. Following incubation, add 3.5-ml $CHCl_3$ to the sample and vortex the entire solution vigorously for 1 min and allow to stand for 30 min.
8. Separate lipid extract from homogenate by centrifugation at 3,000 rpm, 4°C, for 10 min, and remove supernatant from the pellet.
9. Divide lipid extract into two equal volumes for determination of total oxolipids and for determination of lipid class distribution.

3.1.2.2. Endarterectomy Lesions; Kamido et al. *(27)*

1. Wash carotid endarterectomy lesions (~1 mg wet weight) in cold deoxygenated PBS (50 mM, pH 7.4) containing 0.54-mM EDTA and 10-μM BHT.
2. Resect, finely mince and homogenize intima in cold, deoxygenated sodium carbonate buffer (100 mM), pH 11 (4–5 ml).

3. Centrifuge the homogenates at 2,000 × *g* and 5°C for 10 min.
4. Use supernates directly for the isolation of lesion lipoproteins by density gradient ultracentrifugation at 417,000 × *g* and 15°C for 4 h.
5. For aortic lesions, resect thoracic and abdominal aorta samples. Weigh the resected intima and pool the tissues from the same lesion stage to yield 10 g wet tissue.
6. Freeze the tissue in liquid N_2, pulverize, and reconstitute in 10-mM phosphate/0.15-M NaCl buffer, pH 7.4 (50 ml), containing 0.3-mM EDTA, 100-μM diethylenetriamine pentaacetic acid, 50 μg/ml soybean trypsin inhibitor, 100-μM BHT, and 10-mM 3-aminotriazole.
7. Agitate the samples overnight (4°C) and centrifuge at 2,750 × *g* and 4°C for 30 min. Ultracentrifuge the resulting tissue supernate at 100,000 × *g* and 4°C for 30 min. Discard the pellet and top layer and collect the supernate containing all lesion lipoproteins.
8. Add the carotid and aortic lesion supernates and samples of lesion lipoproteins (~200 μl) to hexane/MeOH (5:1, v/v; 12 ml).

3.1.2.3. Rabbit Hearts; Bergvist and Kuksis ***(28)***

1. Kill rabbits under deep anesthesia induced by Nembutal.
2. Dissect hearts on ice to obtain a total lipid extract by adding 4 ml of ice cold $CHCl_3$/MeOH 3:1 (v/v) containing 0.005% BHT to approximately 1 g of heart tissue and homogenize under ice-cold conditions.
3. Mix homogenate with another 4 ml of $CHCl_3$/MeOH 3:1 (v/v) and 1 ml distilled water, and agitate vigorously for 1 min, then centrifuge at 800 × *g* for 20 min.
4. Aspirate $CHCl_3$ layer and concentrate in a rotary evaporator. Dry residue under a stream of dry N_2.
5. Isolate phospholipid fraction by solid phase extraction.
6. Use a silica column (Sep-Pak, Waters, Milford, MA) of 3 ml capacity packed with aminopropyl-derivatized silica ($-NH_2$).
7. Dissolve total lipid sample in a small amount of $CHCl_3$ and layer on top of the column, and flush with a mixture of 2-ml $CHCl_3$ and 1 ml of IPA.
8. Flush the column next with MeOH containing 0.005% BHT to obtain mainly phospholipids.
9. Concentrate phospholipid eluate on a rotary evaporator, dry under a stream of N_2 and redissolve residue in 150 μl of MeOH.

10. Inject a 1 μl portion into the LC/MS system. (*See* **Note 4**).

3.1.3. Autoxidized Seed Oils

3.1.3.1. Products of Corn and Sunflower Oils; Kuksis et al. ***(30)***; Sjovall et al. ***(3)***

1. Purify sunflower oil TAGs by normal phase TLC using conventional Silica gel H (Merck) plates and heptane/$(IP)_2O$/HOAc 60:40:4 (by vol) as the developing solvent.
2. Locate resolved components by iodine staining of a narrow strip of the developed plate.
3. Recover purified TAGs by scraping and extracting gel with $CHCl_3$/MeOH 2:1 (v/v).
4. Wash extracts with distilled H_2O and dry with anhydrous Na_2SO_4.
5. Evaporate solvents to dryness under N_2 and dissolve residue in 1 ml of $CHCl_3$.
6. Transfer sample to 15-ml centrifuge tube and evaporate solvent under N_2 by rotating the tube to deposit a thin film of TAGs covering the inside of the tube up to 3 cm from bottom.
7. Flush tube with air and incubate it in the open at room temperature for 60 days.
8. Stop the reaction by dissolving contents in 5 ml of $CHCl_3$/MeOH (2:1, v/v), 100 μl of 2% EDTA in water, and 10 μl of 2% BHT.
9. Wash extracts three times with H_2O (3 × 1 ml) and evaporate solvents under N_2 at 38°C, and add 1-M DNPH·HCl (3.6 mg/ml) to an aliquot of dry sample.
10. Shake mixture vigorously and keep in dark at room temperature for 4 h and overnight at 4°C.
11. Extract lipids with 5 ml of $CHCl_3$/MeOH 2:1 (v/v), evaporate solvent under N_2, and take up residue in an appropriate solvent for TLC or LC-MS.
12. Perform TLC as outlined for purification of original TAG sample.
13. Locate yellow bands on the chromatoplates (in daylight).
14. Recover oxo-TAGs from the silica gel scrapings of individual TLC bands by extraction with $CHCl_3$/MeOH 2:1 (v/v).
15. Wash extracts with distilled H_2O, dry over anhydrous Na_2SO_4 and save for LC/ELSD and LC/ESI-MS analysis *(3)*.

3.1.3.2. Formation of TAG Core Aldehydes During Rapid Oxidation of Corn and Sunflower Oils with TBHP/ Fe^{2+}; Sjovall et al. ***(31)***

1. Purify sunflower oil TAGs by normal phase TLC using conventional Silica gel H (Merck) plates and heptane/IP_2O/HOAc 60:40:4 (by vol) as the developing solvent.
2. Locate resolved components by I_2 staining of a narrow strip of the developed plate.
3. Recover purified TAG by scraping the gel from appropriate areas of the plate and extracting it with $CHCl_3$/MeOH 2:1 (v/v).

4. Wash extracts with distilled water and dry with anhydrous Na_2SO_4.
5. Evaporate solvents to dryness under N_2 and dissolve residue in 1 ml of $CHCl_3$.
6. Add 1 ml of 35% TBHP in H_2O to 10 mg of purified TAGs in the presence of 10-μM $FeSO_4$ and 100 μl of 0.2% taurocholic acid.
7. Incubate reaction mixture on a mechanical agitator in dark for 30 min to 30 h at 37°C.
8. Stop reaction by diluting with 5 ml of $CHCl_3$/MeOH 2:1 (v/v), 100 μl of 2% EDTA in H_2O and 10 μl of 2% BHT in MeOH.
9. Wash extracts three times with H_2O (3 × 1 ml) and evaporate solvent to dryness under N_2 at 38°C.
10. Add freshly prepared DNPH reagent in 1-M HCl (3.6 mg/ml) to an aliquot of the dried sample.
11. Shake mixture vigorously in the dark at room temperature for 4 h and overnight at 4°C.
12. Extract lipids with 5 ml of $CHCl_3$/MeOH 2:1 (v/v) and collect the $CHCl_3$ phase.
13. Dry down the $CHCl_3$ under N_2 and take up the residue in an appropriate solvent for chromatography and MS.
14. For LC/ESI-MS analyses follow **Protocol 2.1.4.2**.
15. Identify major core aldehyde species (50–60% of total TAG core aldehydes) as the mono [9-oxo]nonanoyl- and mono[12-oxo]-9,10-epoxy dodecenoyl- or [12-oxo]-9-hydroxy-10,11-dodecenoyl-diacylglycerols (DAGs).
16. Note that as many as 113 molecular species of TAG core aldehydes may be specifically identified, accounting for 32–53% of the DNPH-reactive material of high molecular weight, and representing 25–33% of the total oxidation products.

3.1.3.3. Shark Liver Oil and Human Milk Fat; Hartvigsen et al. *(32)*

1. Obtain a total lipid extract of freeze-dried human milk with $CHCl_3$/MeOH 2:1 (v/v).
2. Recover diacylalkylglyceryl ethers (DAGE) from the milk lipid extract and the shark liver oil by preparative double one-dimensional TLC as follows.
3. Use Silica gel H TLC plates (200 × 200 × 0.25 mm) activated for 2 h at 110°C before use.
4. Develop plates in hexane/Et_2O 90:10 (v/v).
5. Locate lipids by spraying plates with 0.2% 2,7-dichlorofluorescein in EtOH and Schiff base reagent.
6. Extract lipids and core aldehydes from gel scrapings with $CHCl_3$/MeOH 2:1 (v/v).

7. Subject the purified DAGE to regiospecific analysis to reveal the *sn*-1-*O*-alkyl-, *sn*-2-acyl, and *sn*-3-acyl-chain composition and distribution.
8. Perform mild peroxidation of 2-monoacylglycerol (2-MAGE) from either human milk fat or shark liver oil in a tube with oxygen, capping, and heating at 80°C for 3 h.
9. Recover and analyze 2-MAGE by reversed phase LC/ESI-MS.

3.2. Analytical Methods

3.2.1. Lipid Extraction

3.2.1.1. Folch et al. *(19)* Used $CHCl_3$/MeOH 2:1 (v/v) with Endogenous Water in the Tissue as the Ternary Component of the System

1. Add MeOH (1.5 ml) to 200 μl of the sample aliquot and vortex.
2. Add 3 ml of $CHCl_3$, and incubate mixture for 1 h at room temperature in a shaker.
3. Induce phase separation by adding 1.25 ml of water.
4. Leave extract for 10 min at room temperature, then centrifuge at 1,000 × *g* for 10 min.
5. Collect lower ($CHCl_3$) layer and wash upper layer with 2 ml of a solvent mixture calculated to represent the lower phase ($CHCl_3$/MeOH/H_2O, by vol).
6. Combine organic phases and evaporate solvents in a vacuum centrifuge and dissolve residue in 200 μl of $CHCl_3$/MeOH/H_2O (60:30:4.5 by vol) for storage.

3.2.1.2. Bligh and Dyer *(20)* Used $CHCl_3$/MeOH 1:1 (v/v) for Extraction of Tissues That Contain Relatively Little Lipid but a High Proportion of Water

1. Dilute 20 μl of plasma to 800 μl with H_2O.
2. Add 3 ml of MeOH/$CHCl_3$ (1:1, v/v) and leave standing for 1 h at room temp.
3. Induce phase separation by adding 1 ml of $CHCl_3$ and 1 ml of H_2O.
4. Complete extraction according to **Protocol 3.2.1.1**.

3.2.1.3. Hara and Radin *(33)* Used Hexane/IPA 3:2 (v/v) as a Solvent of Low Toxicity

1. Add 1.0 ml of IPA containing 0.1 g/l BHT to 0.5 ml of EDTA plasma.
2. Add 2 ml of hexane and perfuse vial with N_2, cap and vortex for 1 min and centrifuge for 3 min at 3,000 × *g*.
3. Collect upper phase by aspiration and repeat extraction three times and pool and evaporate extracts under dry N_2.
4. Dissolve residue in $CHCl_3$/MeOH (2:1, v/v) and apply to a TLC plate for examination of the extract. (*See* **Note 5**.)

3.2.2. Purification and Prefractionation

3.2.2.1. Kamido et al. *(11)* Used TLC to Isolate and Purify Oxo-Cholesteryl Esters Resulting from Peroxidation of Cholesteryl Linoleate and Archidonate

1. Extract total lipids with $CHCl_3$/MeOH, collect lower layer, evaporate to dryness under N_2, and take up residue in $CHCl_3$ for purification and identification of core aldehydes.
2. Resolve chloroform extracts of reaction mixture by TLC on silica gel H plates (20 × 20 cm, 250-μm-thick layer) using heptane/$(IP)_2O$/HOAc 60:40:4 (by vol) as the developing agent.
3. Locate aldehydes by spraying strips of plate with Schiff's reagent.

4. Scrape silica gel from plate corresponding to aldehyde standards and extract with $CHCl_3$/MeOH 2:1 (v/v).

3.2.2.2. Kamido et al. (34) Used TLC to Isolate and Purify the Cholesteryl Ester Core Aldehydes as the DNPH Derivatives from Total Lipid Extracts of Atheroma Samples

1. Apply samples to silica gel H plates (20 × 20 cm, 250-μm-thick layer).
2. Use double development with CH_2Cl_2 (to a height of 10 cm) and, after solvent evaporation, with toluene (to a height of 17 cm).
3. Scrape off yellow zones corresponding to DNPH derivatives of standard [5-oxo]valeroyl (R_f 0.34), [9-oxo]nonanoyl (R_f 0.41) cholesterol, and the aldehyde esters of 7-ketocholesterol (R_f 0.10), and extract with $CHCl_3$/MeOH (2:1, v/v).
4. Similarly recover the DNPH derivatives of the glycerophospholipid-bound core aldehydes from the origin of the TLC plate (R_f 0.00–0.05).

3.2.2.3. Sjovall et al. (35) Used TLC to Separate the Oxidation Products of Corn Oil TAGs Following Conversion of Core Aldehydes to DNPH Derivatives

1. Apply an aliquot of freshly prepared DNPH derivatives of oxidized corn oil as a band to a TLC place (20 × 20 cm) coated with Silica Gel H (Merck & Co.) and develop it with heptane/ $(IP)_2O$/HOAc 60:40:4 (by vol).
2. Note the presence of nine yellow bands (day-light) and a band for residual TAGs.
3. Locate other bands by staining the TLC plate with 2,7-dichlorofluorescein and viewing under UV light (254 nm).
4. Recover compounds from gel scrapings with $CHCl_3$/MeOH 2:1 (v/v).
5. Wash extracts with distilled water, dry with Na_2SO_4, evaporate solvent under N_2, and save residue for subsequent LC/ELSD and LC/ESI-MS analysis *(35)*.

3.2.2.4. Sjovall et al. (3) Used Reversed Phase LC Columns and Cartridges to Purify and Isolate the TBHP Oxidation Products of Synthetic TAGs

1. Equip a Hewlett-Packard Model 1050 liquid chromatograph with a Varex ELSD II light scattering detector (Varex, MD, USA) and a reversed phase Supelcosil LC-18 HPLC column (250 × 4.6 mm ID).
2. Use a linear gradient of 20–80% IPA in MeOH (0.85 ml/min) in 30 min to obtain the elution profile of the oxidation products of 18:0/18:0/18:2 following a 45 min exposure at 37°C to 7.8-M TBHP.
3. Use N_2 as nebulization gas at an evaporation temperature of 85°C.

3.2.2.5. Bergqvist and Kuksis (28) Isolated Total Glycerophospholipid Fraction from Rat Heart Tissue by Solid Phase Extraction and Purified the Sample on a Sep-Pak Cartridge

1. Use a silica column (Sep-Pak, Waters) of 3-ml capacity packed with aminopropyl-derivatized silica ($-NH_2$).
2. Dissolve total lipid sample in a small amount of $CHCl_3$ and layer on top of the column, and flush with a mixture of 2-ml $CHCl_3$ and 1 ml of IPA.
3. Elute the column next with MeOH containing 0.005% BHT to obtain mainly phospholipids.

4. Concentrate phospholipid eluate on a rotary evaporator, dry under a stream of N_2 and redissolve residue in 150 μl of MeOH.
5. Save sample for LC/ESI-MS *(28)*.
6. Separate lipid classes by TLC on silica gel G (Merck, Germany) using a mobile phase of light petroleum/Et_2O/ MeOH/HOAc 170:30:2:2 (by vol).
7. Visualize lipids by I_2 vapor and collect bands corresponding to authentic standards by eluting with 2 ml of ice-cold $CHCl_3$/MeOH 2:1 (v/v).
8. Complete analysis of base hydrolysates as in **Protocol 2.1.1.1**.

3.2.3. Derivatization

3.2.3.1. Triphenylphosphine Reduction *(6)*

1. To an aliquot of dry methyl esters of fatty acid hydroperoxides in a screw cap vial add TPP (0.7 mg in 700 μl MeOH/Et_2O). (*See* **Note 6**).
2. Close vial, vortex and let the mixture stand at room temperature for 2 h.
3. Evaporate reaction mixture to dryness under N_2 and redissolve residue in hexane/IPA/HOAc 93:6:1 (by vol) for HPLC analysis.

3.2.3.2. Sodium Borohydride Reduction *(7)*

1. To a methanol solution containing 500 μg of oxolipid ester add $NaBH_4$ (500 μg) in PBS or MeCN for 30 min to achieve 90% reduction below 15°C.
2. Stir reaction mixture for 15 min, add water, and extract product into CH_2Cl_2.
3. Evaporate solvent to dryness, dissolve residue in EtOH, and hydrogenate using PtO_2 as catalyst.
4. After 5 min, filter and evaporate to dryness *(36)*.

3.2.3.3. Methoxymation (MOX); Kamido et al. *(11)*

1. Prepare MOX derivatives by heating core aldehydes with 100 μl of the methoxylamine·HCl reagent at 60°C for 3 h. Perform methoxymation prior to trimethylsilylation.
2. Purify *O*-methyloxime derivatives using C18 Sep-Pak cartridges.

3.2.3.4. Preparation of PFB Derivatives; Li et al. *(37)*

1. Treat oxolipid sample with 200 μl of PFB-Br in CH_3Ac (1:19, v/v) followed by 200 μl of *N,N*-diisopropylethylamine in CH_3Ac (1:9, v/v) and heat at 60°C for 60 min.
2. Allow solution to cool and evaporate solvents to dryness under N_2 at room temperature.
3. Redissolve residue in 100 μl of hexane/EtOH (97:3, v/v) and save for normal-phase chiral LC/MS analysis. (*See* **Note 7**).

3.2.3.5. Trimethylsilylation and *tert*-Butyldimethylsilylation *(11)*

1. Prepare TMS ethers by treating the lipid fractions or their MOX derivatives with a silylating reagent made up of one part BSTFA + 1% trimethylsilyl chloride and one part pyridine for 30 min at room temperature.

2. Evaporate reaction mixture to dryness under N_2 and take up residue in petroleum ether.
3. Prepare TBDMS ethers by treating the lipid fractions of their MOX derivatives (0–10 mg) with 150 µl of TBDMS/imidazole reagent by heating at 80°C for 20 min.
4. After cooling, mix reaction mixture with 5 ml light petroleum spirit and wash three times with 0.5-ml H_2O and dry petroleum extracts over Na_2SO_4.

3.2.3.6. 2,4-Dinitrophenylhydrazone Formation *(38)*

1. Treat aldehydes or aldehyde esters with 0.5 ml of freshly prepared 2,4-dinitrophenylhydrazine in 1-N HCl (0.5 mg/ml), mix vigorously, and keep the reaction mixture in the dark for 2 h at room temperature and overnight at 4°C.
2. Extract reaction mixture with $CHCl_3$/MeOH 2:1 (v/v) and isolate and purify hydrazones by TLC using a double development with CH_2Cl_2 (up to a height of 10 cm) and, after solvent evaporation, the same direction with toluene (up to 17 cm).

3.2.3.7. Preparation of Hydrazones of Cholesteryl Ester Core Aldehydes *(39)*

1. Treat an aliquot of core aldehyde fraction with 2,4-dinitrophenylhydrazine (0.5 mg in 1 ml of 1-N HCl for 2 h at 20°C and at 4°C overnight. (*See* **Note 8**).
2. Extract DNPH derivatives with $CHCl_3$/MeOH 2:1 (v/v) and collect the $CHCl_3$ layer.
3. Analyze the DNPH derivatives by LC/ESI-MS *(39)*.

3.2.4. Chemical and Enzymic Hydrolyses

3.2.4.1. Hydrolysis with Alkali *(41)*

1. To release oxo-fatty acids, hydrolyze oxolipid esters with 1-M KOH or 1-M NaOH (2.0 ml) at room temperature for 90 min.
2. Acidify fractions (pH 3) with 2-M HCl and add an internal standard.
3. Extract lipids with hexane (5 ml) and evaporate to dryness with a stream of N_2.
4. Dissolve residue in 100 µl of MeOH/MeCN/6.5-mM NH_4Ac (pH 5.7) (24.5:45.5:30, by vol) in preparation for LC/MS analysis. *(41)*.

3.2.4.2. Transmethylation *(42)*

1. To the lipid sample (0.1–0.5 mg) add 150 µl of a solution of NaOMe (0.5 N) in anhydrous MeOH.
2. Mix thoroughly and allow to stand for 5 min at room temperature.
3. Add one drop (10 µl) of AcCl to acidify solution and evaporate solvents to dryness.
4. Recover fatty acid methyl esters by extracting twice with 2 ml of hexane.

5. Transmethylation may also be performed in presence of P_2O_5-dried scrapings of silica gel with 2 ml of 0.1-N NaOMe in dry MeOH for 1 h at room temperature *(43)*.

3.2.4.3. Hydrolysis with Phospholipase C *(11, 43)*

1. Hydrolyze (1 h at 37°C) the core aldehydes of PtdCho and PtdEtn and their DNPH derivatives with phospholipase C (*Bacillus cereus*) in order to release the polar head groups which interfere with GLC of the aldehydes *(43)*.
2. Recover hydrolysis products by extraction with $CHCl_3$ and immediately trimethylsilylate for GLC. *(11)*.
3. Alternatively, place individual lipoprotein fractions (equivalent to 0.1–0.2 ml of plasma) or whole plasma into 18-ml centrifuge tubes with screw caps.
4. Dilute contents with 1-ml water, add 1 ml of diethyl ether followed by 2 ml of a solution (0.1 mg/ml of Tris buffer, pH 7.3) of phospholiopase C (α-toxin of *Clostridium welchii*, Sigma, St. Louis, MO).
5. Add 1.3 ml of 10% $CaCl_2$ solution and mix the solution well.
6. Incubate mixture at 33°C for 2 h with shaking.
7. Extract enzyme digest with 10-ml $CHCl_3$/MeOH 2:1 (v/v) containing 200 μg of tridecanoylglycerol as internal standard, and centrifuge solution to break emulsion.
8. Collect the lower $CHCl_3$ layer and pass it through a Pasteur pipet containing 2 cm length of anhydrous Na_2SO_4.
9. Evaporate solvent to dryness and trimethylsilylate for GC.

3.2.4.4. Hydrolysis with Pancreatic Lipase *(44)*

1. Hydrolyze purified DAGE with Et_2O preextracted pancreatic lipase in presence of gum arabic for 30 min.
2. Extract digestion products with Et_2O).
3. Resolve and recover degradation products by TLC using silica gel G containing 5% boric acid and develop plates with hexane/IP_2O/HOAc (50:50:4, by vol).

3.3. Results

3.3.1. Autoxidation of Cholesteryl Esters

GC and GC-MS analyses are extensively utilized for the analysis of fatty acids and fatty acid hydroxides. The methodology is well known and is not considered here in detail. However, GC and GC-MS analysis of intact lipid esters and especially intact oxolipid esters has remained largely unknown although effective methods exist for the conversion of various oxolipid esters into volatile derivatives suitable for GC and for GC-MS.

3.3.1.1. Lipoproteins

1. Prepare oxo-LDL as per **Protocol 3.1.1**.
2. Obtain total lipid extract as per **Protocol 3.2.1.1**.

*3.3.1.1.1. Kamido et al. (**27**) Performed Extensive GC Analyses on Intact Oxocholesteryl Esters in Autoxidized LDL and HDL and Identified Both [5-Oxo]valeroyl and [9-Oxo]nonanoyl Esters of Cholesterol and Oxocholesterol*

3. To an aliquot of the lipoprotein extract equivalent to 0.2 ml of plasma, add 0.2–0.4 mg Phospholipase C (*Bacillus cereus*) in 4 ml of 17.5-mM Tris buffer, pH 7.3, along with 1.3 ml of 1% $CaCl_2$ and 1 ml of Et_2O, and incubate mixture with shaking for 2 h at 30°C.
4. Treat reaction mixture with 5 drops of 0.1-N HCl and extract once by vigorous shaking with 10-ml $CHCl_3$/MeOH (2:1, v/v) containing 150–250 μg tridecanoylglycerol as internal standard.
5. Separate solvent phases by centrifugation at 200 × *g*.
6. Collect clear $CHCl_3$ phase, dry it by passing through a Pasteur pipet containing 2 g of anhydrous Na_2SO_4.
7. Evaporate solvents under N_2 and dissolve residue in TRISIL-BSA (150–250 μl) and transfer the solution to a sampling vial and then seal the vial.
8. Inject automatically 1 μl of the solution onto a gas chromatographic column without the benefit of a flash evaporator. *See* **Protocol 2.1.1**.
9. Program temperature from 175 to 350°C at 8°C/min and identify oxolipid peaks by reference to standards.
10. Confirm oxolipid peak identities by LC/ESI-MS of appropriate TLC fractions and DNPH derivatives. *See* **Protocol 2.1.4.5**. (*See* **Note 9**.)

*3.3.1.1.2. Kamido et al. (**45**) Used Reversed Phase LC-MS to Isolate and Identify Cholesteryl Ester Core Aldehydes Following Copper-Catalyzed Peroxidation of Human LDL (see Scheme 1)*

1. Collect whole blood by venipuncture from a healthy volunteer into tubes containing EDTA (1 mg/ml) after a 12 h fast.
2. Separate plasma by low speed centrifugation (1,500 × *g* for 20 min).
3. Add BHT (20 μM).
4. Prepare LDL (d = 1.019–1.063 g/ml) by conventional ultracentrifugation.
5. For dialysis and oxidation follow **Protocol 3.1.1**.
6. To 1 ml of oxidized LDL solution (1.5 mg protein/ml), add 0.1 ml of 1% EDTA, 10 μl of 2% BHT, and 1 ml freshly prepared 2,4-dinitrophenylhydrazine in 1-N HCl (0.5 mg/ml), mix vigorously, and keep in dark for 2 h at room temperature and then overnight at 4°C.
7. Extract reaction mixture with $CHCl_3$/MeOH 2:1 (v/v).
8. Apply lipid extract to Silica gel H plates (20 × 20 cm), and subject them to double development with CH_2Cl_2 (to a height of 10 cm) and, after solvent evaporation, with toluene (to a height of 17 cm).

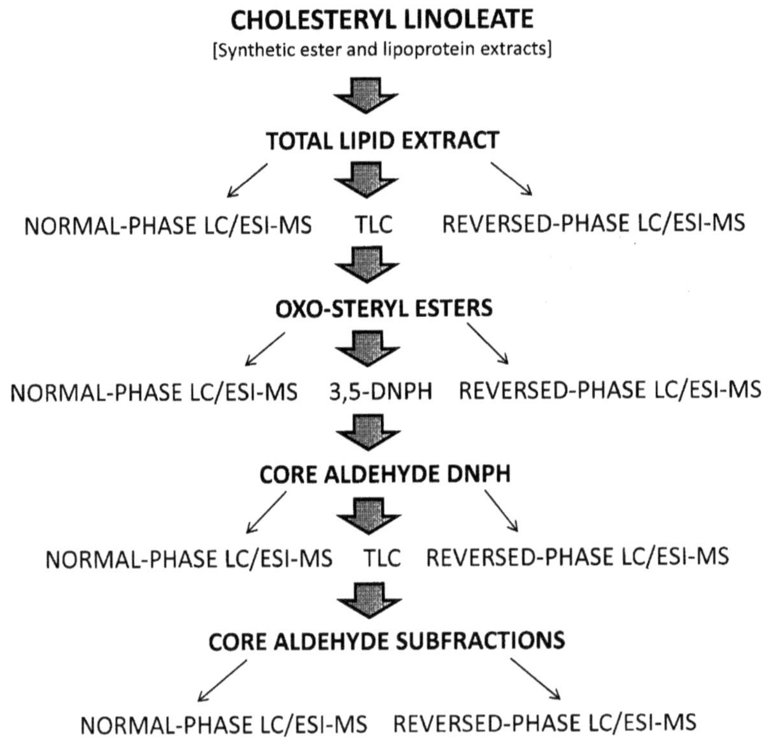

Scheme 1. Diagram for identification of cholesteryl linoleate oxidation products (original drawing).

9. Scrape off the yellow zones corresponding to standard DNPH derivatives of 5-oxovaleroyl (R_f = 0.26) and 9-oxononanoyl (R_f = 0.33) cholesterol, and extract with $CHCl_3$/MeOH.
10. Remove the core aldehydes of 7-ketocholesteryl esters (R_f = 0.10) in a similar manner.
11. Separate DNPH derivatives of the aldehydes by reversed phase LC on a Supelcosil LC-18 column (250 mm × 4.6 mm ID, Supelco) using MeCN/IPA 4:1 (v/v) or a linear gradient of 30–90% EtCN in MeCN as the eluting solvents.
12. Install column into a Hewlett-Packard Model 1084B Liquid chromatograph and operate with a flow rate of 0.1–1.5 ml/min, while monitoring peaks at 358 nm.
13. Admit about 1% of the HPLC column effluent to a Hewlett-Packard Model 5985B quadrupole mass spectrometer via DLI interface.
14. Record NICI spectra every 5 s over the entire chromatogram in the mass range 200–900.
15. Extract single ion mass chromatograms of the DNPH derivatives from the total ion spectra corresponding to C_4

(m/z 650), C_5 (m/z 664), C_6 (m/z 678), C_7 (m/z 692), C_8 (m/z 706), C_9) (m/z 720), and C_{10} (m/z 734) cholesteryl ester core aldehydes.

16. Obtain similar single ion mass chromatograms for the C_4 to C_{10} 7-ketocholesterol core aldehydes.
17. Calculate relative quantities of core aldehydes from peak areas.

*3.3.1.1.3. Kamido et al. (**11**) Used TBHP/Fe^{2+} to Peroxidize Liposomal Cholesteryl Esters in Order to Obtain Aldehydes for Biochemical and Metabolic Studies in Amounts Larger than Those Provided by $CuSO_4$ Catalyzed Autoxidation*

1. Dissolve 1–2 mg of cholesteryl ester (linoleate or arachidonate) in 100 μl of $CHCl_3$/MeOH 2:1 (v/v) in a 15-ml test tube and evaporate solvent under N_2.
2. To the dry sample add 1 ml of 35% aqueous TBHP and 100-μM $FeSO_4$, and shake the tube mechanically for 3 h at 37°C in the dark.
3. Stop reaction by diluting reaction mixture with five volumes of $CHCl_3$/MeOH 2:1 (v/v) and wash with H_2O to remove excess TBHP.
4. Collect lower layer, evaporate to dryness under N_2, and take up residue in $CHCl_3$ for purification and identification of core aldehydes.
5. Resolve chloroform extracts of reaction mixture by TLC on silica gel H using heptane/$(IP)_2O$/HOAc 60:40:4 (by vol) as the developing agent.
6. Locate aldehydes by spraying strips of plate with Schiff's reagent.
7. Scrape silica gel from plate corresponding to aldehyde standards and extract with $CHCl_3$/MeOH 2:1 (v/v).
8. Reduce core aldehydes and derived oxocholesterols with $NaBH_4$ to convert any 7- keto to 7-hydroxycholesterol derivatives.
9. Convert an aliquot of cholesteryl ester core aldehyde extracts to hydrazones by treatment with 0.5 ml of freshly prepared DNPH reagent (2,4-dinitrophenylhydrazine in 1-N HCl at a concentration of 0.5 mg/ml).
10. Mix vigorously and keep in the dark for 2 h at room temperature and overnight at 4°C.
11. Extract reaction mixture with $CHCl_3$/MeOH 2:1 (v/v) and purify the recovered hydrazones by TLC using the double development described above *see* **Protocol 3.3.1.1.2.8.**
12. Resolve DNPH derivatives of core aldehydes by reversed phase LC using a Supelcosil LC-18 column (250 mm × 4.6 mm ID, Supelco) using MeCN/IPA 4:1 (v/v) at 1 ml/min or a linear gradient of 30–90% EtCN in MeCN (1.5 ml/min) as the eluting solvents.
13. Admit about 1% of the LC column effluent to a Hewlett-Packard Model 5985B single quadrupole mass spectrometer

via a DLI interface. Take NCI mass spectra every 5 s over the entire chromatogram, in the m/z range 200–900.

14. Identify cholesteryl ester and 7-ketocholesteryl ester core aldehydes as above.

*3.3.1.1.4. Ahmed et al. (**47**) Reported Formation of Core Aldehydes of Cholesteryl Esters and TAGs During Oxidation of Plasma LDL with Peroxynitrate Generator (SIN-1)*

1. For isolation of LDL, *see* **Protocol 3.1.1.1**.
2. Oxidize LDL (1 mg/ml protein) by the peroxynitrite donor, SIN-1 (1 mM) for up to 20 h at 37°C in presence of DTPA (100 μM), a metal ion chelator.
3. Take aliquots of oxidation mixture at desired times and stop oxidation by addition of 100-μM BHT.
4. Extract LDL lipid-soluble oxidation products with $CHCl_3$/MeOH 2:1 (v/v) after addition of an internal standard 15:0/15:0 GroPCho.
5. Evaporate organic phase to dryness under N_2 and convert any core aldehydes to DNPH derivatives. For method of conversion, *see* **Protocol 3.2.3.6**.
6. Perform LC/ESI-MS using a normal phase silica column (4.6 mm × 250 mm, Alltech Associates) installed in Hewlett-Packard Model 1050 liquid chromatograph, connected to a Hewlett-Packard Model 5989A Quadrupole Mass Spectrometer, and equipped with a nebulizer-assisted ESI interface.
7. Elute column with a linear gradient of 100% A ($CHCl_3$/MeOH/30% NH_4OH 80:19.5:0.5, by vol) to 100% B ($CHCl_3$/MeOH/H_2O/30% NH_4OH, 60:34:5.5:0.5, by vol) for 14 min, followed by 100% B for 10 min at a flow rate of 1 ml/min.
8. Split the effluent 1:50 to give a 20 μl/min flow into the mass spectrometer.
9. Set capillary exit voltage at 150 V, with electron multiplier at 1,795 V.
10. Record positive ion ESI spectra in the mass range 450–1,100 amu.
11. Identify molecular species of the oxidation products based on single ion mass chromatograms of the putative ions corresponding to the DNPH derivatives of azelaic acid ester of 7-ketocholesterol (*m/z* 735), [8-oxo]octanoyl aldehyde ester of cholesterol (*m/z* 705), [9-oxo]nonanoyl aldehyde ester of cholesterol (*m/z* 719), and of azelaic acid ester of cholesterol (*m/z* 705).
12. Identify the DNPH derivatives of TAG core aldehydes corresponding to *m/z* 945, 926, 928, and 902.

3.3.1.2. Atheromas

*3.3.1.2.1. Kamido et al. (**27**) Used GLC of Intact Oxolipids to Identify Oxo-Cholesteryl Esters and Oxo-Glycerolipids as Minor Components of Total Lipid Extracts of Atherosclerotic Plaques (See* **Fig. 1***)*

1. Obtain plaque material in the operating room from patients undergoing endarterectomy for arteriosclerosis obliterans or aortic reconstruction for aneurysm.
2. Place atheroma samples in 0.01-M phosphate-buffered saline, pH 7.4, containing EDTA (1 mg/ml) and BHT (200 μg/ml), wash, mince, and homogenize using a Waring blender.
3. Extract total lipids using $CHCl_3$/MeOH (2:1, v/v) and subject to dephosphorylation with phospholipase C as per **Protocol 3.2.4.3**.
4. Trimethylsilylate the resulting neutral lipid mixture in presence of appropriate internal standards and perform High temperature GC as per **Protocol 2.1.1**.
5. Note presence of oxolipid peaks and confirm oxolipid peak identities by LC/ESI-MS of appropriate TLC fractions and DNPH derivatives. *See* **Protocol 2.1.4.4**. (*See* **Note 10**.)

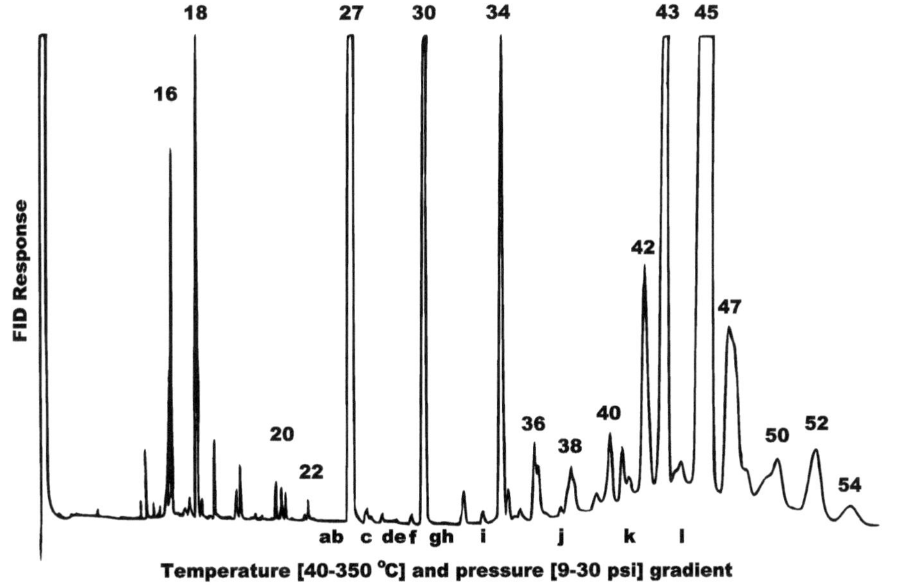

Fig. 1. Total lipid profile of human atheroma as obtained by nonpolar capillary GC following dephosphorylation with PLC and trimethylsilylation (unpublished original from *27*). Peak identification: 16–18, free fatty acids; 20–22, monoacylglycerols derived from lysoPtdCho; 27, free cholesterol; 30, tridecanoylglycerol internal standard; 34, d18:1/16:0 ceramide derived from SM; 34–40, diacylglycerols with 34–38 acyl carbons; 42, d18:1/24:1 ceramide derived from SM; 43–47, cholesteryl esters with 16–20 acyl carbons; 50–534, triacylglycerols with 50–54 acyl carbons. Tentatively: (**a**) 1-*O*-hexadecyl-2-[5-oxo]valeroylglycerol derived from alkylacyl GroPCho; (**b**) 1-palmitoyl-2-[5-oxo]valeroylglycerol derived from diacyl GroPCho; (**c**) cholest-5-ene-3β,7β-diol,5,6 – epoxy-5α-cholestan-3β-ol, 5,6β-epoxy-5β-ol; (**d**) 3β-hydroxycholest-5-en-7-one; (**e**) 1-*O*-hexadecyl-2-[9-oxo]nonanoylglycerol derived from alkylacyl GroPCho; (**f**) 1-palmitoyl-2-[9-oxo]nonanoylglycerol derived from alkylacyl GroPCho; (**g**) 1-stearoyl-2-[9-oxo]nonanoylglycerol derived from diacyl GroPCho; (**h**) cholesteryl [5-oxo]valerate; (**i**) cholesteryl [9-oxo]nonanoate; (**j**) 16:0/18:1/[9-oxo]9:0Gro; (**k**) 18:1/18:1/[9-oxo]9:0Gro. GC instrumentation is given under Protocol 2.1.1.

*3.3.1.2.2. Hoppe et al. (**39**) Demonstrated the Presence of Cholesteryl Ester Core Aldehydes in Human Atheroma Using Reversed Phase LC/ESI-MS of the DNPH Derivatives*

1. Obtain atherosclerotic lesions from carotid endarterectomy samples and immediately place them into PBS, pH 7.4, containing 0.1% EDTA, 0.15% ε-aminocaproic acid, 40-μM BHT, and 1-mM phenylmethylsulfonfluoride (PMSF).
2. Separate plaque material (0.4–1.0 g) from media and adventitia, mince into small (0.5–1.0 mm^2) pieces.
3. Extract total lipids from atherosclerotic tissue (250 μl) by adding 3.5 ml of $CHCl_3$/MeOH 1:1 (v/v), which forms a one-phase solvent system.
4. After 10–30 min extraction at room temperature with occasional shaking, add 1.25-ml H_2O and vortex mixture vigorously. Centrifuge to break emulsion into two phases.
5. Blow down the $CHCl_3$ extracts under N_2 and treat the residue with 0.5 ml of freshly prepared 2,4-dinitrophenylhydrazine in 1-N HCl (0.5 mg/ml).
6. Mix vigorously, keep in dark for 2 h at room temperature, then overnight at 4°C.
7. Extract with $CHCl_3$/MeOH 2:1 (v/v) and subject DNPH derivatives along with unreacted lipids to reversed phase LC/ESI-MS.
8. Install an HP reversed phase C_{18} column (100 mm × 2.1 mm ID) into a liquid chromatograph and connect it to a mass spectrometer via the ESI interface.
9. Develop column with a linear gradient of 100% solvent system A (MeOH/H_2O/30% NH_4OH 80:12:0.5, by vol) by changing to 100% solvent B (MeOH/hexane/30% NH_4OH 80:12:0.5, by vol) in 30 min, after initially holding for 1 min at 100% system A, and finally at 100% system B for 5 min.
10. For fraction collection, adjust a split ratio 1:50 and maintain the flow rate to the mass spectrometer at 25 μl/min.
11. Record a total negative ion current profile and note the presence of major peaks at *m/z* 571 (7β-hydroxycholesteryl 9-carboxynonanoate), *m/z* 557 (5,6-epoxycholesteryl 8-carboxyoctanoate), *m/z* 735 (7α-hydroxycholesteryl 9-oxononanoate DNPH), *m/z* 555 (cholesteryl 9-carboxynonanoate), *m/z* 705 (cholesteryl 8-oxooctanoate DNPH), and *m/z* 719 (cholesteryl 9-oxononanoate DNPH).

*3.3.1.2.3. Kamido et al. (**27**) Used LC/ESI-MS to Identify Oxo-Cholesteryl Esters and Oxo-Glycerolipids as Minor Components of Total Lipid Extracts of Atherosclerotic Plaques*

1. *See* **Protocol 2.1.1** and **Protocol 3.2.3.7** for preparation of total lipid extracts of atheromas and the preparation of DNPH derivatives of any aldehydes, respectively.
2. Perform a C18 reversed phase LC and LC/thermospray ionization(TSI)-MS of both native and DNPH modified lipid extracts using a linear gradient of 10–30% IPA in

MeCN or 20–80% IPA in MeOH (30 min) on a reversed phase column.

3. Alternatively, perform reversed phase LC/ESI-MS on unreacted and DNPH modified total lipid extracts using a linear gradient of 100% Solvent A ($MeOH/H_2O/NH_4OH$, 80:19.5:0.5, by vol) over 30 min, after initially holding for 1 min at 100% A, followed by 100% B ($CHCl3/MeOH/H_2O$/30% NH_4OH, 60:34:5.5:0.5, by vol) for 5 min.
4. Generate both positive and negatively charged ions by thermospray and ESI.
5. Record single ion mass chromatograms to represent the DNPH derivatives of the major cholesteryl ester core aldehydes: C_9 (*m/z* 719); C_8 (*m/z* 705); C_7 (*m/z* 691; C_6 (*m/z* 677); C_5 (*m/z* 663); and C_4 (*m/z* 649).
6. Record single ion mass chromatograms for the mono- (*m/z* 677) and di- (*m/z* 875) DNPH derivative of 7-ketocholesteryl [5-oxo]valerate.

3.3.1.3. Liposomes

*3.3.1.3.1. Hoppe et al. (**39**) Utilized the LC-MS Method of Kamido et al. (**11**) to Demonstrate the Production of Cholesteryl Ester Core Aldehydes Along with the Core Acids, During Autoxidation of Cholesteryl Linoleate*

1. For preparation of the cholesteryl ester core aldehydes follow **Protocol 3.3.1.1.3**.
2. Use LC/ESI-MS in the negative ion mode to identify 7α-hydroxycholesteryl 9-carboxynonanoate, 7β-hydroxycholesteryl 9-carboxynonanoate, 5,6-epoxycholesteryl 8-carboxyoctanoate, 5,6-epoxycholesteryl 9-carboxynonanoate, 7α-hydroxycholesteryl 9-oxononanoate DNPH, 7β-hydroxycholesteryl 9-oxononanoate DNPH, 7-ketocholesteryl 9-carboxy nonanoate DNPH, cholesteryl 9-carboxynonanoate, cholesteryl 8-oxo-octanoate DNPH, and cholesteryl 9-oxononanoate DNPH.

*3.3.1.3.2. Herrera et al. (**48**) Reported Formation of Core Aldehydes of Highly Toxic 7-Hydroperoxycholesterol During Autoxidation of Cholesteryl Linoleate*

1. Oxidize purified cholesteryl linoleate (5 mg) as a thin film in air at 25–75°C for up to 24 h. *(39)*.
2. Prepare DNPH derivatives by treating aliquots of the oxidation mixture with 2,4-dinitrophenylhydrazine (0.5 mg in 1 ml of 1-N HCl) for 2 h at 20°C and at 4°C overnight *(49)*.
3. Resolve underivatized autoxidation products by TLC with hexane/Et_2O/HOAc 60:70:1.5 (by vol).
4. Resolve the DNPH derivatives similarly using heptane/diisopropyl ether/HOAc 60:40:4 (by vol). View the plates under UV before and after staining with iodine vapor.
5. Perform reversed phase HPLC of the underivatized oxidation products using an LC system consisting of a Supelcosil LC-18 column (25 mm × 4.6 m ID, Supelco, Bellefonte, PA) interfaced with a UV (235 nm) and a mass detector (ELSD) eluting with a linear gradient of 20–80% IPA in MeOH over 30 min at a flow rate of 0.8 ml/min.

6. Resolve DNPH derivatives in a similar LC system, except monitor column effluents in UV at 358 nm.
7. Perform normal phase LC on the underivatized oxidation products on a Spherisorb column (3 µm, 100 mm × 4.6 mm ID, Altech Guelph, Ontario) installed in a Hewlett-Packard (Palo Alto, CA) Model 1090 liquid chromatograph interfaced with a Hewlett-Packard Model 5985B quadrupole mass spectrometer (*see* below).
8. Elute the column with a linear gradient of 100% solvent system A ($CHCl_3$/MeOH/30% NH_4OH, 80:19.5:0.5 by vol) by changing to 100% solvent B ($CHCl_3$/MeOH/H_2O/30% NH_4OH (60:34:5.5:0.5 by vol) in 14 min, followed by 100% system B for 10 min.
9. Perform normal phase LC/ESI-MS on a Hewlett-Packard Model 1090 liquid chromatograph interfaced with a Hewlett-Packard Model 5985B quadrupole mass spectrometer equipped with a nebulizer assisted ESI interface in the positive and negative ion mode *(31)*.
10. Analyze DNPH derivatives of oxo-steryl esters by normal phase LC/ESI-MS and by LC/CID/ESI-MS *(49)*.
11. Note that the major peaks in the positive ion current profile of the DNPH derivatives obtained by normal phase LC/ESI-MS for the cholesteryl linoleate oxidation products recovered after 60 min of exposure to air at 100°C correspond to the DNPH derivatives of cholesterol C_9 aldehyde (*m/z* 719), 7-hydroxycholesterol C_{13} aldehyde hydroperoxide (*m/z* 841), 7-hydroperoxycholesterol C_{13} hydroperoxy hydroxide (*m/z* 857), 7-hydroperoxycholesterol C_9 aldehyde (*m/z* 750), 7-ketocholesterol C_9 aldehyde (*m/z* 734), while the smaller peaks are due to further degradation products (*m/z* 417, 513, and 525). The total yield was 1–2% of original cholesteryl ester. (*See* also **ref.** ***42)***.

3.3.2. Autoxidation of Glycerophospholipids

3.3.2.1. Lipoproteins

*3.3.2.1.1. Kamido et al. (**49**) Isolated PtdCho Core Aldehydes from Human Plasma Lipoproteins Following Copper-Catalyzed Peroxidation*

1. Prepare LDL and HDL by conventional ultracentrifugation from fresh human plasma containing EDTA (1 mg/ml) and BHT (20 µM) **Protocol 3.1.1.3.**
2. Dialyze both for 24 h at 4°C in the dark against vacuum-degassed 0.011-M phosphate buffer, pH 7.4, containing 10-µM EDTA, 0.15-M NaCl, and 0.1 mg/ml chloramphenicol.
3. Use a dialysis bag (Spectra/Por 15.9 diameter tubing) sterilized by boiling in distilled deionized H_2O for 30 min prior to use.
4. Oxidize dialyzed lipoprotein solution (1.5 mg protein/ml LDL and 4.5 mg protein/ml HDL) in a dialysis bag immersed in a 100 fold volume of the dialysis buffer containing 5-µM $CuSO_4$. *See* **Protocol 3.1.1.3.**

5. Keep the system in dark at room temperature for 24 h while oxygen is being bubbled continuously through the external buffer.
6. Isolate ester bound fatty aldehydes as hydrazones by adding 1 ml of freshly prepared 2,4-dinitrophenylhydrazine in 1-N HCl (0.5 mg/ml) to 1 ml of the oxidized lipoprotein solution (1.5 and 4.5 mg protein/ml of LDL and HDL, respectively.
7. Shake vigorously and allow to stand in the dark for 2 h at room temperature and then overnight at 4°C.
8. Extract reaction mixture with $CHCl_3$/MeOH 2:1 (v/v), blow down solvent to a small volume and apply to Silica gel H plates for TLC.
9. Separate DNPH derivatives by a double development with CH_2Cl_2 (to a height of 10 cm) and, after solvent evaporation, with toluene (to a height of 17 cm).
10. Recover the DNPH derivatives of PtdCho-bound core aldehydes from the origin of the plate (R_f = 0.0–0.05) by eluting the gel scrapings with MeOH.
11. Digest aliquots of the extracts with phospholipase C (*Bacillus cereus*) for 2 h at 37°C. Extract the released DNPH derivatives of the DAG core aldehydes and purify by TLC using $CHCl_3$/MeOH 95:5 (v/v) as the developing solvent.
12. Resolve the DNPH derivatives (R_f 0.20) from residual DAGs (R_f 0.30) and ceramides (R_f 0.25) and recover separately by extraction with $CHCl_3$/MeOH 2:1 (v/v) after spraying with fluorescein location by UV.
13. Analyze the recovered DNPH derivatives of the aldehydes by reversed phase HPLC on a Supelco LC-18 column (250 × 4.6 mm ID) using a linear gradient of 30–90% EtCN in MeCN as the eluting solvent and a UV detector.
14. Install the column into a HP Model 1084 Liquid Chromatograph operated at a flow rate of 1–1.5 ml/min and monitor peaks at 358 nm.
15. Admit about 1% of the HPLC effluent to a HP Model 5985B quadrupole mass spectrometer via a DLI interface.
16. Obtain NICI spectra every 5 s over the entire chromatogram in the mass range 200–900.
17. Obtain single ion plots of ions of interest by recalling data stored in the computer.
18. Note the presence of major peaks for 16:0/5:0ALD (*m/z* 608), 18:0/5:0ALD (*m/z* 636), 16:0/9:0ALD (*m/z* 664), 18:1/9:0ALD (*m/z* 690) and 18:0/9:0ALD (*m/z* 692) glycerols as DNPH derivatives.

*3.3.2.1.2. Pruzanski et al. (**50**) Reported the Release of Mono- and Di-Hydroperoxides and Mono- and Di-Hydroxides of 18:2 and 20:4 During an Unprotected Hydrolysis of PtdCho of Human Plasma HDL by Group IIA Secretory PLA_2*

1. Obtain normal HDL from blood of healthy volunteers and acute phase HDL from patients 34–38 h after surgical procedures.
2. Use identical conditions of conventional ultracentrifugation (1.06–1.25 g/ml) for HDL from both sources as above.
3. Obtain total lipids from normal and acute phase HDL by extraction with $CHCl_3$/MeOH 2:1 (v/v) without acidification to avoid plasmalogen decomposition.
4. Blow down $CHCl_3$ phase to dryness under a stream of N_2 and redissolove residue in 2 ml of $CHCl_3$/MeOH 2:1 (v/v).
5. Perform normal phase LC separations of total lipid extracts using Spherisorb 3 μ columns (100 mm × 4.6 mm ID, Alltech Associates) installed into a Hewlett Packard Model 1060 Liquid Chromatograph connected to a Hewlett-Packard Model 5988B quadrupole MS equipped with a nebulizer assisted ESI interface (HP 59987A).
6. Identify oxygenated fatty acids among the free fatty acids released by group IIA $sPLA_2$ in the total lipid profile obtained by normal phase LC/ESI-MS as follows.
7. Record selected ion mass chromatograms at m/z 311 (18:2 monohydroperoxide); m/z 335 (20:4 monohydroperoxide); m/z 343 (18:2 dihydroperoxide); m/z 359 (22:6 monohydroperoxide); m/z 295 (18:2 monohydroxide); and m/z 319 (20:4 monohydroxide) (*see* **Fig. 2**).

*3.3.2.1.3. Pruzanski et al. (**51**) Reported the Presence of PtdCho Isoprostanes in Normal and Acute Phase HDL*

1. Obtain normal HDL from blood of healthy volunteers and acute phase HDL from patients 34–38 h after surgical procedures.
2. Use above conditions of conventional ultracentrifugation (1.06–1.25 g/ml).
3. Obtain total lipids from normal and acute phase HDL by extraction with $CHCl_3$/MeOH 2:1 (v/v) without acidification to avoid plasmalogen decomposition.
4. Blow down chloroform phase to dryness under a stream of N_2 and redissolve residue in 2 ml of $CHCl_3$/MeOH 2:1 (v/v).
5. Perform normal phase HPLC separations of total lipid extracts using Spherisorb 3 μ columns (100 mm × 4.6 mm ID, Alltrech Associates) installed into a Hewlett Packard Model 1060 Liquid Chromatograph connected to a Hewlett-Packard Model 5988B quadrupole mass spectrometer equipped with a nebulizer assisted ESI interface (HP 59987A).
6. Set capillary exit voltage at 150 V with electron multiplier at 1,795 V.

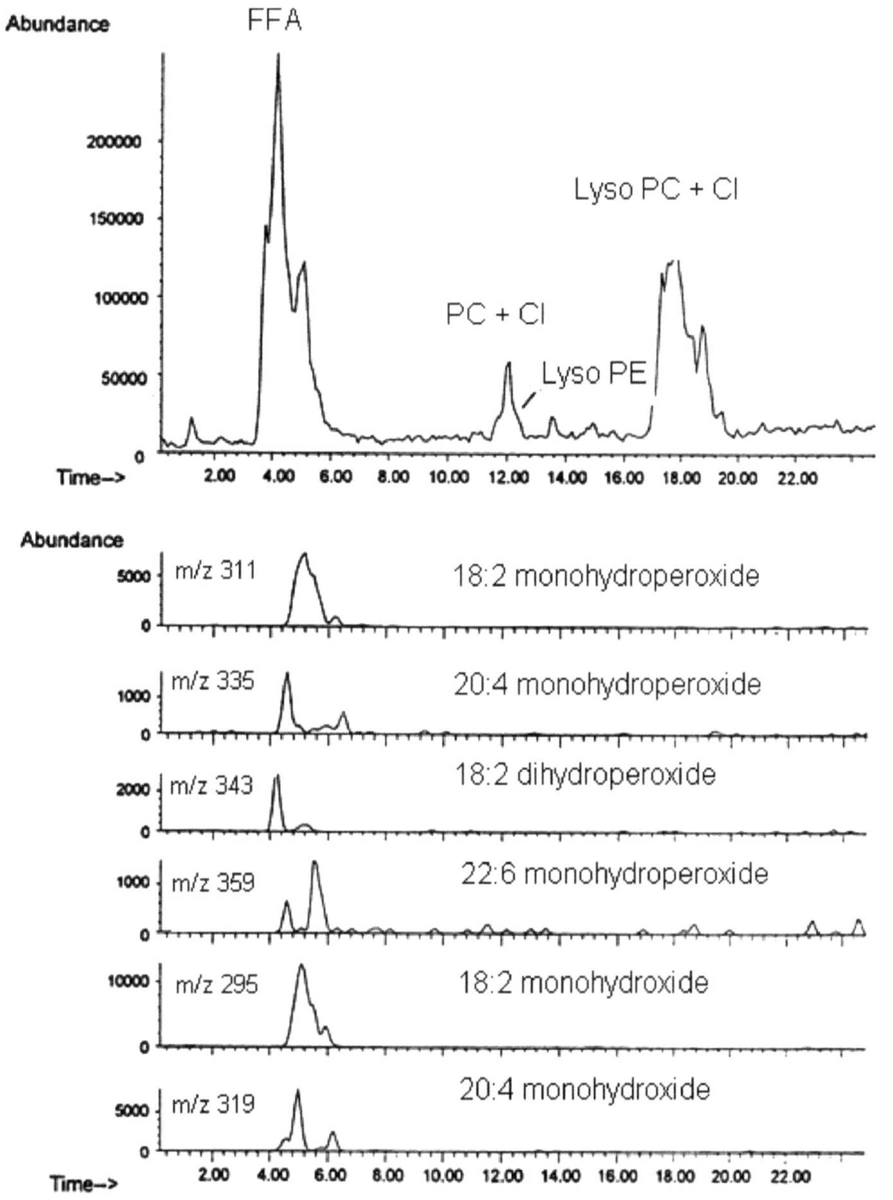

Fig. 2. Mass chromatogram of oxofatty acids released from APHDL during a 2-h hydrolysis with group IIA $sPLA_2$ as obtained by normal phase LC/ESI-MS (redrawn and updated from *50*). *Upper panel*, total negative ion current profile; *lower panel*, single negative ion chromatograms. Peak identification is indicated in the figures. Mono- and dihydroperoxy and hydroxy derivatives of the fatty acids are indicated by 1× and 2× the oxygen function. The peak doublets and multiplets are due to separation of positional and geometric isomers of the oxygenated fatty acids.

7. Elute column with a linear gradient of 100% A ($CHCl_3$/MeOH/30% NH_4OH 80:19.5:0.5 by vol) to 100% B ($CHCl_3$/MeOH/H_2O/30% NH_4OH, 60:34:5.5:0.5 by vol) for 14 min, followed by 100% B for 10 min, at a flow rate OF 1 ml/min.

8. Obtain positive ESI and negative ESI spectra by injecting each sample twice.
9. Obtain positive spectra in the mass range 350–1,100 and NI spectra in the mass range 300–1,100.
10. Identify isoprostanoic acid-containing GroPChos on basis of masses provided by ESI-MS, the knowledge of the fatty acid composition of the pospholipid classes, and the retention times of standards prepared by peroxidation of 16:0/20:4 GroPCho.
11. Note the following $[M + H]^+$ ions at *m/z* 810 (5,6-epoxy Iso-PGA_2GroPCho), *m/z* 828 (5,6-epoxy Iso-PGE_2GroP-Cho, *m/z* 830 (Iso-PGE_2/D_2GroPCho), *m/z* 832 (Iso-PGE_2GroPCho), *m/z* 858 (Iso-PGE_2/D_2GroPCho, and *m/z* 860 (Iso-PGE_2GroPCho).

*3.3.2.1.4. Ahmed et al. (**47**) Used LC/ESI-MS to Demonstrate the Formation of PtdCho Isoprostanes During Peroxidation of HDL with a Peroxynitrite Generating System*

1. Isolate HDL from serum of subjects fasted for 12–14 h by ultracentrifugation between densities 1.063 and 1.21 g/ml and dialyze exhaustively in 10-mM PBS, pH 7.4.
2. Oxidize HDL (1 mg/ml protein) by the peroxynitrite donor, SIN-1 (1 mM) for up to 20 h at 37°C in presence of DTPA (100 μM), a metal ion chelator.
3. Extract lipid-soluble HDL oxidation products after addition of 15:0/15:0 GroPCho as internal standard.
4. Evaporate organic phase under N_2 and redissolve lipids in 500 μl $CHCl_3$/MeOH 2:1 v/v.
5. Perform LC/ESI-MS analysis using normal phase silica column (4.6 × 250 mm, Alltech Associates, Deerfield, IL) in a Hewlett-Packard Model 1050 liquid chromatograph, connected to a Hewlett-Packard Model 5989A Quadrupole Mass Spectrometer, equipped with a nebulizer –assisted ESI interface (HP 59987A).
6. Elute column with a linear gradient of 100% A ($CHCl_3$/MeOH/30% NH_4OH 80:19.5:0.5 by vol) to 100% B ($CHCl_3$/MeOH/H_2O/30% NH_4OH, 60:34:5.5:0.5 by vol) for 14 min, followed by 100% B for 10 min, at a flow rate of 1 ml/min.
7. Split effluent 1:50, resulting in 20 μl/min being admitted to the mass spectrometer.
8. Set capillary exit voltage at 150 V, with electron multiplier at 1,795 V. Record positive ESI spectra in the mass range of 450–1,100 amu.

9. Identify molecular species on the basis of the masses provided by ESI-MS, the knowledge of the fatty acid composition of HDL PtdCho, and the relative retention time of the PtdCho standard (*see* **Fig. 3**).

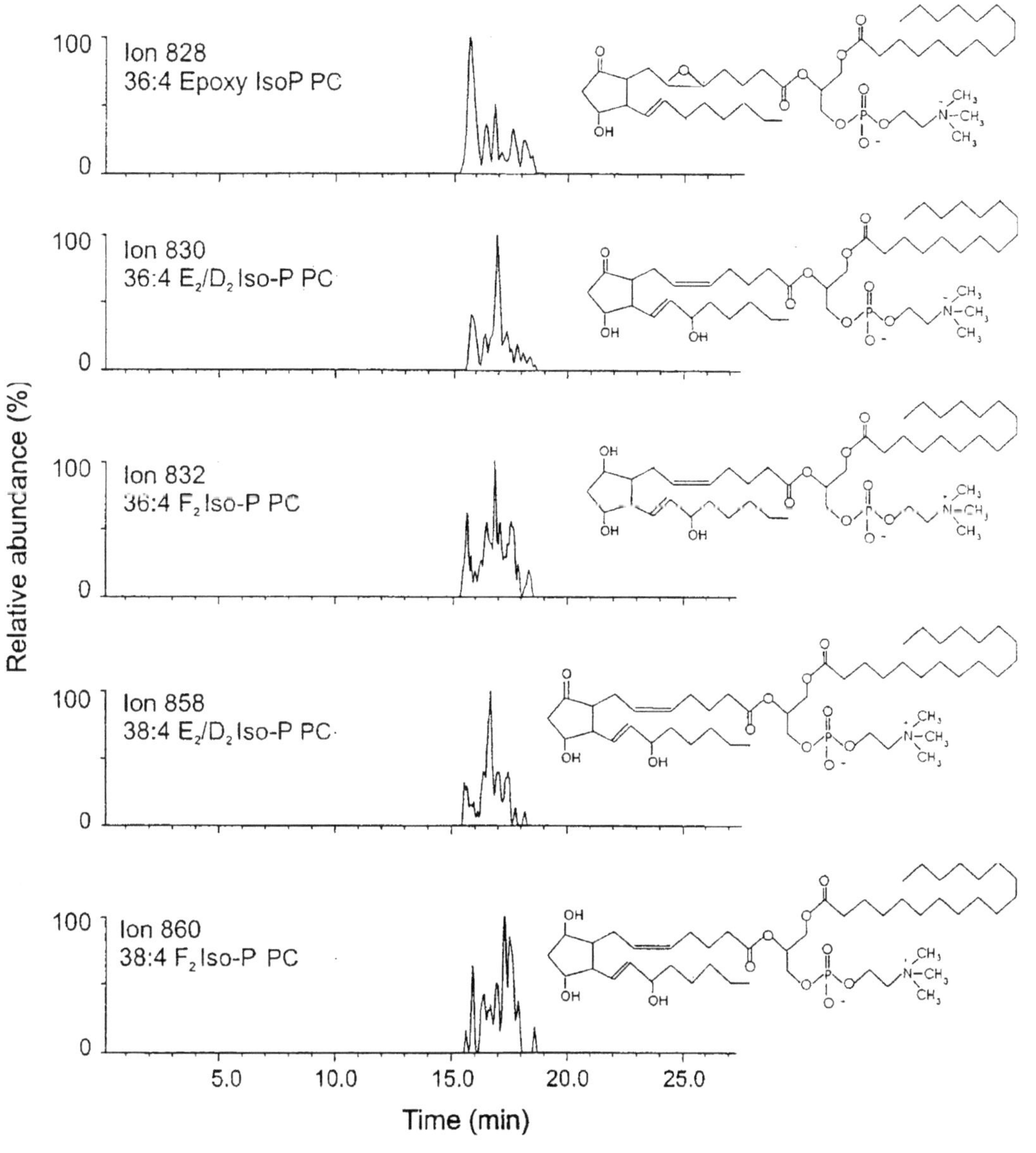

Fig. 3. Single ion mass chromatograms of the isoprostane GroPCho in HDL after 2 h of oxidation with SIN-1 (redrawn and updated from *47)*. Peaks are identified and structural formulas are given in the figure. Epoxy IsoP PC; epoxy-isoprostane GroPChos; IsoP PC, isoprostane GroPChos; 36:4 and 38:4 represent total carbon number of isoprostane diacylglycerol:number of double bonds.

3.3.2.2. Atheroma

*3.3.2.2.1. Kamido et al. (**27, 34**) Identified Glycerophospholipid Core Aldehydes in Human Atherosclerotic Lesions Using Protocols Previously Developed for Identification of the Core Aldehydes of Cholesteryl Esters. See* **Protocols 3.3.1.2.1** *and* **3.3.1.2.2**

1. Thaw aortic lesions, blot and weigh individually, freeze in liquid N_2 and pulverize in mortar.
2. Suspend fine powder in 2 ml of cold PBS buffer, homogenize, and extract with $CHCl_3$/MeOH 2:1 (v/v) following addition (10% of total phospholipid) of 15:0/15:0 GroPCho as internal standard.
3. Use 10 ml of solvent per 0.5 g of tissue.
4. Follow **Protocol 3.3.2.1.3** for normal phase LC/ESI-MS of isoprostane and core aldehyde derivatives of glycerophospholipids.
5. Record positive and negative ion spectra in the m/z range of 400–1,100 amu; use lower range for scanning for FA and nitrogen bases.
6. Prepare DNPH derivatives from another aliquot of sample and record corresponding ions for PtdCho core aldehydes.
7. Record single ion mass chromatograms for the choline, ethanolamine, and inositol glycerophospholipids and their hydroperoxy, hydroxy and core aldehyde derivatives.
8. Confirm presence of major oxolipid species by pseudo-MS/MS by raising capillary exit voltage from 150 to 300 V and reinjecting the sample.
9. Estimate lipid and lysolipid content of each phospholipid class along with the corresponding isoprostane and core aldehyde derivatives of PtdCho.

*3.3.2.2.2. Ravandi et al. (**52**) Reported an LC Isolation and Identification of PtdCho Core aldehydes from Human Atherosclerotic Lesions of Various Stages of Development*

3.3.2.3. Other Tissues

*3.3.2.3.1. Bergqvist and Kuksis (**28**) Reported Normal Phase LC/ESI-MS Profile of Bovine Heart Cardiolipin After Treatment with TBHB*

1. Suspend CL (1–3 mg) in 1 ml of 7% TBHP solution in water containing 100 μl of 2% taurocholic acid and 10-μM Fe^{2+} ions.
2. Shake reaction mixture on a mechanical agitator at 35°C in the dark for 1–4 h.
3. At the end of oxidation add 100 μl of 2% EDTA in water and 10 μl of 2% BHT in MeOH.
4. Extract an aliquot of reaction mixture with $CHCl_3$/MeOH (2:1, v/v) and save for TLC and LC/ESI-MS.
5. React the remainder of the reaction mixture with DNPH in 2 ml of 1-N HCl at room temperature (4 h) and overnight at 4°C before lipid extraction.
6. Note a lack of resolution between native cardiolipin and its hydroperoxides on the normal phase LC column (100 × 4.6 mm ID, Alltech Associates, Deerfield, IL) using a gradient elution with $CHCl_3$/MeOH/NH_4OH. *See* **Protocol of 2.1.2**.

7. Record single ion mass chromatograms corresponding to the major oxidation products of bovine heart cardiolipin. *See* **Protocol 2.1.4.4**.
8. Note the following major species of oxo-CL identified: *m/z* 1,448, $[M-1]^-$ (LLLL); *m/z* 1,480, $[M-1]^-$ (LLLL-OOH); *m/z* 1,482, $[M-1]^-$ (LLLO-OOH); *m/z* 1,512, $[M-1]^-$ (LLLL-di-OOH); *m/z* 1,514, $[M-1]^-$ (LLLO-di-OOH); *m/z* 1,544, $[M-1]^-$ (LLLL-tri-OOH); *m/z* 1,576, $[M-1]^-$ (LLLL-tetra-OOH). (*See* **Note 11**).

3.3.3. Autoxidation of TAGs

3.3.3.1. Lipoproteins, Chylomicrons, and Digesta

3.3.3.1.1. Suomela et al. (24, 53) Have Identified Hydroxy, Epoxy and Keto-Acid Containing TAGs Along with Core Aldehydes in Pig Lipoproteins Following Consumption of Autoxidized Sunflower Seed Oil

1. Collect blood samples in EDTA tubes from the jugular vein at selected time intervals following a test meal.
2. Separate plasma from cells by centrifugation.
3. Obtain a chylomicron-rich fraction from plasma by overlaying 1.8 ml plasma (2 × 0.9 ml) with 1.6-ml NaCl solution (d = 1,006 g/l, including 11.4-g NaCl and 0.1-g disodium EDTA dihydrate) and ultracentrifuge with a TFT 45.6 rotor (Kontron Instruments, Italy) at 38,000 × *g* (18,000 rpm) and 16°C for 30 min.
4. Aspirate the top 1.1 ml to collect the chylomicron-rich fraction.
5. Overlay the infranate again with NaCl solution (0.9 ml) and ultracentrifuge the samples to separate the VLDL-rich fraction at 160,000 × *g* (37,000 rpm) and 16°C for 15 h.
6. Extract lipids from plasma and lipoprotein suspensions with $CHCl_3$/MeOH 2:1 (v/v) using the following relative volumes of $CHCl_3$/MeOH/H_2O 8:4:3. Wash lower, organic phase by adding MeOH/0.88% KCl 1:1 (v/v) corresponding to one fourth of the volume of the organic phase.
7. Extract lipids from plasma and lipoprotein suspensions using relative volumes *:4:3 (by vol) of $CHCl_3$/MeOH/H_2O. (*See* **Note 12**).
8. Scrape silica gel from plates representing the fractions of oxo-TAGs.
9. Recover oxo-TAGs from the gel by extraction with $CHCl_3$/MeOH and simultaneously washing with distilled H_2O in a ratio of $CHCl_3$/MeOH/H_2O 8:4:3.
10. Evaporate solvents under N_2 and take up residue in 1–4 ml of freshly prepared DNPH reagent (1 mg DNPH in 2 ml of 1-M HCl) and keep in dark at 60°C for 30 min or at 6°C overnight.
11. Extract DNPH derivatives with $CHCl_3$/MeOH 2:1 (v/v) using a relative $CHCl_3$/MeOH/H_2O volume ratio of 8:4:3.
12. Evaporate extracts to dryness under N_2 and redissolve residue in 200 μl of IPA.

13. Heat the sample quickly to 60°C immediately before injection into the LC system with Discovery HS C18 column (250 mm × 4.6 mm, 5 μm, Supelco) and elute column at 0.85 ml/min with a linear gradient of 20–80% IPA in MeOH over 20 min and holding at final concentration for 10 min.
14. Lead 85% of the eluate to an ELSD through a UV/VIS detector that is set to register DNPH derivatives at 358 nm.
15. Lead 15% of the eluate to Finnigan MAT TSQ 700 triple quadrupole mass spectrometer equipped with a nebulizer assisted ESI interface (Finnigan, CA, USA).
16. Record full-scan mass spectra (*m/z* 450–1,100) in positive ionization mode and full scan negative ionization mode (*m/z* 600–1,200) for DNPH derivative analysis.
17. Use elution factors of standards for the identification of a large variety of TAG oxidation products, including hydroxyl, epoxy, and keto acid esters, as well as C_8, C_9, and $C_{12:1}$ core aldehydes of TAGs and mixed function oxo-TAGs *(3)*.

*3.3.3.1.2. Tarvainen et al. (**54**) Used Reversed Phase LC/ESI-MS/MS to Identify Oxoli-pids in an Artificial Model of Digestion and absorption of Dietary Fat*

1. Prepare total lipid extracts ($CHCl_3$/MeOH) from separate parts of artificial lipid digestion model (mouth, stomach, small intestine).
2. Separate total lipid extracts by reversed phase LC/LSD into native and oxidized lipid fractions by reference to oxolipid standards.
3. Elute column with a gradient of solvent A (CH_3CN/H_2O/HCOOH, 50:50:0.1 by vol) and solvent B (IPA/$(CH_3)_2CO$/HCOOH, 90:10:0.1, by vol) as follows: 0 min, 10% B; 15 min, 35% B; 40 min, 50% B; 50 min, 65% B; 60 min, 65% B; 60.1 min, 70% B; 75 min, 70% B; 76 min, 90% B; 86 min, 90% B.
4. Evaluate different food preparations with respect to generation of oxolipids in the presence and absence of the antioxidant activity of α-tocopherol acetate.
(*See* **Note 13**).

3.3.3.2. Atheroma

Kamido et al. *(27)* used high temperature GLC to demonstrated the presence of TAG core aldehydes,16:0/18:1/9:0ALD and 18:1/18:1/9:0ALD, in human atheroma samples as small peaks emerging between major peaks 40 and 42, and major peaks 43 and 45, respectively (*see* **Fig. 1**). The relative retention times of these TAG core aldehydes corresponded to those of standards prepared from seed oil TAGs by ozonization *(30)*. The hydroperoxides of TAGs also present in the atheroma samples were not recovered on high temperature GLC.

3.3.3.3. Seed Oils, Fish Oils, and Milk Fats

*3.3.3.3.1. Kuksis et al. (**30**) and Sjovall et al. (**3**) Used Reversed Phase LC/ESI-MS to Identify Mixed Hydroperoxide, Epoxide, Hydroxide and Core Aldehyde Derivatives Among the Autoxidation Products of Corn and Sunflower Oils*

1. Purify sunflower oil TAGs by normal phase TLC using conventional Silica gel H (Merck) plates and heptane/$(IP)_2O$/HOAc 60:40:4 (by vol) as the developing solvent.
2. Locate resolved components by iodine staining of a narrow strip of the developed plate.
3. Recover purified TAGs by scraping the gel and extracting it with $CHCl_3$/MeOH 2:1 (v/v).
4. Wash extracts with distilled H_2O and dry with anhydrous Na_2SO_4.
5. Evaporate solvents to dryness under N_2 and dissolve residue in 1 ml of $CHCl_3$.
6. Transfer sample to 15 ml centrifuge tube and evaporate solvent under N_2 by rotating the tube to deposit a thin film of TAGs covering the inside of the tube up to 3 cm from bottom.
7. Flush tube with air and incubate it in the open at room temperature for 60 days.
8. Stop the reaction by dissolving contents in 5 ml of $CHCl_3$/MeOH (2:1, v/v), 100 μl of 2% EDTA in water and 10 μl of 2% BHT.
9. Wash extracts three times with H_2O (3 × 1 ml) and evaporate solvents under N_2 at 38°C, and add 1-M DNPH·HCl (3.6 mg/ml) to an aliquot of dry sample.
10. Shake mixture vigorously and keep in dark at room temperature for 4 h and overnight at 4°C.
11. Extract lipids with 5 ml of $CHCl_3$/MeOH 2:1 (v/v), evaporate solvent under N_2, and take up residue in an appropriate solvent for TLC or LC-MS.
12. Perform TLC as outlined for purification of original TAG sample.
13. Locate yellow bands on the chromatoplates (in daylight).
14. Recover oxo-TAGs from the silica gel scrapings of individual TLC bands by extraction with $CHCl_3$/MeOH 2:1 (v/v).
15. Wash extracts with distilled H_2O, dry over anhydrous Na_2SO_4 and save for LC/LSD and LC/ESI-MS analysis.
16. Resolve individual TLC fractions of oxo-TAGs as DNPH derivatives by LC on a Supelcosil LC-18 column (250 mm × 4.6 mm ID) using a linear gradient of 20–80% IPA in MeOH (0.85 ml/min) in 30 min.
17. Use a Hewlett-Packard Model 1050 liquid chromatograph coupled to a Varex ELSD II light scattering detector (Varex, MD, USA) using N_2 as nebulization gas and evaporation temperature of 85°C.

18. Perform LC/ESI-MS with normal phase LC column using a linear gradient of $CHCl_3$/MeOH/30% NH_4OH for analyses of oxo-TAGs.
19. Set the ionization (capillary exit) voltage of this instrument at 70 V but increase to 300 V to obtain fragment ions from any clearly resolved components (pseudo MS-MS).
20. Use same LC conditions as above, except add 1% NH_4OH in IPA post-column at a flow rate of 0.15 ml/min in order to enhance ionization.
21. Acquire mass spectra in the 400–1,600 mass range over the entire elution profile.
22. Identify mixed acid hydroperoxides, oxides, epoxides and core aldehydes of TAGs based on normal phase TLC, reversed LC, and $[M - H]^-$ ions obtained by normal phase LC/ESI-MS in negative ionization mode.

1. Purify sunflower oil TAGs by normal phase TLC using conventional Silica gel H (Merck) plates and heptane/IP_2O/HOAc 60:40:4 (by vol) as the developing solvent.
2. Locate resolved components by I_2 staining of a narrow strip of the developed plate.
3. Recover purified TAG by scraping the gel from appropriate areas of the plate and extracting it with $CHCl_3$/MeOH 2:1 (v/v).
4. Wash extracts with distilled water and dry with anhydrous Na_2SO_4.
5. Evaporate solvents to dryness under N_2 and dissolve residue in 1 ml of $CHCl_3$.
6. Add 1 ml of 35% TBHP in H_2O to 10 mg of purified TAGs in the presence of 10-μM $FeSO_4$ and 100 μl of 0.2% taurocholic acid.
7. Incubate reaction mixture on a mechanical agitator in dark for 30 min to 30 h at 37°C.
8. Stop reaction by diluting with 5 ml of $CHCl_3$/MeOH 2:1 (v/v), 100 μl of 2% EDTA in H_2O and 10 μl of 2% BHT in MeOH.
9. Wash extracts three times with H_2O (3 × 1 ml) and evaporate solvent to dryness under N_2 at 38°C.
10. Add freshly prepared DNPH reagent in 1-M HCl (3.6 mg/ml) to an aliquot of the dried sample.
11. Shake mixture vigorously in the dark at room temperature for 4 h and overnight at 4°C.
12. Extract lipids with 5 ml of $CHCl_3$/MeOH 2:1 (v/v) and collect the $CHCl_3$ phase.

13. Dry down the $CHCl_3$ under N_2 and take up residue in an appropriate solvent for chromatography and MS.
14. For LC/ESI-MS analyses follow **Protocol 3.3.3.3.1**.
15. Identify major core aldehyde species (50–60% of total TAG core aldehydes) as the mono [9-oxo]nonanoyl- and mono[12-oxo]-9,10-epoxy dodecenoyl- or [12-oxo]-9-hydroxy-10,11-dodecenoyl-DAGs.
16. Note that as many as 113 molecular species of TAG core aldehydes may be specifically identified, accounting for 32–53% of the DNPH-reactive material of high molecular weight, and representing 25–33% of the total oxidation products.

3.3.3.3.3. Sjovall et al. (57) Identified and Quantified TAG Core Aldehydes as DNPH Derivatives in Autoxidized Sunflower Seed Oil Using Reversed Phase LC/ESI-MS

1. Use commercial sunflower seed oil (Kultasula, Toijala, Finland) without further purification.
2. Expose three aliquots of the sunflower seed oil (10 ml) to air in open bottles kept in the dark.
3. Take regular samples and store under nitrogen at −18°C until analyzed.
4. Convert hydroperoxide functional groups to hydroxide groups by a treatment with TPP (1 mg/ml benzene) to prevent core aldehyde formation during workup.
5. Remove excess TPP by TLC in a neutral lipid solvent system after derivatization of the core aldehydes to 2,4-dinitrophenylhydrazones **Protocol 3.2.3.6**.
6. Apply the DNPH derivatives of the TAG core aldehydes to an LC Supelcosil LC-18 column (250 × 4.6 mm ID, Supelco) and develop column with a linear gradient of 20–80% IPA in MeOH (0.85 ml/min) in 30 min.
7. For LC/ESI-MS analysis, follow **Protocol 3.3.3.3.1**.
8. Note that the autoxidized samples differ from those oxidized by TBHP by the presence of 12:1 and 13:2 core aldehydes, which apparently decomposed under the more harsh conditions of chemical oxidation.
9. Calculate that the core aldehydes make up 2–12 g/kg of oil on basis of UV detection and 2–9 g/kg by ESI-MS detection, whereas the hydroperoxides measured in the unreduced state by LC with ELSD can be estimated at 200 g/kg after 18 day autoxidation.

3.3.3.3.3. Hartvigsen et al. (32) Have Reported Formation of Core Aldehydes During Peroxidation of Alkyldiacylglycerols (DAGE) Isolated from Shark Oil

1. Isolate DAGE from shark liver oil by preparative double one-dimensional TLC (silica gel H plate (20 × 20 cm, 0.25 mm) developed with hexane/Et_2O 90:10 (v/v).
2. Visualize TLC bands by spraying with 0.2% 2,7-dichlorofluorescein in EtOH.

3. Recover lipids from TLC plates by extracting with $CHCl_3$/MeOH 2:1 (v/v), washing with H_2O, drying with Na_2SO_4, evaporating solvents under N_2, and redissolving residue in $CHCl_3$/MeOH 2:1 (v/v).
4. Subject purified DAGE from milk fat and shark oil to hydrolysis with pancreatic lipase see **Protocol 3.2.4.4.**
5. Extract digestion products with Et_2O and recover the DAGE, MAGE and 3-MAGE fractions.
6. Perform mild peroxidation of 2-MAGE by flushing the purified 2-MAGE from either shark liver oil in a tube with O_2, capping, and heating it at 80°C for 3 h.
7. Analyze peroxidized 2-MAGE by LC/ESI-MS. *See* **Protocol 3.3.3.3.1**.
8. Prepare DNPH derivatives of purified 2-MAGE by reaction with DNPH·HCl in the dark (0.5 mg in 1-ml 1-N HCl) for 2 h at room temperature and overnight at 4°C.
9. Extract DNPH derivatives with $CHCl_3$/MeOH 2:1, dry over Na_2SO_4, evaporate solvents under N_2, dissolve residue in $CHCl_3$/MeOH 2:1 (v/v), and analyze by reversed phase LC/ESI-MS. *See* **Protocol 2.1.2**.
10. Characterize the various core aldehydes by chromatographic retention time and diagnostic ions obtained by on-line ESI-MS.
11. Recognize a total of 23 molecular species of 1-*O*-alkyl-*sn*-glycerols with *sn*-2-oxoacyl group from the oxidized shark liver oil (*see* **Fig. 4**).

3.3.3.3.4. Hartvigsen et al. ***(32)*** Have Reported Formation of Core Aldehydes During Peroxidation of Alkyldiacylglycerols (DAGE) Isolated from Human Milk Fat

1. Isolate DAGE from human milk lipid extract by preparative double one-dimensional TLC (silica gel H plate (20 × 20 cm, 0.25 mm) developed with hexane/Et_2O 90:10 (v/v).
2. Visualize TLC bands by spraying with 0.2% 2,7-dichlorofluorescein in EtOH.
3. Recover lipids from TLC plates by extracting with $CHCl_3$/MeOH 2:1 (v/v), washing with H_2O, drying with Na_2SO_4, evaporating solvents under N_2, and redissolving residue in $CHCl_3$/MeOH 2:1 (v/v).
4. Subject purified DAGE from milk fat to hydrolysis with pancreatic lipase.
5. Extract digestion products with Et_2O and recover the DAGE, MAGE and 3-MAGE fractions.
6. Perform mild peroxidation of 2-MAGE by flushing the purified 2-MAGE from human milk in a tube with O_2, capping, and heating it at 80°C for 3 h.
7. Analyze peroxidized 2-MAGE by LC/ESI-MS. *See* **Protocol 3.3.3.3.1**.

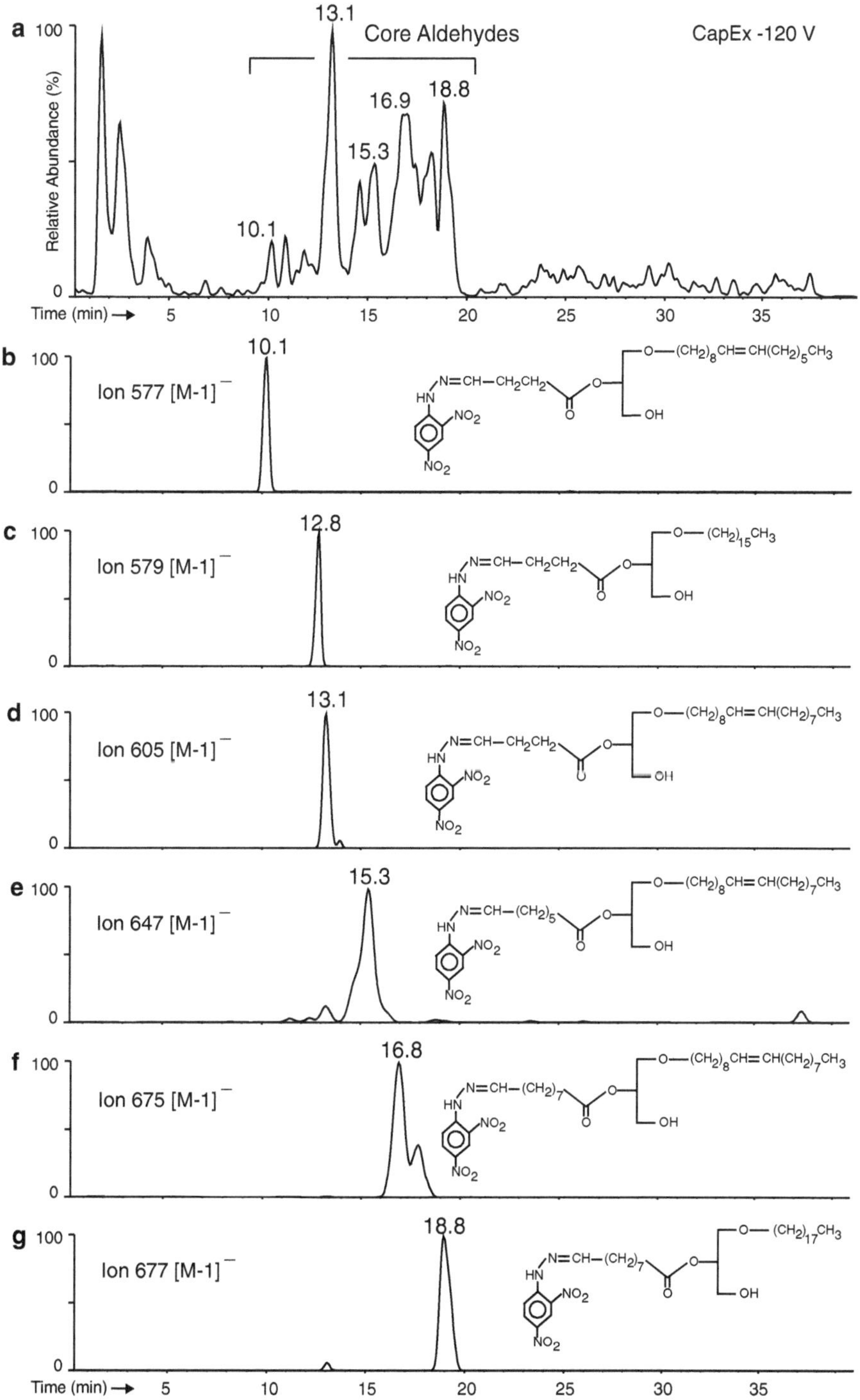

Fig. 4. Reversed phase LC/ESI-MS analysis of autoxidized and DNPH-derivatized 1-*O*-alkyl-2-oxoacyl-*sn*-glycerols from shark liver oil (with permission from *32)*. An aliquot (50 μl) of the extract of the reduced adduct in MeOH/H_2O/HOAc (50:50:1, by vol) was admitted to the ESI-MS interface via the LC column and the column eluted with a linear gradient of 100% A ($CHCl_3$/MeOH/30% NH_4OH, 80:19.5:0.5, by vol) to 100% B ($CHCl_3$/MeOH/H_2O/30% NH_4OH, 60:34:5.5:0.5, by vol) in 14 min, then at 100% B for 10 min. (**a**) Total negative ion current profile of 1-*O*-alkyl-2-[DNPH-oxo]acyl-*sn*-glycerols. (**b–g**) reconstructed single-ion mass chromatograms of *m/z* 577, 579, 605, 647, 675, and 677 $[M-1]^-$ diagnostic ions of 1-*O*-hexadecenyl-2-[DNPH-4-oxo]butyroyl-*sn*-glycerol, 1-*O*-hexadecyl-2-[DNPH-4-oxo]butyroyl-*sn*-glycerol, 1-*O*-octadecenyl-2-[DNPH-4-oxo]butyroyl-*sn*-glycerol, 1-*O*-octadecenyl-2-[DNPH-7oxo]heptanoyl-*sn*-glycerol, 1-*O*-octadecenyl-2-[DNPH-9-oxo]nonanoyl-*sn*-glycerol, and 1-*O*-octadecyl-2-[DNPH-9-oxo]nonanoyl-*sn*-glycerol, respectively.

8. Prepare DNPH derivatives of purified 2-MAGE by reaction with DNPH·HCl in the dark (0.5 mg in 1-ml 1-N HCl) for 2 h at room temperature and overnight at 4°C.
9. Extract DNPH derivatives with $CHCl_3$/MeOH 2:1, dry over Na_2SO_4, evaporate solvents under N_2 and dissolve residue in $CHCl_3$/MeOH 2:1 (v/v), analyze by reversed phase LC/ESI-MS. *See* **Protocol 3.3.3.3.1**.
10. Characterize the various core aldehydes by chromatographic retention time and diagnostic ions obtained by on-line ESI-MS.
11. Recognize 1-*O*-octadecyl-2-(9-oxo)nonanoyl-*sn*-glycerol as a readily detected lipid ester aldehyde in oxidized human milk fat.

3.3.4. Oxolipid Adducts of Amino Lipids, Amino Acids and Proteins

3.3.4.1. Cholesteryl Ester Core Aldehyde Adducts

3.3.4.1.1. Hoppe et al. (39) Reported Binding of Cholesteryl Ester Core Aldehydes to Amino Acids

1. Dissolve cholesteryl 9-oxononanoate (1 mg) in IPA/MeOH 1:1 (2 ml) and blow off the solvent in a test tube.
2. Add H_2O (2 ml) containing Nα-*tert*-butoxycarbonyl (N-BOC) -Lys (2 mg) followed by MeOH (1 ml).
3. Sonicate mixture (1 min) using short bursts at maximum power (Bronson 1200 Sonicator).
4. Leave sonicated mixture standing (4 h) at room temperature.
5. Reduce the formed Schiff base by adding $NaCNBH_3$ in MeOH to a final concentration of 70 mM and keep mixture at room temperature (1 h).
6. Extract reduced base with $CHCl_3$/MeOH 2:1 (v/v) and blow down solvent under N_2.
7. Redissolve residue in a small volume of $CHCl_3$/MeOH 2:1 for injection into the LC/ESI-MS system.
8. Admit an aliquot (50 μl) of the extract of the reduced adduct in MeOH/H_2O/HOAc (50:50:1, by vol) to the ESI-MS interface via the LC column and elute the column with a linear gradient of 100% A ($CHCl_3$/MeOH/30% NH_4OH, 80:19.5:0.5, by vol) to 100% B ($CHCl_3$/MeOH/H_2O/30% NH_4OH, 60:34:5.5:0.5, by vol) in 14 min, then at 100% B for 10 min.
9. Note that the free ε-amino group readily reacts with the core aldehyde to yield a Schiff base of a correct molecular mass of m/z 769 $[M + 1]^+$ on flow injection/ESI-MAS at a capillary exit (CapEx) of 120 V.
10. Note that reduction of the base with $NaCNBH_3$ increases its mass to m/z 771 and permits LC/ESI-MS recording of the total negative ion current profile of the product (LC peak; 7.21–8.52 min).

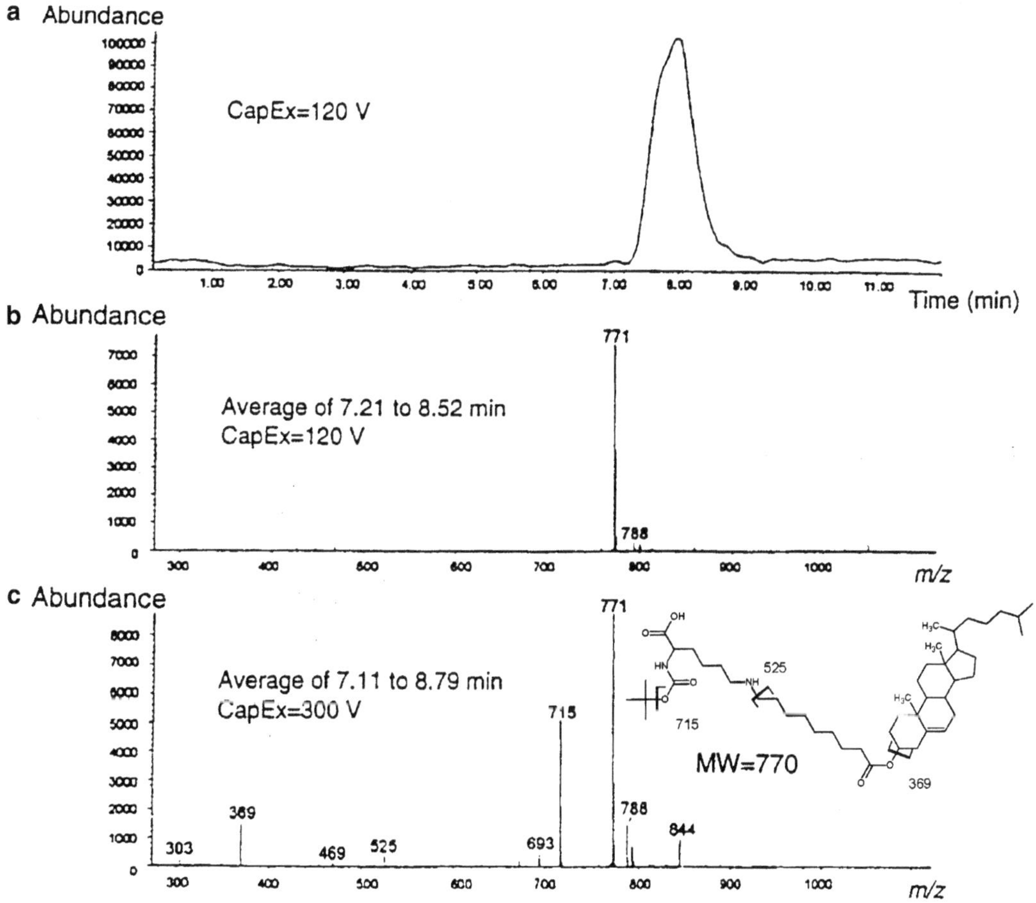

Fig. 5. Demonstration of Schiff base formation between cholesteryl [9-oxo]nonanoate and the ε-amino group of N-BOC-Lys (redrawn and updated from *39*). (**a**) Total positive ion current profile of $NaCNBH_3$-reduced base as obtained by LC/ESI-MS at CapEx voltage of 120 V. (**b**) Full mass spectrum at CapEx of 300 V averaged over the entire peak eluted in (**a**). (**c**) Full mass spectrum at CapEx 300 V averaged over the entire peak eluted in (**a**). Ion identification: *m/z* 771, $[M + 1]^+$; *m/z* 788, $[M + NH_4]^+$; *m/z* 715, $[M\text{-}tert\text{-butyl} + 1]^+$; *m/z* 845, $[M + 57 + 18]^+$; *m/z* 525, $[M - N\text{-BOC-Lys}]^+$; *m/z* 369, cholesterol ring; *m/z* 303, reduced Schiff base of N-BOC-Lys and [9-oxo]nonanoate.

11. Note that increasing the exit voltage from 120 to 300 V yields the fragment ions anticipated from a loss of the *tert*-butyl group $[M - 57]^+$ at *m/z* 715, $[M - N\text{-BOC-Lys}]^+$ at *m/z* 525, and $[M - \text{N-BOC-Lys} + \text{[9-Oxo]nonanoate}]^+$ at *m/z* 369 (cholesterol ring) (*see* **Fig. 5**).

3.3.4.1.2. Hoppe et al. (39) Reported Binding of Radiolabeled Cholesteryl Ester Core Aldehydes to Mouse Peritoneal Macrophages and Serum Proteins

1. Complex cholesteryl ester core aldehydes with cellular proteins by incubating mixed [^{3}H]cholesteryl ester core aldehydes (25 μl in PS-liposomes; 200,000 dpm/well) with mouse peritoneal macrophage (MPM) cultures (20 h).
2. Collect conditioned media and cell monolayers separately and subject to lipid extraction using $CHCl_3$/MeOH 2:1 (v/v).

3. Wash protein interface extensively with $CHCl_3$ (3 × 3 ml).
4. Wash cellular protein repeatedly with hexane/IPA.
5. Dissolve protein residues in 0.1-N NaOH, let stand, and neutralize samples with an equimolar amount of HCl for measurement of ^{3}H label.
6. Mix with scintillation fluid and count protein bound [^{3}H]-label as % of total recovered from cell or media.
7. In parallel, dry aliquots of [^{3}H]CL and [^{3}H]cholesteryl ester core aldehydes (10^6 dpm) in $CHCl_3$ on a 2 × 2 mm glass filter and place on bottom of boroslicate glass tubes.
8. Add 50 μl of 10% lipoprotein deficient serum (LPDS) in PBS containing EDTA and 40 μM of BHT.
9. Incubate at 37°C for 24 h and extract unbound lipid with Et_2O (3 × 1 ml).
10. For SDS/PAGE, mix aqueous phases with sample buffer containing 63-mM Tris-HCl (pH 6.8), 10% glycerol, 2% SDS, and 0.00025% bromophenol blue.
11. Resolve sample electrophoretically on precast 4–12% Tris-glycine gels (NOVEX, San Diego, CA).
12. Fix gels, stain and soak in EN'HANCE according to manufacturer's specifications.
13. Note binding of [^{3}H] core aldehyde but not [^{3}H]CL in cell free medium.

3.3.4.2. PtdCho Core Aldehyde Adducts

3.3.4.2.1. Ravandi et al. (58, 59) Demonstrated a Schiff Base Formation Between the PtdCho Core Aldehydes and Aminoglycerophospholipids

1. Dissolve dioleoyl GroPEtn (2 mg) and the 1-palmitoyl (stearoyl)-2-[9-oxo]nonanoyl GroPCho (1 mg) in $CHCl_3$/MeOH 2:1 and keep the mixture at room temperature for 1 h.
2. After this time add freshly prepared $NaCNBH_3$ in MeOH solution to final concentration of 70 mM and keep the reaction mixture at 4°C for 30 min.
3. At end of reaction, remove excess reagent by washing with H_2O.
4. Reduce the Schiff bases by $NaBH_4$ or PtO_2.
5. Identify the reduced Schiff base of PtdCho-core aldehyde-PtdEtn by normal phase LC/ESI-MS.
6. Prepare and identify the reduced Schiff base of PtdSer using identical methods.
7. Estimate the yields of the Schiff bases by LC/ESI-MS to range from 30% for PtdSer to 60% for the PtdEtn adduct.
8. Note that normal phase LC/ESI-MS of the $NaCNBH_3$-reduced reaction products of dioleoyl GroPEtn and 1-palmitoyl(stearoyl)-2-[9-oxo]nonanoyl-GroPCho give two major ions, one at m/z 1,378 $[M + 1]^+$ corresponding to 18:1/18:1 GroPEtn plus 16:0/9:0 GroPCho core aldehyde and another at m/z 1,406

$[M+1]^+$ corresponding to 18:1/18:1 GroPEtn plus 18:0/9:0 GroPCho core aldehyde.

9. Note that the total mass spectrum recorded at CapEx +300 V gives two other peaks at *m/z* 1,400 and 1,428 corresponding to the monosodium salts of the two Schiff's bases.
10. Identify other peaks in the LC/ESI/CAD-MS (CapEx 300 V) as the carboxy (*m/z* 666 and *m/z* 694) and hydroxy (*m/z* 652 and *m/z* 680) derivatives of the 16:0 and 18:0 GroPCho core aldehydes.
11. Note that ionization of the PtdSer-PtdCho core aldehyde Schiff base at CapEx of 300 V results in fragment ions, which closely resembles the pattern just established for the PtdEtn adduct.

3.3.4.2.2. Ravandi et al. (59) Subsequently Prepared Reduced Schiff Bases of Amino Acids and PtdCho Core Aldehydes

1. Dissolve PtdCho core aldehyde (16:0/[5-oxo]5:0) and 16:0/[9-oxo]9:0) (1 mg) in MeOH (2 ml) and add two-fold molar excess of the amino acid (valine, isoleucine, lysine, and lysine methyl ester) in saturated solution in H_2O (2 ml).
2. Shake reaction mixture at room temperature for 1 h, and then reduce the imine by adding $NaCNBH_3$ in MeOH solution to a final concentration of 70 mM and keeping the reaction mixture at 4°C for 30 min.
3. At the end of the reaction, recover the Schiff bases and the residual PtdCho core aldehydes by extraction with $CHCl_3$/MeOH 2:1 (v/v).
4. Evaporate solvents and take up residues in $CHCl_3$/MeOH/30% NH_4OH 80:19.5:0.5 (by vol).
5. Isolate Schiff bases by normal phase LC (Spherisorb, 3 μm, 100 × 4.6 mm ID, Alltech) installed in Hewlett-Packard Model 1050 Liquid Chromatograph. Elute column with a linear gradient of 100% A ($CHCl_3$/MeOH/30% NH_4OH 80:19.5:0.5, by vol) to 100% B ($CHCl_3$/MeOH/H_2O/30% NH_4OH 60:34:5.5:0.5, by vol) in 14 min, then at 100% B for 10 min.
6. Identify Schiff bases by normal phase LC/ESI-MS by admitting the liquid chromatograph effluent to a Hewlett-Packard Model 5988B single quadrupole mass spectrometer equipped with a nebulizer-assisted ESI interface.
7. Note that the $NaCNBH_3$ reduced reaction products from incubation of valine with 1-palmitoyl(stearoyl)-2-[9-oxo]nonanoyl-*sn*-3-GroPCho give major ions at *m/z* 751 $[M+1]^+$ representing valine plus 16:0/9:0Ald GroPCho, and at *m/z* 779 $[M+1]^+$ representing valine plus 18:0/9:0 Ald GroPCho.
8. Admit an aliquot (50 μl) of the extract of the reduced adduct in MeOH/H_2O/HOAc (50:50:1, by vol) to the ESI-MS interface via the LC column and elute the column with a

linear gradient of 100% A ($CHCl_3$/MeOH/30% NH_4OH, 80:19.5:0.5, by vol) to 100% B ($CHCl_3$/MeOH/H_2O/30% NH_4OH, 60:34:5.5:0.5, by vol) in 14 min, then at 100% B for 10 min.

9. Perform an LC/ESI/CAD-MS (Cap Ex +120 V) of the total ion current to show additional major ions at *m/z* 773 and *m/z* 801 corresponding to the monosodium adducts of the two Schiff bases, respectively. Similarly, reduce reaction products of isoleucine and the PtdCho core aldehydes to yield major ions at *m/z* 765 [*M* + 1]+ representing isoleucine plus 16:0/9:0 Ald GroPCho and at *m/z* 793 [*M* + 1]+ representing isoleucine plus 18:0/9:0 AldGroPCho.
10. Use similar methods to demonstrate that the PtdCho core aldehydes react effectively with the methyl ester of lysine to yield *m/z* 794 [M+1]+ for the 16:0–9:0 and *m/z* 822 for the 18:0–9:0 species. The spectrum also includes ions at *m/z* 816 and *m/z* 844, which correspond to the monosodium adducts of the palmitoyl and stearoyl species.
11. Likewise, show that the free lysine adduct of 16:0–9:0 Ald GroPCho on reversed phase LC/ESI-MS yields an ion at *m/z* 778 for the nonreduced molecule (*see* **Fig. 6**).

3.3.4.2.3. Ravandi et al. (59) Identified [5-oxo] valeroyl- and [9-oxo]-nonanoyl GroPCho Adducts of Myoglobin

1. Dissolve horse skeletal muscle myoglobin (0.5 mg; Sigma) in distilled H_2O (1 ml).
2. To this solution add PtdCho core aldehyde mixture (2 mg) in EtOH (2 ml) to give an approximate 100:1 ratio of aldehyde to protein.
3. Keep reaction mixture at room temperature for 1 h, then add $NaCNBH_3$ to a final concentration of 70 mM and keep at 4°C for 30 min.
4. Dialyze reaction mixture against distilled H_2O for 24 h with five changes of the solvent in order to remove excess reducing agent.
5. Lyophilize dialyzed sample and keep at –20°C until further analysis.
6. Analyze Schiff bases of PtdCho core aldehyde and myoglobin by means of Hewlett-Packard Model 5988B single quadrupole mass spectrometer equipped within ESI interface using a flow injection mode.
7. Dissolve lyophilized sample in 1-ml MeOH/H_2O/HOAc (50:50:1, by vol) and inject 50 μl of the sample, representing 1.5 nmol protein, into the ESI interface at 100 μl/min.
8. Record positive ion spectra in the m/z range 300–2,000.
9. Note that the original apomyoglobin gives a multicharged ion spectrum that can be deconvoluted to give a MW of 16,948.73, which corresponds to the literature value of 16,950. (*See* **Note 14**.)

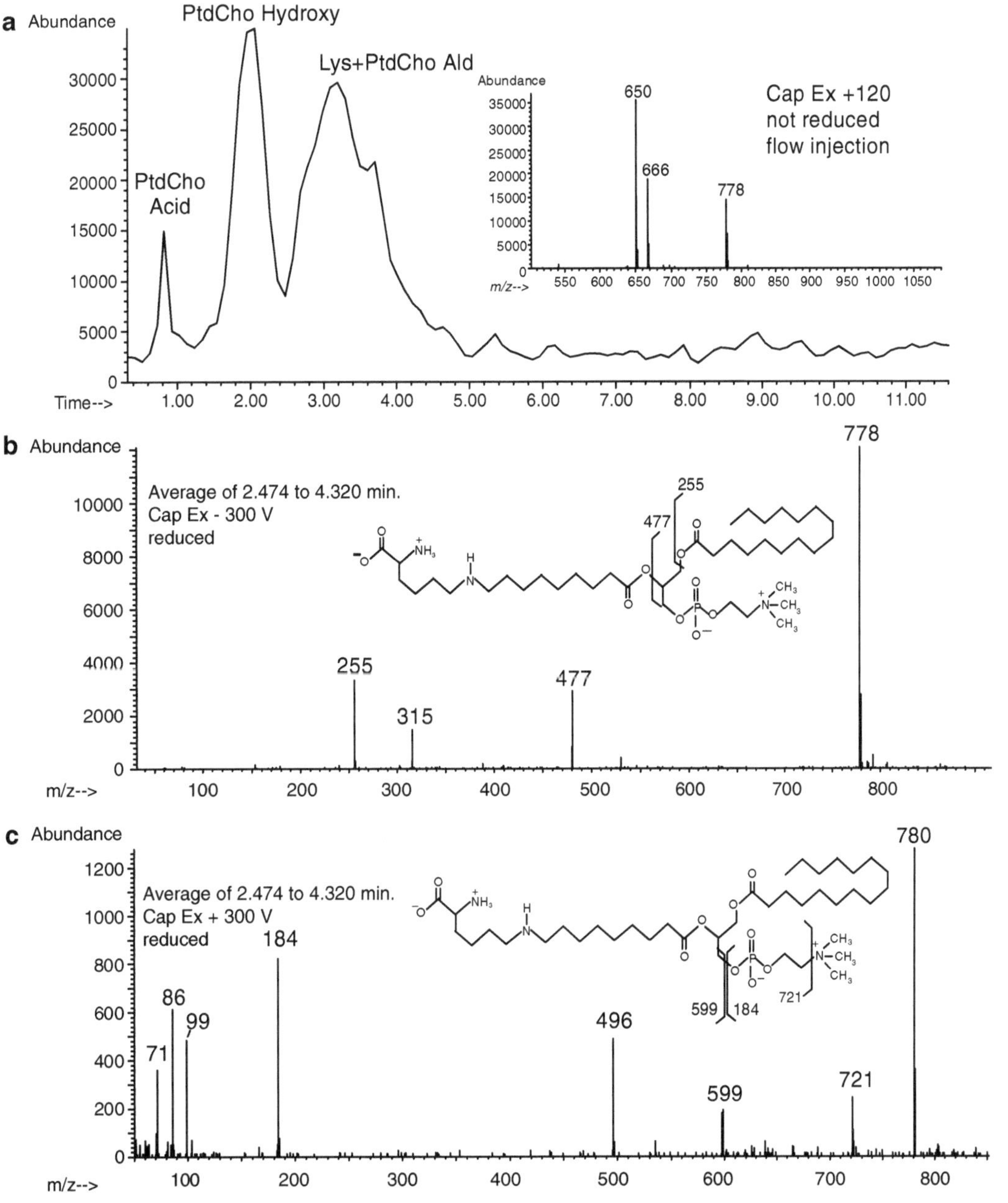

Fig. 6. Reversed phase LC/ESI-MS of the reduced reaction products of free lysine and 16:0/9:0Ald GPC (with permission from (*59*)). Total ion current (**a**) and fragmentation spectra of the reduced Schiff base of free lysine and 16:0/9:0Ald PC at negative CapEx voltage of −300 V (**b**) and at positive CapEx voltage of +300 V (**c**). In both (**b**) and (**c**), all the major ions are accounted for by the characteristic fragmentation of the PC moiety indicated in the structural formulae are given in the figure.

10. Note that other incubations gave evidence of the formation of Schiff bases with the mono-16:0/9:0 core aldehyde and mono-18:0/9:0 core aldehyde, the di-16:0/9:0 core aldehyde, and the mixed 16:0/9:0 core aldehyde, and the 18:0/9:0 core aldehyde adducts (*see* **Fig. 7**).

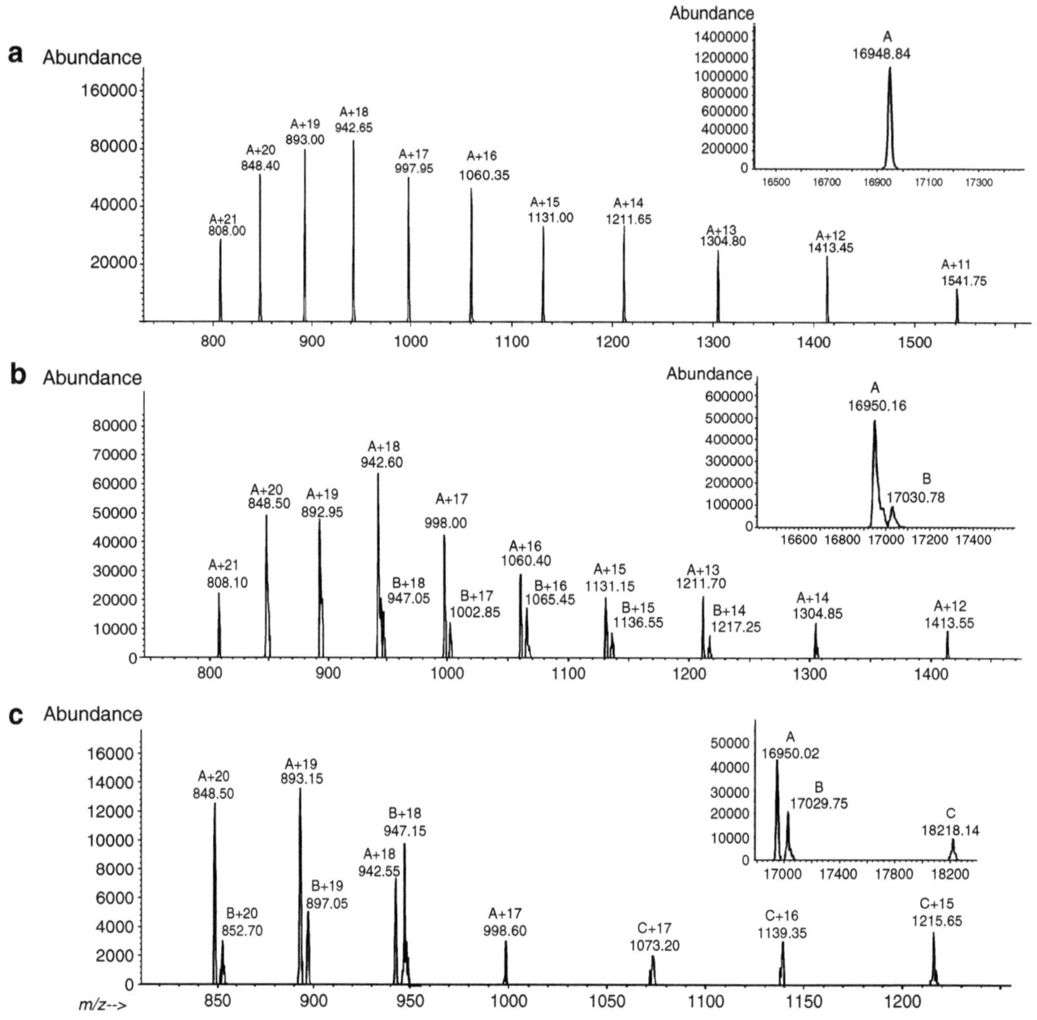

Fig. 7. Flow ESI-MS spectra and the deconvoluted MW of the horse skeletal muscle apomyoglobin (MW 16,948.84 kDa) (**a**), its sodium cyanoborohydride reduction product (MW 17,030.78 kDa) (**b**), and the product of interaction of myoglobin with two molecules of 16:0/9:0Ald GPC (MW 18,218.14 kDa) (**c**). Positive ion spectra were taken in the *m/z* range 300–2,000 (with permission from *59*). (*See* Note 14).

3.3.4.2.4. Ahmed et al. (61) Complexed Oxo-PtdCho Derived from Peroxynitrite Oxidation (SIN-1) with Apoprotein AI

1. Prepare ApoAI proteoliposomes using apoAI-palmitoyl/(5-oxo)valeroylGroPCho (palmitoyl/(9-oxo)nonanoylGro PCho-dimyristoylGroPCho-cholesterol POVPC (PONPC)-DMPC-cholesterol in a molar ratio 1:18:82:5. (*See* **Note 15**).
2. Incubate PtdCho core aldehyde proteoliposomes at 37°C for 6 h to allow Schiff base formation.
3. Reduce Schiff base adducts with 30-mM $NaCNBH_3$ for 1 h at 4°C and dialyze overnight.
4. Analyze apoAI adducts by sodium dodecyl sulfate-polyacrylamide gel electrophoresis (SDS-PAGE) by adding to each

well on an 8% polyacrylamide precast mini-gel (ICS Bioscience, Kaysville, UT).

5. Fix gels with a solution of 7% HOAc and 15% MeOH fro 30 min, stain with SYRO Ruby and visualize in a Gel Doc EQ System (Bio-Rad).
6. View protein stained SDS-PAGE gels of ApoAI proteoliposomes containing apoAI-14:0/14:0GroPCho (DMPC), apoAI-DMPC-POVPC and apoAI-DMPC-PONPC. (*See* **Note 16**).

3.3.4.3. Acylglycerol Core Aldehyde Adducts

1. Reduce Schiff base of 2-MAG-Ald and PtdEtn as for aminophospholipids *(59)*.

3.3.4.3.1. Kurvinen et al. (65) Prepared and Characte-rized Monoacylglycerol Core Aldehyde Adducts of Amino Lipids (PtdEtn and PtdSer)

2. Dissolve 2-MAG-Ald (1 mg) and a 2-fold excess of PtdEtn (6 mg) in 4 ml of $CHCl_3$/MeOH (2:1, v/v) and allow to react at room temperature for 16 h.
3. After reaction, add freshly prepared $NaCNBH_3$ in MeOH for a final concentration of 70 mM and keep the mixture at 4°C for 30 min to allow reduction of the imine bond of the Schiff base.
4. Remove excess reducing agent by passing the reaction mixture through anhydrous Na_2SO_4.
5. Identify reduced Schiff base by LC/ESI-MS see Protocol number.

3.3.4.3.2. Kurvinen et al. (65) Prepared and Characterized Monoacylglycerol Core Aldehyde Adducts of Valine, Nα-acetyl-L-Lysine Methyl Ester, and the Tripeptides (GGG, GGH, and GHK)

1. Dissolve 2-[9-oxo]nonanoylglycerol (1 mg) in MeOH (2 ml) and add two-fold molar excess of amino acid (valine, Nα-acetyl-L-lysine methyl ester) or peptide (GGG, GGH and GHK) dissolved in water (2 ml).
2. Shake reaction mixture at room temperature (1 h).
3. Add freshly prepared $NaCNBH_3$ in MeOH to a final concentration of 70 mM and keep reaction mixture at 4°C for 30 min.
4. Add chloroform (5 ml) and allow phases to separate.
5. Recover Schiff base adducts of amino acids from the organic phase and the Schiff bases of peptides from the aqueous phase.
6. Identify reduced Schiff base by normal phase LC/ESI-MS. *See* **Protocol 2.1.4**.
7. Dissolve 2-[9-oxo]nonanoylglycerol (1 mg) in MeOH (2 ml) and add two-fold molar excess of PtdEtn dissolved in $CHCl_3$/MeOH 2:1 (v/v) (4 ml).
8. Allow reaction to proceed at room temperature (16 h).

9. Add freshly prepared $NaCNBH_3$ in MeOH to a final concentration of 70 mM and keep mixture at 4°C for 30 min to reduce the imine bond of Schiff base.
10. Remove excess reagent by washing the organic phase with water and passing it through anhydrous Na_2SO_4.
11. Identify reduced Schiff base by normal phase LC/ESI-MS. *See* **Protocol 2.1.4.**

3.4. Physiological Activity of Some Oxolipids

Lipidomics has been essential for establishing the structure of the oxolipids used for demonstration of their potential physiological activity.

3.4.1. PtdCho Core Aldehydes

3.4.1.1. Kamido et al. *(66)* Demonstrated Platelet Aggregation by Physiologically Relevant Concentrations of Alkyl GroPCho Core Aldehydes (PAF mimics)

1. Prepare C_5-alkyl GroPCho core aldehydes by subjecting 1-*O*-hexadecyl-2-arachidonoyl-*sn*-GroPCho to ozonization and reduction with TPP *(58)*.
2. Prepare C_5-acyl GroPCho core aldehydes by subjecting 1-palmitoyl-2-arachidonoyl *sn*-GroPCho to similar ozonization and reduction with TPP *(58)*.
3. Draw blood from a carotid artery of male Japanese white rabbits anaesthetized with pentabarbiturate, using 3.8% sodium citrate as anticoagulant (9:1), and prepare platelet-rich plasma by centrifugation of blood at 225 × *g* for 10 min.
4. To the platelet rich plasma, add a 1:1,000 volume of prostacyclin (0.1 mg/ml) and centrifuge at 1,200 × *g* for 10 min. Wash platelets with citrate-buffered saline (pH 6.5) and suspend in HEPES-Tyrode buffer (pH 7.38) at density of 3.0 × 10 8 platelets/ml.
5. Mix 400 µl of the platelet suspension with 4 µl of a solution of core aldehyde in 99.5% EtOH and determine platelet aggregating activity in HEPES-Tyrode buffer containing 1-mM $CaCl_2$ and fibrinogen (0.5 mg/ml) at 37°C.
6. Monitor platelet aggregation by continuous recording of light transmission using C550 whole blood aggregator (ChronoLog, Havertown, PA).
7. Perform quantitative analysis of the shape change of the platelets by turning the stirrer on and off, alternatively, every 30 s *(67)*.
8. Observe that in contrast to C_5 alkyl GroPCho core aldehyde, C_5 acyl GroPChol core aldehyde did not induce aggregation up to 10 µM, above which platelet lysis occurred.
9. Observe, however, that C_5 acyl core aldehyde induced platelet shape change, which was suppressed in the presence of protein kinase inhibitor (staurosporine).
10. By contrast, 10-µM C_5 alkyl and C_5 acyl GroPCho core aldehydes inhibited endothelium dependent relaxation of rabbit

artery by 50%, while endothelium independent relaxation was not affected.

3.4.1.2. Ahmed et al. *(68)* Demonstrated that Formation of PtdCho Core Aldehyde-Apoprotein AI Adducts Enhanced Uptake of Oxidized HDL by THP-1 Macrophages

1. Prepare apoAI proteoliposomes according to Sorci-Thomas et al. *(62)* using apoAI/palmitoyl/(5-oxo)valeroylGroPCho (palmitoyl/(9-oxo)nonanoylGroPCho/dimyristoylGroP-Cho in a molar ratio of 1:18:82:5.
2. Incubate proteoliposomes for 6 h at 37°C to allow Schiff base adduct formation.
3. Reduce complexes with 30-mM $NaCNBH_3$ for 1 h at 4°C and dialyze overnight. Confirm formation of adduct by phosphorus assay and by sodium-dodecyl sulfate-polyacrylamide gel electrophoresis (SDS-PAGE).
4. Prepare Dil (1,1′-dioctadecyl-3,3,3′,3′-tetramethylindioarbocyanine perchlorate) or DiO (3,3′-dihexadecyloxa-carbocyanine perchlorate) labeled HDL(or oxidized HDL) and PtdCho core aldehyde proteoliposomes by adding slowly 300 μl of Dil solution in dimethyl sulfoxide (3 mg/ml) to 5-ml HDL (2 mg protein/ml), predialized in 0.1-M PBS, pH 7.4, and incubate for 8 h at 37°C under nitrogen in the dark *(69)*.
5. Interact oxidized HDL with THP-1 cells following transformation of monocytes to macrophages by adding 164-nM phorbol myristate acetate for 36 h, prior to incubation with labeled lipoproteins.
6. Incubate THP-1 macrophages for 2 h at 37°C with Dil-labeled lipoproteins or proteoliposomes on slides in culture plates.
7. Wash plates extensively with formalin-PBS, pH 7.4 (4:96, v/v) as described *(69)*.
8. Remove slides from culture plates and perform fluorescent confocal microscopy using an argon laser with excitation at 488 nm and fluorescent emission at 501 nm for detection of Dio-labeled complexes and using a helium neon laser with excitation at 553 nm and fluorescence emission at 570 nm of Dil-labeled complexes.
9. Detach cells with rubber policeman and resuspend in 0.5% paraformaldehyde fixative solution prepared in PBS, ph 7.4, and evaluate by flow cytometry.
10. Perform flow-cytometry experiments by measuring 10 000 individual cells gated for macrophages, based on their forward and side light-scatterer profiles (Coulter FASCORT and CELL Quest software, Coulter Becton Dickinson, San Jose, CA).
11. Note that binding and uptake of PtdCho core aldehyde-apoprotein AI proteoliposomes by THP-1 macrophages was similar to that observed for oxidized HDL and oxidized LDL.

3.4.1.3. Hartvigsen et al. (32) Have Pointed that Hydrolysis and Peroxidation of Alkylacyl Glycerols Containing Docosahexaenoic Acid in the Intestine May Yield Short-Chain Monoalkylglycerol Core Aldehydes, Which Upon Absorption and Further Simple Metabolic Transformation May Yield γ-Hydroxy Butyrate

(*See* Scheme 2). Absorption of 2-C_4- acyl PtdCho core aldehydes as 1-alkyl or 1-acyl esters, followed by reduction to alcohol and lipolysis would release γ-hydroxy butyrate (GHB), a physiologically active oxolipid and a common street drug. The potential formation of GHB is supported by previous studies of Kanazawa and Ashida *(55, 56)*, who demonstrated core aldehyde formation in the intestine, and Suomela et al. *(24, 53)*, who recovered a large variety of oxolipids including lipid ester core aldehydes in pig chylomicrons after feeding peroxidized sunflower seed oil.

1. The individual transformation steps are well established metabolic events and need not be limited to intestinal tissue or glycerolipids *(70, 71)*.
2. The aldo-keto reductases are wide-spread tissue enzymes and readily convert aldehyde groups into hydroxyl groups.
3. Short-chain alcohol esters are readily released from the *sn*-2-position of PtdCho by PAF acetyl hydrolase or by various other

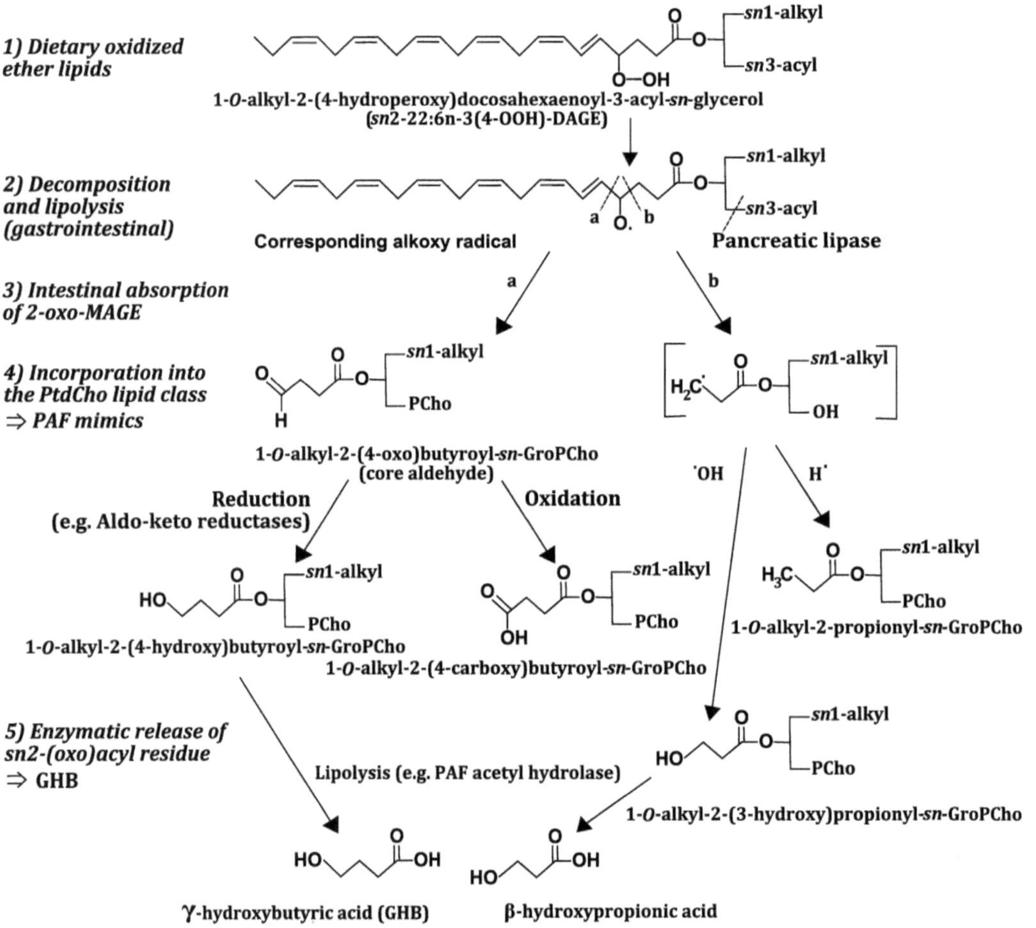

Scheme 2. Proposed mechanism of formation of PAF mimics and GHB from diacylalkylglyceryl ether (DAGE) containing docosahexaenoic acid (with permission from *32*).

short-chain acid ester hydrolases to yield short-chain hydroxy fatty acids.

4. It can be readily seen that GHB would be most directly derived from the docosahexaenoic acid esters, while the transformation of other core aldehyde esters to GHB would require additional oxidation steps.

4. Notes

1. Voyksner and Pack *(2)* provide a detailed description of CID process and spectra in the transport region of an ESI-MS.
2. Zhang et al. *(12)* synthesized a large number of doubly allylic dihydroperoxides as putatitve intermediates in the generation of biologically active aldehydes in vivo.
3. Chung et al. *(25)* have described methods for the preparation of various lipoprotein classes, including LDL and HDL.
4. Adachi et al. *(29)* have used a similar method of preparation of homogenates for the isolation of the glycerophospholipids from rat heart tissue.
5. Rapid TLC analysis of total lipid extracts may be done on commercially coated silica gel plates (EM Labs) using $CHCl_3/MeOH/H_2O$ (65:25:4, by vol) *(33)*. The plates are sprayed with iodine in MeOH, ninhydrin, or a phospholipid detecting reagent.
6. Schmitt et al. *(36)* reduce the hydroperoxides to their corresponding hydroxides during extraction utilizing a modified Dole procedure in which the reducing agent, TPP, is present.
7. Lee et al. *(37)* recommend the use of trifluoroacetate derivatives of hydroxy fatty acids for sensitive targeted lipidomics analysis with electron capture.
8. Buldt and Karst *(40)* recommend 1-methyl-1-DNPH for similar applications.
9. Hui et al. *(46)* reported that in patients with advanced liver disease, hydroperoxide levels of plasma cholesteryl ester and TAG are 11,903 nM and 3,318 nM, respectively, indicating involvement of lipid peroxidation. These levels of plasma lipid hydroperoxides are much higher than the about 3–4 nM reported previously without the use of hydroperoxide internal standard.
10. The cholesteryl oxoacyl esters including core aldehydes were resolved by conventional TLC using heptanes/diisopropyl ether/glacial acetic acid (60:40:4, by vol). Lipid fractions

were located by spraying with 2,7-dichlorofluorescein and Schiff's reagent *(45)*.

11. The rabbit heart cardiolipins appeared to posses a much wider distribution of molecular species than the commercial preparation of the bovine heart cardiolipins reported in detail elsewhere *(29)*.
12. The core aldehydes were detected by spraying with a Schiff's reagent.
13. Kanazawa et al. *(55, 56)* have reported that dietary hydroperoxides of linoleic acid decompose to aldehydes in stomach before being absorbed into the body.
14. The molecular weight of apomyoglobin of horse skeletal muscle has been determined by mass spectrometry previously *(60)*.
15. Proteoliposomes to be used for oxidation experiments, including controls, were prepared by cholate dialysis of emulsions of apoA-I/16:0/18:2 GroPCho/14:0/14:0 GroPCho/Cholesterol in a molar ratio 1:18:82:5 *(62)*.
16. Lipid ester core aldehyde adducts have also been prepared by Itabe et al. *(63)* and Gugiu et al. *(64)*.

Acknowledgments

The various research projects were funded in whole or in part by The Medical Research Council of Canada, Ottawa, Canada, The Heart and Stroke Foundation of Ontario, Toronto, Ontario and The Academy of Finland, Helsinki, Finland.

References

1. Wolf, C., and Quinn, P. J. (2008). Lipidomics: Practical aspects and applications. *Progr. Lipid Res.* **47**, 15–36.
2. Voyksner, R. D., and Pack, T. (1991). Investigation of collision-activation decomposition process and spectra in the transport region of an electrospray single-quadrupole mass spectrometer. *Rapid Commun. Mass Spectrom.* **5**, 1263–1268.
3. Sjovall, O., Kuksis, A., Marai, L., and Myher, J. J. (1997). Elution factors of synthetic oxotriacylglycerols as an aid in identification of peroxidized natural triacylglycerols by reverse-phase high-performance liquid chromatography with electrospray mass spectrometry. *Lipids* **32**, 1211–1218.
4. Terao, J, and Matsushita, S. (1981). Analysis of photosensitized oxidation products of unsaturated triglycerides and vegetable oils by gas chromatography mass spectrometry. *Agric. Biol. Chem.* **45**, 601–608.
5. Terao, J., Asano, I., and Matsushita, S. (1984). High performance liquid chromatographic determination of phospholipid peroxidation products of rat liver after carbon-tetrachloride administration. *Arch. Biochem. Biophys.* **235**, 326–333.
6. Chan, H. W. S., Prescott, F. A. A., and Svoboda, P. A. T. (1976). Thermal decomposition of individual positional isomers of methyl linoleate hydroperoxide: Evidence of carbon-oxygen bond scission. *J. Am. Oil Chem. Soc.* **53**, 572–576.
7. Chiba, T., Takazawa, M., and Fujimoto, K. (1989). A simple method for estimating carbonyl content in peroxide-containing oils. *J. Am. Oil Chem. Soc.* **66**, 1588–1592.

8. Deffense, E. (1993). Nouvelle method d-analyse pour separer, via HPLC, les isoimeres de position 1–2 et 1–3 des triglycerides mono-insatures des graisses vegetales. *Rev. Fr. Corps Gras* **40**, 33–39.
9. Privett, O. S., and Blank, M. L. (1963). A method for the structural analysis of triglycerides and lecithins. *J. Am. Oil Chem. Soc.* **40**, 70–75.
10. Kuksis, A., Myher, J. J., Ravandi, A., and Cocks, N. (1994). Identification of polyoxyglycerophospholipids by normal phase LC/MS with electrospray. *Inform* **5**, 478–479.
11. Kamido, H., Kuksis, A., Marai, L., and Myher, J. J. (1992). Identification of cholesterol-bound aldehydes in copper-oxidized low density lipoprotein. *FEBS Lett.* **304**, 269–272.
12. Zhang, W., Sun, M., and Salomon, R. G. (2006). Preparative singlet oxygenation of linoleate provides doubly allylic dihydroperoxides: Putative intermediates in the degeneration of biologically active aldehydes in vivo. *J. Org. Chem.* **71**, 5607–5615.
13. Podrez, E. A., Poliakov, E., Shen, Z., Zhang, R., Deng, Y., Sun, M., Finton, P. J., Shan, L., Gugiu, B., Fox, P. L., Hoff, H. F., Salomon, R. G., and Hazen, S. L. (2002). Identification of a novel family of oxidized phospholipids that serve as ligands for the macrophage scavenger receptor CD36. *J. Biol. Chem.* **277**, 38503–38516.
14. Podrez, E. A., Poliakov, E., Shen, Z., Zhang, R., Deng, Y., Sun, M., Finton, P. J., Shan, L., Febbraio, M., Hajjar, D. P., Silverstein, R. L., Hoff, H. F., Salomon, R. G., and Hazen, S. L. (2002). A novel family of atherogenic oxidized phospholipids promotes macrophage foam cell formation via the scavenger receptor CD36 and is enriched in atherosclerotic lesions. *J. Biol. Chem.* **277**, 38517–38523.
15. Han, X., and Gross, R. W. (2003). Global analyses of cellular lipodomes directly from crude extracts of biological samples by ESI mass spectrometry: a bridge to lipidomics. *J. Lipid Res.* **44**, 1071–1079.
16. Garscha, U., Nilsson, T., and Oliw, E. H. (2008). Enantiomeric separation and analysis of unsaturated hydroperoxy fatty acids by chiral column chromatography-mass spectrometry. *J. Chromatogr. B* (In press).
17. Havrilla, C. M., Hachey, D. L., and Porter, N. A. (2000). Coordination (Ag+) ion spray-mass spectrometry of peroxidation products of cholesterol linoleate and cholesterol arachidonate: high performance liquid chromatography-mass spectrometry analysis of peroxide products from polyunsaturated lipid autoxidation. *J. Am. Chem. Soc.* **122**, 8042–8055.
18. Byrdwell, W. C. (2005). Dual parallel liquid chromatography/dual mass spectrometry for lipid analysis. In *Modern Methods for Lipid Analysis* (Byrdwell, W. C., editor), AOCS Press, Champaign, IL, pp. 510–576.
19. Folch, J., Lees, M., and Stanley G. H. S. (1957). A simple method for the isolation and purification of total lipids from animal tissues. *J. Biol. Chem.* **226**, 497–509.
20. Bligh, E. G., and Dyer, W. J. (1959). A rapid method of total lipid extraction and purification. *Can. J. Biochem. Physiol.* **37**, 911–917.
21. Arborati, M., Benchorba, D., Lesieur, I., Bizot-Esiard, J. G., Guardiola-Lemaitre, B., Chapman, M. J., and Ninio, E. (1997). Oxidative degradation of cholesteryl esters in low density lipoproteins: analysis by liquid chromatography-light scattering and protection by a new synthetic antioxidant, S20478. *Fundam. Clin. Pharmacol.* **11**, 68–77.
22. Anonymous (1974). Manual of Laboratory Operations. *Lipid Research Clinics Program.* Volume 1. Lipid and Lipoprotein Analysis. pp. 51–56.
23. Kuksis, A., Myher, J. J., Breckenridge, W. C., and Little, J. A. (1979). Lipid profiles of human plasma high density lipoproteins. In *Report of the High Density Lipoprotein Methodology Workshop* (K. Lippel, editor) NIH Publication No. 79–1661. Bethesda, MD, pp. 142–163.
24. Suomela, J.-P., Ahotupa, M., and Kallio, H. (2005). Triacylglycerol oxidation in pig lipoproteins after a diet rich in oxidized sunflower seed oil. *Lipids* **40**, 437–444.
25. Chung, B. H., Segrest, J. P., Ray, M. J., Brunzell, J. D., Hokanson, J. E., Krauss, R. M., Baudrie, K., and Cone, J. T. (1986). Single vertical spin-density gradient ultracentrifugation. *Methods Enzymol.* **128**, 181–209.
26. Lenz, M. L., Hughes, H., Mitchell, J. R., Via, D. P., Guyton, J. R., Taylor, A. A., Gotto, Jr., A. M., and Smith, C. V. (1990). Lipid hydroperoxy and hydroxy derivatives in copper-catalyzed oxidation of low density lipoprotein. *J. Lipid Res.* **31**, 1043–1050.
27. Kamido, H., Nonaka, K., Kuksis, A., Marai, L., and Ravandi, A. (1996). Identification of lipid ester core aldehydes and hydroperoxides in human atherosclerotic lesions. American Heart Association 69th Annual Meeting, New Orleans, LA, *Circulation* **94** (Suppl.: 1), 707.
28. Bergqvist, M., and Kuksis, A. (1997). Liquid chromatography with on-line electrospray mass spectrometry of oxidized diphosphatidylglycerol. In *New Techniques and Applications in Lipid Analysis* (R. E. McDonald and M. M. Mossoba, eds.), AOCS Press, Champaign, IL, pp. 81–99.

29. Adachi, J., Yoshioka, N., Sato, M., Nakagawa, K., Yamamoto, Y., and Ueno, Y. (2005). Detection of phosphatidylcholine oxidation products in rat heart using quadrupole time-of-flight mass spectrometry. *J. Chromatogr. B* **823**, 37–43.
30. Kuksis, A., Myher, J. J., Marai, L., and Geher, K. (1993). Analysis of hydroperoxides and core aldehydes of triacylglycerols. In *17th Nordic Lipid Symposium, Lipidforum* (Malkki, Y., ed.), Bergen, Norway, pp. 230–238.
31. Sjovall, O., Kuksis, A., and Kallio, H. (2002). Formation of triacylglycerol core aldehydes during rapid oxidation of corn and sunflower oils with *tert*-butyl hydroperoxide/Fe^{2+}. *Lipids* **37**, 81–94.
32. Hartvigsen, K., Ravandi, A., Harkewicz, R., Kamido, H., Bukhave, K, Holmer, G., and Kuksis, A. (2006). 1-O-Alkyl-2-(ω-oxo)acyl-sn-glycerols from shark oil and human milk fat are potential precursors of PAF mimics and GHB. *Lipids* **41**, 679–693.
33. Hara, A., and Radin, N. S. (1978). Lipid extraction of tissues with low toxicity solvent. *Anal. Biochem.* **90**, 420–426.
34. Kamido, H., Koyama, K., Nonaka, K., Yamana, K., Kosuga K., Kuksis, A., and Marai, L. (1994). Cholesteryl ester core aldehydes (components of oxidized low density lipoproteins) exist in human atherosclerotic lesions and are cytotoxic to cultured human endothelial cells. In *Abstracts, Xth InternationalSynmposium on Atherosclerosis*, Montreal, Quebec, October 9–14.
35. Sjovall, O., Kuksis, A., and Kallio, H. (2001). Analysis of molecular species of peroxide adducts of triacylglycerols following treatment of corn oil with *tert*-butyl hydroperoxide. *Lipids* **36**, 1347–1356.
36. Schmitt, D., Shen, Z., Zhang, R., Colles, S. M., Wu, W., Saloman, R. G., Chen, Y., Chisolm, G. M., and Hazen, S. L. (1999). Leukocytes utilize myeloperoxidase-generated nitrating intermediates as physiological catalysts for the generation of biologically active oxidized lipids and sterols in serum. *Biochemistry* **38**, 16904–16915.
37. Lee, S. H., Williams, M. V., and Blair, I. A. (2005). Targeted chiral lipidomics analysis. *Prostglandins & Other Lipid Mediators* 77, 141–157.
38. Esterbauer, H., and Cheeseman, K. H. (1990). Determination of aldehydic lipid peroxidation products: malonaldehyde and 4-hydroxynonenal. *Methods Enzymol.* **186**, 407–421.
39. Hoppe, G., Ravandi, A., Herrera, D., Kuksis, A., and Hoff, H. F. (1997). Oxidation products of cholesteryl linoleate are resistant to hydrolysis in macrophages, form complexes with proteins, and are present in human atherosclerotic lesions. *J. Lipid Res.* **38**, 1347–1360.
40. Buldt, A., and Karst, U. (1997). 1-Metryl-1-(2,4-dinitriophenyl)hydrazine as a new reagent for HPLC determination of aldehydes. *Anal. Chem.* **69**, 3617–3622.
41. Ravandi, A., Kuksis, A., and Shaikh, N. A. (2000). Glucosylated glycerophosphoethanolamines are the major LDL glycation products and increase LDL susceptibility to oxidation. *Arterioscler. Thromb. Vasc. Biol.* **20**, 467–477.
42. Kenar, J. A., Havrilla, C. M., Porter, N. A., Guyton, J. R., Brown, S. A., Klemp, K. F., and Selinger, E. (1996). Identification and quantification of the regioisomeric cholesteryl linoleate hydroperoxides in oxidized human low density lipoprotein and high density lipoprotein. *Chem. Res. Toxicol.* **9**, 737–744.
43. Kuksis, A. (1972). New developments in determination of structure of glycerides and phosphoglycerides. *In Progress in the Chemistry of Fats and Other Lipids* (Holman, R. T. editor). Pergamon Press, Oxford, **12**, 1–163.
44. Yang, L.-Y., Kuksis, A., and Myher, J. J. (1990). Lipolysis of menhaden oil triacylglycerols and corresponding fatty acid alkyl esters by pancreatic lipase in vitro. A reexamination. *J. Lipid Res.* **31**, 137–148.
45. Kamido, H., Kuksis, A., Marai, L., Myher, J. J., and Peng, H. (1992). Preparation, chromatography, and mass spectrometry of cholesteryl ester and glycerolipid-bound aldehydes. *Lipids* 27, 645–650.
46. Hui, S. P., Murai, T., Yoshimura, T., Chiba, H., Nagasaka, H., and Kurosawa, T. (2005). Improved HPLC assay for lipid peroxides in human plasma using the internal standard of hydroperoxide. *Lipids* **40**, 515–522.
47. Ahmed, Z., Ravandi, A., Maguire, G. F., Emili, A., Draganov, D., La Du, B. N., Kuksis, A., and Connelly, P. W. (2001). Apolipoprotein A-I promotes the formation of phosphatidylcholine core aldehydes that are hydrolyzed by paraoxonase (PON-1) during high density lipoprotein oxidation with a peroxynitrite donor. *J. Biol. Chem.* **276**, 24473–24481.
48. Herrera, D., Ravandi, A., and Kuksis, A. (2007). Formation of core aldehydes of 7-hydroperoxycholesterol during autoxidation of cholesteryl linoleate. In *Abstracts, 98th AOCS Annual Meeting & Expo*, Quebec City, PQ, May 13–16, p. 117.
49. Kamido, H., Kuksis, A., Marai, L., and Myher, J. J. (1995). Lipid ester-bound aldehydes among copper-catalyzed peroxidation products of human plasma lipoproteins. *J. Lipid Res.* **36**, 1876–1886.

50. Pruzanski, W., Stefanski, E., de Beer, F. C., de Beer, M. C., Vadas, P., Ravandi, A., and Kuksis, A. (1998). Lipoproteins are substrates for human secretory group IIA phospholipase A_2: preferential hydrolysis of acute phase HDL. *J. Lipid Res.* **39**, 2150–2160.
51. Pruzanski, W., Stefanski, E., deBeer, F. C., de Beer, M. C., Ravandi, A., and Kuksis, A. (2000). Comparative analysis of lipid composition of normal and acute-phase high density lipoproteins. *J. Lipid Res.* **41**, 1035–1047.
52. Ravandi, A., Babaei, S., Leung, R., Monge, J. C., Hoppe, G., Hoff, H., Kamido, H., and Kuksis, A. (2004). Phospholipids and oxophospholipids in atherosclerotic plaques at different stages of plaque development. *Lipids* **39**, 97–109.
53. Suomela, J.-P., Ahotupa, M., Sjovall, O., Kurvinen, J.-P., and Kallio, H. (2004). Diet and lipoprotein oxidation: analysis of oxidized triacylglycerols in pig lipoproteins. *Lipids* **39**, 639–647.
54. Tarvainen, M., Suomela, J.-P., Kallio, H., Ahotupa, M., and Kuksis, A. (2008). Critical lipid oxidation products in an artificial digestion model. In Abstracts, 99th Annual AOCS Meeting and Expo, Seattle, Washington, p. 20.
55. Kanazawa, K., and Ashida, H. (1998). Catabolic fate of dietary trilinoleoylglycerol hydroperoxides in rat gastrointestines. *Biochim. Biophys. Acta* **1393**, 336–348.
56. Kanazawa, K., and Ashida, H. (1998). Dietary hydroperoxides of linoleic acid decompose to aldehydes in stomach before being absorbed unto the body. *Biochim. Biophys. Acta* **1393**, 349–361.
57. Sjovall, O., Kuksis, A., Kallio, H. (2003). Tentative identification and quantification of TAG core aldehydes as dinitrophenylhydrazones in autoxidized sunflower seed oil using reversed phase HPLC with electrospray ionization MS. *Lipids* **38**, 1179–1190.
58. Ravandi, A., Kuksis, A., Myher, J., and Marai, L. (1995). Determination of lipid ester ozonides and core aldehydes by high performance liquid chromatography with on-line mass spectrometry. *J. Biochem. Biophys. Methods* **30**, 271–285.
59. Ravandi, A., Kuksis, A., Shaikh, N., and Jackowski, G. (1997). Preparation of Schiff base adducts of phosphatidylcholine core aldehydes and aminophospholipids, amino acids, and myoglobin. *Lipids* **32**, 989–1001.
60. Biemann, K., Mass spectrometry of peptides and proteins. (1992). *Ann. Rev. Biochem.* **61**, 977–1010.
61. Ahmed, Z., Ravandi, A., Maguire, G., Kuksis, A., and Connelly, P. W. (2000). Comparative study of formation of neutral lipid oxidation products in high- and low density lipoproteins by peroxynitrite donor, SIN-1. In *Abstracts, 91st AOCS Annual Meeting & Expo*, San Diego, CA, April 25–28, page S117.
62. Sorci-Thomas, M. G., Parks, J. S., Kearns, M. W., Pate, G. N., Zhang, C., and Thomas, M. J. (1996). High level secretion of wild-type and mutant forms of human proapoA-I using baculovirus-mediated Sf-9 cell expression. *J. Lipid Res.* **37**, 673–683.
63. Itabe, H., Yamamoto, H., Suzuki, M., Kawai, Y., Nakagawa, Y., Suzuki, A., Imanaka, T., and Takano, T. (1996). Oxidized phosphatidylcholines that modify proteins. *J. Biol. Chem.* **271**, 33208–33217.
64. Gugiu, B. G., Mouillesseaux, K., Duong, C., Herzog, T., Hekimian, A., Koroniak, L., Vondriska, T. M., and Watson, A. D. (2008). Protein targets of oxidized phospholipids in endothelial cells. *J. Lipid Res.* **49**, 510–520.
65. Kurvinen, J.-P., Kuksis, A., Ravandi, A., Sjovall, O., and Kallio, H. (1999). Rapid complexing of oxoacylglycerols with amino acids, peptides and aminophospholipids. *Lipids* **34**, 299–305.
66. Kamido, H., Eguchi, H., Ikeda, H., Imaizumi, T., Yamana, K., Hartvigsen, K., Ravandi, A., and Kuksis, A. (2002). Core aldehydes of alkyl glycerophosphocholines in atheroma induce platelet aggregation and inhibit endothelium-dependent arterial relaxation. *J. Lipid Res* **43**, 158–166.
67. Zhao, B., Dierichs, B., Liu, B., and Holling-Rauss, M. (1994). Functional morphological alterations of human blood platelets induced by oxidized low density lipoprotein. *Thromb. Res.* **74**, 293–301.
68. Ahmed, Z., Ravandi, A., Maguire, G. F., Kuksis, A., and Connelly, P. W. (2003). Formation of apolipoprotein A-I-phosphatidylcholine core aldehyde Schiff base adducts promotes uptake by THP-1 macrophages. *Cardiovasc. Res.* **58**, 712–720.
69. Stephan, Z. F., and Yurachek, E. C. (1993). Rapid fluorometric assay of LDL receptor activity by DiI-labeled LDL. *J. Lipid Res.* **34**, 325–330.
70. Kuksis, A., and Mamelak, M. (2006). Gammahydroxybutyrate and lipid peroxidation. *INFORM* **17** *(9)*, 607–610.
71. Luci, S., Konig, B., Giemsa, B., Huber, S., Hause, G., Kluge, H., Stangl, G. I., and Eder, K. (2007). Feeding of a deep-fried fat causes PPARα activation in the liver of pigs as non-proliferating species. *Br. J. Nutr.* **97**, 872–882.

Chapter 5

Lipid Raft-Redox Signaling Platforms in Plasma Membrane

Fan Yi, Si Jin, and Pin-Lan Li

Summary

Membrane lipid rafts (LRs) have been demonstrated to be importantly involved in transmembrane signaling in a variety of mammalian cells. Many receptors can be aggregated within the LR clusters to form signaling platforms. Currently, LRs were reported to be clustered to aggregate, recruit, and assemble NADPH oxidase subunits and related proteins in various cells in response to various stimuli, forming redox signaling platforms. These LR signaling platforms may play important roles in the regulation of cellular activity and cell function, and also in the development of cell dysfunction or injury associated with various pathological stimuli. This LRs clustering-mediated mechanism is considered to take a center stage in redox signaling associated with death receptors. In this chapter, some basic methods and procedures for characterization of LR-redox signaling platforms formation and for determination of the function of these signaling platforms are described in detail, which include identification of LR-redox signaling platforms in cell membrane by using fluorescent or confocal microscopy of LR-redox signaling platforms and fluorescent resonance energy transfer analysis, isolation of LR-redox signaling platforms by flotation of detergent-resistant membranes, and function measurement of LR-redox signaling platforms by electron spin resonance spectroscopy. It is expected that information provided here will help readers to design necessary experiments in their studies on LR signaling platforms and redox regulation of cell function.

Key words: Lipid microdomains, Reactive oxygen species, Molecular trafficking, Sphingolipids

1. Introduction

There is increasing evidence that clustering of distinct cholesterol- and sphingolipid-rich membrane microdomains or lipid rafts (LRs) is importantly involved in transmembrane signaling in a variety of mammalian cells. Many receptors including tumor necrosis factor-α receptors, Fas, DR3, -4, -5, insulin receptors, and integrins as well as other postreceptor signaling molecules may

Donald Armstrong (ed.), *Lipidomics*, Methods in Molecular Biology, vol. 580,
DOI 10.1007/978-1-60761-325-1_5, © Humana Press, a part of Springer Science+Business Media, LLC 2009

be aggregated within the LR clusters to form signaling platforms *(1, 2)*. It has been indicated that LRs clustering or platform formation is implicated in the regulation of a number of biological processes in different cells including cell growth, differentiation and apoptosis, T-cell activation, tumor metastasis, neutrophil and monocyte infiltration, as well as infection of different pathogen organisms such as bacteria, viruses, and parasites *(3)*. Among different LRs, a ceramide-enriched membrane platform has been extensively studied. Evidence is increasingly accumulated that this ceramide-enriched membrane platform plays an essential role in the regulation of cellular signaling. The mechanism mediating this cellular signaling is associated with establishment of a proximity of many receptor molecules, facilitation of transactivation of signaling molecules associating or interacting with a receptor, and amplification of the specific signaling of the activated receptors *(4, 5)*. Therefore, this ceramide-enriched membrane platform is also referred as to LR signaling platform. It has been established that the formation of the LR signaling platforms with aggregation of different signaling molecules is an important mechanism, determining the variety of transmembrane signaling that could robustly amplify signals from activated receptors on the cell membrane *(6, 7)*.

Studies in our laboratory and by others have reported that in response to different stimuli such as activation of death receptors, carcinogenic factors, and degenerative stimuli LRs may be clustered due to ceramide production by activation acid sphingomyelinase, where various redox molecules like superoxide ($O_2^{\cdot -}$), H_2O_2, or peroxinitrite ($ONOO^-$) can be produced. It has been demonstrated that in ceramide-enriched membrane domains or platforms, different enzymes or factors associated with production or metabolism of redox molecules are clustered by translocation, recruitment, or aggregation such as NADPH oxidase, $O_2^{\cdot -}$ dismutase, and thioreductase *(8, 9)*. Since this LR platform produces redox molecules in response to different stimuli and thus regulates cellular activity or cell function, it is now called the LR-redox signaling platform. In this regard, the NADPH oxidase subunits have been demonstrated to be clustered with ceramide-enriched membrane domains, and this LR clustering-mediated redox signaling has been commented as taking center stage in signaling of death receptors *(10)*. This LR clustering mechanism may provide a driving force to cause NADPH oxidase assembling and activation.

It is well know that NADPH oxidase is a multicomponent enzyme complex that consists of the membrane-bound cytochrome b_{558} (gp91^{phox} and p22^{phox}) and cytoplasmic proteins (p40^{phox}, p47^{phox}, p67^{phox}, and Rac GTPase) that translocate to the membrane to form an assembled complex following cellular stimulation to produce $O_2^{\cdot-}$. The p47*phox* translocation is considered as a key step, to some extent, a marker event, for the assembly and activation of NADPH oxidase, which is assumed to be initiated by the phosphorylation of this subunit at various phosphorylation sites by PKC, PKA, or MAPK *(11)*. In addition, the catalytic subunits of this enzyme are termed NOX proteins, which include several known members, namely, NOX1, NOX2 (gp91*phox*), NOX3, NOX4, and NOX5, DUOX1, and DUOX2 *(12)*. However, for a long time, it is unknown what is the precise mechanism that drive p47*phox* translocation and subsequent assembly of other NADPH oxidase subunits so efficiently in the cell membrane *(13, 14)*. Demonstration of LRs clustering of these NADPH oxidase may shift a paradigm in understanding the activation of NADPH oxidase and redox signaling *(8, 15–17)*.

In this chapter, the methods and procedures for characterization of LR-redox signaling platform formation and related protocols for functional studies of LR signaling platforms are described in detail. These basic procedures and methods include identification of LR-redox signaling platforms in cell membrane by using fluorescent or confocal microscopy of LR-redox signaling platforms and fluorescent resonance energy transfer (FRET) analysis, isolation of LR-redox signaling platforms by flotation of detergent-resistant membranes (DRMs), and function measurement of LR-redox signaling platforms by electron spin resonance (ESR) spectroscopy. The authors hope that these protocols would help readers design experiment to understand the physiological or pathological relevance of LR-redox signaling platforms, to explore the molecular mechanisms underlying the formation of LR-redox signaling platforms, and to develop new therapeutic strategies for treatment of diseases or pathological processes related to this LR signaling platform.

It should be noted that besides these methods in this chapter, other general visualization techniques for LRs may also be used for further studies on such LR-redox signaling platforms. For example, total internal reflection microscopy allows us to get information of the diffusivity of particles in the membrane as well as to reveal membrane corrals, barriers, and sites of confinement. Fluorescence correlation and cross-correlation spectroscopy can be used to gain information of fluorophore mobility in the membrane. In addition, atomic force microscopy, scanning ion conductance microscopy, nuclear magnetic resonance, and superresolution microscopy such as stimulated emission depletion may also be used, if related equipment or instruments are available. **Figure 1a** summarizes all commonly used methods for studies of LRs or LR-redox signaling platforms. The rationales of methods that we introduce in this chapter are described in following text.

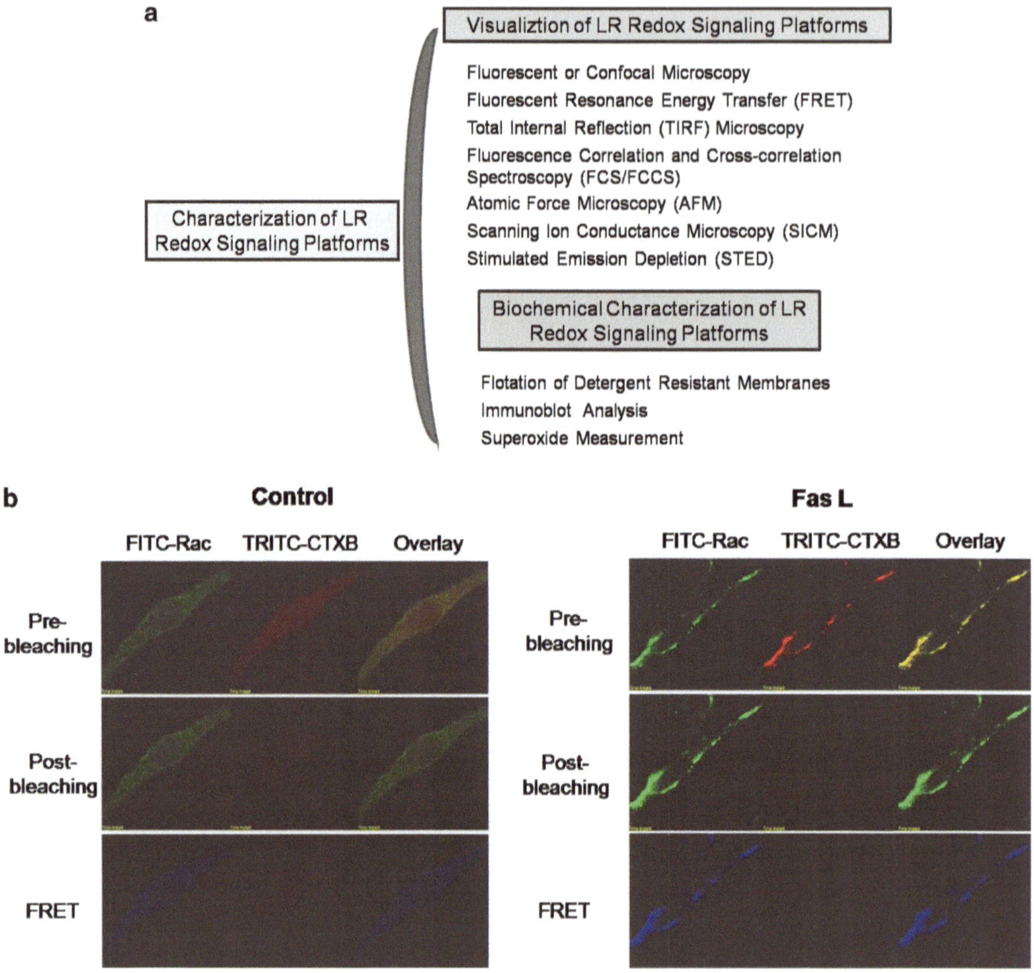

Fig. 1. Characterization of lipid raft redox signaling platforms in plasma membrane. (**a**) Methods commonly used to characterize of the formation of lipid raft redox signaling platforms. (**b**) Representative images of FRET analysis between FITC-Rac1 and TRITC-CTXB in BCAECs. The left group of images shows a control cell costained with FITC-Rac1 and TRITC-CTXB that underwent an acceptor bleaching protocol. Both the pre- and postbleaching images were presented on the top and middle panels. FRET image (in blue) was generated by subtraction of fluorescent intensity in the prebleaching image from that in the postbleaching image of FITC-Rac1 labeling. As shown in FRET image (blue in the *bottom image*), there was very low FRET detected under control condition. The right group of images shows a FasL-stimulated cell that underwent the same FRET protocol. In addition to detected patch formation (lipid raft clustering and Rac aggregation) and colocalization of both molecules seen in the overlaid images (*top panel*) in response to FasL, a more intense FRET image (blue one in the *bottom*) was detected in this FasL-treated BCAEC, demonstrating that energy transfer occurs between a Rac1 and LR component-GM1 ganglioside.

1.1. Identification of LR-Redox Signaling Platforms in Cell Membrane: Fluorescent or Confocal Microscopy and FRET Analysis

These methods are used to detect a colocalization of LRs components and aggregated or recruited NADPH oxidase subunits or other molecules related to redox signaling on the cell membrane. Although individual LRs are too small to be resolved on the cell surface by standard light microscopy, clustered LRs could be visualized by fluorescence or other staining techniques if their components are cross-linked with antibodies or lectins. Therefore, fluorescent or confocal microscopy of LR patches or

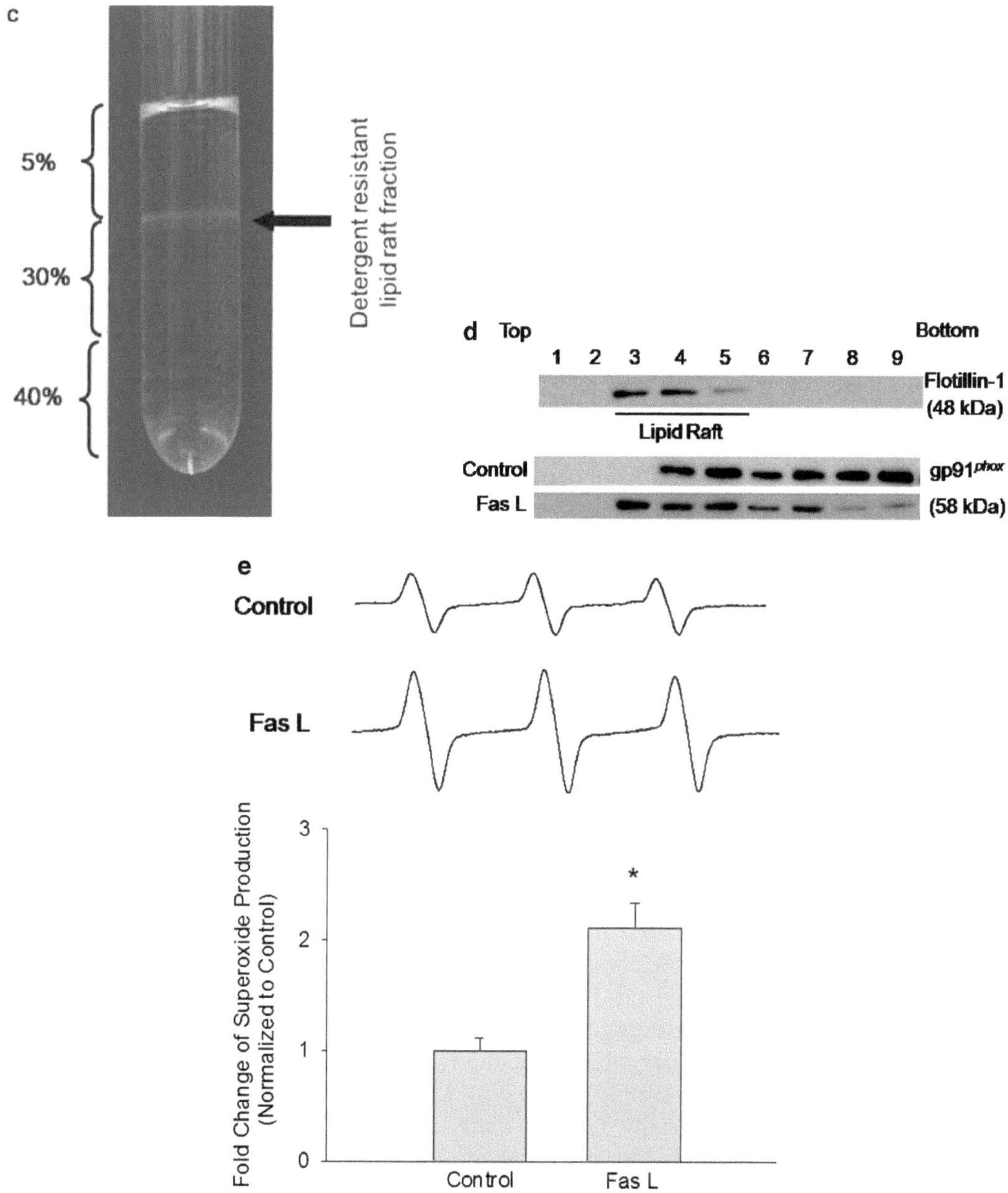

Fig. 1. (continued) (**c**) Isolation of detergent-resistant lipid raft fractions from BCAECs based on their detergent insolubility and low density. (**d**) After density gradient ultracentrifugation, nine fractions from top to bottom were fractionated and then analyzed by immunoblotting. Fractions #3–5 were designated as LRs indicated by flotillin-1 (one marker for lipid rafts). The blot pattern for gp91^{phox} indicated that gp91^{phox} subunit was aggregated to membrane lipid raft fractions of FasL-treated BCAECs. (**e**) Representative ESR spectra showing SOD-inhibitable $O_2^{\cdot -}$ signals (*upper*) and summarized data depicting $O_2^{\cdot -}$ production indicating that FasL significantly increased $O_2^{\cdot -}$ production in lipid raft fractions.

spots on the cell membrane is widely used as a common method currently. One of LRs markers is fluorescent labeled-cholera toxin (CTX), which is used based on its capacity of binding to the raft constituent ganglioside GM_1, a glycosphingolipid that consists of a ceramide backbone with four sugars esterified, one of these being N-acetylneuraminic acid, galactose, and glucose *(18)*. Since this LR signaling platform is ceramide-enriched domain, ceramide can also be used as a marker to detect this LR signaling platform or ceramide-enriched microdomains by fluorescent or confocal microscopy.

The current advances in fluorescence microscopy, coupled with the development of new fluorescent probes, make FRET analysis widely being used as a powerful technique for studying molecular interactions in cells. This analysis will reveal molecular proximity of various molecules in a variety of cells with improved spatial (angstrom) and temporal (nanosecond) resolution, distance range, and sensitivity as well as a broader range of biological applications *(19)*. FRET is a phenomenon that occurs between a fluorophore pair, donor and acceptor (e.g., FITC and TRITC). The fluorophore pair shares the character that the emission wavelength of the donor can overlap with the excitation wavelength of the acceptor, in which energy can transfer from the donor to the acceptor. Two key factors determine the occurrence of FRET, molecular orientation and distance between the molecules. It is proposed that FRET can only take place between two molecules within 7–10 nm range. Detected FRET generally indicates that the two molecules are closely located, which may generate energy transfer to each other or interact to lead to molecular reaction. Here we present some methods specific to detect FRET between LR components and redox producing or regulatory enzymes, in particular NADPH oxidase subunits and GM_1.

1.2. Isolation and Analysis of LR-Redox Signaling Platforms: Flotation of DRMs

This method is used to identify redox signaling molecules or other associated proteins and receptors in isolated LRs fractions. DRM complex or detergent-insoluble glycolipid-enriched domains (DIG) can float to low-density fractions during sucrose gradient centrifugation. These LR fractions are rich in raft proteins and therefore analyzing the raft proteins in DRM provides a reliable and simple means of identifying possible LR components, especially LRs-associated proteins *(2, 4, 20)*. Furthermore, in combination with proteomic techniques developed recently, this membrane flotation technique can help demonstrate all unidentified molecules including receptors, enzymes, and adaptors if large-scale proteomic analyses could reach enough resolutions and sensitivity. Currently, only some targeted proteomic analysis can be done given technical limitations *(21)*.

1.3. Functional Measurement of LR-Redox Signaling Platforms: ESR Spectroscopy

Among methods to characterize the activity and modulation of redox molecules or redox-related enzyme activity such as NADPH oxidase including lucigenin-enhanced chemiluminescence, dihydroethidium (DHE) fluorescent spectrometric assay, HPLC analyses, fluorescent dye intracellular trapping detection, and ESR, the most direct and definitive method is ESR spectrometric analysis. ESR, also called electron paramagnetic resonance spectroscopy, is a technique for studying chemical species that have one or more unpaired electrons, such as organic and inorganic free radicals or inorganic complexes possessing a transition metal ion. ROS are often free radicals with unpaired electrons. Because they are very short-lived, it has been challenging to measure ROS from biological samples. The recently developed ESR has made highly specific, quantitative and reproducible measurements of ROS possible. Now ESR is commonly used for measurements of nitric oxide, $O_2^{\cdot -}$, and other ROS from live cells, organelles, and tissues *(16, 22)*.

2. Materials

2.1. Equipment

1. Centrifuge (.maximum force ~17,000 × *g*).
2. The Optima™ Series ultracentrifuges and related rotors (Beckman, Fullerton, CA).
3. Olympus FV-300 FluoView Confocal Microscope Workstation (Shinjuku-kuTokyo, Japan) or Leica TCS-SP2 AOBS inverted confocal laser scanning microscope and workstation (Wetzlar, Germany).
4. Miniscope 200 ESR spectrophotometer (Magnettech, Berlin, Germany).

2.2. Reagents

2.2.1. Cell Culture

1. RPMI 1640 medium supplemented with 15% (v/v) fetal bovine serum, 1% antibiotic solution (Invitrogen, Carlsbad, CA).
2. Sterile phosphate-buffered saline (PBS) solution, pH = 7.4 (Sigma, St. Louis, MO).
3. 0.25% Trypsin-EDTA (Invitrogen, Carlsbad, CA).

2.2.2. Slide Preparations

1. Alexa Fluor 488-conjugated CTX B subunit (Molecular Probes, Carlsbad, CA) or anti-ceramide antibody (Alexis Biochemicals, Farmingdale, NY).
2. gp91*phox* monoclonal antibody (BD Bioscience, San Jose, CA) or other related antibodies for NADPH oxidase subunits.
3. Texas Red-conjugated anti-mouse secondary antibody (Santa Cruz, Santa Cruz, CA).
4. 4% Paraformaldehyde (PFA) solution in PBS (*see* **Note 1**).
5. 100 ml 0.05% (v/v) Tween-20 in PBS (PBT).
6. 10 ml 1% (w/v) BSA in PBT.
7. 20 ml 0.1% BSA in PBT (diluted from 1% (w/v) BSA in PBT).
8. All conjugated, primary, and secondary antibodies diluted in 0.1% BSA in PBT.
9. Vectashield Mounting Media (Vector Labs, Burlingame, CA).

2.2.3. FRET Analysis

1. TRITC-conjugated CTX B subunit (acceptor).
2. FITC-labeled primary antibodies (donor).
3. TRITC-anti-mouse-IgG (as positive control).

2.2.4. Flotation of DRMs

1. MES-buffered saline (MBS) buffer: 25-mM 2-(*N*-morpholino) ethanesulphonic acid (MES), 150-mM NaCl, and 1-mM EDTA, pH = 7.4.
2. Triton X-100 detergent (Sigma, St. Louis, MO).
3. 60% OptiPrep® density gradient medium (Sigma, St. Louis, MO) (*see* **Note 2**).
4. Phenylmethylsulfonyl fluoride (Roche, Branchburg, NJ).
5. Na_3VO_4 (Sigma, St. Louis, MO).
6. One tablet "complete" protease inhibitors dissolved in 1-ml dH_2O (Roche, Branchburg, NJ)
7. MBS buffer containing 1-mM Na_3VO_4, 1-mM phenylmethylsulfonyl fluoride, and "complete" protease inhibitors (1:50 dilution) (solution A).
8. Solution A containing 1% (v/v) Triton X-100 (solution B) (*see* **Note 3**).
9. Thirty percent and 5% density gradient solutions: Dilute the Optiprep density gradient medium (60%) 1:1 and 1:11 in solution A for 30% and 5% solutions, respectively.

2.2.5. Superoxide Measurement by ESR

1. 10-mM 1-hydroxy-3-methoxycarbonyl-2,2,5,5-tetramethylpyrrolidine (CMH) (Noxygen, Elzach, Germany) (*see* **Note 4**).
2. Polyethylene glycol superoxide dismutase (PEG-SOD; 1000 U/ml) (Sigma, St. Louis, MO).
3. Diethyldithiocarbamate (DETC) (Sigma, St. Louis, MO).
4. Deferoamine (Sigma, St. Louis, MO).
5. Kreb/HEPES buffer (KHB).
6. Modified Kreb's/HEPEs buffer (containing 25 μM deferoxamine and 5 μM DETC) (*see* **Note 5**).
7. 10-mM CMH in modified Kreb's/HEPEs buffer.

3. Methods

3.1. Confocal Microscopy

1. On day 1, plate Bovine coronary arterial endothelial cells (BCAECs) in T-75 flasks at 5×10^6 cells/flask (*see* **Note 6**).
2. On day 2, split and plate cells at about 70% confluence on a four-chamber glass slide in fresh medium for at least 2 h at 37°C (*see* **Note 7**).

3. Add stimuli such as FasL (final concentration is 10 ng/ml) to treat cells for 15 min.
4. Aspirate medium and wash each chamber twice quickly with 0.5 ml PBS.
5. Fix the cells by adding 0.5 ml of 4% PFA to each chamber and incubating for 15 min.
6. Wash cells in PBT three times (each time for 5 min) on the shaker and then incubate cells for 30 min in 0.5 ml of 1% BSA.
7. Wash cells in PBT three times (each time for 5 min) on the shaker and then incubate cells in 0.5 ml of the working concentration of anti-gp91 antibody diluted in 0.1% BSA for 1 h.
8. Repeat step 7 using Texas Red-conjugated anti-mouse secondary antibody working solution (*see* **Note 8**).
9. Repeat step 7 using the Alexa 488-conjugated CTX working solution (*see* **Note 9**).
10. Wash cells in PBT three times (each time for 5 min) on the shaker.
11. Allow the slide to dry and remove the plastic chamber piece and sealer holding in place completely.
12. Place one drop of Vectashield Mounting Media on each sheet of cells and cover with a No. 1.5 thickness coverslip (Warner Instruments, Hamden, CT). Gently push out any air bubbles that form underneath the coverslip and seal the edges with clear nail polish.
13. Store slides at 4°C in the dark before and during viewing under fluorescence.
14. Staining is visualized using a conventional fluorescence microscope or a Leica TCS SP2 scanning confocal microscope (*see* **Note 10**).

3.2. FRET Analysis

1. Same procedures as steps 1–6 in **Subheading 3.1**.
2. Wash cells in PBT three times (each time for 5 min) on the shaker and then incubate cells in 0.5 ml of the working concentration of the FITC-conjugated donor antibody diluted in 0.1% BSA for 1 h.
3. Repeat step 2 using the TRITC-conjugated acceptor antibody.
4. Same procedures as steps 10–13 in **subheading 3.1**.
5. Capture digital images of donor, acceptor, and FRET fluorescent patterns (*see* **Note 11**).
6. Effectively and irreversibly bleach acceptor fluorescence by continuous excitation for 2 min at the acceptor wavelength.
7. Capture digital images of donor, acceptor, and FRET fluorescent patterns after bleaching.

8. FRET images are obtained by the subtraction of the prebleaching images from the postbleaching images (*see* **Note 12**).

3.3. Flotation of DRMs

1. Culture BCAECs in 10 cm^2 dishes at 5 × 10^6 cells/dish.
2. Aspirate medium. Culture cells with fresh medium. Then treat BCAECs with or without FasL (10 ng/ml, 15 min, Upstate, Billerica, MA).
3. Wash each dish three times in cold PBS, aspirate PBS, and then add 1 ml PBS to scrape the cells using a rubber, flexible scraper, transfer cells of each dish into 1.7 ml eppendorf tube individually.
4. Centrifuge the cell suspension at 3,000 × *g* for 5 min and discard the supernatant.
5. Resuspend cell pellets in 1 ml solution B (*see* **Note 13**)
6. Homogenize cell extracts by 10–15 passages through a 25-gauge needle and then incubate 60 min on ice.
7. Dilute each sample to 1.5 ml by adding the appropriate amount of solution B. Combine this with 3 ml 60% density gradient solution and mix well by pipetting up and down. Transfer this mixture (final including 40% density gradient solution) to a Beckman ultracentrifuge tube.
8. Add 4.5 ml 30% density gradient solution carefully to the ultracentrifuge tube, not disturbing the barrier between this and the solution already sitting in the tube (*see* **Note 14**).
9. Add 4.5 ml 5% density gradient solution carefully to the centrifuge tube, not disturbing the barrier between this and the solution already sitting in the tube.
10. Equalize the masses of all samples by weighing and identifying the heaviest sample, then adding dropwise 5% density gradient solution to the others so that all masses are exactly equal.
11. Spin in a Beckman SW 32 Ti rotor ultracentrifuge at 32,000 rpm at 4°C for 20 h (*see* **Note 15**).
12. Remove each sample from the ultracentrifuge and, by pipetting from the very top of the sample, aliquot the whole sample into nine 1.5 ml fractions in 1.7 ml ependorff tubes. Nine fractions are collected from the top to the bottom (fraction numbers 1–9) (*see* **Note 16**). Store samples in –80°C until ready to proceed.
13. For immunodetection of LR-associated proteins, 30 μl of each fraction (see **Note 17**) are subjected to SDS-PAGE, transferred onto a nitrocellulose membrane, and prepared for Western blot analysis using related antibodies such as LR marker flotinllin-1 and NADPH oxidase subunits ($p47^{phox}$, $p67^{phox}$, $gp91^{phox}$, and Rac GTPase et. al) (see **Note 18**).

14. Analysis: All films with immunoreactive blots are scanned by a densitometer and the intensity of corresponding protein bands was quantitated using UN-SCAN-IT software (Silk Scientific Corporation, Orem, Utah).

3.4. Superoxide Measurement by ESR Spectroscopy

1. LR isolation and identification (*see* **Subheading 3.3**).
2. Add 10 μl 10 mM CMH to 90 μl modified Kreb's/HEPEs buffer and incubate for 10 min (*see* **Note 19**).
3. 50 μl from step 2 is transferred to a 50-μl capillary tube and analyzed in an ESR spectrometer by time scan to quantify $O_2^{\cdot-}$ (*see* **Note 20**).
4. Incubate 20 μg LR fractions and 10 μl 3mM NADPH in modified Kreb's/HEPEs buffer for 10 min (total volume is adjusted to 90 μl using Kreb's/HEPEs buffer) (*see* **Note 21**).
5. After incubation, add 10 μl 10 mM CMH to 90 μl reaction mixture from step **4**.
6. 50 μl from step 5 is transferred to a 50 μl capillary tube immediately and analyzed in an ESR spectrometer by time scan to quantify $O_2^{\cdot-}$.
7. Incubate 20 μg LR fractions, 10 μl 1000 U/ml polyethylene glycol superoxide dismutase (PEG-SOD) and 10 μl 3mM NADPH in modified Kreb's/HEPEs buffer for 10 min (total volume is adjusted to 90 μl using Kreb's/HEPEs buffer).
8. After incubation, add 10 μl 10 mM CMH to 90 μl reaction mixture from step 7.
9. 50 μl from step 8 is transferred to a 50 μl capillary tube immediately and analyzed in an ESR spectrometer by time scan to quantify $O_2^{\cdot-}$.
10. A time scan of CMH oxidation is recorded and normalized to the protein content of the sample. $O_2^{\cdot-}$ production from LRs is calculated as the SOD-inhibitable fraction of CMH oxidation.

3.5. Results

To better help readers to understand these methods and design necessary experiments in their studies on LR signaling platforms and redox regulation of cell function, we used FasL-induced formation of LR-redox signaling platforms as example to illustrate some representative results.

3.5.1. FasL-Induced FRET between Rac1 and LR Component

FRET can be detected by confocal microscopy between a fluorophore, FITC as donor and TRITC as acceptor, which shares the character to allow FRET. Acceptor (TRITC) bleaching protocol was applied to calculate the FRET efficiency. Representative images of FRET analysis between FITC-Rac1 (one of NADPH oxidase subunits) and TRITC-CTXB in BCAECs are shown in **Fig. 1b**. The left group of images shows a control cell costained with FITC-Rac1 and TRITC-CTXB that underwent an acceptor bleaching protocol. Both the pre- and postbleaching images were presented on the top and middle panels. FRET image (in blue) was generated by subtraction of fluorescent intensity in the prebleaching image from that in the postbleaching image of FITC-Rac1 labeling. As shown in FRET image (blue in the bottom image), there was very low FRET detected under control condition. The right group of images shows a FasL-stimulated cell that underwent the same FRET protocol. In addition to detected patch formation (LR clustering and Rac aggregation) and colocalization of both molecules seen in the overlaid images (top panel) in response to FasL, a more intense FRET image (blue one the bottom) was detected in this FasL-treated BCAEC, demonstrating that energy transfer occurs between a Rac1 and LR component-GM1 ganglioside.

3.5.2. Isolation of LR fractions by Flotation Assay

Flotation of DRMs is used to identify redox signaling molecules or other associated proteins and receptors in isolated LRs fractions. The LR fraction was located at the interface between 5% and 30% density gradient fractions in the ultracentrifuge tube after centrifugation (**Fig. 1c**).

3.5.3. FasL-Induced gp91^{phox} Aggregation in LR fractions

In **Fig. 1d**, Western blot analysis showed a positive expression of flotillin-1 in fractions 3–5 (from top to bottom), which was referred to LR fractions. NADPH oxidase subunit, gp91^{phox} can be detected in most of the membrane fractions from BCAECs. However, there was a distribution change among these fractions with a marked increase in gp91^{phox} protein in LR fractions when BCAECs were stimulated by FasL.

3.5.4. FasL-Enhanced NADPH Oxidase Activity

NADPH oxidase activity was detected by measurement of $O_2^{\cdot-}$ production in isolated LR-enriched fractions using ESR. Representative ESR spectra were shown in **Fig. 1e** indicating SOD-inhibitable $O_2^{\cdot-}$ signals (upper) and summarized data for $O_2^{\cdot-}$ production (bottom) indicating that FasL significantly increased $O_2^{\cdot-}$ production in LR fractions.

4. Notes

1. Preparation: Mix 0.4 g PFA in 1.0 ml dH_2O and add 100 µl 1 M NaOH. Then heat this mixture until the PFA is dissolved. Finally, dilute the solution to 10 ml with PBS.
2. OptiPrep® Density Gradient Medium is used for the isolation of cells and cell organelles, which contains 60% (w/v) solution of iodixanol in sterlized water. Gradient sucrose buffer is also commonly used for LR isolation.

3. The concentration of detergent Triton X-100 can be variable (1–2%) based on different cells or tissues. Other detergents such as Brij-96 may be also used.
4. Other spin trap agents specific for superoxide such as DMPO (5,5-dimethyl-1-pyrolline-*N*-oxide), DEPMPO (5-diethoxyphosphoryl-5-1-pyrroline-*N*-oxide), DPPMDPO (5-(diphenylphosphinoyl)-5-methyl-4,5-dihydro-3H-pyrrole *N*-oxide) are also used in this study.
5. Deferoxamine and DETC are used as metal chelators to decrease CMH background.
6. Isolation of BCAECs is based on the reference *(23)*.
7. Split and culture cells based on common laboratory procedures.
8. All remaining steps are performed in the dark to protect fluorescent signals.
9. Green fluorescence (Alexa 488/FITC) should be paired with Red/Orange fluorescence (Texas Red/TRITC/Alexa 555), and vice versa.
10. The patch formation of Alexa 488-labeled CTX and gangliosides complex represents the clusters of LRs. Clustering is defined as one or several intense spots of fluorescence on the cell surface, whereas unstimulated cells display a homogenous distribution of fluorescence throughout the membrane. In each experiment, the presence or absence of clustering of 200 cells in each sample is scored by three independent observers. The results are given as the percentage of cells showing a cluster after the indicated treatment as described. Similar analysis is also used to summarize target proteins with LRs colocalization.
11. Starting with this step, the followings are FRET visualization-acceptor bleaching procedures. Donor control images are observed under donor excitation/emission wavelengths; Acceptor control images are observed under acceptor excitation/emission wavelengths; FRET control images are observed under donor excitation/acceptor emission wavelengths.
12. The FRET efficiency is calculated through the following formula:

$$E = \frac{\text{FITC}_{\text{post}} - \text{FITC}_{\text{pre}}}{\text{FITC}_{\text{post}}} \times 100\%$$

13. All steps except cell culture and treatment should be performed on ice, preferably in a cold room set at or below 4°C.
14. In this protocol, gradient solution is 40%, 30%, and 5%, each fraction volume is 4.5 ml. This gradient concentration and fraction volume may be adjusted based on different cells and other factors.

15. Alternative centrifuge rotors may be used, the rotor speed RPM need to be adjusted based on different rotors.
16. The number of aliquot fractions may be adjusted.
17. For immunoblot analysis, if protein concentration of each fraction is too low to detect the signal, the proteins of each fraction may be precipitated by mixing with equal volume of 30% trichloroacetic acid and 30 min of incubation on ice. Proteins then spin down by centrifugation at 13,000 × *g* at 4°C for 15 min. The protein pellet is carefully washed with cold acetone twice, air dried, and then resuspended in suitable volume of 1 M Tris-HCl (pH 8.0), which are ready for immunoblot analysis.
18. Western blot analysis is based on common laboratory procedures.
19. Total reaction volume is 100 μl.
20. This measurement is considered as CMH background or baseline. The ESR settings are as follows: biofield, 3,350; field sweep, 60 G; microwave frequency, 9.78 GHz; microwave power, 20 mW; modulation amplitude, 3 G; 4,096 points of resolution; receiver gain, 100; and kinetic time, 10. These settings may be adjusted based on the signals.
21. From steps 4 to 9, LR fractions are prepared and incubated with NADPH and the spin probe CMH in the presence or absence of PEG-SOD.

Acknowledgments

The studies cited in this chapter were supported by grants from the National Institutes of Health (HL-57244, HL-70726, and HL-51055). The content of this publication does not necessarily reflect the views or policies of the Department of Health and Human Services, nor does mention of trade names, commercial products, or organization imply endorsement by the US Government.

References

1. Kabouridis, P.S. (2006) Lipid rafts in T cell receptor signalling. *Mol Membr Biol* **23**: 49–57.
2. Simons, K., and Toomre, D. (2000) Lipid rafts and signal transduction. *Nat Rev Mol Cell Biol* **1**:31–39.
3. Hawkes, D.J., and Mak, J. (2006) Lipid membrane; a novel target for viral and bacterial pathogens. *Curr Drug Targets* **7**:1615–1621.
4. Gulbins, E., and Grassme, H. (2002) Ceramide and cell death receptor clustering. *Biochim Biophys Acta* **1585**:139–145.

5. Li, P.L., Zhang, Y., and Yi, F. (2007) Lipid raft redox signaling platforms in endothelial dysfunction. *Antioxid Redox Signal* **9**: 1457–1470.
6. Jacobson, K., Mouritsen, O.G., and Anderson, R.G. (2007) Lipid rafts: at a crossroad between cell biology and physics. *Nat Cell Biol* **9**:7–14.
7. Allen, J.A., Halverson-Tamboli, R.A., and Rasenick, M.M. (2007) Lipid raft microdomains and neurotransmitter signalling. *Nat Rev Neurosci* **8**:128–140.
8. Zhang, A.Y., Yi, F., Zhang, G., Gulbins, E., and Li, P.L. (2006) Lipid raft clustering and redox signaling platform formation in coronary arterial endothelial cells. *Hypertension* **47**:74–80.
9. Zuo, L., Ushio-Fukai, M., Ikeda, S., Hilenski, L., Patrushev, N., and Alexander, R.W. (2005) Caveolin-1 is essential for activation of Rac1 and NAD(P)H oxidase after angiotensin II type 1 receptor stimulation in vascular smooth muscle cells: role in redox signaling and vascular hypertrophy. *Arterioscler Thromb Vasc Biol* **25**:1824–1830.
10. Touyz, R.M. (2006) Lipid rafts take center stage in endothelial cell redox signaling by death receptors. *Hypertension* **47**:16–18.
11. Bedard, K., and Krause, K.H. (2007) The NOX family of ROS-generating NADPH oxidases: physiology and pathophysiology. *Physiol Rev* **87**:245–313.
12. Ushio-Fukai, M. (2006) Redox signaling in angiogenesis: role of NADPH oxidase. *Cardiovasc Res* **71**:226–235.
13. Bokoch, G.M., and Zhao, T. (2006) Regulation of the phagocyte NADPH oxidase by Rac GTPase. *Antioxid Redox Signal* **8**:1533–1548.
14. Takeya, R., and Sumimoto, H. (2006) Regulation of novel superoxide-producing NAD(P)H oxidases. *Antioxid Redox Signal* **8**:1523–1532.
15. Jin, S., Yi, F., and Li, P.L. (2007) Contribution of lysosomal vesicles to the formation of lipid raft redox signaling platforms in endothelial cells. *Antioxid Redox Signal* **9**:1417–1426.
16. Jin, S., Zhang, Y., Yi, F., and Li, P.L. (2008) Critical role of lipid raft redox signaling platforms in endostatin-induced coronary endothelial dysfunction. *Arterioscler Thromb Vasc Biol* **28**: 485–490.
17. Schmitz, G., and Grandl, M. (2007) Role of redox regulation and lipid rafts in macrophages during Ox-LDL-mediated foam cell formation. *Antioxid Redox Signal* **9**:1499–1518.
18. Grassme, H., Jekle, A., Riehle, A., Schwarz, H., Berger, J., Sandhoff, K., Kolesnick, R., and Gulbins, E. (2001) CD95 signaling via ceramide-rich membrane rafts. *J Biol Chem* **276**: 20589–20596.
19. Sekar, R.B., and Periasamy, A. (2003) Fluorescence resonance energy transfer (FRET) microscopy imaging of live cell protein localizations. *J Cell Biol* **160**:629–633.
20. Sowa, G., Pypaert, M., and Sessa, W.C. (2001) Distinction between signaling mechanisms in lipid rafts vs. caveolae. *Proc Natl Acad Sci U S A* **98**:14072–14077.
21. MacLellan, D.L., Steen, H., Adam, R.M., Garlick, M., Zurakowski, D., Gygi, S.P., Freeman, M.R., and Solomon, K.R. (2005) A quantitative proteomic analysis of growth factor-induced compositional changes in lipid rafts of human smooth muscle cells. *Proteomics* **5**:4733–4742.
22. Hwang, J., Kleinhenz, D.J., Lassegue, B., Griendling, K.K., Dikalov, S., and Hart, C.M. (2005) Peroxisome proliferator-activated receptor-gamma ligands regulate endothelial membrane superoxide production. *Am J Physiol Cell Physiol* **288**:C899–905.
23. Zhang, D.X., Yi, F.X., Zou, A.P., and Li, P.L. (2002) Role of ceramide in TNF-alpha-induced impairment of endothelium-dependent vasorelaxation in coronary arteries. *Am J Physiol Heart Circ Physiol* **283**:H1785–1794.

Chapter 6

Mass Spectrometry Analysis of Polyisoprenoids Alcohols and Carotenoids via ESI(Li^+)-MS/MS

Fabio Luiz D'Alexandri, Renata Tonhosolo, Emilia A. Kimura, and Alejandro Miguel Katzin

Summary

Direct analysis of polyisoprenoid alcohols by electrospray ionization mass spectrometry (ESI-MS) often produces poor results requiring off-line time- and sample-consuming derivatization techniques. In this chapter, we describe a simple ESI-MS approach for the direct analysis of polyisoprenoid alcohols from biological samples. Lithium iodide is used to promote cationization by intense formation of $[M+Li]^+$ adducts. Detection of polyisoprenoids with mass determination can thus be performed with high sensitivity (LOD near 100 pM), whereas characteristic collision-induced dissociations observed for both dolichols and polyprenols permit investigation of their structure. We also describe a simple ESI-MS approach for the direct analysis of carotenoids in biological samples using lithium iodide to promote their ionization and the analysis of several carotenoids as proof-of-principle cases. Finally, we applied ESI(Li^+)-MS and ESI(Li^+)-MS/MS to investigate the presence of carotenoids in *Plasmodium falciparum*.

Key words: Polyisoprenoids, Electrospray ionization mass spectrometry (ESI-MS), Lithium iodide, Carotenoids and *Plasmodium falciparum*

1. Introduction

Prenols, also known as isoprenoids, are the most numerous and diverse group of natural products, covering more than 30,000 different compounds *(1)*. These lipids are synthesized from the five carbon precursor isomers isopentenyl diphosphate and dimethylallyl diphosphate *(2)*, and exhibit diverse functions in many different organisms including viruses, bacteria, plants, fungi, yeasts, and mammals. Cholesterol, steroid hormones, bile

Donald Armstrong (ed.), *Lipidomics*, Methods in Molecular Biology, vol. 580,
DOI 10.1007/978-1-60761-325-1_6, © Humana Press, a part of Springer Science+Business Media, LLC 2009

acids, retinoids, heme a, ubiquinone, prenylated proteins, isopentenylated tRNAs, and long-chain isoprenoids such as dolichols, represent some of the known isoprenoid derivatives present in mammalian cells *(3)*. Other isoprenoids with important biological functions are found in insects (juvenile hormone, ecdysone), plants (gibberellic acid, abscissic acid, trisporic acids, rubber, and carotenoids), and fungi/yeasts (prenylated mating hormone) *(3)*.

Animals, bacteria, and some parasites such as *Leishmania amazonensis* are known to biosynthesize isoprenoids via the mevalonic acid pathway *(4, 5)*, whereas plants, some bacteria, and parasites such as *Plasmodium falciparum* use the 2-*C*-methyl-D-erythritol 4-phosphate pathway *(6, 7)*. These two pathways have been found to occur concurrently in organisms such as bacteria and fungi *(8)*.

Many isoprenoids are linear polymers and receive the generic name polyisoprenoid alcohols, which are divided into two main groups: *(1)* dolichols (**scheme 1A**), which are either α-saturated isoprenoid alcohols found in all animal cells and some bacteria, parasites, fungi, and plants, and *(2)* polyprenols (**scheme 1B**), which are α-unsaturated isoprenoid alcohols found in the green tissues of many plants, bacteria, yeast, and parasites.

Polyprenols and dolichols are found in cells as free alcohols and esters *(9)*. Their biological roles are related to the ability to increase the permeability and fluidity of cell membranes and in the transport of vacuolar proteins *(10)*. They also take part in the transport mechanisms involving the endoplasmic reticulum *(11)*. Dolichols also participate in posttranslational modification of proteins *(12, 13)*, an event that is involved in tumor cell growth and differentiation and cellular signaling *(3, 14)*. Phosphorylated dolichols can also be found and play a role as carrier of oligossacharides in the biosynthesis of glycoproteins and glycosylphosphatidyl inositol anchors *(15, 16)*. Phosphorylated polyprenols have been postulated to serve as donors of isoprenoid groups for protein prenylation *(12)*.

The role of polyprenol-like substances in the biosynthesis of glycoconjugates has been studied extensively, and many mass spectrometric methods have been used for the analysis of these compounds in biological samples.

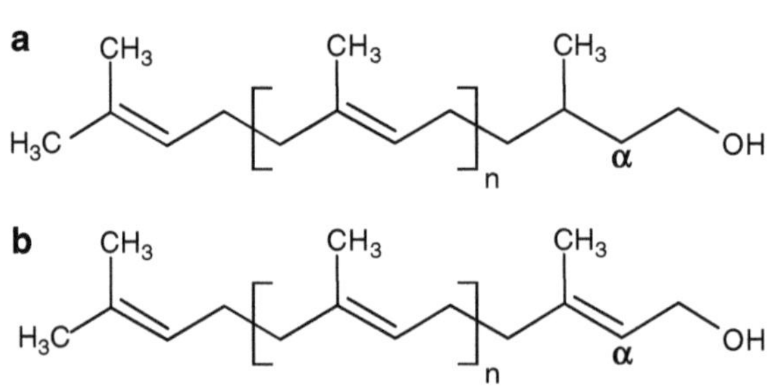

Scheme 1. Structure of dolichols (**a**) and polyprenols (**b**). α-Terminal isoprene unit.

Fast atom bombardment mass spectrometry (FAB-MS) of glycosyl monophosphopolyprenols produces intense $[M-H]^-$ ions, and tandem mass spectrometry of these ions allowed the determination of masses of both glycosyl and lipid moieties *(17)*. By desorption electron impact ionization, an intense $[M-H_3PO_4]^+$ ion as well as fragments corresponding to the successive loss of isoprene units of 68 Da were observed in analyses of prenyl phosphates. Alternatively, desorption chemical ionization yields ions corresponding to the loss of 66, 78, and 98 Da (i.e., of a part or the entire phosphate moiety) of a prenyl phosphate molecule *(17)*.

In the case of polyisoprenoid alcohols, electron ionization mass spectrometry (EI-MS) promotes an efficiently ionization of underivatized polyisoprenoids *(18, 19)*, and the EI mass spectra contain a large amount of structural information. However, this may be masked by the presence of impurity peaks, especially in biological samples. EI-induced dissociation occurs extensively, forming numerous fragment ions, and makes the mass spectra difficult to interpret in terms of the structural characterization and identification of the molecular ion.

Normally off-line derivatization to *tert*-butyl-dimethylsilyl (TBDMS) ethers is used for the analysis of polyisoprenoids with EI-MS *(18)*. The hydroxyl groups of polyisoprenoids are commonly converted to TBDMS ethers, improving sensitivity and the quality of structural information from EI-MS.

Analysis of underivatized polyisoprenoid alcohols is also difficult with soft ionization techniques such as FAB-MS *(17, 18, 20)* and field desorption mass spectrometry *(21)*. These molecules fail to ionize efficiently, and the mass spectra present poor structural information with a lack of molecular ions, and the multiple fragments observed often are not structurally characteristic. Off-line derivatization in sulfates *(18)* or phosphates *(17)* is necessary before MS analysis to increase polarity and facilitate protonation or deprotonation of the molecules. Because of the high hydrophobicity of the polyisoprenoid alcohols, the same difficulties are found in electrospray ionization mass spectrometry (ESI-MS).

Although good-quality spectra normally are obtained after derivatization, with easy detection of molecular ions and fragment ions that reflect structural aspects, these preliminary off-line steps involving sample manipulation are time- and sample-consuming. We considered performing the ESI-MS analysis of polyisoprenoids in samples of parasites such as *Plasmodium* and *Leishmania*, for which normally very small amounts are employed, but we realized that the use of derivatization would make this approach unviable.

The same problem can be observed for another class of prenols: the carotenoids. Carotenoids are the widespread lipophilic pigments, with over than 600 compounds, synthesized by all photosynthetic organisms and some nonphotosynthetic fungi and bacteria. Their chemical characteristic and physical properties are responsible for their light absorption as well as for inactivation of free radicals and antioxidant properties, being essential in photosynthesis events. They can also interfere in the fluidity of biomembranes *(22–25)* and can act in intracellular signaling *(26)*.

Their biosynthesis starts with the condensation of two molecules of geranylgeranyl diphosphate to form phytoene, the initial C40 carotenoid skeleton. Different carotenoids are derived essentially by modifications in the base structure by cyclization of the end groups and by introduction of oxygen functions giving them their characteristic colors and antioxidant properties. Lycopene, β- and α-carotene, lutein, xantophyls, retinoids (vitamin A), and abscisic acid are examples of carotenoid compounds.

Becase of their importance in plant metabolism and human alimentation, the identification and characterization of these compounds by mass spectrometry is widely used. Nowadays, the technique used to identify carotenoids is high-performance liquid chromatography mass spectrometry (HPLC-MS), where many carotenoids can be identified and even quantified at the same time *(27–29)*. The common ionization technique used is the atmospheric pressure chemical ionization (APCI), once its characteristics allow the ionization of the most of the known carotenoid compounds. However, high sample amounts are necessary, and sometimes these amounts are not possible to be achieved. A good alternative is the use of ESI technique where small amounts of sample are necessary and there are no problems with thermal instability of some carotenoids. However, this technique is more susceptible to mobile phase interference and usually low detection limits are achieve when compared with APCI for carotenoid analysis.

More recently, efficient ionization of low-polar compounds such as lipids and steroids using ESI has been done with prior derivatization as their Na^+ and Li^+ adducts via the addition of sodium or lithium salts *(30–33)*. Our group also demonstrated that polyisoprenoids alcohols can be analyzed by ESI-MS as Li^+ adducts with detection limits in pM *(34)*.

In this chapter, we present a detailed method for analyze of polyisoprenoid and carotenoid compounds using ESI(Li^+)-MS/MS. We also present methods for the extraction and prepurification of polyisoprenoid and carotenoid compounds from biological samples. We also show that this method can be used for characterization of carotenoids in biological samples. Finally, we applied ESI(Li^+)-MS and ESI(Li+)-MS/MS to screen for the presence of carotenoids in *P. falciparum*.

2. Materials

2.1. Equipment

1. LCQ Duo ion trap (Thermo Finnigan) working with an ESI source.
2. Gilson HPLC 322 pump (Gilson, Villiers-le-Bel, France) and a gradient module connected to a 152 UV visible detector, a temperature regulator 831, and a fraction collector FC203B (Gilson, Villiers-le-Bel, France). UNIPOINT™ Software (Gilson, Inc.) was used as the operational and analytical system.
3. Harvard syringe pump (model 11, Harvard, Hollinston, MA, USA).
4. Omni-Fit N_2 pressure system (OmniFit, Cambridge, UK).
5. Optiplex 755 computer (Dell, Round Rock, TX, USA), 1 Mb RAM, equipped with a Windows XP operating system. The system software Xcalibur™ 2.0 SR2 (Thermo Finnigan) was used for data processing.

2.2. Reagents

1. Albumax I was purchased from Gibco (Carlsbad, CA, USA). RPMI 1640 medium, D-sorbitol and saponin were purchased from Sigma (St. Louis, MO, USA). HPLC grade methanol, propan-2-ol and chloroform were purchased from J. T. Baker (Phillipsburg, NJ, USA), 1-butanol and *n*-pentane were purchased from Carlo Erba (Rodano, MI, France) and hexane was purchased from Mallinckrodt (Phillipsburg, NJ, USA). Lithium iodide Hydrate (≥98.0%) was purchased from Fluka (Buchs, Switzerland).
2. Authentic standards of dolichols of 11, 19, and 24 isoprene units and polyprenols of 8, 9, 10, 11, and 12 isoprene units were kindly provided by Prof. Tadeusz Chojnacki (Institute of Biochemistry and Biophysics, Polish Academy of Sciences, Warsaw, Poland).
3. β-Carotene, lycopene, and lutein standards were obtained from Sigma Chemical Company.

2.3. Supplies

1. Luna C_{18} column (4.6 × 250 mm, 5 μm) (Phenomenex, Allerod, Denmark).
2. C_{18} security guard cartridges (4 × 2 mm) (Phenomenex).
3. 2 ml amber glass vials with plastic crew caps (8 mm) and Teflon-faced seals (8 mm) (Chromacol, Thermo Fisher Scientific).
4. Glass μl syringes (10, 100, and 500 μl) with PTFE-tipped plungers (Hamilton Company, Bonaduz, Switzerland).
5. Fluoropore™ PTFE membrane filters (47 mm, 0.2 μm) (Millipore, Billerica, MA, USA).
6. 0.2-μm Syringe nylon filter (Advantec MFS, Dublin, CA, USA).

3. Methods

3.1. Standards Preparation (Notes 1 and 2)

1. For HPLC analysis of polyisoprenoids, solutions of 1 mM of each polyisoprenoid standards were prepared in methanol. Equal volumes of each stock solution (10 μl) were mixed together and injected into the HPLC columns for analysis.
2. To perform MS analysis of the polyisoprenoid standards, solutions of 1 μM of polyprenols of 8, 9, 10, 11, and 12 isoprene units and dolichols of 11, 19, and 24 isoprene units were prepared in chloroform/methanol (1:1, v/v), 2 mM of lithium iodide.
3. For HPLC analyses of carotenoids, extracts from *Anacardium occidentale* or *Physalis angulata* that had their carotenoid contend previously identified by LC-PDA-APCI-MS were used *(35)*. They were resuspended in methanol and 20 μl were used for HPLC analyses.
4. For MS analyses of carotenoids, commercial standards of lutein, β-carotene, and lycopene were prepared in chloroform/methanol (1:1, v/v), 2 mM of lithium iodide to a final concentration of 1 μM.

3.2. Parasite Culture

1. *P. falciparum* 3D7 clone was cultured *in vitro* according to Trager and Jensen *(36)*, using tissue culture flasks (75 cm^2) with a gas mixture of 5.05% CO_2, 4.93% O_2, and 90.2% N_2. Parasite development and multiplication were monitored by microscopic evaluation of Giemsa-stained thin smears.
2. Cultures (~15% parasitemia) were initially synchronized in ring stage (1–10 h after invasion) by two treatments with 5% (w/v) D-sorbitol *(37)* solution in water and maintained in culture until the trophozoite stage (20–24 h after reinvasion) or the schizont stage (30–35 h after reinvasion). The cultures were centrifuged at 2,000 × *g* and the parasites were isolated from erythrocytes by treatment with 0.1% (w/v) saponin for 5 min and washed three times with phosphate-buffered saline (0.007 M Na_2HPO_4, 0.01 M NaH_2PO_4, pH 7.4, 0.15 M NaCl) at 10,000 × *g* for 10 min. The isolated parasites were then lyophilized and stored at −80°C.

3.3. Polyisoprenoid Extraction (Note 1)

1. For the extraction of the polyisoprenoids from biological sample, the first step is to lyophilize the samples. The lyophilization reduces the volume of the sample by the evaporation of the liquid in lower temperature, keeping the lipids intact. After lyophilization, the sample can be stored at room temperature. However, it should be stored at −80°C to avoid degradation.

2. Before the extraction, the samples should be thawed first. The pellet is extracted by adding three volumes of hexane, and then it is sonicated for 5 min and centrifuged at 8,000 × *g* for 5 min. If the volume of the pellet is high (more than 2 ml), the use of a cell homogenizator is recommended. This procedure should be repeated three times. The solvent can be transferred to a glass tube and dried under nitrogen stream. Volumes higher than 10 ml can also be dried using lyophilization or speedvac. After that, the samples should be stored at −80°C.
3. To analyze the polyisoprenoids from *P. falciparum*, freeze-dried pellets of schizont parasites (~2 × 10^{12}) were extracted three times with 1 ml of hexane. The hexane extracts were dried under a nitrogen stream and stored at −80°C for subsequent HPLC analysis.

3.4. Carotenoid Extraction (Note 1)

1. For the extraction of the carotenoids from biological samples, the first step is also lyophilization of the samples. Once lyophilized, the samples should be maintained at −80°C or low temperature and safety from light to avoid carotenoid degradation.
2. After thawing the samples, they should be successively extracted with four volumes of ice-cold acetone, sonicated for 5 min, and centrifuged at 8,000 × *g* for 5 min. If the volume is high (more than 2 ml) or the tissues are hard, the use of a cell homogenizer is recommended. This procedure should be repeated three times. The pooled extracts can be dried under a nitrogen stream or speedvac and stored at −80°C.
3. In the case of *P. falciparum* samples, lyophilized parasites were successively extracted with four volumes of ice-cold acetone and centrifuged at 8,000 × *g* for 5 min. The pooled extracts were dried under a nitrogen stream and stored at −80°C, for posterior analysis.

3.5. Polyisoprenoid and Carotenoid Prepurification

1. Usually no prepurification before HPLC analysis is necessary. However, for some kind of sample such as blood and liver, a prepurification by solid phase extraction may be necessary. For this, C_{18} solid phase cartridges can be used following the manufacture's instructions.
2. Before HPLC analysis, the extracts must be filtered using nylon filter cartridges (0.2 μm) to avoid column damage and clogging.

3.6. HPLC Analysis (Note 2)

1. For HPLC analysis of polyisoprenoids, the hexane fraction was first resuspended in 300 μl of methanol and analyzed using an Phenomenex Luna C_{18} column (4.6 mm × 25 cm, 5 μm). The gradient elution system used was methanol/water

(9:1, v/v; solvent A) and hexane/propan-2-ol/methanol (1:1:2, v/v/v; solvent B). A linear gradient was run from 5% to 100% of B over a period of 25 min. The flow rate was 1.5 ml/min. The eluent was monitored at 210 nm. Fractions were collected at 0.5-min intervals *(38)*.

2. For HPLC analysis of carotenoids, the acetone extracts of each parasite stages were resuspended in 300 μl of acetonitrile, filtered through a 0.2 μm nylon filter, and analyzed using a Phenomenex Luna C_{18} column (250 mm × 4.6 mm × 5 μm). A gradient elution system was used, with acetonitrile/ethyl acetate/water (88:2:10, v/v/v) as solvent A and acetonitrile/ethyl acetate/water (85:15:0) as solvent B. The gradient program applied was as follows: 0–15 min, 0–100% B; 15–45 min, 100% B; 45–50 min, 100% to 0% B; 50–55 min, 0% B. The flow rate was 1 ml/min and the column was maintained at 29°C. The UV detector was set at 450 nm, and fractions were collected at 1-min intervals.

3.7. Mass Spectrometry Analysis (Notes 3 and 4)

1. To optimize the MS analysis, the polyisoprenoid and carotenoid standards in chloroform/methanol (1:1, v/v) containing 2 mM of lithium iodide were continuously injected directly into the ESI source using a syringe pump operating at a 10-μl/min. The parameters for each standard were tuned with the autotune operation in the LCQ Tune-Plus software.
2. For all sample analysis, they were resuspended in 10 μl chloroform/methanol (1:1, v/v) containing 2 mM of lithium iodide and were loaded into the 10-μl loop of the inject valve of the mass spectrometer. The solvent (chloroform/methanol [1:1, v/v], 2 mM of lithium iodide) was pumped continuously using the Omni-Fit pressure system.
3. All ESI(Li^+)-MS spectra were acquired in the positive ion mode, with spray voltage, capillary voltage, and capillary temperature set at 4.52 kV, 17 V, and 250°C, respectively. For ESI(Li^+)–MS/MS and ESI(Li^+)–MS^3, relative collision energy of 40% (2 eV) was applied in all of the analyses, and the sheath (N_2) and collision (He) gas pressure settings were 80 and 20 arbitrary units, respectively. No in-source dissociation was attempted. ESI(Li^+)-MS spectra were acquired in both full-ion mode and selective ion monitoring (SIM) mode over the *m/z* ranges presented in **Table 1**. The smoothing filter and background subtraction were used for data processing. For qualitative analysis, the Quan Browser software (version 2.0) was used.

3.8. Results

3.8.1. Mass Spectrometry Analysis of Polyisoprenoids

1. **Figure 1** shows the ESI(Li^+)–MS (in the *m/z* region near that of the expected Li^+ adduct) for 1 μM standard solutions of dolichol 11 (**Fig. 1a**) and polyprenol 11 (**Fig. 1b**)

Table 1
Molecular formulas, molecular weights, masses of $[M+Li]^+$ ions (Da), and *m/z* range used for MS analyses

Standard	Molecular formula	Molecular mass	Single-charged lithium ad duct ion mass	*m/z* range (full-Ion mode)	*m/z* range (selective ion monitoring
Geranylgeraniol	$C_{20}H_{34}O$	290.48	297.38	50–500	297 ± 3
Polyprenol 8	$C_{40}H_{66}O$	592.95	569.85	400–600	570 ± 3
Polyprenol 9	$C_{45}H_{74}O$	631.06	637.96	550–700	637 ± 3
Polyprenol 10	$C_{50}H_{82}O$	699.18	706.08	650–900	706 ± 3
Polyprenol 11	$C_{55}H_{90}O$	767.30	774.20	700–850	774 ± 3
Dolichol 11	$C_{55}H_{92}$	769.31	776.21	700–850	776 ± 3
Polyprenol 12	$C_{60}H_{98}O$	835.42	842.32	650–900	842 ± 3
Dolichol 12	$C_{60}H_{100}O$	837.43	844.33	750–900	844 ± 3
Dolichol 19	$C_{95}H_{156}O$	1314.24	1321.14	700–1400	1321 ± 3
Dolichol 24	$C_{120}H_{196}O$	1654.51	1659.71	1500–1800	1660 ± 3
β-Carotene	$C_{40}H_{56}$	536.87	543.81	520–560	536 ± 3[a]
Lycopene	$C_{40}H_{56}$	536.87	543.81	520–560	536 ± 3[a]
Lutein	$C_{40}H_{56}O_2$	568.87	575.81	540–580	551 ± 3[b]

[a] *m/z* range relative to the $[M]^{\cdot+}$ ion
[b] *m/z* range relative to the $[M+H-H_2O]^+$ion
Adapted with permission from *(34)*

spiked with 2 mM of lithium iodide. In these spectra, the singly charged lithium adducts $[M+Li]^+$ at *m/z* 776 and 774 are clearly observed for dolichol and polyprenol of 11 isoprene units, respectively. These spectra indicate, therefore, that ESI(Li^+)-MS of polyisoprenoids is indeed feasible and efficient, and this "single-ion" detection improves sensitivity and facilitates mixture analysis. It should also be noted that the protonated molecules $[M+H]^+$ in both ESI(Li^+)-MS are absent.

2. ESI(Li^+)-MS spectra were acquired along the *m/z* range shown in **Table 1**, and ESI(Li^+)-MS spectra over a more restricted range (*m/z* 800 ± 5) were also acquired to measure background noise. All ESI(Li^+)-MS spectra were acquired using LiI at a concentration of 2 mM (***Note* 4**).
3. To evaluate the efficiency of ESI(Li^+)-MS in analyzing polyisoprenoids with different isoprene chain sizes, polyisoprenoid standards (geranylgeraniol, dolichol 24 and a

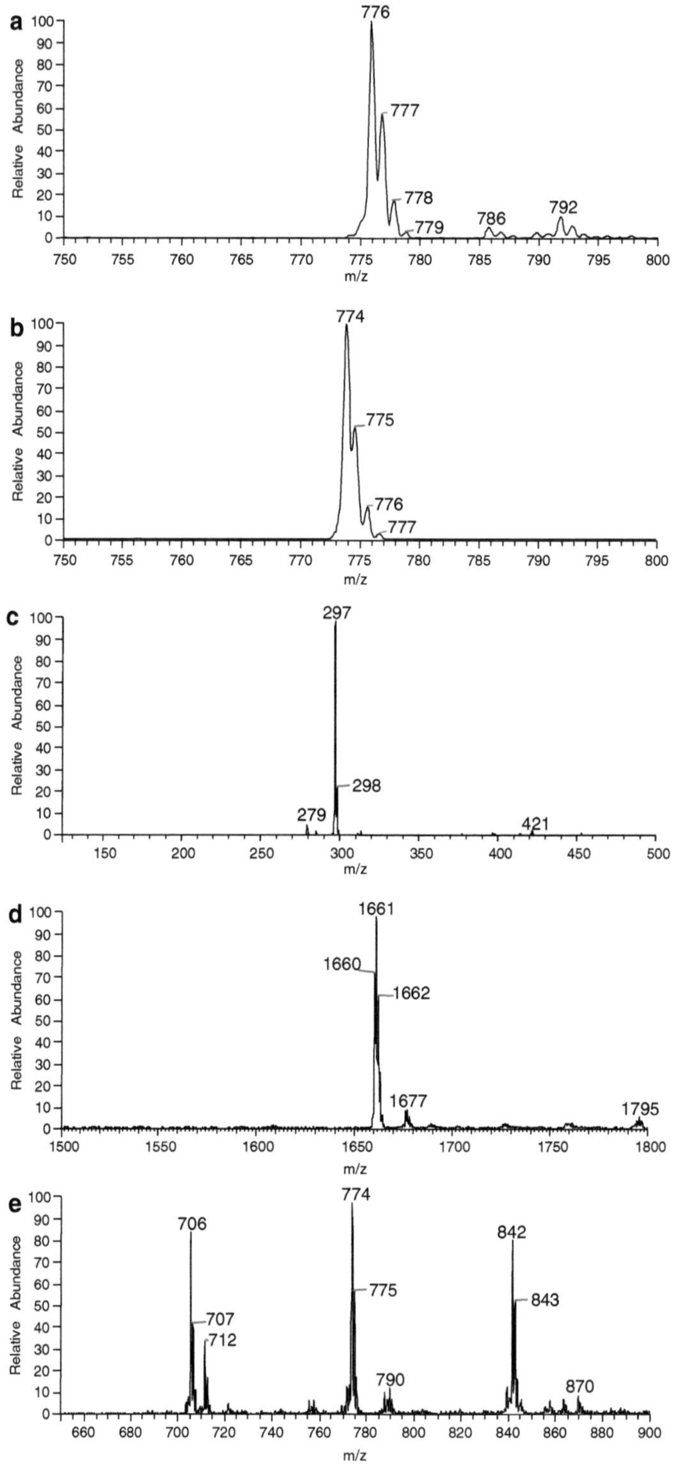

Fig. 1. ESI(Li+)-MS spectra of dolichol 11 (**a**), polyprenol 11 (**b**), geranylgeraniol (**c**), dolichol 24 (**d**), and a mixture of standards of polyprenols of 10, 11, and 12 isoprene unit standards (**e**). Reprinted with permission from ***(34)***.

mixture of polyprenols of 10, 11, and 12 isoprene units) at 1 μM concentrations were tested (**Fig. 1c**, **d**, and **e**, respectively). ESI(Li+)-MS was found to be efficient regardless of the chain size. As **Fig. 1e** shows, the three isoprenoids in a mixture (polyprenols of 10, 11, and 12 isoprene unit standards) were ionized with a similar efficiency to that of their $[M+Li]^+$ adducts.

4. **Figure 2** shows ESI(Li+)-MS/MS of $[M+Li]^+$ ions at *m/z* 776 from dolichol 11 (**Fig. 2a**) and *m/z* 1661 from dolichol 24 (**Fig. 2b**). A series of very structurally diagnostic fragment ions corresponding to sequential loss of 68 Da isoprene units (C_5H_8) are evident ($[M+Li-(C_5H_8)_n]^+$). A series of $[M+Li-(C_5H_8)_n-H_2O]^+$ fragments initiated by water loss (the $[M+Li-H_2O]^+$ at *m/z* 758 and *m/z* 1643) are also detected, but in lower abundances.
5. For polyprenols, unlike dolichols, ESI(Li+)-MS/MS of $[M+Li]^+$ detects mainly a single-fragment ion as a result of water loss (**Fig. 2c**). As **Fig. 2c** shows for the polyprenol of 11 isoprene units, although water loss from the $[M+Li^+]$ adduct at *m/z* 774 forming $[M+Li-H_2O]^+$ at *m/z* 756 is the main process, nevertheless, a minor $[M+Li-CH_2O]^+$ at *m/z* 744 is also detected. The same dissociation pattern was observed for all other polyprenols tested. This contrasting dissociation behavior therefore differentiates between dolichols and polyprenols. Although ESI(Li+)-MS/MS of the $[M+Li]^+$ adducts of polyprenols shows water loss as the major process and thus provides limited structural information, ESI(Li+)-MS^3 of the $[M+Li-H_2O]^+$ fragment retrieves the missing structural information. This is because ESI(Li+)-MS^3 of polyprenols displays the same series of structurally diagnostic fragment ions as a result of sequential loss of 68 Da isoprene units (**Fig. 2d**), which is $[M+Li-(C_5H_8)_n-H_2O]^+$, as does ESI(Li+)-MS/MS of dolichols, which is $[M+Li-(C_5H_8)_n]^+$.
6. **Table 2** shows that concentrations as low as 100 pM can be detected in the full-ion mode for dolichols 19 and 24, and as expected, even better results are obtained using SIM (50 pM for dolichols 19 and 24). ESI(Li+)-MS is therefore a sensitive method for polyisoprenoid detection (***Note* 5**).
7. This method was applied for the identification and characterization of polyisoprenoid different biological sources: characterization of polyisoprenoids from different stages of *L. amazonensis (39, 40)*, characterization of dolichols of 11 and 12 isoprene units from *P. falciparum (34)*, characterization of octaprenyl pyrophosphate synthase products in *P. falciparum (41)*, characterization of polyisoprenoid released from proteins in P. falciparum *(13)*.

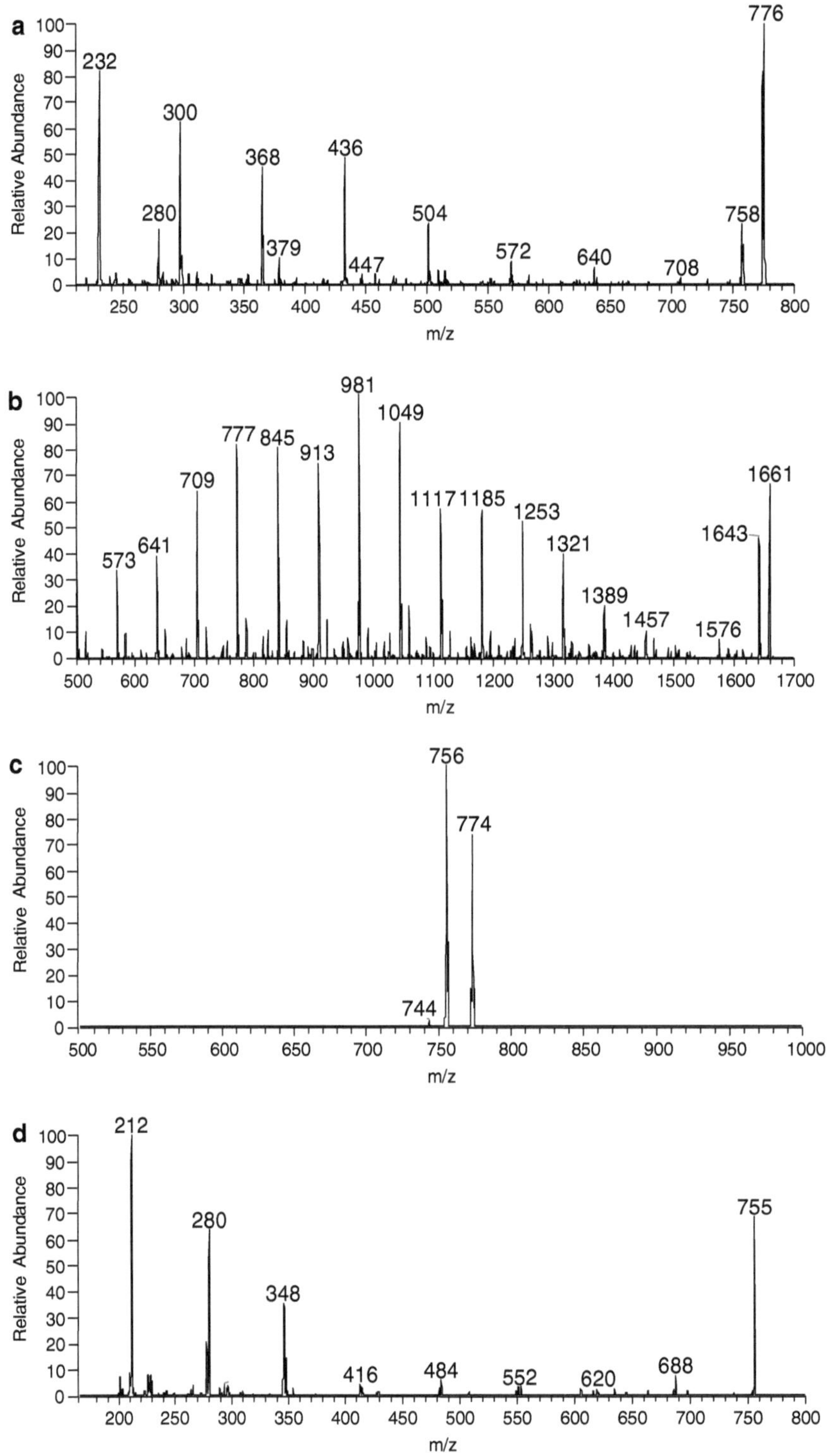

Fig. 2. (**a,b**) ESI(Li+)-MS/MS spectra of [M+Li]+ ion at *m/z* 776 and *m/z* 1661 from 1μM solution of dolichols of 11 and 24 isoprene units, respectively. (**c, d**) ESI(Li+)-MS/MS and ESI(Li+)-MS^3, respectively, of $[M+Li]^+$ ion at *m/z* 774 and $[M+Li-H_2O]^+$ ion at *m/z* 756 from 1 μM solution of polyprenol standard of 11 isoprene units. Reprinted with permission from ***(34)***.

Table 2
Detection limits (3:1 S/N ratio) in nM of polyisoprenoid standards in the different analytical modes

Standard	Full-ion mode	Selective ion monitoring	MS/MS	MS^3
Geraniol	0.25[a]	0.10[a]	1.00[b]	—
Geranylgeraniol	0.20[a]	0.10[a]	1.00[b]	—
Dolichol 11	0.15[a]	0.10[a]	1.00[b]	—
Dolichol 19	0.10[a]	0.05[a]	0.10[b]	—
Dolichol 24	0.10[a]	0.05[a]	0.10[b]	—
Polyprenol 11	0.15[a]	0.10[a]	1.00[c]	10.00[d]
Polyprenol 12	0.15[a]	0.10[a]	1.00[c]	10.00[d]
Polyprenol 13	0.15[a]	0.10[a]	1.00[c]	10.00[d]

[a] $[M + Li]^+$ ion was detected
[b] $[M + Li\text{-}(C_5H_8)_n]^+$ fragment ions were detected
[c] $[M + Li\text{-}H_2O]^+$ ion was detected
[d] $[M + Li\text{-}H_2O\text{-}(C_5H_8)_n]^+$ fragment ions were detected
Reprinted with permission from *(34)*

3.8.2. Mass Spectrometry Analysis of Carotenoids

1. **Figure 3** shows ESI(Li$^+$)-MS (in the *m/z* region near that of the expected Li$^+$ adduct) for 1 μM standard solutions of β-carotene (A), lycopene (B), and lutein (C) spiked with 2 mM of lithium iodide. In these spectra, the singly charged lithium adducts $[M+Li]^+$ at *m/z* 543.7 and *m/z* 575.8 are observed for β-carotene, lycopene and lutein, respectively. However, the intensity of these ions is very low, corresponding to approximately 10%. The main ion observed for β-carotene and lycopene was the molecular ion $[M]^{\cdot+}$ at *m/z* 536.7. For lutein, the main ion observed was the ion $[M+H\text{-}H_2O]^+$ at *m/z* 551.6.
2. Although these spectra indicate that the lithium adducts $[M+Li]^+$ cannot be used for mass spectrometry identification of carotenoids, these ions appear to be very important in the ionization process of carotenoid compounds. Without addition of lithium iodide, the molecular ions $[M]^{\cdot+}$ and the $[M+H\text{-}H_2O]^+$ could be detected with a high limit of detection, even when other additives like formic acid and acetic acid were used. Rentel et al. *(42)* verified the ionization of carotenoids as stable silver adducts $[M+Ag]^+$ when $AgClO_4$ solution was used as a postcolumn additive in LC/MS analysis. In their experiments, they also identified the molecular ions

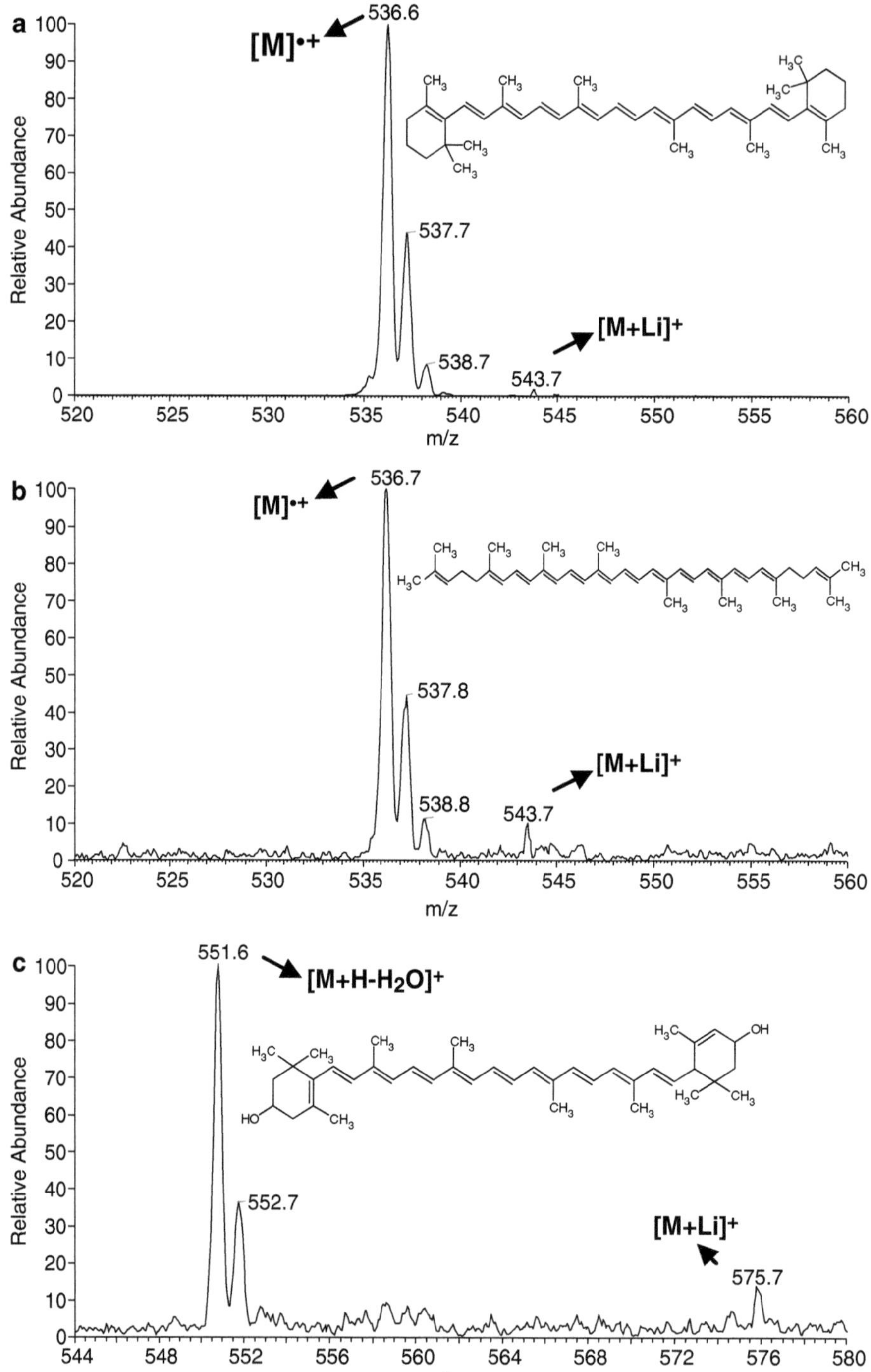

Fig. 3. ESI(Li⁺)-MS spectra of β-carotene (**a**), lycopene (**b**), and lutein (**c**) standards.

$[M]^{\cdot+}$ for all these compounds, but this ion was in lower abundance if compared with the silver adducts. In our case, our results shown that probable the first ionization step of the carotenoid compounds is as $[M+Li]^+$ adduct ions. However, these ions are very instable and the elemental lithium is lost, resulting in the $[M]^{\cdot+}$ molecular ions.

3. **Figure 4** shows the ESI-MS/MS analysis of the β-carotene (A), lycopene (B), and lutein (C) standards. The fragmentation pattern is not different from that present in the literature.

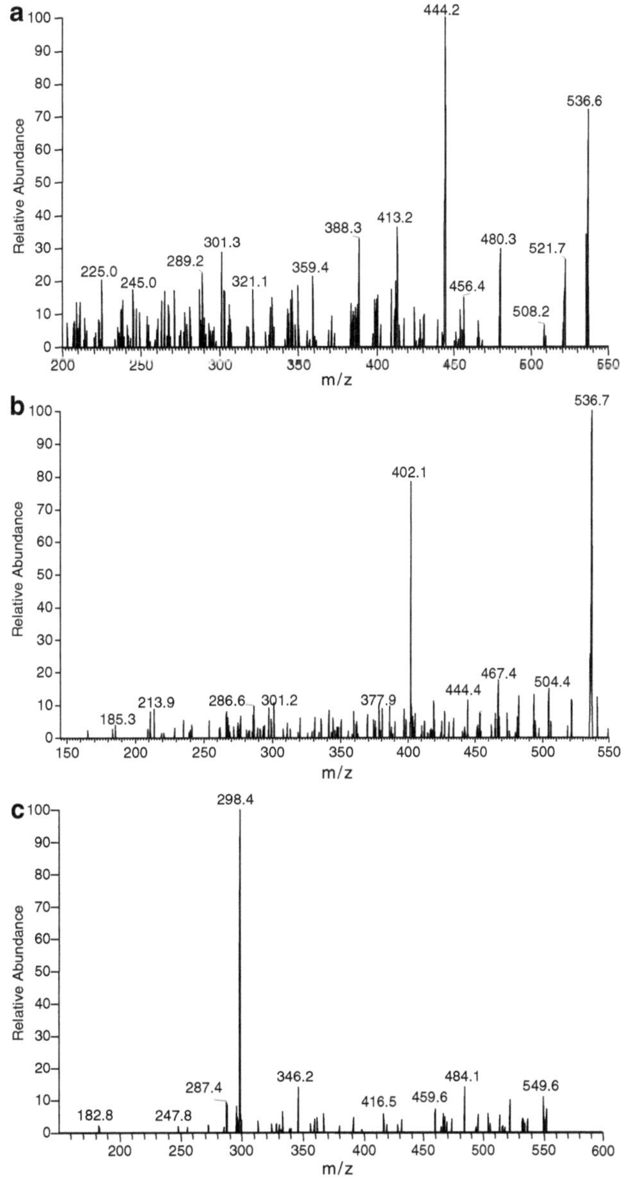

Fig. 4. ESI(Li$^+$)-MS/MS spectra of $[M]^{\cdot+}$ ion at *m/z* 536.6 of β-carotene (**a**) and lycopene (**b**) and of $[M+H-H_2O]^+$ ion at *m/z* 551.6 of lutein standards.

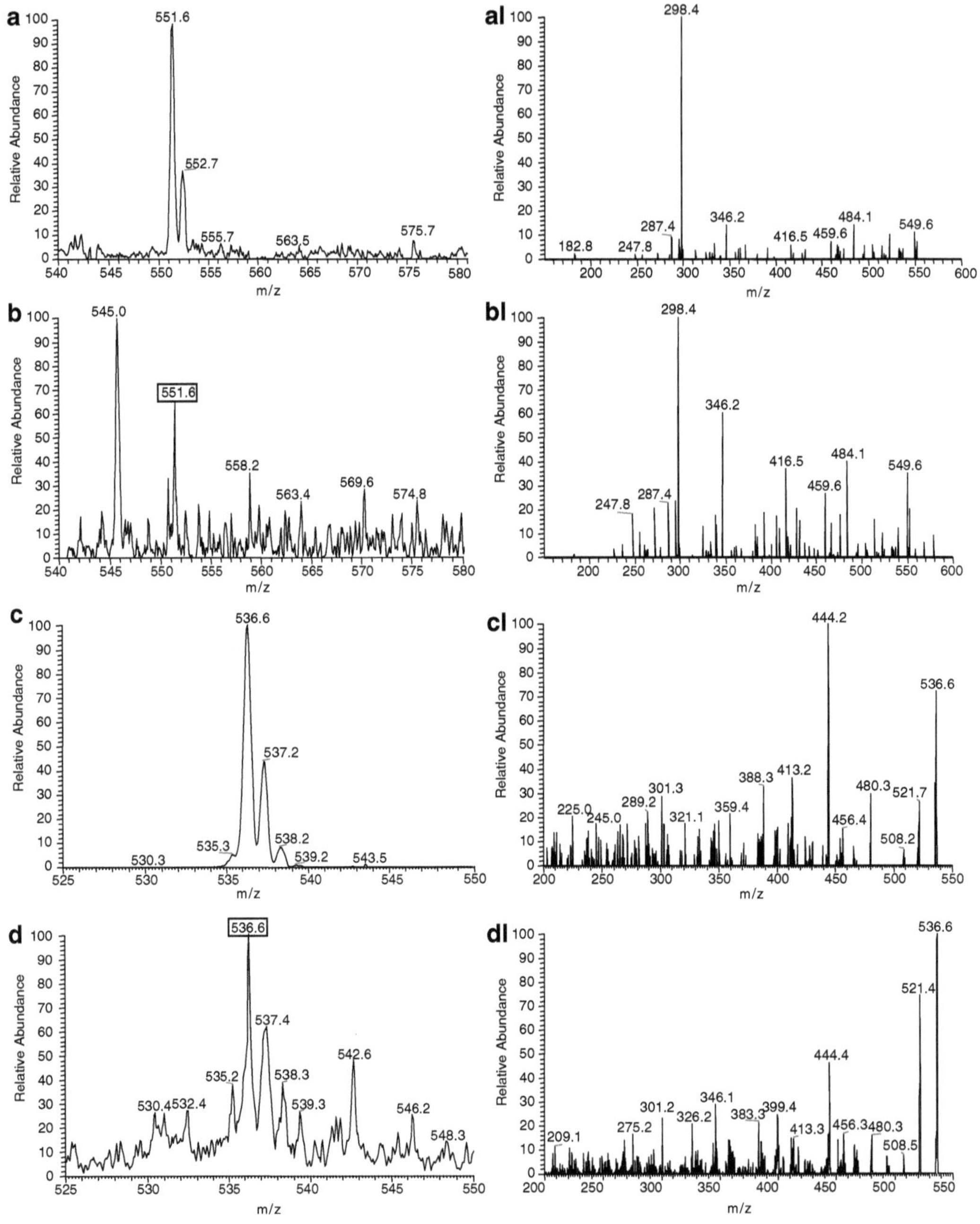

Fig. 5. (**a, c**) ESI(Li+)-MS spectra of lutein and β-carotene standards, respectively. (**b, d**) ESI(Li+)-MS spectra of the hexane extracts of *P. falciparum* schizont parasites with HPLC retention times coincident with the lutein and β-carotene standards, respectively. (al and cl) ESI(Li+)-MS/MS spectra of the ion $[M+H-H_2O]^+$ at *m/z* 551.6 and the ion $[M]^{\cdot+}$ at *m/z* 536.6, respectively, from standards. (bl, dl) ESI(Li+)-MS/MS spectra of the ion $[M+H-H_2O]^+$ at *m/z* 551.6 and the ion $[M]^{\cdot+}$ at *m/z* 536.6, respectively, from *P. falciparum* samples.

In the case of β-carotene, the fragment ion spectrum of the $[M]^{\cdot+}$ ion shows that the most abundant fragment ion is $[M-92]^+$ at *m/z* 444.2 (loss of toluene). In the case of lycopene, the most abundant fragment ion is $[M-135]^+$ at *m/z* 402.1, and for lutein, it is the ion $[M+H\text{-}H_2O\text{-}253]^+$ at *m/z* 298.4.

4. In order to demonstrate that the ESI(Li^+)-MS can be used for identification and characterization of carotenoids in biological samples, we applied this method in the characterization of carotenoid compounds in the malaria parasite *P. falciparum*. For that, the carotenoid compounds present in the schizont forms of *P. falciparum* were extracted and purified by HPLC as described above. After HPLC analysis, the fractions with the same retention time of the lutein (12 min) and β-carotene (38 min) standards were analyzed by mass spectrometry (**Fig. 5**). The same ionization profile of the samples (**Fig. 5a** and **c**) and the standards (**Fig. 5b** and **d**) were observed, where the ion at *m/z* 551.6 represents a lutein molecule ($[M+H\text{-}H_2O]^+$) and the ion at *m/z* 536.6 represents a β-carotene molecule that ionize as a molecular ion ($M^{\cdot+}$). The molecular structures were confirmed by comparing the ESI-MS/MS spectra of the ions at *m/z* 551.6 (**Fig. 5aI**) and *m/z* 536.6 (**Fig. 5cI**) from *P. falciparum* with that of the standard of lutein (**Fig. 5bI**) and β-carotene (**Fig. 5dI**), respectively, and the same fragmentation profile was observed.

4. Notes

1. Polyisoprenoids and especially carotenoids are sensitive to oxidation and consequently standards and samples should be stored at low temperatures (e.g., ≤–80°C) under an inert atmosphere (e.g., argon) in the absence of light, moisture, and active surfaces. They also should be stored in glass vials to avoid plastic contamination.
2. It is important to use HPLC or MS grade solvents for the preparation of chromatographic mobile phases and standards to reduce detected background noise. Filtering the solvents is strongly recommended to avoid column and equipment clogging with eventual impurities.
3. Different solvents for diluting samples were tested, and chloroform:methanol (1:1, v/v) provided the best results. The solvent systems most often employed for chromatography of polyisoprenoids, however, provide inferior ESI(Li^+)-MS sensitivity compared with chloroform:methanol (1:1, v/v).

4. ESI(Li^+)-MS efficiency is dependent on the LiI concentration, since high concentrations promote ion suppression. Highest sensitivity was obtained with LiI at a concentration of 2–5 mM.
5. ESI(Li^+)-MS is therefore a sensitive method for polyisoprenoid detection. Since the measurements were performed in an ion-trap instrument, we expect that even better LODs can be achieved using SIM in triple quadrupole and hybrid mass spectrometers. Nevertheless, we need to keep in mind that this method is a qualitative method. The adaptation of this method to LC methods could allow the quantitation of polyisoprenoids in biological samples, an important step to integrate the polyisoprenoid data to lipidomics.

Acknowledgments

We thank Prof. Tadeusz Chojnacki (Institute of Biochemistry and Biophysics, Polish Academy of Sciences, Warsaw, Poland). R.T is a recipient of an FAPESP fellowship. F.L.D is a recipient of Karolinska Institute fellowship, Sweden. A.M.K is supported by the Brazilian research agencies FAPESP and CNPq.

References

1. Bach TJ. (1995) Some new aspects of isoprenoid biosynthesis in plants – a review. *Lipids* 30, 191–202.
2. Kuzuyama T, Seto H. (2003) Diversity of the biosynthesis of the isoprene units. *Nat Prod Rep* 20, 171–183.
3. Edwards PA, Ericsson J. (1999) Sterols and isoprenoids: signaling molecules derived from the cholesterol biosynthetic pathway. *Annu Rev Biochem* 68, 157–185.
4. Grunler J, Ericsson J, Dallner G. (1994) Branch-point reactions in the biosynthesis of cholesterol, dolichol, ubiquinone and prenylated proteins. *Biochim Biophys Acta* 1212, 259–277.
5. Goldstein JL, Brown MS. (1990) Regulation of the mevalonate pathway. *Nature* 343, 425–430.
6. Wanke M, Skorupinska-Tudek K, Swiezewska E. (2001) Isoprenoid biosynthesis via 1-deoxy-D-xylulose 5-phosphate/2-C-methyl-D-erythritol 4-phosphate (DOXP/MEP) pathway. *Acta Biochim Pol* 48, 663–672.
7. Lichtenthaler HK. (1999) The 1-deoxy-D-xylulose-5-phosphate pathway of isoprenoid biosynthesis in plants. *Annu Rev Plant Physiol Plant Mol Biol* 50, 47–65.
8. Kuzuyama T. (2002) Mevalonate and nonmevalonate pathways for the biosynthesis of isoprene units. *Biosci Biotechnol Biochem* 66, 1619–1627.
9. Rip JW, Rupar CA, Ravi K, Carroll KK. (1985) Distribution, metabolism and function of dolichol and polyprenols. *Prog Lipid Res* 24, 269–309.
10. Valtersson C, van Duyn G, Verkleij AJ, Chojnacki T, de Kruijff B, Dallner G. (1985) The influence of dolichol, dolichol esters, and dolichyl phosphate on phospholipid polymorphism and fluidity in model membranes. *J Biol Chem* 260, 2742–2751.
11. Chojnacki T, Dallner G. (1988) The biological role of dolichol. *Biochem J* 251, 1–9.
12. Hjertman M, Wejde J, Dricu A, et al. (1997) Evidence for protein dolichylation. *FEBS Lett* 416, 235–238.

13. D'Alexandri FL, Kimura EA, Peres VJ, Katzin AM. (2006) Protein dolichylation in Plasmodium falciparum. *FEBS Lett* 580, 6343–6348.
14. Schafer WR, Rine J. (1992) Protein prenylation: genes, enzymes, targets, and functions. *Annu Rev Genet* 26, 209–237.
15. Spiro RG. (2002) Protein glycosylation: nature, distribution, enzymatic formation, and disease implications of glycopeptide bonds. *Glycobiology* 12, 43R–56R.
16. Burda P, Aebi M. (1999) The dolichol pathway of N-linked glycosylation. *Biochim Biophys Acta* 1426, 239–257.
17. Wolucka BA, de Hoffmann E. (1994) Mass spectrometric analysis of prenyl phosphates and their glycosylated forms. *Acta Biochim Pol* 41, 345–349.
18. Griffiths WJ, Hjertman M, Lundsjö A, Wejde J, Sjövall J, Larsson O. (1996) Analysis of dolichols and polyprenols and their derivatives by electron impact, fast atom bombardment and electrospray ionization tandem mass spectrometry. *Rapid Commun Mass Spectrom* 10, 663–675.
19. Gough DP, Hemming FW. (1970) The characterization and stereochemistry of biosynthesis of dolichols in rat liver. *Biochem J* 118, 163–166.
20. Xu P, Widmer G, Wang Y, et al. (2004) The genome of *Cryptosporidium hominis*. *Nature* 431, 1107–1112.
21. Ibata K, Mizuno M, Takigawa T, Tanaka Y. (1983) Long-chain betulaprenol-type polyprenols from the leaves of Ginkgo biloba. *Biochem J* 213, 305–311.
22. Subczynski WK, Markowska E, Sielewiesiuk J. (1993) Spin-label studies on phosphatidylcholine-polar carotenoid membranes: effects of alkyl-chain length and unsaturation. *Biochim Biophys Acta* 1150, 173–181.
23. Strzalka K, Gruszecki WI. (1994) Effect of beta-carotene on structural and dynamic properties of model phosphatidylcholine membranes. I. An EPR spin label study. *Biochim Biophys Acta* 1194, 138–142.
24. Jezowska I, Wolak A, Gruszecki WI, Strzalka K. (1994) Effect of beta-carotene on structural and dynamic properties of model phosphatidylcholine membranes. II. A 31P-NMR and 13C-NMR study. *Biochim Biophys Acta* 1194, 143–148.
25. McNulty H, Jacob RF, Mason RP. (2008) Biologic activity of carotenoids related to distinct membrane physicochemical interactions. *Am J Cardiol* 101, 20D–29D.
26. Duester G. (2008) Retinoic acid synthesis and signaling during early organogenesis. *Cell* 134, 921–931.
27. Kane MA, Folias AE, Wang C, Napoli JL. (2008) Quantitative profiling of endogenous retinoic acid in vivo and in vitro by tandem mass spectrometry. *Anal Chem* 80, 1702–1708.
28. Zhu D, Wang Y, Pang Y, et al. (2006) Quantitative analyses of beta-carotene and retinol in serum and feces in support of clinical bioavailability studies. *Rapid Commun Mass Spectrom* 20, 2427–2432.
29. Gundersen TE, Bastani NE, Blomhoff R. (2007) Quantitative high-throughput determination of endogenous retinoids in human plasma using triple-stage liquid chromatography/tandem mass spectrometry. *Rapid Commun Mass Spectrom* 21, 1176–1186.
30. Ham BM, Jacob JT, Keese MM, Cole RB. (2004) Identification, quantification and comparison of major non-polar lipids in normal and dry eye tear lipidomes by electrospray tandem mass spectrometry. *J Mass Spectrom* 39, 1321–1336.
31. Hsu FF, Turk J, Stewart ME, Downing DT. (2002) Structural studies on ceramides as lithiated adducts by low energy collisional-activated dissociation tandem mass spectrometry with electrospray ionization. *J Am Soc Mass Spectrom* 13, 680–695.
32. Levery SB, Toledo MS, Straus AH, Takahashi HK. (2001) Comparative analysis of glycosylinositol phosphorylceramides from fungi by electrospray tandem mass spectrometry with low-energy collision-induced dissociation of Li(+) adduct ions. *Rapid Commun Mass Spectrom* 15, 2240–2258.
33. Skorupinska-Tudek K, Bienkowski T, Olszowska O, et al. (2003) Divergent pattern of polyisoprenoid alcohols in the tissues of Coluria geoides: a new electrospray ionization MS approach. *Lipids* 38, 981–990.
34. D'Alexandri FL, Gozzo FC, Eberlin MN, Katzin AM. (2006) Electrospray ionization mass spectrometry analysis of polyisoprenoid alcohols via Li+ cationization. *Anal Biochem* 355, 189–200.
35. de Rosso VV, Mercadante AZ. (2007) Identification and quantification of carotenoids, by HPLC-PDA-MS/MS, from Amazonian fruits. *J Agric Food Chem* 55, 5062–5072.
36. Trager W, Jensen JB. (1997) Continuous culture of Plasmodium falciparum: its impact on malaria research. *Int J Parasitol* 27, 989–1006.

37. Lambros C, Vanderberg JP. (1979) Synchronization of Plasmodium falciparum erythrocytic stages in culture. *J Parasitol* 65, 418–420.
38. Low P, Dallner G, Mayor S, Cohen S, Chait BT, Menon AK. (1991) The mevalonate pathway in the bloodstream form of Trypanosoma brucei. Identification of dolichols containing 11 and 12 isoprene residues. *J Biol Chem* 266, 19250–19257.
39. Arruda DC, D'Alexandri FL, Katzin AM, Uliana SR. (2005) Antileishmanial activity of the terpene nerolidol. *Antimicrob Agents Chemother* 49, 1679–1687.
40. Arruda DC, D'Alexandri FL, Katzin AM, Uliana SR. (2008) Leishmania amazonensis: biosynthesis of polyprenols of 9 isoprene units by amastigotes. *Exp Parasitol* 118, 624–628.
41. Tonhosolo R, D'Alexandri FL, Genta FA, et al. (2005) Identification, molecular cloning and functional characterization of an octaprenyl pyrophosphate synthase in intra-erythrocytic stages of Plasmodium falciparum. *Biochem J* 392, 117–126.
42. Rentel C, Strohschein S, Albert K, Bayer E. (1998) Silver-plated vitamins: a method of detecting tocopherols and carotenoids in LC/ESI-MS coupling. *Anal Chem* 70, 4394–4400.

Chapter 7

Detection of a Lipid-Lysine Adduct Family with an Amide Bond as the Linkage: Novel Markers for Lipid-Derived Protein Modifications

Yoji Kato and Toshihiko Osawa

Summary

An amide-type adduct, hexanoyl-lysine (HEL) is generated from the reaction between n-6 fatty acid (FA)-derived lipid peroxide and lysine. Immunochemical and chemical methods can be used to detect the formation of HEL. For example, an ELISA kit using the monoclonal antibody to HEL is now commercially available. We recently identified propanoyl-lysine (propionyl-lysine, PRL) from the reaction of an n-3 FA and a lysine residue. The antibody to PRL has been prepared and characterized. Using these monoclonal antibodies, the localization of adducts in tissues has been confirmed. Moreover, both amide-type adducts, HEL and PRL, can be simultaneously measured using liquid chromatography mass spectrometry (LC/MS/MS) with isotope dilution methods. The LC/MS/MS analysis reveals the rigid amounts of the adducts in human urine. Both the chemical and immunochemical methods are useful for the estimation of amide-type adducts in vivo.

Key words: Hexanoyl-lysine, Propanoyl-lysine, Amide-type adducts, Lipid hydroperoxide, Protein modification

1. Introduction

Lipid peroxidation generates various oxidation products. Among them, aldehydes have high reactivity against biomolecules such as proteins and nucleotides *(1)*. Aldehyde-modified biomolecules have been immunochemically identified in various tissues. On the contrary, our group has also found the formation of novel adducts that have amide bonds forming the linkage between oxidized lipid and lysine. One of these adducts,

Donald Armstrong (ed.), *Lipidomics,* Methods in Molecular Biology, vol. 580,
DOI 10.1007/978-1-60761-325-1_7,

hexanoyl-lysine (HEL), has been detected in atherosclerotic plaques using polyclonal and monoclonal antibodies to HEL *(2, 3)*. Several amide-type adducts, azelayl-lysine (AZL), succinyl-lysine (SUL), glutaroyl-lysine (GLL), and propanoyl-lysine (PRL), have also been recently identified *(4–6)*. Because these adducts are probably generated from oxidation products of n-3 and/or n-6 FAs with lysine residues, the simultaneous determination of all adducts may prove a useful marker to determine which kinds of lipids are oxidized in vivo (**Fig. 1**). Since SUL, AZL, and GLL have a C-terminal moiety (COOH) derived from parent fatty acids (FAs), these should be esterified with cholesterols and phospholipids in vivo. To immunochemically detect the carboxyalkylamide-type adducts in a tissue, the scission of lipid ester bond has to be done by phospholipase A2 or alkaline hydrolysis of the ester bond *(6)*. On the contrary, the alkylamide-type adducts HEL and PRL do not need such pretreatments for detection. This chapter describes how to prepare and characterize the monoclonal antibodies to HEL and PRL.

Immunochemical data often need to be confirmed using other methods because the result might be due to nonspecific binding of the antibody. Chemical approaches are also beneficial to estimate levels of some biomarkers. However, some oxidative stress biomarkers are generated during sample preparation for chemical derivatization or during other process. This chapter shows the methods used for simultaneous quantitation of PRL and HEL in human urine using LC/MS/MS without derivatization. The methods use an isotope dilution method *(7)* using chemically synthesized deuterated HEL and PRL as internal standards (IS). Note that this method requires less than 100 μl of urine to determine the amounts of both adducts.

1.1. Advantages and Disadvantages

The detection of PRL and HEL by immunochemical approaches has three primary disadvantages relative to other methods:

1. It is difficult to detect the HEL and PRL simultaneously, although double-staining techniques are available to overcome this limitation.
2. It is hard to evaluate the accurate amount of adducts, whereas ELISA is a useful technology for estimation of the levels of adducts.
3. The "true" specificity of an antibody is hard to prove. Similarly, it is not easy to prove that the positive immunoreactivity is derived from the reaction of the antibody with the adduct in a sample.

Determination of the levels of PRL and HEL using immunochemical approaches also has advantages relative to other techniques such as HPLC and MS:

1. Crude samples, such as urine, can be used in immunochemical methods without a complicated precleaning process.

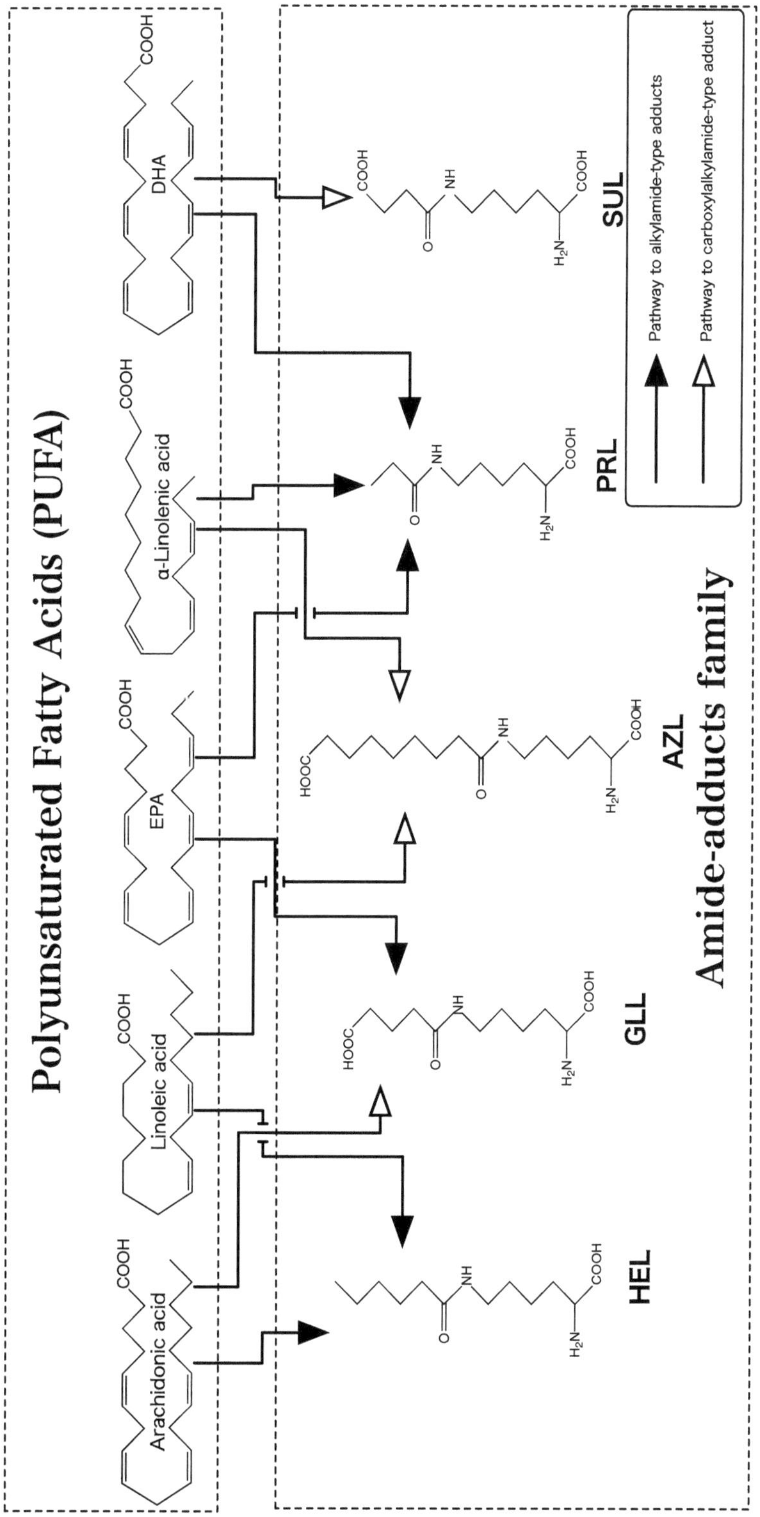

Fig. 1. Formation of the amide-type adduct family from the reaction of oxidized PUFA with lysine.

2. The cost of analysis is inexpensive. For example, to quantify the adduct/antigen by ELISA, only a microplate reader is needed. Moreover, in most cases, a microplate reader should already be available. Nevertheless, the price of a microplate reader is much less than that of MS.
3. Many samples can be measured simultaneously by ELISA.
4. The localization of adducts in tissue can be visualized by immunohistochemical staining.

The detection of PRL and HEL by LC/MS/MS has four disadvantages relative to other methods including immunochemical methods:

1. Partial purification of biological specimens (urine) is often necessity for LC/MS/MS. The intact sample may be usable, but the mass spectrometer may become contaminated, which causes a decreased sensitivity, necessitating more frequent maintenance of the machine.
2. The analysis itself takes around 30 min (or more) per sample. Therefore, a large number of samples can not be examined simultaneously in contrast with ELISA.
3. The cost of investment for LC/MS/MS is much higher relative to most other methods including ELISA.
4. To accurately measure the amount of PRL and HEL, the standards as well as stable isotopic IS have to be prepared by the investigator. The method for synthesis of PRL and HEL (including IS) is not difficult, but some reagents are needed for the preparation and, importantly, it is time consuming. A deuterated compound is usually expensive. In this case, one must purchase hexanoic-d11 acid and propionic-d4 acid prior to preparation of IS.

The detection of PRL and HEL by LC/MS/MS also has many advantages relative to other instruments, such as ELISA and HPLC:

1. Precise quantitation of HEL and PRL is possible, particularly with IS-containing stable isotopes, with which the quantitation is highly reliable. There is no need to worry about coelution of adduct with the IS because they have different molecular weights.
2. Structural information is available. Because the technique monitors the charged ion of molecular mass as well as fragmented ions caused by collision-induced dissociation, the detection itself provides structural information. This means that the information obtained provides reliable quantitation.
3. The sensitivity is considerably higher compared with other methods. The detection limits are typically 1 nM or less.

2. Materials

2.1. Equipment and Supplies (Immunochemical Analysis)

1. Microplate reader: Bio-Rad Model 550 (Bio-Rad).
2. Microscope.

2.2. Equipment and Supplies (LC/MS/MS)

1. LC/MS/MS system: the tandem mass spectrometry system includes API3000 (Applied Biosystems) and Agilent HP1100 HPLC (Dual pump, autosampler, UV-Vis detector, degasser).
2. Solid-phase extraction system: GL-Science Vacuum Manifold (12-port model) with a vacuum pump (DIVAC 1.2 L).
3. Centrifugal evaporator system: EYELA CVE-3100 with a cold-trap (UNITRAP UT2000) and a vacuum pump (ULVAC GCD-051X).
4. Rotary Evaporator system for chemical synthesis: EYELA N-N series with a vacuum pump (DIVAC 0.6 L).

2.3. Reagents

All reagents are of the highest grade available.

3. Methods

3.1. Immunochemical Analysis

3.1.1. Preparation of Antibodies

3.1.1.1. Anti-HEL Antibody

The monoclonal antibody to HEL can be prepared as previously described *(2, 8)* and is now commercially available. Briefly, hexanoyl keyhole limpet hemocyanin (KLH) and its analogue, hexanoyl bovine serum albumin (BSA), are prepared from hexanoic acid and proteins with 1-ethyl-3-(3-dimethylaminopropyl) carbodiimide (EDC) and sulfo-*N*-hydroxysuccinimide (sulfo-NHS) *(9)*. Balb/c mice are primed intraperitoneally with the modified KLH emulsified in Freund's complete adjuvant. Mice are given booster doses twice at a 2-week interval. Three days after the final intraperitoneal injection of the antigen without adjuvant, the animals are sacrificed, and a myeloma cell line and the spleens of immunized mice are fused using polyethylene glycol. The selection of hybridomas is carried out using an ELISA. The positive clones are selected using hexanonyl BSA as the antigen for the ELISA. The hybridomas obtained are injected intraperitoneally into Balb/c mice, which have been injected with pristane (2, 6, 10, 14-tetramethylpentadecane) 1 week before. About 7–10 days later, the ascites are collected and then partially purified by ammonium sulfate fractionation and a conventional protein G column. Alternatively, the hybridomas are cultivated in nonserum medium and the antibody purified from the conditioned medium using a protein G column.

3.1.1.2. Anti-PRL Antibody

The detailed method for the preparation of the antibody to PRL has been shown in detail (10). Briefly, propionic acid, EDC, and sulfo-NHS are mixed and reacted in dimethylformamide at room temperature for 24 hr. The mixture is then reacted with KLH or BSA in 0.1 M phosphate buffer (pH 7.4) for 4 h at room temperature. The modified proteins (PRL-KLH and PRL-BSA) are dialyzed against phosphate-buffered saline (PBS). Balb/c mice are injected with the immunogen (PRL-KLH) with complete Freund's adjuvant. Four weeks later, mice are given a booster dose of the immunogen with incomplete Freund's adjuvant. The final booster dose (the conjugate in PBS) is injected into the vein and spleen cells are fused with myeloma cells according to the methods for HEL as described earlier.

3.1.1.3. Preparation of the Competitors

Benzoyl(Bz)-Gly-HEL is prepared as described previously *(2)*. The reaction of hexanoic acid with Bz-Gly-Lys (BGK) by a carbodiimide method [EDC and *N*-hydroxysuccinimide (NHS)] or the treatment of BGK with hexanoic anhydride (Aldrich) can be used for the preparation. In this study, the carbodiimide-method is used as follows. Hexanoic acid (110 mg), EDC (180 mg), and NHS (110 mg) are mixed and incubated for 24 h at room temperature in dimethyl formamide. To the solution, 2.5 ml BGK (at an equal molar amount with the hexanoic acid added) dissolved in 0.1 M phosphate buffer (pH 7.4) is added and reacted for 4 h. The solution is then mixed with an equal volume of ethyl acetate and 1 M HCl (1:1). The ethyl acetate layer is collected and washed with water, 5% aqueous $NaHCO_3$, and then water again. The ethyl acetate layer is passed through a Na_2SO_4 column to dehydrate. The elution is concentrated and purified by HPLC. The separation is done using a Develosil Combi-RP-5 (20 × 100 mm) with 0.1% trifluoroacetic acid (TFA)/CH_3CN (3:2) at a flow rate of 5 ml/min. Bz-Gly-PRL is synthesized from the reaction of BGK with propionic anhydride (WAKO) or propionic acid with EDC/NHS according to the method for Bz-Gly-HEL (2, 10). Bz-Gly-SUL and Bz-Gly-GLL are prepared. Bz-Gly-SUL and Bz-Gly-GLL are prepared from succinic anhydride (WAKO), or glutaric anhydride (WAKO), as previously described *(6)*.

3.1.2. Characterization of Antibodies

The ELISA is done as previously described *(2, 9)*. Briefly, PRL-BSA (HEL-BSA) is coated onto wells and incubated at 4°C overnight. The plate is washed with 0.05% Tween containing PBS (TPBS) and then blocked with 1% Block Ace for 1 h. The plate is washed and the 50 μl sample (competitors) is added to the well. Then, 50 μl of anti-PRL(HEL) antibody (typically 0.5 μg/ml) is added. After incubation for 2 h at 37°C, the plate is washed by TPBS, and anti-mouse immunoglobulin antibody labeled with peroxidase (1/5,000) is added and reacted for 1 h at 37°C.

The plate is washed with TPBS and 3,3′-5,5′-tetramethylbentidine reagent (TMB, KPL) [or *o*-phenylenediamine (OPD)] with hydrogen peroxide is added to the well. The color development is terminated by the addition of 1 N phosphoric acid for TMB or 2 N H_2SO_4 for OPD. The absorbance at 450 nm for TMB or 490 nm for OPD is measured using a microplate reader.

3.1.3. Immunohistochemical Staining of Antibodies

The staining method has been described in a previous report *(3)*. In brief, the specimens are fixed in formalin for histological examinations. After deparaffinization, the sections are blocked in horse serum for 30 min. After washing with PBS three times (5 min × 3 times), the slide glass is treated with 20 μg/ml antibody at room temperature for 1 h. As a control, PBS is used instead of the antibody. For absorption experiments, the antibody is preincubated with 200 μg/ml antigen (PRL-BSA or HEL-BSA) and then used as a primary antibody. After the washing step, biotinylated anti-mouse IgG antibody is applied to the slide glass for 30 min at room temperature. After treatment with alkaline phosphatase (AP)-conjugated avidin for 30 min, Vector Red, a substrate for AP, is applied to the glass slides for 10 min. The slides are then washed with running water, and methyl-green is added as a counter stain for 30 s. After washing with deionized water and dehydration, the glass slides are encapsulated with Canada balsam. The immunoreactive staining is observed using a microscope.

3.2. LC/MS/MS Analysis

3.2.1. Preparation of Standard

HEL and the IS, deuterated HEL (HEL-D), are prepared as follows *(7)*. Nα-Boc-Lys (0.1 g) is dissolved in 1.6 ml of 0.1 M phosphate buffer (pH 7.4)/ethanol (1:3) and hexanoic anhydride (94 μl) is then added. The formed Nα-Boc-HEL is then purified by reversed phase HPLC and reacted with TFA to remove the Boc moiety. To prepare the deuterated HEL (denoted as HEL-D), another approach is used because deuterated hexanoyl anhydride is not commercially available. Briefly, the Nα-Boc-lysine is reacted with hexanoic-d11 acid (Aldrich) in the presence of EDC and NHS, as previously described *(7)*. Purification of resulting Ne-[d11] hexanoylated Nα-Boc-Lys is done as described earlier. HEL-D is then prepared by treating the obtained deuterated Boc-L-HEL with TFA to remove the Boc moiety, as described earlier. The concentration of the purified HELs is determined using the TNBS method, as previously described *(2)*. Similarly, PRL is synthesized from Nα-Boc-Lys with propionic anhydride. As an IS for LC/MS/MS, PRL-D is prepared as follows. Briefly, propionic 2,2-d2 acid (Aldrich), EDC, and sulfo-NHS are mixed in dimethyl formamide. After incubation for 24 h at room temperature, Boc-Lys in ethanol is added and the mixture was further reacted for 4 h. After HPLC purification, the sample containing deuterated Boc-PRL is dried in a glass flask. To remove the Boc moiety, an aliquot of TFA is added to the flask and then evaporated. The free PRL-D is further purified by HPLC according to the method for PRL purification.

3.2.2. Sample Preparation

Human urine is frozen at –70°C until use. After centrifugation to remove insoluble materials in urine, the mixture of deuterated PRL and HEL (PRL-D and HEL-D) is added as an IS (final concentration, 50 nM). CH_3CN (900 μl) is then added to 100 μl of human urine and kept for 1 h below 4°C. The urine is then centrifuged (14,000 ×g, 4°C, 10 min) and the supernatant is concentrated. The samples are reconstituted with 0.01 M hydrochloride (HCl) and applied to a Spelclean SCX column (500 mg, 3 ml volume), which have been preconditioned with 2 ml methanol, 2 ml of methanol plus 0.1 M HCl (1:1), and 2 ml of 0.01 M HCl. After applying the sample, 2 ml of 0.01 M HCl is added to wash the column. The elution is then done with 2.25-ml 1-M aqueous ammonium solution. The samples obtained are evaporated and then reconstituted with 100 μl of 2 mM ammonium formate, and 10 μl samples are applied twice to the API3000 mass spectrometer connected to an HPLC.

3.2.3. Measurements of Adducts in Urine

Separation by HPLC is done using an Agilent 1100 with a Develosil ODS-SR-5 (2 × 150 mm) column. Solvent A is 2 mM ammonium formate and solvent B is CH_3CN. The elution is performed at a flow rate of 0.2 ml/min. The gradient program is as follows: initial (A 100%), 10 min (A90%), 15 min (A80%), 16 min (A50%), 17 min (A50%), 18 min (A100%), and 28 min (A100%). A switching valve is used to prevent contamination of the ion source. The valve is operated as follows: initial (closed), 4.5 min (open), 10 min (closed), 16 min (open), and 22 min (closed). Under these conditions, PRL and HEL are eluted at 7 and 19 min, respectively. The combinations of multiple-reaction monitoring (MRM) are shown in **Table 1**. The multiple combination of MRM for one adduct contributes to the reliable estimation of amide adducts.

Table 1
Combination of MRM for estimation of adducts

Adducts	Sample/Standard		Internal standard	
	Q1	Q3	Q1	Q3
HEL	245.2	182.2	256.2	193.4
	245.2	84.1	256.2	85.1
PRL	203.0	140.4	205.0	142.1
	203.0	130.3	205.0	130.0
	203.0	84.1	205.0	85.1

3.3. Results

3.3.1. Immunochemical Analysis

HEL has often been used as a biomarkers *(7, 11, 12, 13)*, probably because an antibody to HEL and an ELISA kit are commercially available. On the contrary, PRL is a novel biomarker, and therefore, it is currently more difficult to measure PRL with simple techniques. To improve this situation, we have prepared a monoclonal antibody to PRL. The serum from PRL-KLH-immunized mice showed not only strong reactivity to PRL-BSA but also weak reactivity to HEL-BSA (**Fig. 2**). However, the obtained antibody specifically recognized PRL but not other amide adducts such as HEL, SUL, and GLL (**Fig. 3**). Positive staining of PRL in rabbit atherosclerotic plaques has been observed *(10)*. We have also observed positive staining of PRL, but not HEL, in the hippocampus of old rat brain (unpublished observations). This may means that n-3 FAs are specifically oxidized in the brain during aging. In this manner, PRL staining reveals the presence of oxidative stress in specific loci but may also indicate the specific oxidation of n-3 FA in comparison to HEL staining, which is a marker of n-6 FA-derived modification.

3.3.2. LC/MS/MS Analysis

Whereas immunochemical approaches have many advantages compared with other methods, it is hard to estimate the actual amount of target molecules using immunochemical techniques.

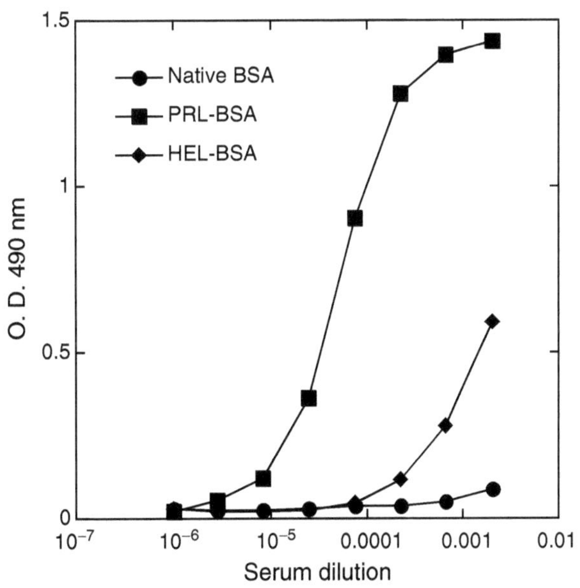

Fig. 2. Native BSA, PRL-BSA, and HEL-BSA were coated onto wells overnight at 4°C. After blocking the plate, the diluted serum was added to the well. After incubation, the plate was washed and anti-mouse IgG antibody-peroxidase labeled was added. The color development was done as described in the text. The antibodies in the serum predominantly reacted with PRL-BSA. However, some antibodies in the serum also appeared to react with HEL-BSA.

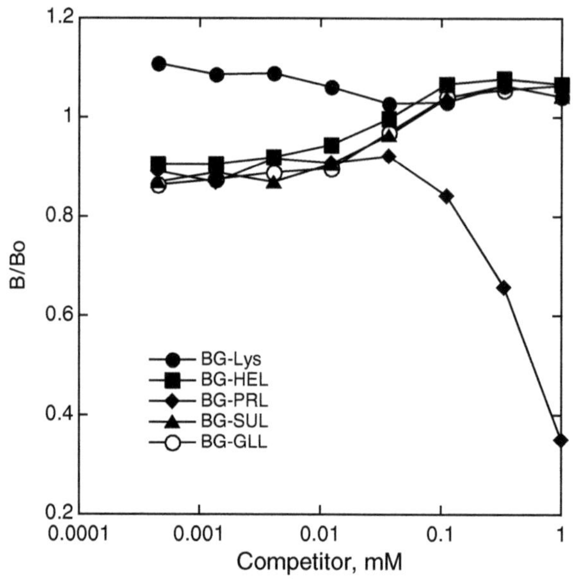

Fig. 3. Competitive ELISA for peptidyl amide-type adducts using the PRL monoclonal antibody. PRL-BSA was coated onto the wells of a microplate. Competitors and the PRL antibody were then added to the wells. After incubation, the plate was washed and anti-mouse IgG peroxidase-labeled antibody was added. The color development was done as described in the text. The antibody specifically recognized PRL but not the other amide adducts. *BG-Lys*, benzoyl-glycyl-lysine; *BG-HEL*, benzoyl-glycyl-HEL; *BG-PRL*, benzoyl-glycyl-PRL; *BG-SUL*, benzoyl-glycyl-SUL; *BG-GLL*, benzoyl-glycyl-GLL.

Chemical approaches can be used to specifically quantify and identify the target molecule. In particular, LC/MS/MS is suitable for the detection of hydrophilic biological molecules at very low levels. We have simultaneously determined urinary PRL as well as HEL using LC/MS/MS with isotope-dilution methods (**Fig. 4**). To achieve high sensitivity, multiple reaction monitoring (MRM) was used for the detection of both PRL and HEL. The relationship between urinary HEL and PRL from healthy people (n = 15) was very high (**Fig. 5**).

4. Notes

1. For HEL and PRL detection by LC/MS/MS, derivatization such as butylation can be used. The butylation method has already been published (*10*). It is noteworthy that the butylation reaction may erode the metals of a centrifugal evaporator, and therefore, an instrument pretreated with antierosion coating (such as Teflon) should be used for the handling/butylation of the samples.

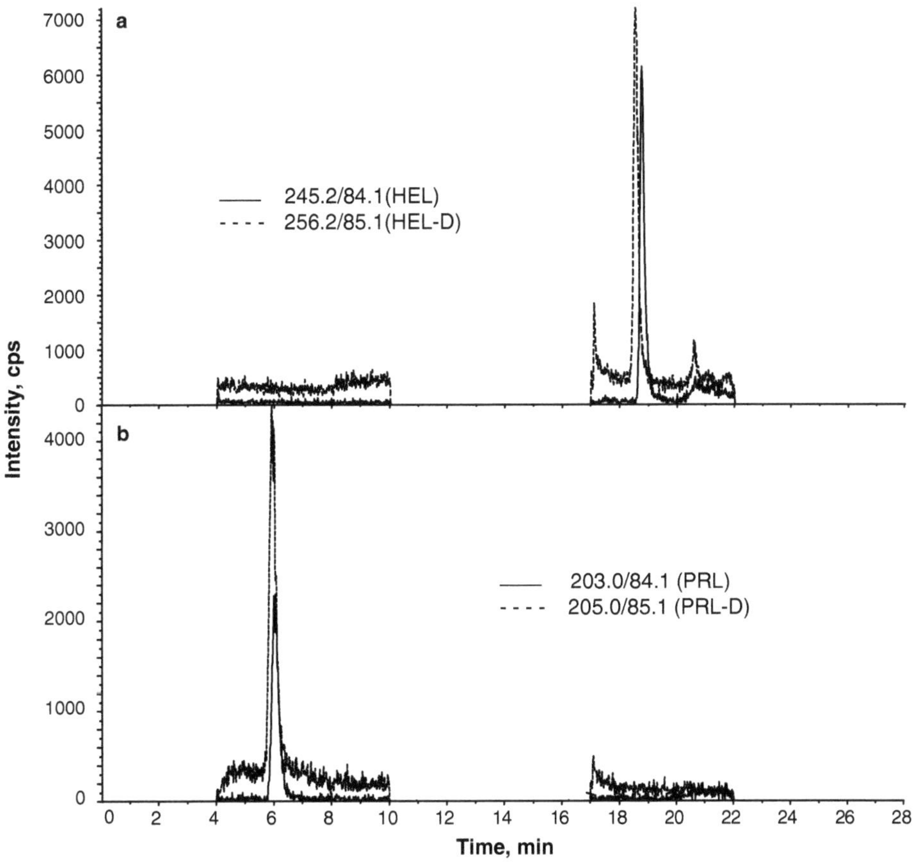

Fig. 4. MRM charts for urinary HEL and PRL along with their internal standards. The *upper panel* (**a**) shows the MRM extracts for urinary HEL and *lower panel* (**b**) shows those for urinary PRL. Both MRM scans were done simultaneously. *Solid lines* show the MRM chart for HEL and PRL, respectively. *Dashed lines* show stable internal standards containing deuterium. A switching valve was used to selectively incorporate the sample into the MS. The valve was opened twice to introduce the samples to the MS from 4 to 10 min for PRL and from 17 to 22 min for HEL.

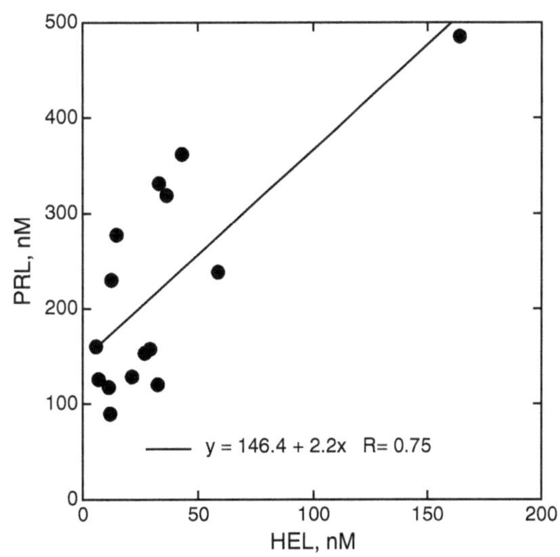

Fig. 5. Correlation between urinary PRL and HEL from healthy people ($n = 15$).

2. Oasis MCX (Waters) may be used instead of the SCX column. The MCX column has dual functions of reversed-phase extraction and cation exchange.
3. To obtain specific monoclonal antibodies, the screening of hybridomas is a critical step because many positive clones can be obtained. For example, during the screening process of the PRL antibody, one should select the clones not only based on those with the high titer against PRL-BSA but also by considering the specificity by comparing other antigens such as HEL-BSA.
4. The amide-type adducts in vivo may be derived from an enzymatic reaction. For example, PRL might be generated by *N*-acetyl transferase enzymes *(13)*. Therefore, the origins of amide adduct must be discussed carefully.

Acknowledgments

We would like to thank Shinsuke Hisaka for his helpful technical assistance.

References

1. Esterbauer, H., Schaur, R. J., and Zollner, H. (1991) Chemistry and biochemistry of 4-hydroxynonenal, malonaldehyde and related aldehydes. *Free Rad. Biol. Med.* **11**, 81–128.
2. Kato, Y., Mori, Y., Makino, Y., Morimitsu, Y., Hiroi, S., Ishikawa, T., and Osawa, T. (1999) Formation of N(epsilon)-(hexanonyl)lysine in protein exposed to lipid hydroperoxide. A plausible marker for lipid hydroperoxide-derived protein modification. *J. Biol. Chem.* **274**, 20406–20414.
3. Fukuchi, Y., Miura, Y., Nabeno, Y., Kato, Y., Osawa, T., and Naito, M. (2008) Immunohistochemical detection of oxidative stress biomarkers, dityrosine and N(epsilon)-(hexanoyl)lysine, and C-reactive protein in rabbit atherosclerotic lesions. *J. Atheroscler. Thromb.* **15**, 185–192.
4. Kawai, Y., Kato, Y., Fujii, H., Makino, Y., Mori, Y., Naito, M., and Osawa, T. (2003) Immunochemical detection of a novel lysine adduct using an antibody to linoleic acid hydroperoxide-modified protein. *J. Lipid Res.* **44**, 1124–1131.
5. Kawai, Y., Fujii, H., Kato, Y., Kodama, M., Naito, M., Uchida, K., and Osawa, T. (2004) Esterified lipid hydroperoxide-derived modification of protein: formation of a carboxyalkylamide-type lysine adduct in human atherosclerotic lesions. *Biochem. Biophys. Res. Commun.* **313**, 271–276.
6. Kawai, Y., Fujii, H., Okada, M., Tsuchie, Y., Uchida, K., and Osawa, T. (2006) Formation of N(epsilon)-(succinyl)lysine in vivo: a novel marker for docosahexaenoic acid-derived protein modification. *J. Lipid Res.* **47**, 1386–1398.
7. Kato, Y., Yoshida, A., Naito, M., Kawai, Y., Tsuji, K., Kitamura, M., Kitamoto, N., Osawa, T. (2004) Identification and quantification of N(epsilon)-(hexanoyl)lysine in human urine by liquid chromatography/tandem mass spectrometry. *Free Radic. Biol. Med.* **37**, 1864–1874.
8. Kato, Y., Miyake, Y., Yamamoto, K., Shimomura, Y., Ochi, H., Mori, Y., and Osawa, T. (2000) Preparation of a monoclonal antibody to N(epsilon)-(hexanonyl)lysine: application to the evaluation of protective effects of flavonoid supplementation against exercise-induced oxidative stress in rat skeletal muscle. *Biochem. Biophys. Res. Commun.* **274**, 389–393.
9. Kato, Y. and Osawa, T. (2002) Detection of lipid hydroperoxide-derived protein modification with polyclonal antibodies. In: *Oxidative*

Stress Biomarkers and Antioxidant Protocols, (Donald Armstrong, ed.), Humana Press, pp. 37–44.

10. Hisaka, S., Kato, Y., Kitamoto, N., Yoshida, A., Kubushiro, Y., Naito, M., and Osawa, T. (2009) Chemical and Immunochemical identification of propanoyllysine derived from oxidized n-3 polyunsaturated fatty acid. *Free Rad. Biol. Med.* **46**, 1463–1471.
11. Shimizu, K., Ogawa, F., Akiyama, Y., Muroi, E., Yoshizaki, A., Iwata, Y., Komura, K., Bae, S., and Sato, S. (2008) Increased serum levels of N(epsilon)-(hexanoyl)lysine, A new marker of oxidative stress, in systemic sclerosis. *J. Rheumatol.* **35**, 2214–2219.
12. Minato, K., Gono, M., Yamaguchi, H., Kato, Y., and Osawa, T. (2005) Accumulation of N(epsilon)-(hexanoyl)lysine, an oxidative stress biomarker, in rice seeds during storage. *Biosci. Biotechnol. Biochem.* **69**, 1806–1810.
13. Kageyama, Y., Takahashi, M., Nagafusa, T., Torikai, E., and Nagano, A. (2008) Etanercept reduces the oxidative stress marker levels in patients with rheumatoid arthritis. *Rheumatol. Int.* **28**, 245–251.
14. Garrity, J., Gardner, J. G., Hawse, W., Wolberger, C., Escalante-Semerena, J. C. (2007) N-Lysine propionylation controls the activity of propionyl-CoA synthetase. *J. Biol. Chem.* **282**, 30239–30245.

Chapter 8

Assessing the Neuroprotective Effect of Antioxidative Food Factors by Application of Lipid-Derived Dopamine Modification Adducts

Xuebo Liu, Naruomi Yamada, and Toshihiko Osawa

Summary

Advances in understanding the neurodegenerative pathologies are creating new opportunities for the development of neuroprotective therapies, such as antioxidant food factors, lifestyle modification, and drugs. However, the biomarker by which to determine the effect of the agent on neurodegeneration is limited. We here address hexanoyl dopamine (HED), one of novel dopamine adducts derived from brain polyunsaturated acid, referring to its in vitro formation, potent toxicity to SH-SY5Y cells, and application to assess the neuroprotective effect of antioxidative food factors. Dopamine is a neurotransmitter and its deficiency is a characterized feature in Parkinson's disease (PD), thereby HED represents a new addition to understanding of dopamine biology and pathophysiology of PD and a novel biomarker for the assessment of neuroprotective therapies. We have established an analytical system using for the detection of HED and its toxicity to the neuroblstoma cell line, SH-SY5Y cells. Here, we discuss the characteristics of the system and its applications to investigate the neuroprotective effect of several antioxidants that originate from food.

Key words: HED, Parkinson's disease, Biomarker, Food factors, Neuroprotective effect

1. Introduction

Increasing evidence suggests that oxidative stress play a crucial role in the majority of neurodegenerative diseases. Parkinson's disease (PD) is a neurodegenerative disorder characterized by a dramatic loss of dopaminergic neurons in the substantia nigra and the subsequent deficiency of dopamine in the brain areas *(1)*. Until now, very little is known about why and how the

Donald Armstrong (ed.), *Lipidomics,* Methods in Molecular Biology, vol. 580,
DOI 10.1007/978-1-60761-325-1_8, © Humana Press, a part of Springer Science+Business Media, LLC 2009

PD neurodegenerative process begins and progresses; however, recent studies indicate that there are high levels of basal oxidative stress in the substantia nigra pars compacta (SNc) in the normal brain and this is increased in PD *(2)*.

Oxidative stress in the brain easily leads to the lipid peroxidation reaction due to a high concentration of polyunsaturated fatty acids, such as docosahexaenoic acid (DHA, C22:6/ω-3) and arachidonic acid (AA, C18:4/ω-6), which are present in the brain *(3)*. We have recently found that lipid hydroperoxides, the primary peroxidative products, can universally react with primary amino groups to form *N*-acyl-type (amide-linkage) adducts *(4–9)*. We have previously described the in vitro and in vivo formation of DHA-derived adducts, N^{ε}-(succinyl) lysine and N^{ε}-(propanoyl) lysine, by LC-MS/MS or immunochemical analysis. In addition, during the reaction of oxidized AA with the lysine residue, the formation of N^{ε}-(hexanoyl) lysine and N^{ε}-(glutaryl) lysine were also detected. The *N*-acyl-type adducts are specific to the peroxidation of polyunsaturated fatty acids; therefore, their formations are the useful markers for the lipid peroxidation, protein modification, and related dysfunction that occur in these fatty acids-enriched tissues.

Dopamine is the endogenous neurotransmitter produced by nigral neurons. Dopamine loss can trigger not only prominent secondary morphological changes, such as density reduction of the dendritic spines, but also changes in the density and sensitivity of dopamine receptors *(1)*; therefore, it is a sign of PD development. The reasons for dopamine loss are attributed to dopamine's molecular instability. Some possible causes of dopamine loss are abnormalities of dopaminergic neurons *(10)*, dopamine degradation by monoamine oxidase A (MAO-A) *(11)*, or autoxidation *(12)* and the reaction with amino acid cysteine *(13)*. Dopamine is a member of catecholamine family. The catechol structure contributes to high oxidative activation of dopamine. Additionally, the N-teminals in dopamine's structure may represent another reactive spot; however, little experimental evidence proves this. On the basis of our previously described reaction between lipid hydroperoxides and NH_2 residues, we examined the reaction of dopamine with reactive LOOH species derived from lipid peroxidation.

We particularly report here on hexanoyl dopamine (HED), a dopamine modified adduct derived from AA, referring to its formation, effect on SH-SY5Y cells, and applications to investigate the neuroprotective effect of antioxidant foods, such as tocopherol *(14, 15)*, curcumin *(16, 17)*, sesamin analogs *(18–21)*, and astaxanthin *(22, 23)*. We have demonstrated that HED was present in rat brain (data not shown) and toxic to the neuroblastoma SH-SY5Y cells, thereby representing a novel biomarker for the assessment of neuroprotective therapies against PD.

2. Materials

2.1. Equipment

2.1.1. Reverse Phase HPLC (Nomura Chemical, Inc., Japan)

Column: Develosil ODS-HG-5 column (20 × 250 mm).

2.1.2. HPLC-Tandem Mass Spectrometry (MS/MS)

Spectrometer: API 2,000 triple quadrupole mass spectrometer (Applied Biosystems) through a TurboIonSpray source.

Chromatography: Develosil ODS-HG-3 column (2.0 × 250 mm) with an Agilent 1,100 HPLC system.

2.2. Reagents

AA and lipoxidase were obtained from the Sigma-Aldrich Co. (St. Louis, MO, USA). Dopamine·HCl was purchased from Nacalai Tesque, Inc. (Kyoto, Japan). Hexanoic anhydride was obtained from Wako Pure Chemical Industries, Ltd. (Osaka, Japan). The antibodies against poly(ADP-ribose) polymerase (PARP) was purchased from the Cell Signaling Technology, Inc. (Beverly, MA). Active caspase-3 rabbit monoclonal antibody was purchased from the Epitomics, Inc. (California).

2.3. Standards

HED was prepared in our laboratory. Briefly, HED was chemically synthesized by incubating dopamine (0.5 mM) with hexanoic anhydride (0.5 mM) in 5 ml of 100-mM sodium phosphate buffer (pH 7.4)-saturated sodium acetate (1:1, v/v) for 60 min at room temperature. The synthesized HED were purified by reverse-phase HPLC using a Develosil ODS-HG-5 column (20 × 250 mm) in an isocratic system of 15% or 50% acetonitrile containing 0.1% trifluoroacetic acid at the flow rate of 6 ml/min. The elution profiles were monitored by absorbance at 280 nm. The amino residues in the dopamine adducts were identified by the ninhydrin reaction. The mass, structure, and formula of the synthesized molecule were identified by HPLC-MS, NMR, and ESI-TOF-MS analyses, respectively.

2.4. Cell Cultures

SH-SY5Y human dopaminergic neuroblastoma cells were kindly donated by Dr. Maruyama (National Institute for Longevity Science, Japan). SH-SY5Y cells and NIH-3T3 cells were grown in Cosmedium-001 (Cosmo-Bio, Tokyo, Japan) containing 5% FBS and Dulbecco's modified Eagle's medium (DMEM) containing 10% FBS, and maintained at 37°C in an atmosphere of 5% CO_2 in air.

2.5. Food Factors

Tocopherol analogues (α-tocopherol, β-tocopherol, γ-tocopherol, and δ-tocopherol), tocotrienol (α-tocotrienol, β-tocotrienol, γ-tocotrienol, and δ-tocotrienol), curcumin anaglogues (crucumin and tetrahydrocrucumin), and sesamin analogues (sesamin, sesamolin, sesaminol, sesaminol-6-catechol (SMLC), and sesaminal triglucoside), and astaxanthin were used in this study. All of them were from our laboratory.

3. Methods

3.1. HED Formation

1. HPLC-MS/MS conditions
   The chromatographic separation was performed by a gradient elution as follows: 0–10 min, linear gradient from 0.1% formic acid to 50% aqueous acetonitrile containing 0.1% formic acid; 10–15 min, hold; 15–20 min, linear gradient to 0.1% formic acid; flow rate = 0.2 ml/min. The instrument response was optimized by infusion experiments with the standard compounds using a syringe pump at the flow rate of 5 μl/min. HED formation was determined using electrospray ionization MS/MS in the multiple reaction monitoring mode.
2. In vitro formation
   The level of AA hydroperoxide (15-HPETE) was determined using a lipid hydroperoxide assay kit (Cayman, Michigan). Dopamine (2 mM) was incubated with 10 mM AA-hydroperoxides in phosphate buffer (pH 7.4) at 37°C for different times. The reaction was terminated by immediate freezing at −80°C. The detection was carried out by HPLC-MS/MS. 15-HPETE was prepared as previously described in a reference *(24)*.

3.2. Cellular Toxicity

1. Assessment of cell viability
   Cell viability was evaluated by an MTT assay. SH-SY5Y cells in 96-well plates were incubated with drugs for different times, followed by further incubation with 500 μg/ml MTT at 37°C for 2 h. Cell viability in some experiments was also measured using PI and Hoechst 33,258 staining.
2. ROS measurement
   Endogenous ROS level was detected by flow cytometry using H_2DCF-DA (Molecular Probes). Briefly, the drug-treated cells were incubated with H_2DCF-DA for 30 min, and the fluorescence of dichlorofluorescein (DCF) was measured using an EPICS Elite Flow Cytometer.
3. Identification of apoptosis induction
   Western blot analysis – The cells were washed twice with phosphate-buffered saline, pH 7.0, and lysed with lysis buffer (50 mM Tris-HCl, pH 7.5, 150 mM NaCl, 1% Triton X-100, 0.5% sodium deoxycholate, 0.1% SDS, 100 μg/ml phenylmethylsulfonyl fluoride). After protein quantification, equal amounts of the protein (total protein, 20–50 μg) were boiled with Laemmli sample buffer for 5 min at 100°C. The samples were run on 10% SDS-polyacrylamide gels, transferred to a nitrocellulose membrane, incubated with 5% skim milk in Tris-buffered saline (TTBS containing 10% Tween 20) for blocking, washed, and treated with the primary antibodies. After washing with TTBS, the blots were further incubated

for 1 h at room temperature with the IgG antibody coupled to horseradish peroxidase in TTBS. Blots were then washed three times in TTBS before visualization. An ECL kit was used for detection.

3.3. Assay for HED In Vitro Formation Inhibition by Food Factors

Ten microliters of 1-mM 15-HPETE (in ethanol) and 20 μl of 0.2-mM dopamine (in H_2O, *see* **Note 1**) were added into 60 μl of PB (pH 7.4) with 10 μl of agent (in DMSO). The reaction was performed at 37°C for 24 h, and stopped immediately by putting into −80°C deep freezer. The HED detection was carried out following the protocols described in **Subheading 3.1.2**.

3.4. Statistics

All data were analyzed using Bonferroni/Dunn's multiple comparison procedure.

3.5. Results

Figure 1 shows the proposed chemical formation scheme and HPLC-MS/MS analysis of HED, including structure (**a**), mass (**b**), and formula (**c**). **Figure 2** shows the HED formation.

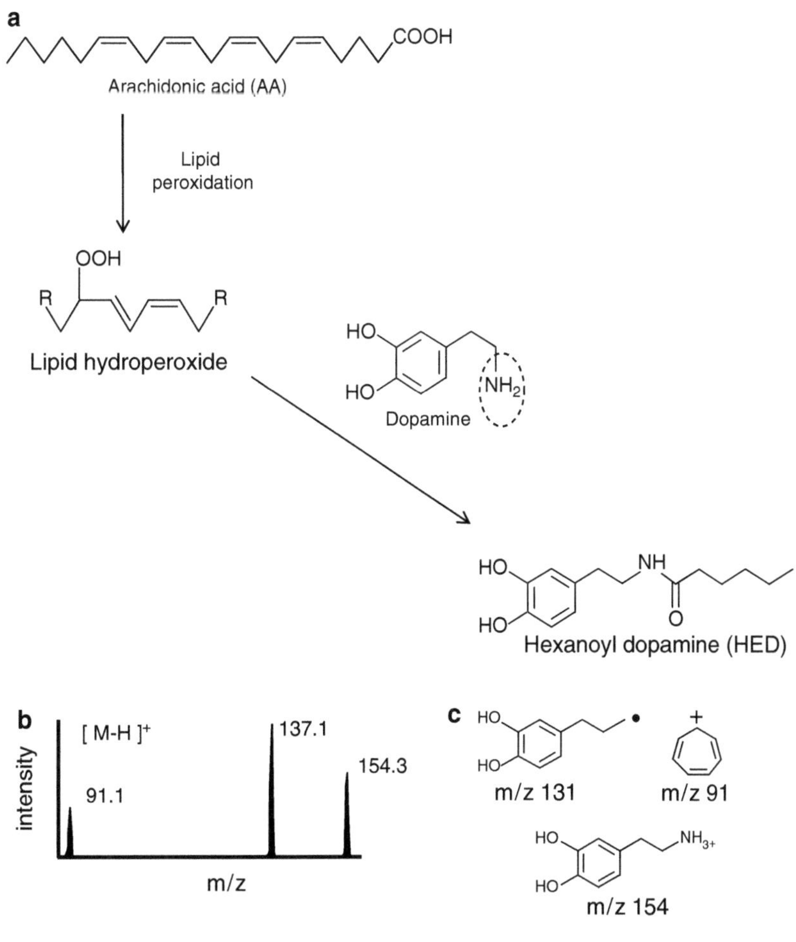

Fig. 1. Proposed chemical formation scheme and HPLC-MS/MS analysis of HED formation. (**a**) Proposed reaction scheme of HED formation. (**b**) The $[MH]^+$ ion *m/z* 252 of HED was subjected to CID, and the daughter ions were scanned. (**c**) The proposed structures of individual ions are shown.

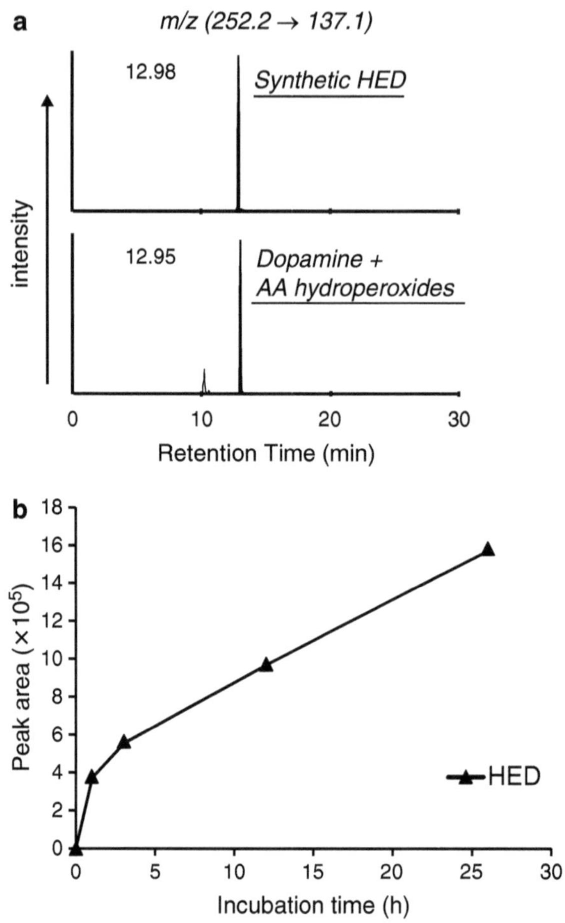

Fig. 2. HPLC-MS/MS analysis of HED formed during the reaction of dopamine with AA hydroperoxides. Dopamine (2 mM) was incubated with AA hydroperoxides (10 mM) in 0.1 M phosphate buffer (pH 7.4) at 37°C. Ion monitoring of HED transition was *m/z* 252. (**a**) Authentic dopamine adduct and reaction mixture of AA hydroperoxides with dopamine. (**b**) HED formation in a time-dependent manner.

Figure 2a shows the in vitro HED detection by LC-MS/MS analysis; **Fig. 2b** shows the dose-dependent formation of HED in the reaction of 15-HPETE with dopamine. We also detected HED formation in vivo (*see* **Note 2**). **Figure 3** shows the effect of HED on SH-SY5Y cellular toxicity. **Figure 3a** shows the effect of HED on cell viability by MTT assay; **Fig. 3b** shows the effect of HED on ROS generation in the cells; **Fig. 3c** shows the effect of HED on apoptosis induction using apoptosis hallmarks including PARP cleavage and accumulation of active caspase-3. In addition, monoamine transporters were needed for HED cytotoxicity (*see* **Note 3**). **Figure 4** shows the chemical structure of antioxidant food factors used in this study. **Figure 5** shows the effect of food factors on the in vitro HED formation. SMLC showed the most significant inhibitive effect on the in vitro HED formation (*see* **Note 4**).

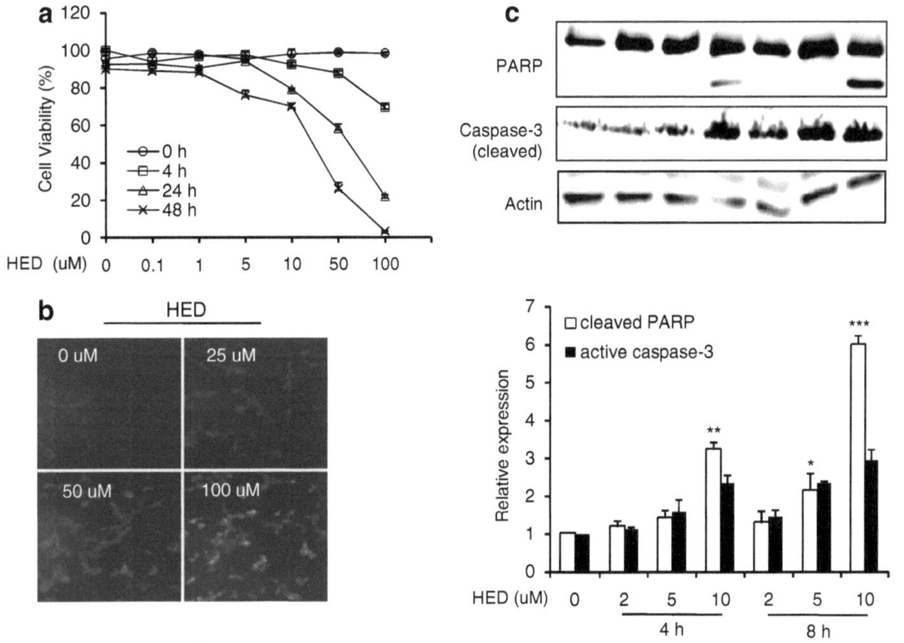

Fig. 3. Effect of HED on SH-SY5Y cellular toxicity. (**a**) Dose- and time-dependent cytotoxicity of HED. SH-SY5Y cells were exposed to 0100 μM HED for different retention times. Cell viability was measured by the MTT assay. (**b**) Dose-dependent ROS generation induced by HED. DCF fluorescence imaging was determined by fluorescence microscope. (**c**) PARP cleavage and active caspase-3 expression in SH-SY5Y cells exposed to 010 μM HED for 4 and 8 h. The cleavage of PARP and expression of active caspase-3 were tested by western blotting and statistically analyzed.

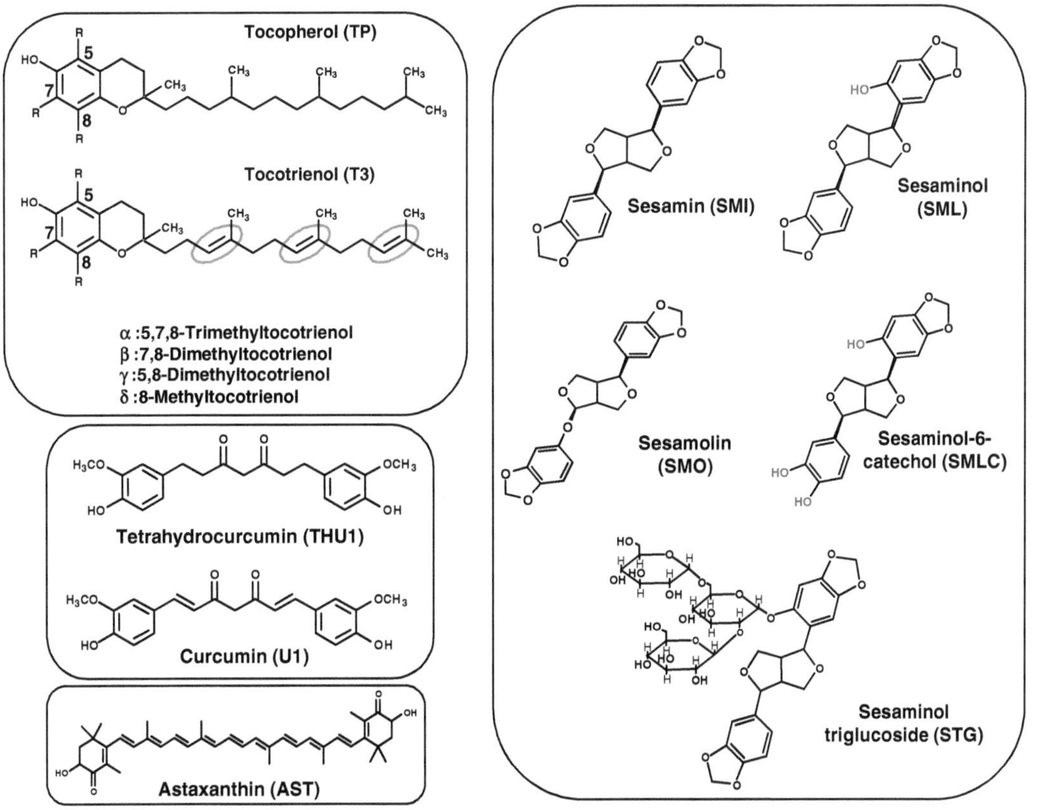

Fig. 4. Chemical structure of antioxidant food factors.

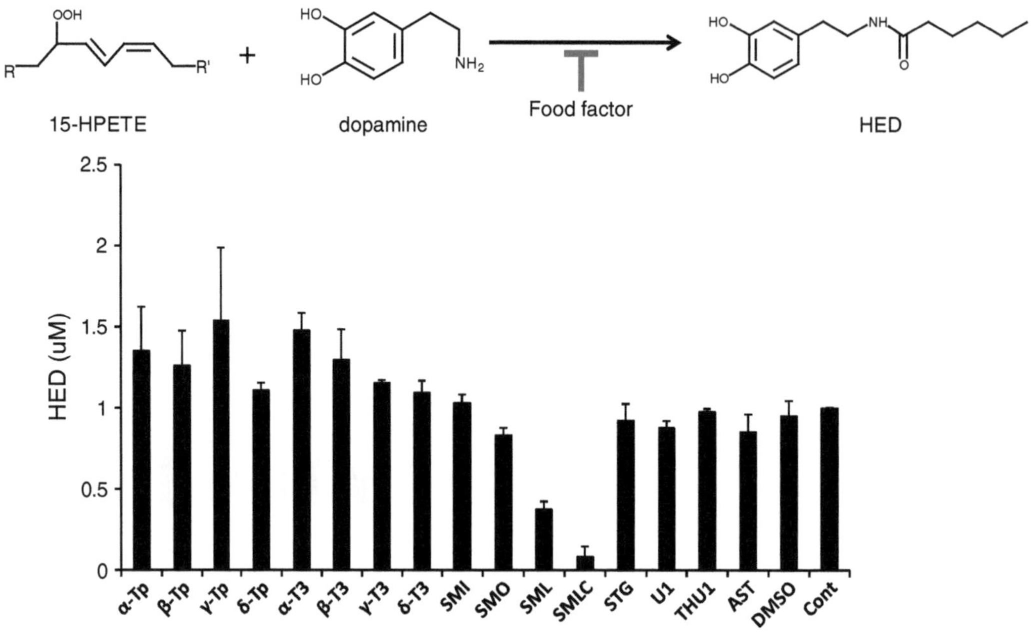

Fig. 5. Effect of food factors on the in vitro HED formation (*see* **Subheading 3.3**).

4. Notes

1. To assess protective effect of antioxidant food factors on HED formation, the method shown in **Subheading 3.3** should be referred. But, it is critical that dopamine solution in H_2O should be prepared on the day because dopamine is easily to be oxidized.
2. We have also demonstrated that HED was present in rat brain (data not shown), in addition to the toxicity to the neuroblastoma SH-SY5Y cells, thereby representing a novel biomarker for the assessment of neuroprotective therapies against PD.
3. HED cytotoxicity has been demonstrated to be selectively to neuronal cells with monoamine transports, such as dopamine transporter, norepinephrine transporter, and 5-HT transporter.
4. In **Fig. 5**, SMLC showed the most significant inhibitive effect on the in vitro HED formation. We also recently found that SMLC markedly prevented HED-induced ROS generation and cell death in SH-SY5Y cells (data not shown). In addition, SMLC was detected in the brain of rats fed with SMLC-contained foods, suggesting that it could cross brain–blood barrier and exhibit the neuroprotective effect in vivo.

References

1. Galvan, A. and Wichmann, T. (2008) Pathophysiology of parkinsonism. *Clin Neurophysiol.* **119**(7), 1459–1474.
2. Jenner, P. (2003) Oxidative stress in Parkinson's disease. *Ann. Neurol.* **53**(Suppl 3), S26–S38.
3. Porter, N.A., Caldwell, S.E., and Mills, K.A. (1995) Mechanisms of free radical oxidation of unsaturated lipids. *Lipids* **30**(4), 277–290.
4. Kato, Y., Makino, Y., and Osawa, T. (1997) Characterization of a specific polyclonal antibody against 13-hydroperoxyoctadecadienoic acid-modified protein: formation of lipid hydroperoxide-modified apoB-100 in oxidized LDL. *J. Lipid Res.* **38**(7), 1334–1346.
5. Kato, Y. and Osawa, T. (1998) Detection of oxidized phospholipid-protein adducts using anti-15-hydroperoxyeicosatetraenoic acid-modified protein antibody: contribution of esterified fatty acid-protein adduct to oxidative modification of LDL. *Arch. Biochem. Biophys.* **351**(1), 106–114.
6. Kato, Y., Mori, Y., Makino, Y., Morimitsu, Y., Hiroi, S., Ishikawa, T., and Osawa, T. (1999) Formation of Nepsilon-(hexanonyl)lysine in protein exposed to lipid hydroperoxide. A plausible marker for lipid hydroperoxide-derived protein modification. *J. Biol. Chem.* **274**(29), 20406–20414.
7. Kawai, Y., Kato, Y., Fujii, H., Mkino, Y., Mori, Y., Naito, N., and Osawa, T. (2003) Immunochemical detection of a novel lysine adduct using an antibody to linoleic acid hydroperoxide-modified protein. *J. Lipid Res.* **44**(6), 1124–1131.
8. Kawai, Y., Fujii, H., Kato, Y., Kodama, M., Naito, N., Uchida, K., and Osawa, T. (2004) Esterified lipid hydroperoxide-derived modification of protein: formation of a carboxyalkylamide-type lysine adduct in human atherosclerotic lesions. *Biochem. Biophys. Res. Commun.* **313**(2), 271–276.
9. Kawai, Y., Fujii, H., Okada, M., Tsuchie, Y., Uchida, K., and Osawa, T. (2006) Formation of Nepsilon-(succinyl)lysine in vivo: a novel marker for docosahexaenoic acid-derived protein modification. *J. Lipid Res.* **47**(7), 1386–1398.
10. Bove, J., Prou, D., Perier, C., and Przedborski, S. (2005) Toxin-induced models of Parkinson's disease. *NeuroRx* **2**(3), 484–494.
11. Gotz, M.E., Kunig, G., Riederer, P., and Youdim, M.B. (1994) Oxidative stress: free radical production in neural degeneration. *Pharmacol. Ther.* **63**(1), 37–122.
12. Hald, A. and Lotharius, J. (2005) Oxidative stress and inflammation in Parkinson's disease: is there a causal link? *Exp. Neurol.* **193**(2), 279–290.
13. LaVoie, M.J. and Hastings, T.G. (1999) Peroxynitrite- and nitrite-induced oxidation of dopamine: implications for nitric oxide in dopaminergic cell loss. *J. Neurochem.* **73**(6), 2546–2554.
14. Yamagishi, M., Osakab, N., Takizawa, T., and Osawa, T. (2001) Cacao liquor polyphenols reduce oxidative stress without maintaining alpha-tocopherol levels in rats fed a vitamin E-deficient diet. *Lipids* **36**(1), 67–71.
15. Atkinson, J., Epand, R.F., and Epand, R.M. (2008) Tocopherols and tocotrienols in membranes: a critical review. *Free Radic Biol Med.* **44** (5), 739–764.
16. Osawa, T. and Kato, Y. (2005) Protective role of antioxidative food factors in oxidative stress caused by hyperglycemia. *Ann. N.Y. Acad. Sci.* **1043**, 440–451.
17. Kitani K., Osawa T., and Yokozawa T. (2007) The effects of tetrahydrocurcumin and green tea polyphenol on the survival of male C57BL/6 mice. *Biogerontology* **8**(5), 567–573.
18. Yokota, T., Matsuzaki, Y., Koyama, M., Hitomi, T., Kawanaka, M., Enoki-Konishi, M., Okuyama, Y., Takayasu, J., Nishino, H., Nishikawa, A., Osawa T, and Sakai T. (2007) Sesamin, a lignan of sesame, down-regulates cyclin D1 protein expression in human tumor cells. *Cancer Sci.* **98**(9), 1447–1453.
19. Sheng HQ, Hirose Y, Hata K, Zheng Q, Kuno T, Asano A, Yamada Y, Hara A, Osawa T, and Mori H. (2007) Modifying effect of dietary sesaminol glucosides on the formation of azoxymethane-induced premalignant lesions of rat colon. *Cancer Lett.* **246**, 63–68.
20. Miyake, Y., Fukumoto, S., Okada, M., Sakaida, K., Nakamura, Y., and Osawa, T. (2005) Antioxidative catechol lignans converted from sesaminol triglucoside by culturing with *Aspergillus*. *J. Agric.Food Chem.* **53**(1), 22–27.
21. Kang, M.-H., Naito, M., Sakai, K., Uchida, K., and Osawa, T. (2000) Mode of action of sesame lignans in protecting low-density lipoprotein against oxidative damage in vitro. *Life Sciences* **66**, 161–171.
22. Liu, X., and Osawa, T. (2007) Cis astaxanthin and especially 9-cis astaxanthin exhibits a higher antioxidant activity in vitro compared to the all-trans isomer. Biochem. *Biophys. Res. Commun.* **357**, 187–193.
23. Aoi, W., Naito, Y., Takanami, Y., Ishii, T., Kawai, Y., Akagiri, S., Kato, Y., Osawa, T., and

Yoshikawa, T. (2008) Astaxanthin improves muscle lipid metabolism in exercise via inhibitory effect of oxidative CPT I modification. Biochem. *Biophys. Res. Comn.* **366**, 892–897.

24. Liu, X. B., Shibata, T., Hisaka, S., and Osawa, T. (2008) DHA hydroperoxides as a potential inducer of neuronal cell death: a mitochondrial dysfunction-mediated pathway. *J. Clin. Biochem. Nutr.* **43**(1), 26–33.

Chapter 9

Mass-Spectrometric Characterization of Phospholipids and Their Hydroperoxide Derivatives In Vivo: Effects of Total Body Irradiation

Yulia Y. Tyurina, Vladimir A. Tyurin, Valentina I. Kapralova, Andrew A. Amoscato, Michael W. Epperly, Joel S. Greenberger, and Valerian E. Kagan

Summary

Combination of electrospray ionization mass spectrometry (ESI-MS), fluorescence high-performance liquid chromatography (HPLC), and 2D-high-performance thin-layer chromatography (2D-HPTLC) is a powerful approach to identify and quantitatively analyze oxidized phospholipids in vivo. We describe application of this methodology in assessments of phospholipid hydroperoxides using as an example their characterization and quantitative determinations in different tissues of mice exposed to total body irradiation (TBI, 10 and 15 Gy). Using ESI-MS, we identified individual molecular species – with particular emphasis on polyunsaturated molecules as preferred peroxidation substrates – in major classes of phospholipids: cardiolipin (CL), phosphatidylcholine (PC), phosphatidylethanolamine (PE), phosphatidylserine (PS), and phosphatidylinositol (PI) isolated from mouse brain, lung, muscles, small intestine, and bone marrow. We show that the pattern of phospholipid oxidation 24 h after TBI is nonrandom and does not follow the phospholipid abundance in tissues. The anionic phospholipids – CL, PS, and PI – are the preferred peroxidation substrates. We identified and structurally characterized individual hydroperoxides in these three classes of phospholipids. The protocols described may be utilized in studies of signaling functions of oxidized phospholipids in cell physiology and pathology.

Key words: Total body irradiation, Oxidative lipidomics, Phospholipid hydroperoxides, 2D-HPLC, HPLC, ESI-MS

1. Introduction

For more than five decades, lipid peroxidation has been viewed as a universal mechanism of tissue damage after acute injury, during chronic disease conditions, and even in the process of

Donald Armstrong (ed.), *Lipidomics*, Methods in Molecular Biology, vol. 580,
DOI 10.1007/978-1-60761-325-1_9,

aging [for review see *(1–3)*]. This perception was based, to a large extent, on relatively nonspecific assessments of lipid peroxidation – mostly its secondary products of aldehydic nature such as malonic dialdehyde and its derivatives *(4–6)*. While these early studies have provided important initial insights into deleterious potential of lipid peroxidation, the methodologies utilized were not sensitive and specific enough to detect, identify, and quantitate individual molecular species of phospholipids involved. The lack of adequate methodologies has hampered appreciation of important signaling functions of phospholipid peroxidation products. With the advent of soft ionization techniques and their utilization in mass-spectrometry, a new era has emerged in analytical biochemistry of lipids and lipid oxidation. This warrants reevaluation of older paradigms relevant to the role of phospholipid peroxidation in cell physiology and pathology. One of recent fascinating illustrations of the new opportunities offered by MS analysis of phospholipid peroxidation products and understanding their significance in cell and tissue damage may be exemplified by studies of the effects of ionizing radiation.

It is a common knowledge that exposure to ionizing radiation results in the formation of free radicals in living systems. The evidence of generation of reactive oxygen species originated from early studies demonstrating the increase in the level of oxidized products after radiation exposure *(7, 8)*. Since the early fifties, lipids have been considered as one of preferred substrates of oxidation by irradiation induced radicals *(9)*. Consequently, lipid peroxidation has been strongly associated with the mechanisms of irradiation injury *(9, 10)*. At high doses, lipid peroxidation has been observed in irradiated model biochemical systems and in cell in vitro *(9)*. In addition, increased contents of (secondary) lipid peroxidation products including thiobarbituric acid-reactive substances and 4-hydroxynonenal were observed in irradiated animals *(6, 11, 12)*. Moreover, accumulation of lipid peroxidation products has been observed in patients after total body irradiation (TBI) *(13, 14)*.

Rapid burst of oxygen radicals, including superoxide and hydroxyl radicals, generated by radiolysis of water *(7, 15)*, is believed to persist for milliseconds and initiate oxidative damage of all intracellular molecules including DNA, proteins, and lipids *(6, 8, 9, 15)*. Although peroxidation of phospholipids yields a plethora of secondary products, the process proceeds via the initial formation of phospholipid hydroperoxides that are the most representative primary peroxidation products *(3, 16)*. However, detailed analysis of oxidation of major classes of phospholipids, their individual molecular species, and relationship between selectivity, and specificity of their oxidation and doses, especially for low-level irradiation, has not been conducted. This is due, to a large extent, the complex nature of lipids and the limitations of tools for analysis. The application of new "soft ionization" techniques such as electrospray ionization (ESI) offers new opportunities in sensitivity and reliability. These technological advancements facilitated the emergence of a new field of research and knowledge – lipidomics and oxidative lipidomics – opening significant opportunities for sensitive quantitative and structural analysis of individual molecular species of phospholipids and their oxidation products and their role in cell and body metabolism. This chapter describes application of this methodology to analysis of phospholipid hydroperoxides in vivo using as an example their characterization and quantitative assessments in different tissues of mice exposed to TBI. Our approach is based on a combination of ESI-MS, fluorescence high-performance liquid chromatography (HPLC), and 2D-high-performance thin-layer chromatography (2D-HPTLC) to identify and analyze phospholipdis and oxidized phospholipids from tissues.

2. Materials

2.1. Total Body Irradiation Procedure and Preparation of Tissue Homogenates

2.1.1. Equipment

1. Shepherd Mark 1 Model 68 cesium irradiator.
2. Tissue terator (Model 985–370, type 2; Biospec products, Inc., Racine, WI).
3. Dounce tissue grinder (Kontes Glass Company, Vernon Hills, IL).

2.1.2. Reagents and Supplies

1. Liquid nitrogen.
2. Ten times phosphate-buffered saline (PBS) (Thermo-Fisher Scientific, Pittsburgh, PA).

2.2. Extraction of Lipids

2.2.1. Equipment

1. Tabletop centrifuge (e.g., CRU-500, GE international Equipment Co., Needham, MA).
2. Vortex (e.g., Daigger Vortex Genie 2, Scientific Industries, Bohemia, NY).
3. 13 × 100-mm Pyrex glass test tubes with screw-cap (Thermo-Fisher Scientific, Pittsburgh, PA).

2.2.2. Reagents and Supplies

1. Chloroform (HPLC grade; Sigma-Aldrich, St. Louis, MO).
2. Methanol (HPLC grade; Sigma-Aldrich, St. Louis, MO).
3. 0.75% KCl (Sigma-Aldrich, St. Louis, MO).

2.3. Determination of Phospholipid Phosphorus

2.3.1. Equipment

1. Fume hood certified to use $HClO_4$.
2. Heating elements to achieve 180–190°C (e.g., Troemner 501 hotplate with aluminum block test tube holders, Thorofare, NJ).

3. Water Bath (e.g., PC-320, Corning, Tonawand, NY).
4. Centrifuge (e.g., CRU-500, GE international Equipment Co., Needham, MA).
5. A spectrophotometer (e.g., Shimadzu UV 160U spectrometer).
6. 12 × 74-mm Borosilicate heavy wall tubes (Thermo-Fisher Scientific, Pittsburgh, PA).

2.3.2. Reagents and Supplies

1. Cell culture disposable glass tubes (Termo-Fisher Scientific, Pittsburgh, PA).
2. Perchloric acid, $HClO_4$ (Aldrich, Milwaukee, WI).
3. MilliQ water.
4. 2.5% Sodium molybdate (ACS grade or higher; Sigma-Aldrich, St. Louis, MO) – water solution.
5. 10% Ascorbic acid (Sigma-Aldrich, St. Louis, MO) – water solution.
6. NaH_2PO_4 (Sigma-Aldrich, St. Louis, MO).

2.4. Separation of Phospholipids by 2D-HPTLC

2.4.1. Equipment

1. Thin-layer chromatography tanks with lids (Thermo-Fisher Scientific, Pittsburgh, PA).
2. HPTLC silica G plates (10 × 10 cm) (Whatman, Schleicher & Schuell, England).
3. Glass microsyringes (from 25 to 100 μl) (Supelco, Sigma-Aldrich, St. Louis, MO).
4. Pasteur pipettes.
5. Iodine chamber.

2.4.2. Reagents and Supplies

1. Diethylene triamine pentaacetic acid (DTPA) (Sigma-Aldrich, St. Louis, MO).
2. Ethylenediaminetetraacetic acid (EDTA) (Sigma-Aldrich, St. Louis, MO).
3. Chloroform (HPLC grade; Sigma-Aldrich, St. Louis, MO).
4. Methanol (HPLC grade; Sigma-Aldrich, St. Louis, MO).
5. Ammonium hydroxide (Sigma-Aldrich, St. Louis, MO).
6. Acetone (HPLC grade; Sigma-Aldrich, St. Louis, MO).
7. Acetic acid (Glacial; Sigma-Aldrich, St. Louis, MO).
8. MilliQ Water.
9. Compressed nitrogen (99.999% purity).

2.5. Assay of Phospholipid Hydrope roxides by HPLC

2.5.1. Equipment

1. Shimadzu SCL-10A vp HPLC system controller.
2. Shimadzu LC-10AT vp liquid chromatograph.
3. Shimadzu SIL-10AD vp Autosampler/Injector.

4. Shimadzu fluorescence detector RF-10Axl.
5. Dell/Pentium®, Intel®, PC.
6. Shimadzu EZstart v. 7.2 chromatography software.
7. A reverse phase C18 column Eclipse XDB-C18, 5 μm, 150 × 4.6 mm (Agilent Technologies, Santa Clara, CA).

2.5.2. Reagents and Supplies

1. Mobile phase: 25 mM KH_2PO_4/K_2HPO_4 (pH 7.0)/methanol (60:40 v/v).
2. Methanol (HPLC grade; Sigma-Aldrich, St. Louis, MO).
3. Phospholipase A_2 (PLA_2) from porcine pancreatic (Sigma-Aldrich, St. Louis, MO).
4. Sodium dodecyl sulfate (electrophoresis purity reagent; Biorad Laboratories, Inc., Hercules, CA).
5. *N*-acetyl-3,7-dihydroxyphenoxazine (Amplex Red) (Molecular Probes, Eugene, OR).
6. Resorufin (Molecular Probes, Eugene, OR).
7. Microperoxidase-11 (Sigma-Aldrich, St. Louis, MO).
8. Ethylene glycol tetraacetic acid (EGTA) (Sigma-Aldrich, St. Louis, MO).
9. Calcium chloride, $CaCl_2$ (Sigma-Aldrich, St. Louis, MO).
10. NaH_2PO_4, Na_2HPO_4 KH_2PO_4 K_2HPO_4 (Sigma-Aldrich, St. Louis, MO).
11. MilliQ Water.
12. Hydrochloric acid, HCl.
13. Butylated hydroxytoluene, BHT.
14. Compressed nitrogen (99.999%).

2.6. ESI-MS

2.6.1. Equipment

1. Finnigan™ LXQ™ quadrupole linear ion-trap mass spectrometer (Thermo-Fisher, San Jose, CA).
2. Dell/Pentium®, Intel®, PC.
3. Xcalibur operating system.

2.6.2 Reagents and Supplies

1. 250 μl syringe for direct infusion (Thermo-Fisher Scientific, Pittsburgh, PA).
2. Phospholipid standards: 1,2-diundecanoyl-*sn*-glycero-3-phos-phocholine–$(C_{11:0})_2$-PC;1,2-diheptadecanoyl-*sn*-glycero-3-phosphoethanolamine – $(C_{17:0})_2$-PE; 1,2-dihep-tadecanoyl-*sn*-glycero-3-[phospho-L-serine] (sodium salt)– $(C_{17:0})_2$-PS; 1,1′,2,2′-tetramyristoyl cardiolipin (ammonium salt) – $(C_{14:0})_4$-CL 1,2-dipalmitoyl-*sn*-glycero-3-phosphoinositol – $(C_{16:0})_2$-PI Avanti Polar lipids, Inc. (Alabaster, AL).

3. Methods

3.1. Exposure of Mice to Total Body Irradiation and Preparation of Tissue Homogenates

Groups of C57BL/6NHsd female mice were irradiated to a dose of 10 and 15 Gy TBI using a Shepherd Mark 1 Model 68 cesium irradiator at a dose rate of 80 cGy/min. The mice were sacrificed 24 h later by CO_2 inhalation. All tissue were isolated and immediately frozen in liquid nitrogen. In the case of small intestine, a 5-cm piece was removed, cut open, washed in PBS to remove fecal material, blotted dry, and frozen in liquid nitrogen. All procedures were preapproved and performed according to the protocols established by the Institutional Animal Care and Use Committee of the University of Pittsburgh.

3.2. Extraction of Lipids from Tissues Is According the Following Protocol

Total lipids are extracted from tissues by Folch procedure *(17)*.

1. Combine tissue homogenates with mixture of chloroform and methanol at the ratio 2:1 v/v (~20 ml chloroform/methanol mixture per 1 g of tissue) (*see* **Note 1**).
2. Add 0.75% KCl (1 ml per 5 ml of chloroform/methanol mixture containing tissue homogenate).
3. Vortex and keep on ice for 1 h to separate phases.
4. Collect lower phase and then evaporate solvent with stream of nitrogen.
5. Dissolve film of lipids by adding 100–250 μl of mixture chloroform with methanol at the ratio 1:1 v/v.

3.3. Determination of Phospholipids in Lipid Extracts

Total lipid phosphorus in total lipid extracts is estimated spectrophotometritally as described by Böttcher et al. *(18)*.

1. Pipet aliquots of lipid extract into test tubes and evaporate solvent to dryness with a stream of nitrogen.
2. Add 70% perchloric acid (125 μl) to each sample, and heat to 175–180°C for 20 min.
3. After cooling, add H_2O (825 μl) to each tube, followed by 2.5% sodium molybdate (125 μl), followed by 10% ascorbic acid (125 μl).
4. Vortex and then heat to 90–100°C for 10 min.
5. After cooling, measure the absorbance of the supernatant at 797 nm. The phosphorus content of the samples is derived from comparison with the standard curve constructed with known amounts of NaH_2PO_4.

3.4. Separation of Phospholipids by 2D-HPTLC

Lipid extracts are separated and analyzed by 2D-HPTLC as described by Rouser et al. *(19)* with a small modifications (*see* **Note 2**).

1. To bind adventitious transition metals from silica, treat plates with methanol containing 1-mM EDTA, 100-μM

DTPA prior to application and separation of phospholipids by 2D-HPTLC.

2. Activate HPTLC plates by heating for 20 min at 120°C to remove all traces of H_2O.
3. Apply lipids (~10–15 μg of phospholipid phosphorus) extract to plate (10 × 10cm). First develop the plates with a solvent system consisting of chloroform:methanol:28% ammonium hydroxide (65:25:5 v/v). After the plates are dried with a forced N_2 blower to remove the solvents, develop the plates in the second dimension with a solvent system consisting of chloroform:acetone:methanol:glacial acetic acid:water (50:20:10:10:5 v/v).
4. Visualize the phospholipids by spraying with water and identify by comparison with authentic phospholipid standards (*see* **Note 2**).
5. Scrape the phospholipid spots into glass tubes and add water (1 ml) and mixture of chloroform with methanol at the ratio of 2:1, v/v and votrex.
6. Keep on ice for 1 h, then collect lower phase. Evaporate solvent and dissolve film of lipids in mixture of chloroform:methanol at a ratio of 1:1, v/v.

3.5. Determination of Phospholipids in Individual Phospholipids

1. Pipet aliquots of individual phospholipid classes extracted from silica into test tubes and evaporate solvent to dryness with a stream of nitrogen and estimate the amount of phospholipids as described in **Subheading 3.3**.

3.6. Assay of Phospholipid Hydroperoxides by HPLC

Lipid hydroperoxides are determined by fluorescence HPLC of resorufin formed in peroxidase-catalyzed reduction of specific phospholipid hydroperoxides (PL-OOH) with Amplex Red as previously described *(20–22)* (*see* **Note 3**).

1. Hydrolyze phospholipids by porcine pancreatic PLA_2 (0.1 U/μl) in 25-mM phosphate buffer containing 1.0-mM Ca, 0.5-mM EGTA, and 0.5-mM sodium dodecyl sulfate (pH 8.0 at RT for 30 min).
2. Add Amplex Red (50 μM) and MP-11 (1.0 μg/μl) to hydrolyzed lipids, and incubate the samples at 4°C for 40 min. Start the reaction by adding 1 μl of solution MP-11 (1.0 μg/μl) and terminate by a stop reagent (100 μl of solution of 10-mM HCl, 4-mM BHT in ethanol).
3. Centrifuge the samples at 10,000 × *g* for 5 min and use the supernatant for HPLC analysis. Inject aliquots (5 μl) into a C-18 reverse phase column (Eclipse XDB-C18, 5 μm, 150 × 4.6 mm) and elute using a mobile phase composed of 25-mM KH_2PO_4 (pH 7.0)/methanol (60:40 v/v) at a flow rate of 1 ml/min. Measure the resorufin fluorescence at 590 nm after excitation at 560 nm using a Shimadzu LC-100AT vp

HPLC system equipped with fluorescence detector (model RF-10Axl) and autosampler (model SIL-10AD vp).

4. The phospholipid hydroperoxdes content of the samples is derived from comparison with the standard curve constructed with the known amounts of resorufin (from 0.06 to 1.3 nmol per injection).

3.7. Identification of Phospholipid Molecular Species and Their Oxidative Products by ESI-MS

Phospholipid hydroperoxides were characterized by ESI-MS *(21, 23)* using LXQ™ with the Xcalibur operating system. MS^n analysis was used for identification of molecular species of phospholipids and their polyunsaturated fatty acid residues (*see* **Notes 4** and **5**).

1. Evaporate organic solvent in samples of individual phospholipids separated by 2D-HPTLC under N_2.
2. Resuspend film of individual phospholipids in chloroform: methanol 1:1 v/v (20 pmol/μl). Samples are ready for ESI-MS analysis by direct infusion. Directly inject samples for acquisition of ESI mass spectra at a flow rate of 5 μl/min.
3. Operate the electrospray probe at a voltage differential of –3.5 to 5.0 kV in the negative or positive-ion modes.
4. Maintain the source temperature at 70–150°C.
5. Before analysis of samples, create a tune file for each class of phospholipids. To achieve this, use the following standards: $(C_{14:0})_4$-CL, $(C_{17:0})_2$-PS, $(C_{17:0})_2$-PE; $(C_{11:0})_2$-PC, $(C_{16:0})_2$-PI. Standards can be obtained from Avanti Polar Lipids.
6. Perform MS^1 analysis using isolation width of 1 m/z, 5 microscans with maximum injection time 200 ms. For MS^2 analysis, maximum injection time is 1.000 ms. Two ion activation techniques can be used for MS^2 analysis: One is a collision-induced dissociation (CID, Q = 0.25, low mass cut-off at 28% of the precursor m/z). Another one is pulsed-Q dissociation technique (PQD), with Q = 0.7, and no low mass cut-off for analysis of low molecular weight fragment ions.
7. Employ full range zoom (200–1800 m/z) in negative or positive-ion modes for MS analysis of individual PL species.
8. Perform isotopic corrections by entering the chemical composition of each species into the Qual browser of Xcalibur (operating system) and using the simulation of the isotopic distribution to make adjustments for the major peaks.

3.8. Results

3.8.1. Separation of Phospholipids by 2D-HPTLC

Five different tissues bone marrow, brain, lung, small intestine, and muscles isolated from control C57BL/6NHsd female mice and mice exposed to TBI at a dose of 10 and 15 Gy were used in the study. Analysis of phospholipids was started with their extraction from mouse tissues. Homogenates were prepared and lipids were extracted by Folch procedure *(17)*. See also **Subheadings 2.1** and **2.2**. Total

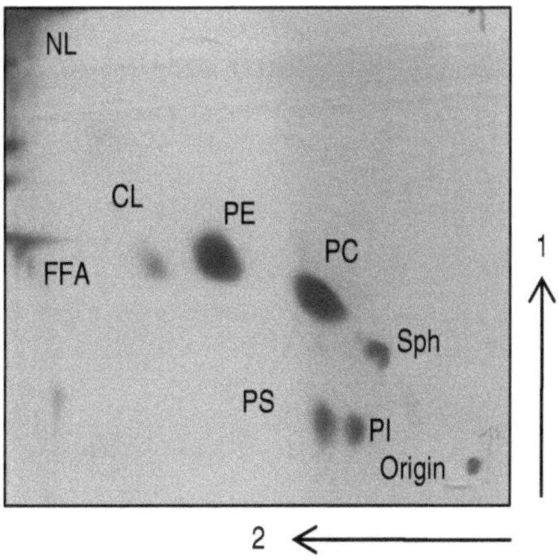

Fig. 1. Typical 2D-HPTLC of total phospholipids extracted from mouse bone marrow. Phospholipids were extracted using Folch procedure and separated by 2D-HPTLC. *NL*, neutral lipids; *FFA*, free fatty acids, *CL*, cardiolipin, *PE*, phosphatidylethanolamine; *PC*, phosphatidylcholine; *PS*, phosphatidylserine, *PI*, phosphatidylinositol; *Sph*, sphingomyelin.

lipid extracts were separated by 2D-HPTLC. **Fig. 1** shows a typical HPTLC chromatogram of total lipids extracted from mouse bone marrow. Six distinct phospholipid spots were detected on the HPTLC plate. The spots were identified as cardiolipin (CL), phosphatidylinositol (PI), phosphatidylethanolamine (PE), phosphatidylserine (PS), phosphatidylcholine (PC), and sphingomyelin (SPH) by comparison with the Rf values measured for authentic standards. Similar HPTLC pattern of phospholipids was observed for lipids extracted from brain, lung, small intestine, and muscles (data not shown). Measurements of lipid phosphorus in samples collected after 2D-HPTLC separation was performed as described by Böttcher et al. *(18)*. Note that TBI caused no changes in phospholipid composition within the tissues (**Table 1**).

3.8.2. Direct Infusion of Phospholipids into Mass Spectrometer

Samples separated by 2D-HPTLC are evaporated under N_2, resuspended in chloroform:methanol 1:1 v/v (20 pmol/μl), and directly utilized for acquisition of negative-ion or positive-ion ESI mass spectra at a flow rate of 5 μl/min (*see* **Subheading 3.7**). Direct infusion was also used in MS^n analysis for identification of molecular species containing highly susceptible to oxidation polyunsaturated fatty acid residues and confirmation of phospholipid structure. Two ion activation techniques were used for MS analysis: CID and PQD *(24)*. Major phospholipids extracted from all five different tissues were analyzed. A typical negative ESI-MS of phospholipids are presented in **Figs. 2–4**.

Table 1
Phosphholipid composition of different tissue of mice exposed to total body irradiation (*TBI*)

Tissue/Treatment (% of total phospholipids)	CL	PS	PI	PC	PE	SPH
Bone marrow						
Control	2.9 ± 0.3	7.9 ± 2.1	8.4 ± 1.3	42.9 ± 5.2	27.0 ± 4.1	10.9 ± 1.3
TBI, 10 Gy	1.2 ± 0.1	9.1 ± 3.0	6.5 ± 1.0	45.4 ± 5.0	28.3 ± 3.4	9.5 ± 1.0
TBI, 15 Gy	1.5 ± 0.2	4.7 ± 2.9	7.5 ± 0.8	48.8 ± 5.9	24.4 ± 3.7	13.1 ± 2.5
Small intestine						
Control	4.7 ± 0.2	7.1 ± 0.4	7.9 ± 0.9	41.2 ± 3.6	30.2 ± 0.5	7.2 ± 1.0
TBI, 10 Gy	4.7 ± 0.9	7.2 ± 1.3	8.4 ± 0.9	42.6 ± 4.3	29.6 ± 0.7	6.8 ± 0.9
TBI, 15 Gy	5.0 ± 0.5	8.1 ± 1.0	8.1 ± 1.0	45.3 ± 0.4	28.6 ± 0.7	4.9 ± 0.9
Lung						
Control	1.1 ± 0.2	6.9 ± 0.2	3.0 ± 0.3	57.7 ± 1.4	23.4 ± 0.8	8.0 ± 0.7
TBI, 10 Gy	1.5 ± 0.6	7.5 ± 0.6	3.9 ± 0.9	55.1 ± 4.0	24.1 ± 2.4	7.8 ± 0.2
TBI, 15 Gy	1.2 ± 0.4	7.3 ± 0.8	3.9 ± 0.9	55.4 ± 3.9	24.0 ± 2.2	7.5 ± 0.3
Brain						
Control	2.1 ± 0.2	12.5 ± 0.6	3.6 ± 0.4	38.2 ± 0.7	39.4 ± 0.5	4.5 ± 0.2
TBI, 10 Gy	2.4 ± 0.3	13.0 ± 0.8	3.4 ± 0.3	38.4 ± 0.5	38.7 ± 0.5	4.4 ± 0.5
TBI, 15 Gy	2.1 ± 0.2	12.2 ± 0.3	4.5 ± 1.0	38.1 ± 0.5	39.6 ± 0.5	4.4 ± 0.6
Muscles						
Control	4.5 ± 1.2	4.1 ± 0.8	6.1 ± 0.5	54.3 ± 2.1	26.6 ± 1.3	4.3 ± 0.9
TBI, 10 Gy	4.3 ± 0.4	4.6 ± 1.0	6.3 ± 0.3	53.6 ± 1.8	26.5 ± 1.8	4.6 ± 0.2
TBI, 15 Gy	4.4 ± 0.2	5.2 ± 0.2	6.7 ± 0.9	53.0 ± 0.3	25.8 ± 1.3	4.9 ± 0.4

C57BL/6NHsd female mice were subjected to 0, 10, and 15 Gy of TBI and sacrificed 24 h thereafter. Lipids from tissues were extracted and separated by 2D-HPTLC. Spots of phospholipids were scraped, phospholipids were extracted from silica and lipid phosphorus was determined. All data are mean ± SD, $n = 3$–4

3.8.3. Identification of PS Molecular Species

ESI-MS analysis of PS isolated from mouse muscles revealed two major molecular clusters with m/z 788.4 and 834.4 (**Fig. 2a**, panel **a**). MS^2 fragmentation of PS species results in the formation of typical product ions characteristic of glycerophospholipids. MS^2 spectra of the major molecular ions $[M-H]^-$ at m/z 788.4 and 834.4 are shown on **Fig. 2a**, panels **b** and **c**, respectively. The loss of serine group of PS yielded the fragments with m/z 701.2 and

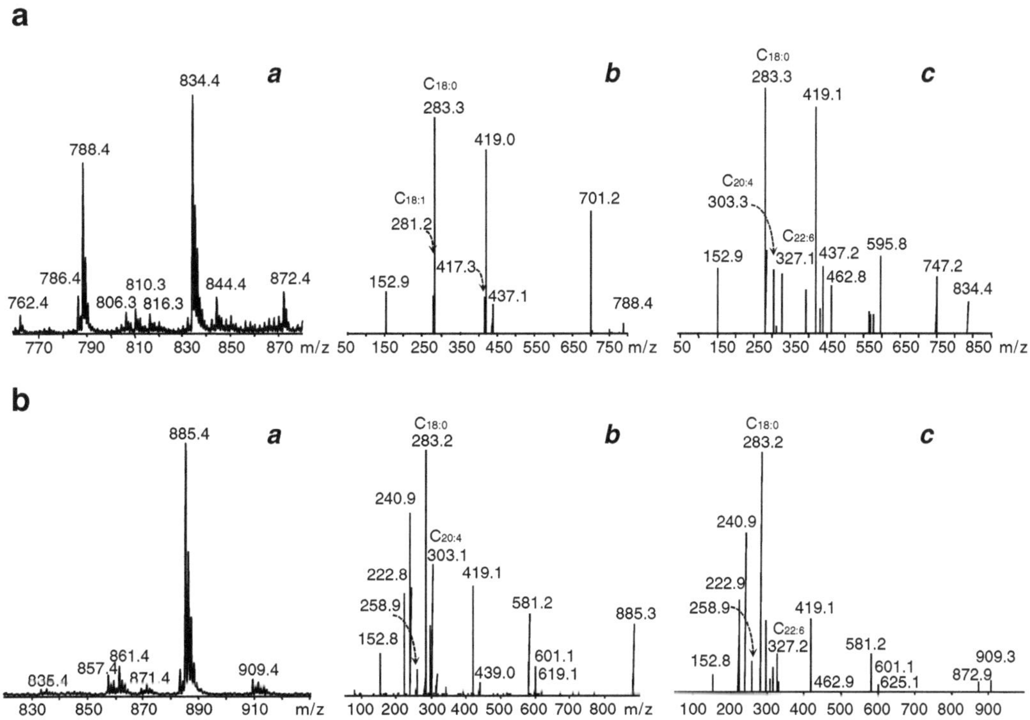

Fig. 2. ESI mass spectra of PS and PI molecular species obtained from mouse muscles and bone marrow. Phospholipids separated by 2D-HPTLC were subjected to MS analysis by direct infusion into mass spectrometer. **a** – Typical negative ESI mass spectra of PS from mouse muscles. Full negative MS spectrum (***a***) and MS^2 negative ESI spectra of molecular species of PS with *m/z* 788.4 (***b***) and 834.4 (***c***). The major molecular species of PS isolated from mouse muscles include $C_{18:0}/C_{18:1}$ at *m/z* 788.4 and $C1_{8:0}/C_{22:6}$ at *m/z* 834.4. **b** – Typical negative ESI mass spectra of PI from mouse bone marrow. Full negative MS spectrum (***a***) and MS^2 negative ESI spectra of molecular species of PI with *m/z* 885.3 (***b***) and 909.3 (***c***). Molecular ions of PI with *m/z* 885.3 and *m/z* 909.3 correspond to $C_{18:0}/_{C20:4}$ and $C_{18:0}/_{C22:6}$ molecular species, respectively. Molecular ion at *m/z* 885.3 is as the major PI molecular species in mouse bone marrow.

747.2. Two product ions with *m/z* 417.3 and 419.3 originated from fragment with *m/z* 701.2 after loss of oleic $C_{18:1}$ and stearic acid $C_{18:0}$, respectively (**Fig. 2a**, panel **b**). Molecular fragments with *m/z* 283.2 and 281.2 detected in MS^2 spectrum correspond to $C_{18:1}$ and $C_{18:0}$. Similarly, product ions with *m/z* 419.1 and 462.8 originated from the fragment with *m/z* 747.2 after loss of docosahexaenoic acid ($C_{22:6}$) and $C_{18:0}$, respectively, and molecular fragments of $C_{22:6}$ (*m/z* 327.3) and $C_{18:0}$ (*m/z* 283.3) were found in MS^2 spectrum (**Fig. 2a**, panel **c**). The ion with *m/z* 152.9 corresponded to glycerophosphate without molecule of water, a common ion formed during phospholipid fragmentation *(25)*. In addition, molecular ions with *m/z* 304.8 corresponding to $C_{20:3}$ fatty acid were also formed during fragmentation of parent ion with *m/z* 834.4. Thus, molecular ion at *m/z* 834.4 can correspond to two molecular species of PS $C_{18:0}/C_{22:6}$ and $C_{20:3}/C_{20:3}$.

Three major molecular clusters of PS with *m/z* 786.7–788.7, 810.7–812.7, and 836.7–838.7 were present in all five studied tissues. Molecular species of PS isolated from bone marrow were represented by major molecular ions with *m/z* 810.3, molecular species at *m/z* 834.3 and 786.3 being the next most abundant molecular species. PS molecular species corresponding to *m/z* 788.7 was detected as a major molecular species in small intestine. Other molecular species in small intestine – in the descending abundance – were those at *m/z* 810.7 and 834.7. Major clusters with *m/z* 786.7–788.7, 810.7–812.7, and 836.7–838.7 were distributed equally in mouse lung. Molecular ion at *m/z* 834.4 corresponding to $C_{18:0}/C_{22:6}$ molecular PS species dominated in mouse brain (**Table 2**).

Table 2
Molecular species of phosphatidylserines extracted from mouse tissues

PS	m/z [M–H]$^-$	Acyl/acyl	Brain	Small intestine	Lung	Muscles	Bone marrow
34:2	758.4	16:0/18:2			*		
34:1	760.5	16:0/18:1	*	*			*
		16:1/18:0		*			*
34:0	762.5	16:0/18:0			*		*
36:4	782.5	16:0/20:4; 18:1/18:3; 18:2/18:2					*
36:3	784.5	16:0/20:3; 18:0/18:3					*
36:2	786.5	18:0/18:2	*	*	*	*	*
		18:1/18:1			*	*	*
		16:0/20:2					*
36:1	788.4	18:0/18:1*	*	*	*	*	*
36:0	790.5	18:0/18/0			*		
38:6	806.5	16:0/22:6			*		*
		18:1/20:5; 18:2/20:4					*
38:5	808.5	18:0/20:5; 18:1/20:4					*
38:4	810.5	18:0/20:4	*	*	*	*	*
		16:0/22:4	*				*
38:3	815.5	18:0/20:3			*	*	*
		18:1/20:2		*	*	*	*
		18:2/20:1			*	*	

(continued)

Table 2 (continued)

PS	m/z [M−H]⁻	Acyl/acyl	Brain	Small intestine	Lung	Muscles	Bone marrow
38:2	814.5	18:0/20:2; 18:2/20:0; 16:0/22:2					*
38:1	816.5	18:1/20:0			*		
39:1	830.5	18:1/21:0					*
40:8	830.5	20:4/20:4; 18:2/22:6			*		*
40:7	832.5	18:1/22:6			*	*	*
		20:3/20:4			*		*
40:6	834.4	18:0/22:6	*	*	*	*	*
		18:1/22:5; 18:2/22:4; 20:3/20:3					*
40:5	836.5	18:0/22:5	*	*	*	*	*
		20:1/20:4			*	*	*
40:4	838.5	18:0/22:4					*
		20:2/20:2; 20:0/20:4					*
40:2	842.5	18:1/22:1			*		
40:1	844.5	18:1/22:0			*	*	
42:10	854.5	20:4/22:6			*		*
42:9	856.5	20:3/22:6; 20:4/22:5			*		*
42:8	858.5	20:3/22:5; 20:2/22:6			*		*
41:1	858.5	23:0/18:1; 23:1/18:0			*		
42:5	864.5	18:0/24:5; 18:1/24:4			*		
		20:0/22:5; 22:1/20:4			*		
866.5		22:0/20:4			*		
42:3	868.5	20:3/22:0; 18:1/24:2; 18:0/24:3			*		
42:1	872.5	18:1/24:0			*	*	
		18:0/24:1				*	

3.8.4. Identification of PI Molecular Species

Molecular ion with *m/z* 885.3 was detected in full MS spectrum of PI as a major PI molecular species in brain, bone marrow, small intestine, muscles, as well as lung (**Table 3**). As an example MS spectra of PI isolated from bone marrow are presented in **Fig. 2b**, panel **a**. Deprotonated ions $[M-H]^-$ with *m/z* 835.4, 857.4, 861.4, 909.4, 911.4, and 913.4 are also presented in full MS spectrum of PI but in relatively lower abundances (**Fig. 2b**, panel **b**). Fragmentation analysis of the major PI molecular ions with *m/z* 885.4 and molecular ion with *m/z* 909.4 are presented in **Fig. 2b**, panels

Table 3
Major molecular species of phosphatidylinositols extracted from mouse tissues

PI	m/z[M-H]$^-$	Acyl/acyl	Brain	Small intestine	Lung	Muscles	Bone marrow
34:2	833.5	16:0/18:2		*	*		*
34:1	835.5	16:1/18:0	*		*		*
		16:0/18:1					*
36:4	857.5	16:0/20:4	*	*	*		*
36:2	861.5	18:0/18:2		*	*		*
		18:1/18:1		*	*		*
36:1	863.5	18:0/18:1				*	*
38:6	881.5	16:0/22:6			*	*	
38:5	883.5	16:0/22:5			*	*	*
		18:1/20:4			*	*	*
		18:0/20:5			*	*	*
38:4	885.5	18:0/20:4	*	*	*	*	*
38:3	887.5	18:0/20:3	*		*	*	*
40:6	909.3	18:0/22:6	*	*	*	*	*
		18:1/22:5			*	*	*
40:5	911.5	18:0/22:5			*	*	*
		18:1/22:4			*	*	*
40:4	913.5	18:0/22:4			*		*
		20:0/20:4			*		*

b and **c**, respectively. Typical daughter fragments formed during fragmentation were observed in MS^2 spectra. Molecular ion with m/z 283.2 corresponding to stearic acid ($C_{18:0}$) was found in both spectra. Molecular ions of arachidonic acid ($C_{20:4}$, m/z 303.1) and docosahexaenoic acid $C_{22:6}$ (m/z 327.2) appeared during fragmentation of the MS^2 spectra of molecular species with m/z 885.3 and 909.3, respectively. Elimination of $C_{20:4}$ or $C_{22:6}$ fatty acids from the parent ions with m/z 885.4 and m/z 909.3 resulted in the product ion with signals at m/z 581.2 $[M\text{-}H\text{-}C_{20:4}]^-$ or $[M\text{-}H\text{-}C_{22:6}]^-$, respectively. Subsequently, a product ion with m/z 419.1 was formed after loosing an inositol group from the fragment with m/z 581.2. Similarly, elimination of $C_{18:0}$ from the parent ions with m/z 885.4 and m/z 909.3 resulted in the product ion $[M\text{-}H\text{-}C_{18:0}]^-$ at m/z 601.1 and m/z 625.1, respectively. Loss of the inositol group from these fragments formed product ions with m/z 439.0 and 462.9, respectively.

Head group-specific ions with *m/z* 259.0, 222.9, and 240.9 and ion corresponds to glycerophosphate without molecule of water (*m/z* 152.8) were detected on MS^2 spectra.

3.8.5. Identification of PE Molecular Species

ESI-MS spectra of muscle PE in negative mode demonstrated the presence of two groups of this glycerophospholipid, diacyl-PE (*m/z* 762.4, 790.4, 800.2, and 810.3) and alkenyl-PE (*m/z* 728. 4, 746.4, and 774.4) (**Fig. 3a**, panel **a**). Fatty acid composition of molecular species of both diacyl-PE and alkenyl-PE was confirmed by MS^2 analysis. After fragmentation of the major molecular ion with *m/z* 790, corresponding to diacyl-PE molecular species, two prominent ions with *m/z* 283.2 and 327.1 were obtained in negative-ionization mode that correspond to $C_{18:0}$ and $C_{22:6}$ fatty acyls, respectively (**Fig. 3a**, panel **b**). Typical

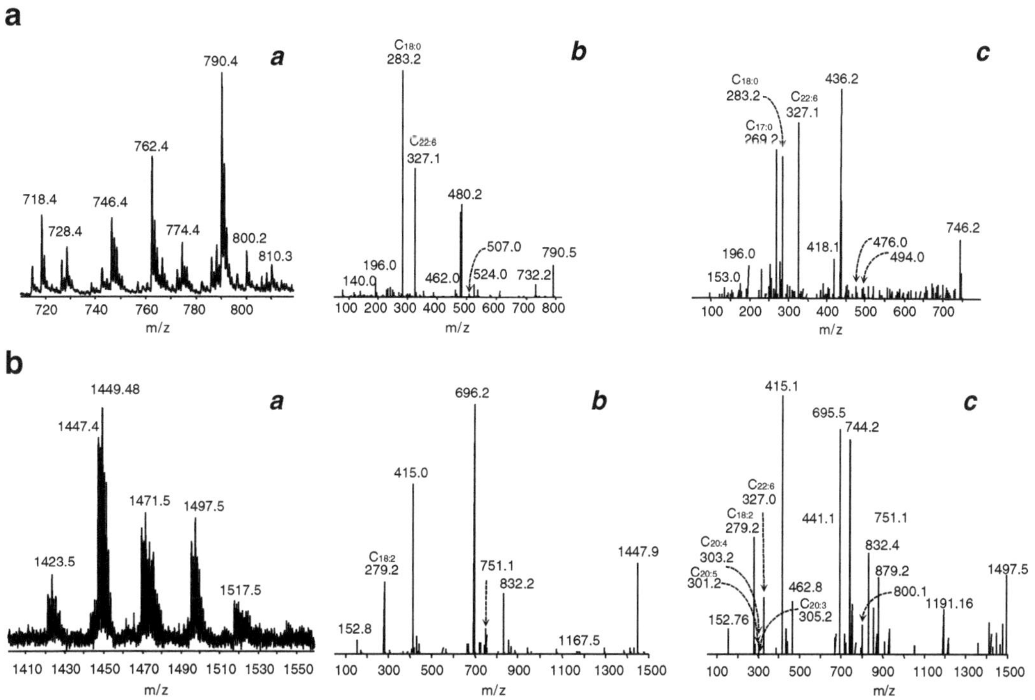

Fig. 3. ESI mass spectra of PE and CL molecular species obtained from mouse muscles. Phospholipids separated by 2D-HPTLC were subjected to MS analysis by direct infusion. **a** – Typical negative ESI mass spectra of PE from mouse muscles. Shown are full negative MS spectrum of PE (***a***) and MS^2 negative ESI spectra of diacyl-molecular species of PE with *m/z* 790.5 (***b***) and alkenyl-molecular species with *m/z* 746.2 (***c***). Molecular species of PE include diacyl-PE (*m/z* 762.4, 790.4, 800.2, and 810.3) and alkenyl-PE (*m/z* 728. 4, 746.4, and 774.4). Molecular ion at *m/z* 790.5 corresponds to a molecular species $C_{18:0}/C_{22:6}$. MS^2 analysis of molecular ion with *m/z* 746.2 revealed the presence of two PE alkenyl molecular species – $C_{16:1p}/C_{22:6}$ and $C_{20:0p}C_{17:0}$. **b** – Typical negative ESI mass spectra of CL from mouse muscles. Shown are full negative MS spectrum of CL (***a***) and MS^2 negative ESI spectra of CL molecular ions with *m/z* 1447.9 (***b***) and 1495.7 (***c***). The major molecular ion of CL with *m/z* 1447.5 corresponds to $C_{18:2}/C_{18:2}/C_{18:2}/C_{18:2}$ molecular species. Three molecular species of CL – $C_{18:2}/C_{18:2}/C_{20:4}/C_{20:4}$, $C_{18:2}/C_{18:2}/C_{18:2}/C_{22:6}$ and $C_{18:0}/C_{18:3}/C_{20:4}/C_{20:5}$ are contributing molecular ions to a signal with *m/z* 1495.7.

fragmentation of PE diacyl species with *m/z* 790.5 was shown in **Fig. 3a**, panel **a**. Daughter ions with *m/z* 480.2 and 524.0 were formed after loosing $C_{18:0}$ and $C_{22:6}$ fatty acids, respectively. Molecular ions with *m/z* 462.0 and 507.3 were derived from these ions after loosing water. Fragmentation of molecular ion with *m/z* 746.2 revealed the presence of two PE alkenyl molecular species – $C_{16:1p}/C_{22:6}$ and $C_{20:0p}C_{17:0}$ (**Fig. 3a**, panel **c**). Molecular ion with *m/z* 327.1 corresponded to $C_{22:6}$ fatty acid. Ions at *m/z* 436.2 and 418.1 were produced from molecular ions with *m/z* 746.2, $C_{16:1p}/C_{22:6}$ after loss of $C_{22:6}$ and $C_{22:6}$ and water, respectively (**Fig. 3a**, panel **c**). Similarly, appearance of molecular ions with *m/z* 269.2, 494.0, and 476.0 corresponded to heptadecanoic acid ($C_{17:0}$), molecular ion without of $C_{17:0}$ and molecular ion after losing $C_{17:0}$ and water, respectively (**Fig. 3ac**). Note that molecular ions with *m/z* 283.2 corresponding to $C_{18:0}$ that belong to PE diacyl molecular species $C_{18:0}/C_{18:0}$ with *m/z* 746.3 overlap with two alkenyl PE species in MS^2 spectrum. PE head group-specific ions with *m/z* 140 and 196 and an ion with *m/z* 152.8 were observed in MS^2 spectra of both diacyl and alkenyl molecular species of PE. All identified molecular species of PE extracted from brain, lung, muscles, bone marrow, and small intestine are shown in **Table 4**.

Table 4
Molecular species of phosphatidylethanolamines extracted from mouse tissues

PE diacyl	*m/z* [M–H]⁻	Acyl/acyl	Brain	Small intestine	Lung	Muscles	Bone marrow
34:2	714.5	16:0/18:2	*				
34:1	716.5	16:0/18:1	*				
35:3	726.5	18:1/17:2; 18:3/17:0				*	
36:4	738.5	16:0/20:4	*		*		*
		18:2/18:2			*		*
		18:1/18:3;					*
		18:0/18:4					
36:2	742.5	18:1/18:1	*	*			*
		16:0/20:2;					*
		18:0/18:2					
36:1	744.5	18:1/18:0	*	*			*
		17:0/19:1					*

(continued)

**Table 4
(continued)**

PE diacyl	m/z [M−H]$^-$	Acyl/acyl	Brain	Small intestine	Lung	Muscles	Bone marrow
36:0	746.5	18:0/18:0				*	*
38:6	762.5	16:0/22:6 18:2/20:4	*		*	* *	*
38:5	764.5	18:1/20:4; 16:0/22:5 18:2/20:3; 16:1/22:4			* *	*	*
38:4	766.5	18:0/20:4; 16:0/22:4	*	*	*	*	*
38:3	768.5	18:0/20:3 18:1/20:2 18:2/20:1		* *		*	* * *
38:1	772.5	18:0/20:1; 18:1/20:0 16:0/22:1				*	* *
38:0	774.5	18:0/20:0 16:0/22:0 17:1/22:6 19:0/19:0			* *	* * *	* *
39:6	776.5	17:0/22:6			*	*	*
40:8	786.5	18:2/22:6				*	
40:7	788.5	20:3/20:4 18:1/22:6			* *	*	
40:6	790.5	18:0/22:6 20:2/20:4; 18:1/22:5; 18:2/22:4	*	*	* *	*	*
40:5	792.5	18:0/22:5; 18:1/22:4 20:1/20:4				*	* *
40:4	794.5	18:0/22:4 20:0/20:4; 18:2/22:2			*		* *
40:1	800.5	18:0/22:1; 18:1/22:0; 20:0/21:0				*	

(continued)

Table 4 (continued)

PE diacyl	m/z [M–H]⁻	Acyl/acyl	Brain	Small intestine	Lung	Muscles	Bone marrow
41:5	806.5	17:1/24:4 19:0/22:5				* *	*
42:11	808.5	20:5/22:6			*	*	
42:10	810.5	20:4/22:6			*	*	*
42:9	812.5	20:3/22:6					*
34:2	700.5	16:1p/18:1	*				
36:5	722.5	16:1p/20:4	*		*		*
36:3	726.5	18:1p/18:2	*				
36:2	728.5	18:1p/18:1	*			*	
38:7	746.5	16:1p/22:6 20:0p/17:0	*		*	* *	*
38:6	748.5	16:1p/22:5 16:0p/22:6			*	* *	*
38:5	750.5	18:1p/20:4; 16:1p/22:4	*	*	*		*
38:4	752.5	16:0p/22:4 18:1p/20:3; 18:0p/20:4					* *
38:3	754.5	18:1p/20:2 18:0p/20:3; 16:1p/22:2	*				*
39:1	772.5	18:1p/21:0					
40:7	774.5	18:1p/22:6	*		*	*	*
40:6	776.5	18:0p/22:6 18:1p/22:5		*	* *	*	
40:5	778.5	18:1p/22:4; 20:1p/20:4			*		*
42:7	802.5	20:1p/22:6		*			
42:4	808.5	20:0p/22:4			*		*

3.8.6. Identification of CL Molecular Species

Molecular species of muscle CL were represented by five molecular clusters (**Fig. 3b**, panel **a**). The major cluster contains molecular ions with m/z 1447.5, 1449.5, and 1450.5. Molecular species at m/z 1423.5, 1474.5, 1497.5, and 1517.5 were also detectable in full MS spectrum but at relatively lower abundance (**Fig. 3b**, panel **a**). Singly charged ions were used for structural identification of CL as described by Hsu and Turk *(26)*. MS^2 analysis of the molecular ion with *m/z* 1447.9 yielded $[a]^-$ and $[b]^-$ ions with *m/z* 696.2 corresponding to $C_{18:2}/C_{18:2}$-phosphatidic acid (–PA), a typical structural fragment of CL. Distinctive fragmentation ions at *m/z* 751.1 ($[a + 56]^-$ and $[b + 56]^-$), *m/z* 832.2 ($[a + 136]^-$ and $[b + 136]^-$), and *m/z* 415.0 ($[a-(C_{18:2})]^-$ and $[b-(C_{18:2})]^-$) were observed in MS^2 spectrum. Molecular ions of $C_{18:2}$ (*m/z* 279.2) were also found in MS^2 spectrum (**Fig. 3b**, panel **b**).

MS^2 analysis of the molecular ion with *m/z* 1497.5 revealed $[a]^-$ and $[b]^-$ ions with *m/z* 695.5 and 744.2 mainly corresponding to $C_{18:2}/C_{18:2}$-PA and $C_{18:0}/C_{22:6}$-PA originated from fragmentation of $C_{18:2}/C_{18:2}/C_{18:2}/C_{22:6}$ molecular ion of CL. Characteristic daughter ions $[a + 56]^-$ (*m/z* 751.1), $[b + 56]^-$ (*m/z* 800.1), $[a + 136]^-$ (*m/z* 832.4), and $[b + 136]^-$ (*m/z* 879.2) were produced during fragmentation process. In addition, ions $[a-(C_{18:2})]^-$ (*m/z* 415.1), $[b-(C_{18:0})]^-$ (*m/z* 462.8), and $[b-(C_{22:6})]^-$ (*m/z* 415.1) were detectable in MS^2 spectrum. Molecular ions of $C_{18:2}$ (*m/z* 279.2) and $C_{22:6}$ (*m/z* 327.0) were presented in MS^2 spectrum as well (**Fig. 2b**). Further, $[b]^-$ ion at *m/z* 744.2 likely corresponding to $C_{20:4}/C_{20:4}$-PA was formed due to fragmentation of $C_{18:2}/C_{18:2}/C_{20:4}/C_{20:4}$ CL molecular ion (with the same *m/z* 1497.5). Consequently, $[b-(C_{20:4})]^-$ (*m/z* 441.1) and molecular ions of $C_{20:4}$ (*m/z* 303.9) were observed in MS^2 spectrum (**Fig. 3b**, panel **c**). Moreover, $C_{18:0}/C_{18:3}/C_{20:4}/C_{20:5}$ molecular species of CL corresponding to molecular ion at *m/z* 1497.5 was also detectable in MS^2 spectrum but in significantly lower abundance (**Fig. 3b**, panel **c**). Molecular ions of $C_{18:1}$ (*m/z* 281.2), $C_{18:0}$ (*m/z* 283.2), $C_{20:4}$ (*m/z* 303.2), $C_{20:5}$ (*m/z* 301.0), and ions formed via loss of one of fatty acids $[M\text{-}H\text{-}C_{20:5}]^-$ (*m/z* 1190.2), $[M\text{-}H\text{-}C_{20:4}]^-$ (*m/z* 1191.2), and $[M\text{-}H\text{-}C_{18:0}]^-$ (*m/z* 1214.5) were observed in MS^2 spectrum along with ions formed after elimination of one of fatty acids (**Fig. 3b**, panel **b**). **Table 5** contains molecular species of CL detected in brain, lung, muscles, bone marrow, and small intestine.

3.8.7. Identification of PC Molecular Species

ESI-MS spectra of PC isolated from mouse small intestine were presented as an example of identification of PC molecular species. In positive-ion mode, the presence of the major molecular species with protonated ions $[M + H]^+$ with *m/z* 758.7, 780.7, 786.7, and 832.7 was evident (**Fig. 4a**, panel **a**). Two additional molecular clusters of PC with *m/z* 734.6 and 832.7 were present in positive full MS spectrum in relatively lower abundance (**Fig. 4a**, panel **a**). A typical fragmentation spectrum in positive-ion mode with *m/z* 758.7 is shown in **Fig. 3a**, panel **b**. The major fragment

Table 5
Major molecular species of cardiolipins extracted from different tissue of mice

	m/z [M–H]⁻	Acyl/acyl	Brain	Small intestine	Lung	Muscles	Bone marrow
68:2	1399.9	$(16:1)_2(18:1)_2$			*		
70:7	1421.5	(16:1)1(18:2)3			*		
70:6	1423.5	$(16:1)_1(18:1)_1(18:2)_2$			*	*	*
70:5	1425.5	$(16:0)_1(18:1)_1(18:2)_2$ $(16:1)_1(18:1)_2(18:2)_1$			* *	*	*
70:4	1427.8	$(16:1)_1(18:0)_1(18:1)_1\ (18:2)_1$ $(16:0)_1(18:1)_2(18:2)_1$	*		*		
72:8	1447.7	$(18:2)_4$ $(18:1)_1(18:3)_1(18:2)_2$	*	*	*	* *	*
72:7	1449.7	$(18:2)_3(18:1)_1$ $(16:1)_1(18:2)_1(18:1)_1\ (20:2)_1$	*	*	*	* *	*
72:6	1451.7	$(18:2)_2(18:1)_2$ $(16:0)_1(16:1)_1(20:3)_1\ (20:2)_1$	*			* *	
72:5	1453.9	$(18:2)_1(18:1)_3$ $(18:2)_2(18:1)_1(18:0)_1$	*		* *		*
72:4	1455.8	$(18:1)_4$	*				
72:3	1457.8	$(18:1)_3(18:0)_1$	*				
73:4	1469.7	$(18:2)_2(17:0)_1(20:0)_1$ $(18:2)_2(19:0)_1(18:0)_1$				* *	
74:11	1469.7	$(18:2)_2(16:1)_1(22:6)_1$ $(18:2)_2(18:3)_1(20:4)_1$				* *	
74:10	1471.9	$(18:2)_3(20:4)_1$ $(18:2)_2(16:0)_1(22:6)_1$ $(18:2)_2(18:1)_1(20:5)$	*	*	*	*	
73:3	1471.9	$(16:0)_1(18:2)_1(18:1)_1\ (21:0)_1$					*
74:9	1473.8	$(18:1)_1(18:2)_2(20:4)_1$ $(18:2)_1(18:1)_1(16:1)_1\ (22:5)_1$ $(16:0)_1(18:2)_2(22:5)_1$ $(16:1)_1(18:2)_1(20:3)_2$	*		*	*	* * *
70:8	1475.8	$(18:1)_2(18:2)_1(20:4)_1$ $(18:2)_3(20:2)_1$	*		*		*
70:7	1477.8	$(18:1)_3(20:4)_1$ $(18:1)_2(18:2)_1(20:3)_1$ $(18:1)_1(18:2)_2(20:2)_1$	*		* * *		

(continued)

Table 5 (continued)

	m/z $[M-H]^-$	Acyl/acyl	Brain	Small intestine	Lung	Muscles	Bone marrow
70:6	1479.8	$(18:0)_1(18:1)_2(20:4)_1$	*				
76:13	1493.7	$(18:2)_2(18:3)_1(22:6)_1$				*	
76:12	1495.7	$(18:2)_2(20:4)_2$ $(18:2)_3(22:6)_1$	*		*	*	*
76:11	1497.7	$(18:1)_1(18:2)_1(20:4)_2$ $(18:1)_1(18:2)_2(22:6)_1$ $(18:2)_2(20:3)_1(20:4)_1$ $(18:2)_3(22:5)_1$	*		* *	* *	*
76:10	1499.7	$(18:1)_2(20:4)_2$ $(18:1)_3(21:0)_1$ $(18:2)_2(20:3)_2$	*		* *		*
76:9	1501.7	$(18:1)_3(22:6)_1$	*				
76:8	1503.6	$(18:1)_3(22:5)_1$	*				*
$(18:1)_2$ $(20:3)_2$							*
78:14	1519.8	$(16:1)_1(20:4)_1(20:3)_1(22:6)_1$	*				
78:13	1521.7	$(18:2)_1(18:1)_1(20:4)_1(22:6)_1$	*				
78:12	1523.7	$(18:1)_1(20:4)_2(20:3)_1$	*				
78:11	1525.7	$(18:2)_1(20:4)_1(20:3)_1(20:2)_1$	*				
78:6	1536.7	$(20:0)_2(16:0)_1(22:6)_1$ $(18:2)_2(18:0)_1(24:2)_1$ $(18:2)_1(18:1)_1(18:0)_1(24:3)_1$			* * *		
$(16:0)_2$ $(23:3)_2$					*		
79:13	1536.7	$(18:2)_1(20:2)_1(22:6)_1(19:3)_1$ $(18:4)_1(20:2)_1(22:4)_1(19:3)_1$ $(18:3)_1(20:2)_1(22:5)_1(19:3)_1$			* * *		
80:20	1536.7	$(20:5)_4$			*		

with *m/z* 575.5 was generated due to loss of the polar head group $[C_5H_{15}NPO_4]^+$. Typical head-group ion at *m/z* 184.0 was detectable in positive MS^2 spectrum as well. In the negative-ionization mode, most of the PC molecular species were observed as chloride adducts $[M + Cl]^-$ (**Fig. 4b**, panel **a**). Fragmentation of molecular ion with *m/z* 792.7 $[M + Cl]^-$ corresponding to *m/z* 758.7 in positive mode is presented in **Fig. 4b**, panel **b**. Typical

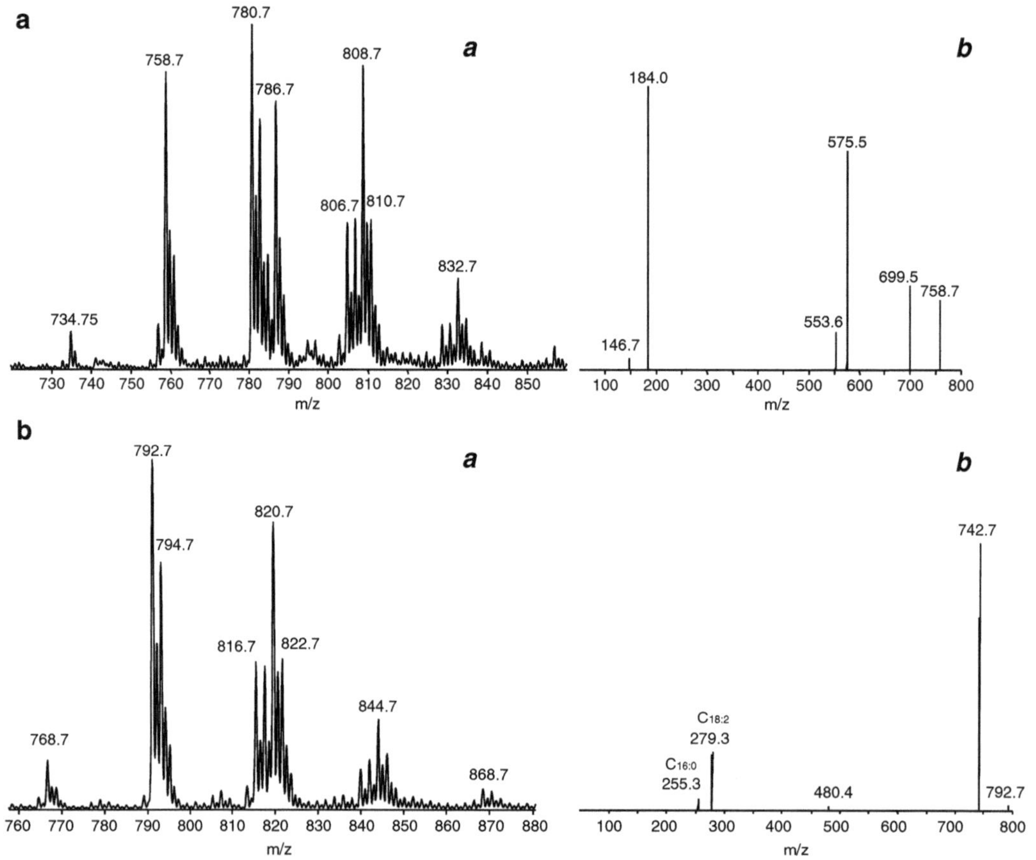

Fig. 4. ESI mass spectra of molecular species of PC obtained from mouse small intestine. PC separated by 2D-HPTLC was directly injected into mass spectrometer. **a** – Typical positive-ion ESI mass spectra of PC. Full ESI MS spectrum of PC molecular species (***a***) and fragmentation (MS^2) profile of molecular species at *m/z* 758.7 ($C_{16:0}/_{C18:2}$) (***b***) performed in a positive-ionization mode. In positive-ionization mode, the major molecular species of small intestine PC were the protonated ions $[M + H]^+$ with *m/z* 758.7 ($C_{16:0}/_{C18:2}$), 780.7 ($C_{16:1}/_{C20:4}$) and 786.7 ($C_{18:0}/_{C18:2}$). **b** – Typical negative-ion ESI mass spectra of PC from mouse small intestine. Full ESI MS spectrum of PC molecular species (***a***) and negative MS^2 profile of PC molecular species (chloride adduct) at *m/z* 792.7 ($C_{16:0}/_{C18:2}$) (***b***). In the negative-ionization mode, most of the PC molecular species were observed as chloride adducts.

ions of fatty acids – palmitic acid ($C_{16:0}$) and $C_{18:2}$ – and fragment [M + Cl-CH_3] with *m/z* 255.3, 279.3, and 742.7 were detectable in negative MS^2 after fragmentation of PC molecular ions *(27)*. Results of identification of PC molecular species present in different mouse tissues are presented in **Table 6**.

3.8.8. Quantitative Assessments of Phospho lipid Hydroperoxdes

In all five tissues studied, MS^2 analysis shows the presence in molecular species of major phospholipids – PC, PE, PS, PI, and CL of highly unsaturated, susceptible to oxidation fatty acids

Table 6
Molecular species of phosphatidylcholines extracted from mouse tissues

PC diacyl	*m/z* [M + H]$^+$	Acyl/acyl	Brain	Small intestine	Lung	Muscles	Bone marrow
30:0	706.5	16:0/14:0			*		
32:2	730.5	16:1/16:1; 14:0/18:2			*		
32:1	732.5	16:0/16:1			*		
32:0	734.5	16:0/16:0	*	*	*		
34:3	756.5	16:1/18:2	*		*		
		16:0/18:3			*		
34:2	758.5	16:0/18:2	*	*	*		
		16:1/18:1			*		
34:1	760.5	16:0/18:1	*		*	*	
		16:1/18:0				*	
34:0	762.6	16:0/18:0	*	*		*	
		14:0/20:0				*	
35:4	768.5	15:0/20:4			*		
35:3	770.5	17:0/18:3					*
		17:1/18:2			*		
36:5	780.5	16:1/20:4		*			*
36:4	782.5	16:0/20:4	*	*	*		
		18:2/18:2					*
36:3	784.5	16:0/20:3	*		*		
		14:1/22:2; 18:0/18:3					*
		18:1/18:2			*	*	*
36:2	786.6	18:0/18:2		*	*		
		18:1/18:1			*		
		16:0/20:2; 17:0/19:2					*
36:1	788.6	18:0/18:1	*				
36:0	790.6	18:0/18:0					*
37:6	792.6	15:0/22:6				*	

(continued)

Table 6 (continued)

PC diacyl	m/z [M + H]$^+$	Acyl/acyl	Brain	Small intestine	Lung	Muscles	Bone marrow
37:4	796.5	17:0/20:4					*
38:7	804.5	16:1/22:6	*		*		
38:6	806.5	16:0/22:6 18:2/20:4 18:4/20:2 18:3/20:3	*	*	*	* * *	* * *
38:5	808.5	18:2/20:3; 18:1/20:4			*		*
38:4	810.6	18:0/20:4 16:0/22:4 14:0/24:4	*		*	*	* *
39:7	818.6	17:1/22:6				*	
39:5	822.6	19:0/20:5				*	
39:1	830.5	19:1/20:0; 18:1/21:0; 17:1/22:0 15:0/24:1				* *	
40:8	830.5	18:2/22:6				*	*
40:7	832.5	18:1/22:6		*	*		*
40:6	834.5	18:1/22:5; 20:2/20:4					*
40:2	842.6	18:1/22:1; 20:0/20:2					*
40:1	844.6	20:0/20:1 16:1/24:0				*	* *
42:4	866.6	22:0/20:4					*
34:0	748.6	18:0o/16:0	*				
36:2	772.5	18:0o/18:2 18:0p/18:1			* *		
38:7	790.5	16:0p/22:6			*		
38:4	796.5	18:0o/20:4		*	*		
41:1	844.6	20:0p/21:0					*

such as $C_{18:2}$, $C_{20:4}$, and $C_{22:6}$. Thus, phospholipids can be readily oxidized by TBI. Quantitative analysis of phospholipid oxidation products was assessed by fluorescence HPLC-based protocol (*see* **Note 6**). The protocol is based on separation of phospholipids using 2D-HPTLC. Phospholipids are hydrolyzed using PLA_2. The FA-OOH released were quantified by a fluorometric assay using Amplex Red reagent (10-acetyl-3,7-dihydroxyphenoxazine) and MP-11. In this system, FA-OOH react with Amplex Red reagent at a stoichiometry of 1:1 to generate a fluorescent product, resorufin (ex = 560 nm, em = 590 nm), analyzed by fluorescence HPLC. Using the protocol, 1–2 pmol of lipid hydroperoxides can be reliably detected in lipid extracts from different tissues. We found that bone marrow, small intestine, and brain accumulated higher levels of oxidized phospholipids than lung and muscles after exposure to TBI. Oxidation of phospholipids was nonrandom and did not follow their abundance (*see* **Table 1**). Only three anionic phospholipids – CL, PS and PI – reveal g-irradiation-induced accumulation of hydroperoxy species whereas more abundant PC and PE remained nonoxidized. These data are summarized in **Table 7**.

3.8.9. Identification of Phospholipid Hydro peroxdes by ESI-MS

Generation of phospholipid hydroperoxides induced by TBI was confirmed by ESI-MS analysis (*see* **Note 6**). As an example, MS^2 spectrum of CL molecular species with *m/z* 1536.8 isolated from irradiated mouse lung is presented on **Fig. 5a**. PQD of the singly charged ion (*m/z* 1536.8) revealed the copresence of oxidized molecular species of CL ($C_{18:2\text{-OOH}}/C_{18:2\text{-OOH}}/C_{18:1}/C_{20:5}$) along with eight nonoxidized CLs ($C_{20:0}/C_{20:0}/C_{16:0}/C_{22:6}$, $C_{18:2}/C_{20:2}/C_{22:6}/C_{19:3}$, $C_{18:3}/C_{20:2}/C_{22:5}/C_{19:3}$, $C_{18:4}/C_{20:2}/C_{22:4}/C_{19:3}$, $C_{16:0}/C_{16:0}/C_{23:3}/C_{23:3}$, $C_{18:2}/C_{18:2}/C_{18:0}/C_{24:2}$, $C_{18:2}/C_{18:1}/C_{18:0}/C_{24:3}$, and $C_{20:5}/C_{20:5}/C_{20:5}/C_{20:5}$). Fragments from these eight nonoxidized molecular species of CL [a], [b], [a + 56], [b + 56], [a + 136], and [b + 136] were formed in the course of PQD fragmentation process. Major molecular fragments with *m/z* 891.7, 893.7, 895.7, 917.4, and 919.7 were identified as [a + 136] ions originating from molecular species $C_{18:2}/C_{20:2}/C_{22:6}/C_{19:3}$, $C_{18:3}/C_{20:2}/C_{22:5}/C_{19:3}$, $C_{18:4}/C_{20:2}/C_{22:4}/C_{19:3}$, $C_{18:2}/C_{18:1}/C_{18:0}/C_{24:3}$, and $C_{18:2}/C_{18:2}/C_{18:0}/C_{24:2}$, respectively (**Fig. 5a**, **c**). In addition, $C_{16:0}$ (*m/z* 255.1), $C_{18:0}$ (*m/z* 283.1), $C_{18:1}$ (*m/z* 281.1), $C_{18:2}$ (*m/z* 279.1), $C_{18:4}$ (*m/z* 274.7), $C_{19:3}$ (*m/z* 290.9), $C_{20:0}$ (*m/z* 311.0), $C_{20:2}$ (*m/z* 307.1), $C_{22:4}$ (*m/z* 331.1), $C_{22:5}$ (*m/z* 329.2), $C_{22:6}$ (*m/z* 326.9), $C_{23:3}$ (*m/z* 347.3), $C_{24:3}$ (*m/z* 360.6), and $C_{24:2}$ (*m/z* 363.1) were detectable in MS^2 spectrum (**Fig. 5b**). Moreover, MS^2 fragmentation of molecular ion with *m/z* 1536.8 yielded $[a]^-$ and $[b]^-$ ions of oxidized CL molecular species with *m/z* 759.7 and 721.5 which corresponded to $C_{18:2\text{-OOH}}/C_{18:2\text{-OOH}}$-PA and $C_{18:1}/C_{20:5}$-PA, originating from molecular

Table 7
Determination of phospholipid hydroperoxides of major phospholipids extracted from different tissue of mice exposed to total body irradiation (*TBI*)

Tissue	Treatment	CL	PS	PI	PC	PE
		(pmol of phospholipid hydroperoxide/nmol of phospholipids)				
Bone marrow	Control	11.6 ± 0.7	2.7 ± 0.1	2.0 ± 0.2	1.0 ± 0.1	2.1 ± 0.1
	TBI, 10 Gy	47.5 ± 1.8*	8.9 ± 0.3*	9.5 ± 0.4*	2.1 ± 0.1	5.1 ± 0.2
	TBI, 15 Gy	28.1 ± 1.5*	2.7 ± 0.3	4.5 ± 0.3	1.5 ± 0.1	4.2 ± 0.1
Small intestine	Control	7.0 ± 1.4	4.0 ± 1.8	6.0 ± 1.4	5.6 ± 1.8	2.7 ± 0.7
	TBI, 10 Gy	34.5 ± 6.2*	25.8 ± 5.5*	14.1 ± 6.0*	4.0 ± 1.3	11.3 ± 4.9
	TBI, 15 Gy	60.1 ± 13.0*	34.7 ± 12.0*	23.2 ± 8.6 *	14.4 ± 3.5*	14.2 ± 3.7*
Lung	Control	6.6 ± 1.9	1.5 ± 0.7	1.5 ± 0.5	0.1 ± 0.1	2.8 ± 0.7
	TBI, 10 Gy	11.5 ± 1.3*	9.5 ± 0.7*	3.4 ± 0.6	0.5 ± 0.1	2.0 ± 1.0
	TBI, 15 Gy	21.9 ± 3.0*	8.4± 1.7*	2.7 ± 0.8	0.4 ± 0.1	3.4 ± 0.1
Brain	Control	20.1 ± 4.0	4.8 ± 1.3	3.1 ± 0.9	0.9 ± 0.3	1.8 ± 0.7
	TBI, 10 Gy	36.5 ± 7.5	4.9 ± 1.5	16.1 ± 4.7*	1.2 ± 0.5	1.5 ± 0.5
	TBI, 15 Gy	50.6 ± 13.0*	6.5 ± 4.0*	11.2 ± 4.8*	5.9 ± 3.6	1.8 ± 0.7
Muscles	Control	13.4 ± 6.8	9.0 ± 0.4	6.5 ± 1.8	0.9 ± 0.3	6.4 ± 2.8
	TBI, 10 Gy	15.1 ± 0.3	12.8 ± 0.6	8.0 ± 1.4	2.0 ± 0.4	5.9 ± 1.9
	TBI, 15 Gy	13.3 ± 1.5	12.3 ± 3.8	3.7 ± 0.1	1.8 ± 0.5	4.9 ± 1.3

C57BL/6NHsd female mice were subjected to 0, 10, and 15 Gy of TBI and sacrificed 24 h thereafter. Lipids from tissues were extracted and separated by 2D-HPTLC. Spots of phospholipids were scraped, phospholipids were extracted from silica and phospholipid hydroperoxides were determined. All data are mean ± SD, $n = 3–4$, $^{*}p < 0.05$ vs Control.

species with m/z 1472.2 ($C_{18:2}/C_{18:2}/C_{18:1}/C_{20:5}$ CL molecular species). Fragments $[a + 56]^-$ (m/z 815.7), $[b + 56]^-$ (m/z 777.3), $[a + 136]^-$ (m/z 895.9), $[b + 136]^-$ (m/z 857.6), $[a\text{-}(C_{18:0\text{-}OOH})]^-$ (m/z 447.8), and $[b\text{-}(C_{20:5})]^-$ (m/z 421.1) were present in MS^2 spectrum (**Fig. 5c**). Consequently, molecular ions of $C_{18:1}$ (m/z 281.4), $C_{20:5}$ (m/z 301.2), and $C_{18:2\text{-}OOH}$ (m/z 311.0) were observed in MS^2 spectrum (**Fig. 5b**). Thus, both $C_{18:2}$ residues were oxidized in $C_{18:2}/C_{18:2}/C_{18:1}/C_{20:5}$ CL molecular species isolated from irradiated mouse lung.

Further, oxidized species of CL with m/z 1480.0, 1511.9, 1543.9, and 1575.8 for singly charged ions (derived from molecular species with m/z 1448.0) corresponding to CL containing

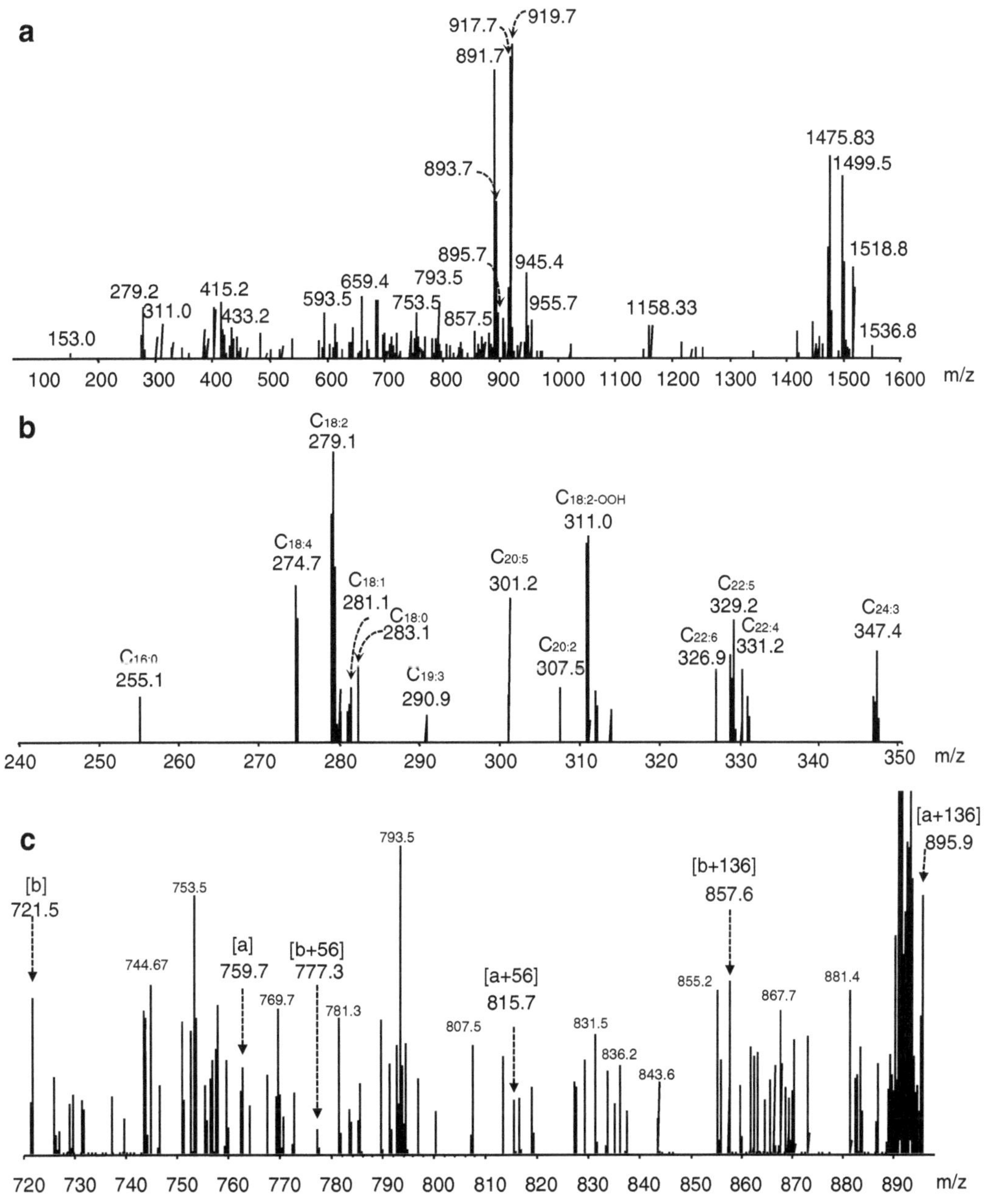

Fig. 5. Typical negative-ion MS[2] ESI mass spectra of oxidized molecular species of CL with *m/z* 1536.8 from lung of mice exposed to TBI at a dose of 15 Gy. (**a**) – ESI mass spectra of CL oxidized molecular species with *m/z* 1536.8. The MS[2] spectrum of the same molecular species in the range of *m/z* from 240 to 350 (**b**) and from 720 to 900 (**c**) are shown. Note that MS[2] analysis revealed co-presence of oxidized molecular species of CL ($C_{18:2\text{-}OOH}/C_{18:2\text{-}OOH}/C_{18:1}/C_{20:5}$) along with eight nonoxidized species of CL ($C_{20:0}/C_{20:0}/C_{16:0}/C_{22:6}$, $C_{18:2}/C_{20:2}/C_{22:6}/C_{19:3}$, $C_{18:3}/C_{20:2}/C_{22:5}/C_{19:3}$, $C_{18:4}/C_{20:2}/C_{22:4}/C_{19:3}$, $C_{16:0}/C_{16:0}/C_{23:3}/C_{23:3}$, $C_{18:2}/C_{18:2}/C_{18:0}/C_{24:2}$, $C_{18:2}/C_{18:1}/C_{18:0}/C_{24:3}$, and $C_{20:5}/C_{20:5}/C_{20:5}/C_{20:5}$). Oxidized species of CL with *m/z* 1536.8 originated from molecular species of CL $C_{18:2}/C_{18:2}/C_{18:1}/C_{20:5}$ with *m/z* 1471.9 after addition of four oxygens and contained two hydroperoxy-linoleate residues 18:2-OOH (*m/z* 311.5).

one, two, three, and four hydroperoxy groups were detected in MS spectrum of CL isolated from the intestines of irradiated mice. In addition, several hydroxyCL species were presented – along with the signals from mono-, di-, tri-, and tetra-hydroperoxy-CLs – in mass spectra of CLs from irradiated small intestine *(21, 28)*. CL doubly charged species at *m/z* 790 corresponding to oxidized molecular species containing $C_{22:6\text{-}OOH}$ (*m/z* 359) along with $C_{22:6\text{-}OH}$ (*m/z* 343) was found in MS^2 spectrum of CL isolated from irradiated brain. The presence of the signal at *m/z* 343 suggests that dihydroxy species in which each of the two $C_{22:6}$ chains contained hydroxy groups was also formed in the brain of irradiated mice (data not shown).

MS analysis of intestinal PS detected the presence of several molecular species with hydroperoxy-PS with *m/z* 866.7, 868.7, and 870.7 originated from PS molecular species at *m/z* 834.7 ($C_{18:0}/C_{22:6}$), 836.7 ($C_{18:0}/C_{22:5}$), and 838.7 ($C_{18:0}/C_{22:4}$). PS oxidation products containing hydroperoxy-docosahexaenoic ($C_{22:6\text{-}OOH}$), hydroperoxy-docosapentaenoic ($C_{22:5\text{-}OOH}$), and hydroperoxy-tetraenoic ($C_{22:4\text{-}OOH}$) fatty acids were found in the spectra *(21, 28)*. Analysis of PS extracted from irradiated lung revealed the presence of molecular ion with *m/z* 866.5. Its MS^2 fragmentation showed overlapping of two molecular species. One of them is nonoxidized PS containing $C_{22:0}/C_{20:4}$ fatty acids (**Fig. 6**). Another one likely corresponds to $C_{18:0}/C_{22:6\text{-}OOH}$ molecular species

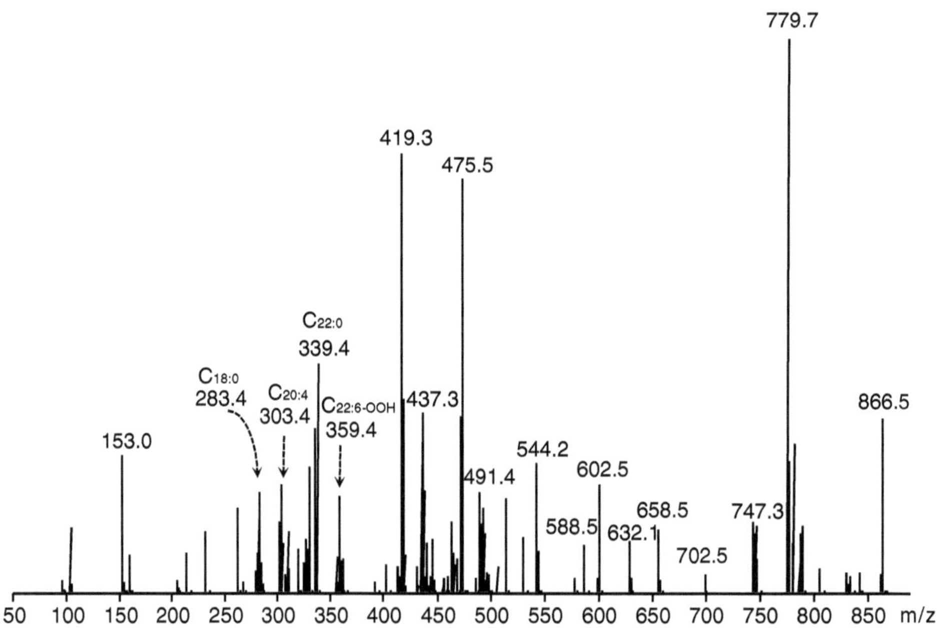

Fig. 6. Typical negative-ion MS^2 ESI mass spectrum of oxidized molecular species of PS from lung of mice exposed to TBI at a dose of 15 Gy. MS^2 fragmentation of molecular ion with *m/z* 866.5 revealed overlapping of nonoxidized PS corresponding to $C_{22:0}/C_{20:4}$ molecular species and oxidized PS corresponding to $C_{18:0}/C_{22:6\text{-}OOH}$ molecular species. Oxidized molecular species of PS (*m/z* 866.5) originated from molecular species with *m/z* 834.3 containing $C_{22:6}$ in *sn*-2 position after addition of two oxygens – $C_{22:6\text{-}OOH}$ (*m/z* 359.4).

originated from molecular species with *m/z* 834 3 containing $C_{22:6}$ in *sn*-2 position. Daughter ions at *m/z* 283.4 and 359.4 corresponding to $C_{18:0}$ and $C_{22:6\text{-OOH}}$ were detectable in MS^2 spectrum (**Fig. 6**). In the brain, PS molecular species containing $C_{22:6}$ ($C_{18:0}/C_{22:6}$) underwent oxidation after TBI. Molecular species at *m/z* 866 corresponding to $C_{18:0}/C_{22:6\text{-OOH}}$ was detected in MS spectrum (data not shown). No oxidation products are found in the most dominant phospholipid classes – PC and PE – in all five examined tissues from irradiated mice.

4. Notes

1. To minimize the risk of oxidation of phospholipids during isolation, extraction of lipids has to be performed on ice. For the same reason, DTPA/EGTA-pretreated HPTLC plated have to be used to prevent catalytic oxidation of lipids by adventitious transition metals.
2. Phospholipid spots can be clearly seen on TLC plates under light. If spots are not readily observable, freezing the plates at −20°C for 10 min with subsequent spraying with water is recommended. Iodine vapors cannot be used for visualization of lipids for ESI-MS analysis or for analysis of hydroperoxy-phospholipids by fluorescence HPLC (Amplex Red protocol).
3. PLA_2-driven reaction results in the production of lyso-phospholipids and release of free fatty acids with hydroperoxide groups (FA-OOH). This treatment is particularly important for assessments of CL oxidation products. While MP-11 can catalyze reduction of some types of PL-OOH without pretreatment with PLA_2, nondeesterified CL-OOH does not effectively react with MP-11/Amplex Red reagents.
4. Analysis of oxidatively modified lipids is complicated by a large variety of peroxidized molecular species of phospholipids as well as their low abundance combined with a relatively low stability and oxidizability of polyunsaturated phospholipids.
5. Additional difficulties of MS analysis of oxidized molecular species of phospholipids are due to overlapping of different molecular species of individual phospholipids as well as their oxidized molecular species within clusters with the same *m/z*. Therefore, detailed fragmentation protocols are required for unequivocal identification and quantitative analysis of phospholipid hydroperoxides.
6. ESI-MS in combination with fluorescence HPLC is a powerful technique permitting quantitation and structural characterization

of individual molecular species of phospholipid hydroperoxides. These protocols are useful for understanding signaling functions of oxidized phospholipids in cell physiology and pathology.

Acknowledgments

This work was supported by grants from NIH U19 AIO68021, HL70755, and Human Frontier Science Program.

References

1. Farooqui AA, Horrocks LA. 1998. Lipid peroxides in the free radical pathophysiology of brain diseases. *Cell Mol. Neurobiol.* 18: 599–608
2. Pamplona R. 2008. Membrane phospholipids, lipoxidative damage and molecular integrity: a causal role in aging and longevity. *Biochim. Biophys. Acta* 1777: 1249–1262
3. Kagan VE. 1988. Lipid Peroxidation in Biomembranes. CRC Press, Boca Raton, FL: pp. 1–181
4. Poli G, Schaur RJ, Siems WG, Leonarduzzi G. 2008. 4-Hydroxynonenal: a membrane lipid oxidation product of medicinal interest. *Med. Res. Rev.* 28: 569–631
5. Nakazawa T, Terayama K, Okuaki H, Yukawa O. 1984. Localization of lipid peroxidation products in liposomes after gamma-irradiation. *Biochim. Biophys. Acta* 769: 323–329
6. Kergonou JF, Bernard P, Braquet M, Rocquet G. 1981. Effect of whole-body gamma irradiation on lipid peroxidation in rat tissues. *Biochimie* 63: 555–559
7. Hatoum OA, Otterson MF, Kopelman D, Miura H, Sukhotnik I, Larsen BT, Selle RM, Moulder JE, Gutterman DD. 2006. Radiation induces endothelial dysfunction in murine intestinal arterioles via enhanced production of reactive oxygen species. *Arterioscler. Thromb. Vasc. Biol.* 26: 287–294
8. Park EC, Yoon JB, Seong JS, Choi KS, Kong ES, Kim YJ, Park YM, Park EM. 2006. Effect of ionizing radiation on rat tissue: proteomic and biochemical analysis. *Prep. Biochem. Biotechnol.* 36: 19–35
9. Stark G. 1991. The effect of ionizing radiation on lipid membranes. *Biochim. Biophys. Acta* 1071: 103–122
10. Belikova NA, Jiang J, Tyurina YY, Zhao Q, Epperly MW, Greenberger J, Kagan VE. 2007. Cardiolipin-specific peroxidase reactions of cytochrome c in mitochondria during irradiation induced apoptosis.. *Int. J. Radiat. Oncol. Biol. Phys.* 69: 176–186
11. Paris F, Fuks Z, Kang A, Capodieci P, Juan G, Ehleiter D, Haimovitz-Friedman A, Cordon-Cardo C, Kolesnick R. 2001. Endothelial apoptosis as the primary lesion initiating intestinal radiation damage in mice. *Science* 293: 293–297
12. Umegaki K, Aoki S, Esashi T. 1995. Whole body X-ray irradiation to mice decreases ascorbic acid concentration in bone marrow: comparison between ascorbic acid and vitamin E. *Free Radic. Biol. Med.* 19: 493–497
13. Dürken M, Herrnring C, Finckh B, Nagel S, Nielsen P, Fischer R, Berger HM, Moison RM, Pichlmeier U, Kohlschütter B, Zander AR, Kohlschütter A. 2000. Impaired plasma antioxidative defense and increased nontransferrin-bound iron during high-dose chemotherapy and radiochemotherapy preceding bone marrow transplantation. *Free Radic. Biol. Med.* 28: 887–894
14. Facorro G, Sarrasague MM, Torti H, Hager A, Avalos JS, Foncuberta M, Kusminsky G. 2004. Oxidative study of patients with total body irradiation: effects of amifostine treatment. *Bone Marrow Transplant* 33: 793–798
15. Radford IR. 1999. Initiation of ionizing radiation-induced apoptosis: DNA damage-mediated or does ceramide have a role?. *Int. J. Radiat. Biol.* 75: 521–528
16. Niki E, Yoshida Y, Saito Y, Noguchi N. 2005. Lipid peroxidation: mechanisms, inhibition,

and biological effects. *Biochem. Biophys.Res. Commun.* 338: 668–676

17. Folch J, Lees M, Sloan-Stanley GH. 1957. A simple method for isolation and purification of total lipids from animal tissue. *J. Biol. Chem.* 226: 497–509
18. Böttcher CSF, Van Gent CM, Fries C. 1961. A rapid and sensitive sub-micro phosphorus determination. *Anal. Chim. Acta* 24: 203–204
19. Rouser G, Fkeischer S, Yamamoto A. 1970. Two dimensional then layer chromatographic separation of polar lipids and determination of phospholipids by phosphorus analysis of spots. *Lipids* 5: 494–496
20. Kagan VE, Tyurin VA, Jiang J, Tyurina YY, Ritov VB, Amoscato AA, Osipov AN, Belikova NA, Kapralov AA, Kini V, Vlasova II, Zhao Q, Zou M, Di P, Svistunenko DA, Kurnikov IV, Borisenko GG. 2005. Cytochrome c acts as a cardiolipin oxygenase required for release of proapoptotic factors. *Nature, Chem. Biol.* 4: 223–232
21. Tyurin VA, Tyurina YY, Kochanek PM, Hamil ton R, DeKosky ST, Greenberger JS, Bayir H, Kagan VE. 2008. Oxidative lipidomics of programmed cell death. *Meth. Enzymol.* 442: 375–393
22. Tyurin VA, Tyurina YY, Feng W, Mnuskin A, Jiang J, Tang M, Zhang X, Zhao Q, Kochanek PM, Clark RSB, Bayir H, Kagan VE. 2008. Mass-spectrometric characterization of phospholipids and their primary peroxidation products in rat cortical neurons during staurosporine-induced apoptosis. *J. Neurochem.* 107: 1614–1633
23. MacMillan DK, Murphy RC. 1995. Analysis of lipid hydroperoxides and long-chain conjugated keto acids by negative ion electrospray mass spectrometry. *J. Am. Soc. Mass Spectrom.* 6: 1190–1201
24. Schwartz JC, Syka JEP, Quarmby ST. 2005. Improving the fundamentals of MSn on 2D linear uon traps: new ion activation and isolation techniques. *in Proccedings Amer. Soc. MS:*, San Antonio, Texas.
25. Hsu FF, Turk J. 2005. Studies on phosphatidylserine by tandem quadrupole and multiple stage quadrupole ion-trap mass spectrometry with electrospray ionization: structural characterization and the fragmentation processes. *J. Am. Soc. Mass Spectrom.* 16: 1510–1522
26. Hsu FF, Turk J, Rhoades ER, Russell DG, Shi Y, Groisman EA. 2005. Structural characterization of cardiolipin by tandem quadrupole and multiple-stage quadrupole ion-trap mass spectrometry with electrospran ionization. *J. Am. Soc. Mass Spectrom.* 16: 491–504
27. Ivanova P, Milne S, Byrne MO, Xiang Y, Brown HA. 2007. Glycerophospholipid identification and quantitation by electrospray mass spectrometry. *Meth. Enzymol.* 432: 21–57
28. Tyurina YY, Tyurin VA, Epperly MW, Greenberger JS 2008. Oxidative lipidomics of gamma-irradiation-induced intestinal injury. *Free Radic. Biol. Med.* 44: 299–314

Part II

Trafficking and Profiling

Chapter 10

Comprehensive mRNA Profiling of Lipid-Related Genes in Microglia and Macrophages Using Taqman Arrays

Richard Mauerer, Yana Walczak, and Thomas Langmann

Summary

Quantitative real-time reverse-transcription (RT)-PCR is a precise and sensitive method to measure mRNA levels over a broad dynamic range. This chapter describes the quantitative transcript analysis of 41 selected lipid-related transcripts in macrophages and microglia using a novel "Lipidomic" Taqman Array. The Taqman Array results show that (1) stimulation with the liver-X-receptor and retinoid-X-receptor ligands T0901317 and 9-*cis* retinoic acid induces several genes of lipid metabolism, (2) lipopolysaccharide (LPS) and interferon-γ (Ifn-γ) strongly repress lipid-related genes, and (3) coincubation with docosahexaenoic acid dampens the repressing effect of LPS. The method described in this chapter can be used to monitor the transcriptional response of 41 dynamic "lipid" genes simultaneously in any cell type.

Key words: Lipid-regulated genes, Lipidomic Taqman Array, Macrophages, Microglia

1. Introduction

Cellular lipidomics is defined as the analysis of metabolism, transport, and localization of lipids species within cells *(1)*. The quantitation of different lipid species from various biochemical pathways and biochemical analysis of lipid metabolism enzymes is an integral part of this concept *(2)*. Recent progress in the field of transcriptomics, mainly the cost reduction of DNA-microarrays and the development of high-throughput real-time reverse-transcription (RT)-PCR systems have also enabled researchers to perform a comprehensive transcriptomic analysis of all lipid-related genes *(3)*.

Donald Armstrong (ed.), *Lipidomics,* Methods in Molecular Biology, vol. 580,
DOI 10.1007/978-1-60761-325-1_10, © Humana Press, a part of Springer Science+Business Media, LLC 2009

Currently available DNA-microarrays cover the complete genomes from various species, including well-annotated human and mouse genes. However, in a given cell type, for example, macrophages or microglia (MG) cells, only a fraction of all lipid-related genes is regulated at the mRNA level. Moreover, DNA-microarrays have a lower dynamic range, are less precise, and are less sensitive than quantitative RT-PCR. Therefore, we used the previously established Taqman Array system *(4)* for the quantitative analysis of selected lipid-regulated genes in primary mouse macrophages and MG as well as the MG cell line BV-2. Our special interest was to study the expression of "lipidomic" genes in these cells under conditions mimicking sterol loading and proinflammatory activation.

We initially identified lipid-related genes in screening experiments with DNA-microarrays *(5, 6)* that (1) displayed a detectable expression in macrophages and MG, (2) showed a differential number of transcripts in lipid-stimulation and cytokine-activation experiments, and (3) were under the control of the key transcription factors of lipid metabolism. We selected 41 genes that fulfilled these criteria and grouped them into four major ontologies, sterol metabolism, fatty acid metabolism, lipid droplet, and transcription factors (*see* **Table 1**). With this approach, we sought to compile a set of genes with a maximum amount of biologically relevant information one can get from mRNA profiling experiments. In addition to these 41 "lipidomic" genes, we added four reference or housekeeping genes [18S ribosomal RNA (18S), beta-actin (Actb), ribosomal protein large 2 (Rplp2), and glyceraldehyde-3-phosphate dehydrogenase (Gadph)] for normalization, and three inflammation marker controls [tumor necrosis factor (Tnf), interleukin 1-beta (Il1b), and interleukin 6 (Il6)] for validation of proinflammatory stimulation experiments.

Our "lipidomic" Taqman Array is based on an Applied Biosystems 7900HT microfluidic card. This method allows simultaneous analysis of 41 lipid-related genes and 7 controls in 2 replicates of 4 different samples per run (*see* **Fig. 1**). The 2-μl small-volume design of each reaction well substantially decreases sample and reagent consumption, and the precaptured primers and probes save time by reducing labor-intensive pipetting steps.

To evaluate this Taqman Array for lipidomic research and to further characterize the mRNA regulation of lipid-related genes in macrophages and MG, we monitored the expression of mRNAs in primary murine bone marrow-derived macrophages (BMM), murine primary MG, and the MG cell line BV-2. The cells were stimulated with the LXR and RXR agonists T0901317 and 9-*cis* retinoic acid (RA) to mimic sterol loading, LPS and Ifn-γ for

Table 1
List of genes represented on the "lipidomic" Taqman Array

No.	Gene Name	Gene Symbol	Taqman Assay ID
Reference genes			
1	Glyceraldehyde 3-phosphate dehydrogenase	Gapdh	Mm99999915_g1
2	18S rRNA	18S	Hs99999901_s1
3	Actin-β	Actb	Mm00607939_s1
4	Ribosomal protein, large P2	Rplp2	Mm00782638_s1
Sterol metabolism			
5	ATP-binding cassette transporter A1	Abca1	Mm01350760_m1
6	ATP-binding cassette transporter G1	Abcg1	Mm01348250_m1
7	Lipoprotein lipase	Lpl	Mm00434770_m1
8	Cd36 scavenger receptor	Cd36	Mm00432403_m1
9	LDL-Receptor	Ldlr	Mm00440169_m1
10	Apolipoprotein E	Apoe	Mm00437573_m1
11	HMG-CoA-reductase	Hmgcr	Mm01282499_m1
12	HMG-CoA-synthase 1	Hmgcs1	Mm00524111_m1
13	Insulin-induced gene 1	Insig1	Mm00463389_m1
14	START domain containing 4, sterol regulated	Stard4	Mm00505395_m1
15	Acetyl-coenzyme A acetyltransferase 1	Acat1	Mm00507463_m1
16	Acetyl-coenzyme A acetyltransferase 2	Soat2	Mm00448823_m1
17	CYP27A1	Cyp27a1	Mm00470430_m1
Fatty acid metabolism			
18	Solute carrier family 16, member 6	Mct6	Mm00506192_m1
19	Solute carrier family 27, member 1	Fatp1	Mm00449511_m1
20	Solute carrier family 27, member 3	Fatp3	Mm01220009_g1
21	Glycerol kinase	GK	Mm00433896_m1
22	Acyl-coenzyme A dehydrogenase, very long chain	Acadvl	Mm00444296_m1
23	Fatty acid-binding protein 4	FABP4	Mm00445878_m1
24	Fatty acid-binding protein 5	FABP5	Mm00783731_s1
25	Fatty acid desaturase 1	FADS1	Mm00507605_m1

(continued)

Table 1 (continued)

No.	Gene Name	Gene Symbol	Taqman Assay ID
26	Fatty acid desaturase 2	FADS2	Mm00517221_ml
27	Fatty acid desaturase 3	FADS3	Mm00517643_ml
28	Stearoyl-CoA desaturase (δ-9-desaturase)	Scd1	Mm00772290_ml
29	Hydroxyacyl-coenzyme A dehydrogenase β subunit	Hadhb	Mm00523880_gl
30	Uncoupling protein 2	Ucp2	Mm00627599_ml
Lipid droplet			
32	Adipophilin	Adfp	Mm00475794_ml
32	Phospholipase A2, group IVA	Pla2g4a	Mm00447040_ml
33	Prostaglandin-endoperoxide synthase 2	Cox2	Mm00478374_ml
34	Leukotriene C4 synthase	Ltc4s	Mm00521864_ml
35	Thromboxane A synthase 1	Tbxas1	Mm00495553_ml
36	Arachidonate 12-lipoxygenase	Alox12	Mm00545833_ml
37	Arachidonate 15-lipoxygenase	Alox15	Mm01250458_ml
38	Arachidonate 5-lipoxygenase	Alox5	Mm01182743_ml
39	Arachidonate 5-lipoxygenase-activating protein	Alox5ap	Mm00802100_ml
40	Leukotriene A4 hydrolase	Lta4h	Mm01246216_ml
Transcription factors			
41	Sterol regulatory element-binding transcription factor 1	Srebf1	Mm01138344_ml
42	Sterol regulatory element-binding transcription factor 2	Srebf2	Mm01306283_ml
43	LXRα	Lxra	Mm00443451_ml
44	PPARδ	Ppard	Mm01305433_ml
45	PPARγ	Pparg	Mm01184322_ml
Inflammatory markers			
46	Tumor necrosis factor	Tnf	Mm00443258_ml
47	Interleukin 1-beta	Il1b	Mm01336189_ml
48	Interleukin 6	Il6	Mm00446191_ml

	1	2	3	4	5	6	7	8	9	10	11	12	13	14	15	16	17	18	19	20	21	22	23	24	Port
A	Abca1	Abcg1	Acat1	Soat2	Actb	Acadvl	Adfp	Apoe	Alox1	Alox15	18S	Alox5	Alox5a	Cd36	Cyp27a	Fabp4	Fabp5	Fads1	Fads2	Fads	Gapdh	Gyk	Hmgcr	Hmgcs	1
B	Hadhb	Insig1	Il1b	Il6	Ldlr	Lta4h	Ltc4s	Lpl	Lxra	Pla2g4a	Ppard	Pparg	Ptgs2	Rplp2	Mc6	Fatp1	Fatp3	Srebf1	Srebf2	Stard	Scd1	Tbxas1	Tnf	Ucp2	
C	Abca1	Abcg1	Acat1	Soat2	Actb	Acadvl	Adfp	Apoe	Alox1	Alox15	18S	Alox5	Alox5a	Cd36	Cyp27a	Fabp4	Fabp5	Fads1	Fads2	Fads	Gapdh	Gyk	Hmgcr	Hmgcs	2
D	Hadhb	Insig1	Il1b	Il6	Ldlr	Lta4h	Ltc4s	Lpl	Lxra	Pla2g4a	Ppard	Pparg	Ptgs2	Rplp2	Mc6	Fatp1	Fatp3	Srebf1	Srebf2	Stard	Scd1	Tbxas1	Tnf	Ucp2	
E	Abca1	Abcg1	Acat1	Soat2	Actb	Acadvl	Adfp	Apoe	Alox1	Alox15	18S	Alox5	Alox5a	Cd36	Cyp27a	Fabp4	Fabp5	Fads1	Fads2	Fads	Gapdh	Gyk	Hmgcr	Hmgcs	3
F	Hadhb	Insig1	Il1b	Il6	Ldlr	Lta4h	Ltc4s	Lpl	Lxra	Pla2g4a	Ppard	Pparg	Ptgs2	Rplp2	Mct6	Fatp1	Fatp3	Srebf1	Srebf2	Stard	Scd1	Tbxas1	Tnf	Ucp2	
G	Abca1	Abcg1	Acat1	Soat2	Actb	Acadvl	Adfp	Apoe	Alox1	Alox15	18S	Alox5	Alox5a	Cd36	Cyp27a	Fabp4	Fabp5	Fads1	Fads2	Fads	Gapdh	Gyk	Hmgcr	Hmgcs	4
H	Hadhb	Insig1	Il1b	Il6	Ldlr	Lta4h	Ltc4s	Lpl	Lxra	Pla2g4a	Ppard	Pparg	Ptgs2	Rplp2	Mct6	Fatp1	Fatp3	Srebf1	Srebf2	Stard	Scd1	Tbxas1	Tnf	Ucp2	
I	Abca1	Abcg1	Acat1	Soat2	Actb	Acadvl	Adfp	Apoe	Alox1	Alox15	18S	Alox5	Alox5a	Cd36	Cyp27a	Fabp4	Fabp5	Fads1	Fads2	Fads	Gapdh	Gyk	Hmgcr	Hmgcs	5
I	Hadhb	Insig1	Il1b	Il6	Ldlr	Lta4h	Ltc4s	Lpl	Lxra	Pla2g4a	Ppard	Pparg	Ptgs2	Rplp2	Mct6	Fatp1	Fatp3	Srebf1	Srebf2	Stard	Scd1	Tbxas1	Tnf	Ucp2	
K	Abca1	Abcg1	Acat1	Soat2	Actb	Acadvl	Adfp	Apoe	Alox1	Alox15	18S	Alox5	Alox5a	Cd36	Cyp27a	Fabp4	Fabp5	Fads1	Fads2	Fads	Gapdh	Gyk	Hmgcr	Hmgcs	6
L	Hadhb	Insig1	Il1b	Il6	Ldlr	Lta4h	Ltc4s	Lpl	Lxra	Pla2g4a	Ppard	Pparg	Ptgs2	Rplp2	Mct6	Fatp1	Fatp3	Srebf1	Srebf2	Stard	Scd1	Tbxas1	Tnf	Ucp2	
M	Abca1	Abcg1	Acat1	Soat2	Actb	Acadvl	Adfp	Apoe	Alox1	Alox15	18S	Alox5	Alox5a	Cd36	Cyp27a	Fabp4	Fabp5	Fads1	Fads2	Fads	Gapdh	Gyk	Hmgcr	Hmgcs	7
N	Hadhb	Insig1	Il1b	Il6	Ldlr	Lta4h	Ltc4s	Lpl	Lxra	Pla2g4a	Ppard	Pparg	Ptgs2	Rplp2	Mct6	Fatp1	Fatp3	Srebf1	Srebf2	Stard	Scd1	Tbxas1	Tnf	Ucp2	
O	Abca1	Abcg1	Acat1	Soat2	Actb	Acadvl	Adfp	Apoe	Alox1	Alox15	18S	Alox5	Alox5a	Cd36	Cyp27a	Fabp4	Fabp5	Fads1	Fads2	Fads	Gapdh	Gyk	Hmgcr	Hmgcs	8
P	Hadhb	Insig1	Il1b	Il6	Ldlr	Lta4h	Ltc4s	Lpl	Lxra	Pla2g4a	Ppard	Pparg	Ptgs2	Rplp2	Mct6	Fatp1	Fatp3	Srebf1	Srebf2	Stard	Scd1	Tbxas1	Tnf	Ucp2	

Fig. 1. Lipidomic Taqman Array map.

proinflammatory stress, and LPS in combination with DHA as anti-inflammatory lipid agonist. We identified a strong impact of treatment with these compounds on the mRNA expression of the majority of selected lipid-related genes studied with our Taqman Array.

The method described in this chapter can be used to rapidly and accurate quantify transcriptionally dynamic "lipid" genes in any cell type. The Lipidomic Taqman Array may be applied to study lipid disorders or to quantify the transcriptional effects of pharmacological treatments on lipid-related genes.

2. Materials

2.1. Equipment

1. NanoDrop 1000 spectrophotometer (PeqLab).
2. 2100 Bioanalyzer (Agilent).
3. Multifuge 3L (Heraeus), centrifuge buckets, and adapters for Taqman Arrays.
4. Applied Biosystems 7900HT Fast Real-Time System with a Taqman Array block (Applied Biosystems).

2.2. Reagents

1. Collagenase type I (Sigma).
2. Hyaluronidase (Sigma).
3. DNase I (Roche).
4. DMEM-medium (PAA).
5. Penicilin/Streptomycin (PAA).
6. 70-μm Cell strainer (Becton Dickinson).
7. Recombinant human macrophage colony stimulating factor (M-CSF) (R&D Systems).
8. Red blood cell lysis buffer (Sigma).
9. LPS from *Escherichia coli* 0111:B4 (Sigma).
10. Recombinant mouse Ifn-γ (R&D Systems).
11. DHA (Biozol).
12. 9-*cis* RA (Sigma).
13. T0901317 (Sigma).
14. Phosphate-buffered saline (PBS).
15. Ethanol (100%, 70%).

2.3. Supplies

1. RNeasy Mini Kit (Qiagen).
2. RNA 6000 Nano LabChip Kit (Agilent).

3. Affinity script multiple temperature cDNA synthesis kit (Stratagene).
4. TaqMan Arrays (Applied Biosystems).
5. Gene Expression Master Mix (Applied Biosystems).

3. Methods

3.1. Cell Isolation

3.1.1. Protocol for BV-2 MG

1. Culture adherent BV-2 cells in DMEM + 10% FCS + 1% Pen/Strep.
2. Plate the cells in 10 cm culture dishes at a density of 1×10^6/ml medium.

3.1.2. Protocol for Mouse BMM

1. Sacrifice 2–3 months old C57/BL6 mice with CO_2 and spray the mouse with 70% ethanol.
2. Using a sharp scissor, cut and dissect tibias and femurs from muscles and tendons.
3. Place the tibias and femurs in 10 cm tissue culture dishes on ice and remove the knee with a scissor.
4. Drill a 27 gauge needle connected to a 5 ml syringe into the end of the femur and tibia and flush 2 ml of DMEM + 10% FCS + 1% Pen/Strep through the femur and tibia.
5. Resuspend cells by pipetting up and down and put the cells into a 15 ml polypropylene tube.
6. Centrifuge the cells at $950 \times g$ for 10 min.
7. Discard the supernatant and add 2 ml red blood cell lysis buffer.
8. Incubate for 7 min and add 5 ml DMEM + 10% FCS + 1% Pen/Strep.
9. Centrifuge at $950 \times g$ for 10 min.
10. Discard the supernatant and resuspend the cell pellet in 10 ml DMEM + 10% FCS + 1% Pen/Strep supplemented with 50 ng/ml M-CSF.
11. Plate cells in culture dishes at a density of 1×10^6/ml medium.
12. Maintain cells in a humidified 37°C incubator for 7 days.

3.1.3. Protocol for Brain MG

1. Cut skin along the midline in a caudal to rostral direction and expose the skull.

2. Using a sharp scissor, cut the skull and run a pair of curved forceps under the hemispheres to scoop out the cortex.
3. Remove meninges and superficial vasculature in a petri dish.
4. Cut 300 mg brain portions into small pieces and incubate for 45 min at 37°C in 1 ml PBS with 1 mg/ml collagenase type I, 0.3 mg/ml DNase I, and 0.2 mg/ml hyaluronidase.
5. Add 10 ml DMEM + 10% FCS + 1% Pen/Strep and filter the cell suspension through a 70 μm cell strainer.
6. Seed the cells in 75 cm^2 flasks containing 10 ml DMEM + 10% FCS + 1% Pen/Strep supplemented with 50 ng/ml M-CSF.
7. Maintain cells in a humidified 37°C incubator for 10 days.

3.2. Cell Stimulation and Harvest

1. Prepare stock solutions for 9-*cis* RA (1 mM), T0901317 (5 mM), LPS (1 mg/ml), Ifn-γ (100 μg/ml), and DHA (250 mg/ml).
2. Stimulate BV-2 cells, BMM, and brain MG with the following final concentrations for 24 h: 9-*cis* RA (10 μM) + T0901317 (10 μM), LPS (20 ng/ml) + Ifn-γ (50 ng/ml), and LPS (20 ng/ml) + DHA (100 μM).
3. Following stimulation, wash the cells twice with PBS and add 600 μl cell lysis buffer RLT (Qiagen) + 6 μl β-mercaptoethanol per 10 cm culture dish.
4. Store RLT-lysate at –80°C before RNA isolation.

3.3. RNA Isolation

Total RNA isolation can be performed with any standard method. However, we recommend to use a method with minimal phenol contamination and genomic DNA fragments. RNase-free conditions should be kept throughout the procedure by using DEPC-treated chemicals. The following procedure is adapted from the RNeasy Mini Kit (Qiagen) (*see* **Note 1**).

1. Pass the RLT-lysate ten times through a blunt 20-gauge needle fitted to an RNase-free syringe.
2. Pipet the lysate directly into a QIAshredder (Qiagen) spin column placed in a 2 ml collection tube, centrifuge for 2 min at full speed in a table top centrifuge.
3. Add one volume of 70% ethanol to the homogenized lysate, and mix well by pipetting up and down.
4. Transfer up to 700 μl of the sample, including any precipitate that may have formed, to an RNeasy spin column placed in a 2 ml collection tube and centrifuge 1 min at 13,000 rpm.
5. Discard the flow-through, add 700 μl buffer RW1 to the RNeasy spin column, and centrifuge for 1 min at 13,000 rpm.

6. Discard the flow-through, add 500 μl buffer RPE to the RNeasy spin column, and centrifuge for 1 min at 13,000 rpm.
7. Discard the flow-through, add 500 μl buffer RPE and centrifuge 2 min at 13,000 rpm.
8. Place the RNeasy spin column in a RNase-free 1.5 ml collection tube and add 50 μl RNase-free water directly to the spin column membrane.
9. Centrifuge for 1 min at 13,000 rpm and collect total RNA.
10. Store total RNA at −80°C.

3.4. RNA Quantification and Quality Check

Total RNA samples can be quantified using spectrophotometry and quality checked by determining the A260/A280 ratio. Optimal RNA preparations have a concentration of >500 ng/ml and an A260/A280 ratio >1.8. We recommend to use the Nanodrop 100 system for quantification and the Agilent 2100 bioanalyzer for quality control (*see* **Note 2**).

3.5. cDNA Synthesis

We routinely use Stratagene's AffinityScript reverse transcriptase, a genetically engineered version of MMLV reverse transcriptase. This RT provides very good cDNA yields and full-length cDNAs (*see* **Note 3**).

1. Prepare the first-strand cDNA synthesis reaction in a RNase-free microcentrifuge tube by adding the following components: 1 μg of total RNA, 3 μl Random Primers (0.1 μg/μl), and RNase-free water to a total volume of 15.7 μl.
2. Incubate the reaction at 65°C for 5 min to denature RNA secondary structures.
3. Cool the reaction at room temperature for 10 min to allow the primers to anneal to the RNA.
4. Add the following components to each reaction, for a final reaction volume of 20μl: 2.0 μl of ten times AffinityScript RT Buffer, 0,8 μl of dNTP mix (25 mM each), 0.5 μl of RNase Block Ribonuclease Inhibitor (40 U/μL), and 1 μl of Affinity Script Multiple Temperature RT.
5. Incubate the reaction at 25°C for 10 min.
6. Mix the reaction components gently, and then place the tube in a temperature-controlled thermal block at 42°C for 30 min (*see* **Note 4**).
7. Terminate the reaction by incubating at 70°C for 15 min.
8. Place the completed first-strand cDNA synthesis reactions on ice or store at −20°C for later use.

3.6. Reaction Setup for Lipidomic Taqman Arrays

The Lipidomic Taqman Array contains an Array of 384 reaction wells for two-step RT-PCR. The wells contain preloaded Taqman Gene Expression Assays for real-time amplification and relative mRNA quantification of 41 lipid-related genes, 4 reference genes, and 3 inflammatory markers (*see* **Table 1** and **Fig. 1**). The Taqman Arrays are shipped with an Array information CD-ROM that contains the Assay Information File (AIF), two Array map files, and a SDS setup file. The AIF contains gene annotation information for each Gene Expression Assay on the Taqman Array, including primers and reporter dye concentrations, the context sequence, and targeted exons. The Array map files show the position of the assays and replicate distribution on the Taqman Array.

1. Thaw cDNA samples and dilute to 20 ng/µl with dH_2O.
2. For each sample, add the following components to a 1.5 ml microfuge tube: 5 µl cDNA sample (100 ng), 45 µl dH_2O, 50 µl Taqman Gene Expression Master Mix (two times).
3. Mix tubes thoroughly by gentle vortexing and spin briefly to eliminate air bubbles.
4. Remove Taqman Arrays from the packaging and bring to room temperature to avoid condensed water at the optical side of the plates.
5. Load 100 µl of each sample-specific PCR reaction mix into a 100 µl pipette, place the tip in the fill port, and dispense the reaction mix into the fill reservoir (*see* **Note 5**).
6. Place the Array holder on a lab bench and insert Taqman Arrays with the fill reservoirs upward and the reaction wells in the same direction as the "this side out" label on the Array holder.
7. Place the filled Array holder in the centrifuge bucket and centrifuge two times for 1 min at 331 × *g*.
8. Examine Taqman Arrays to determine whether filling of the reaction wells is complete (*see* **Note 6**).
9. Use the microfluidic card sealer to isolate the wells of the Taqman Array by sealing the main fluid distribution channels.
10. Trim the fill reservoirs from the Taqman Array using scissors.

3.7. Running Lipidomic Taqman Arrays on a 7900HT Fast Real-Time PCR System

1. Open the Applied Biosystems SDS software version v2.1 or higher. When using version v2.3 import the setup file from the CD-ROM supplied with the Taqman Array into a new plate document.
2. Select relative quantification (RQ) as assay type and 384-well Taqman Array as container.
3. Make sure that the default thermal cycling conditions for Taqman Arrays are set: AmpliTaq Gold DNA polymerase activation at 94.5°C for 10 min, and each of 40 cycles: melting at 97°C for 30 s, and annealing/extension at 59.7°C for 1 min.
4. Save the plate document as SDS 7900HT template file (*.sdt).

5. Enter sample names, define endogenous control (reference gene), and save the plate document as *.sds file.
6. Connect to the SDS 7900HT instrument and select Real-time run.
7. Place the prepared Taqman Array in the instrument tray and start run (*see* **Note 7**).
8. When the run is complete close the Run Complete dialog box and remove the array from the instrument.

3.8. Data Analysis for RQ

Relative gene expression values are obtained from Taqman Arrays using the comparative Ct method for RQ *(7)*. In this method, quantity is expressed relative to a calibrator cDNA sample, which is set to 1 and all other quantities are expressed as an *n*-fold difference relative to this cDNA sample. In our experiments, we have selected cDNA from unstimulated cells (BMM, MG, and BV-2 cells) as calibrators for the individual stimulations.

1. Open the Applied Biosystems Relative Quantification Manager Software v1.2 and create a new RQ study from the SDS file created in **Subheading 3.6**.
2. Load all plates from one experiment into the RQ study document, and mark all wells to be analyzed.
3. Select the endogenous control (Gapdh, 18S rRNA, Actb, or Rplp2).
4. Select the default automatic baseline and threshold option for automatic Ct calculation and check the amplification plots (*see* **Note 8**).
5. Select the cDNA sample to be used as the calibrator.
6. Save the RQ study as *.sdm file.
7. Analyze all samples and check the gene expression plot. This plot shows the fold-difference of the target sample relative to the calibrator as log10 values.
8. Export the results data as *.txt file.
9. Open the *.txt file in an Excel spreadsheet. For better visualization at a linear scale, RQ values <1 are converted to negative RQ values with the simple equation RQ(lin) = −1/ RQ value (*see* **Note 9**).
10. Visualize the data as table or 3D bar graph.

3.9. Results

Using the Lipidomic Taqman Array, the relative expression levels of 41 lipid-related genes was quantified in BMM, primary brain MG, and BV-2 microglial cells (BV2) stimulated with T0901317 + 9-*cis* RA (**Fig. 2a**), LPS + Ifn-γ (**Fig. 2b**), and LPS + DHA (**Fig. 2c**). These results show that (1) stimulation of all three myeloid cell types with T0901317 + 9-cis RA has an inductive effect on several genes of lipid metabolism, (2) the proinflam-

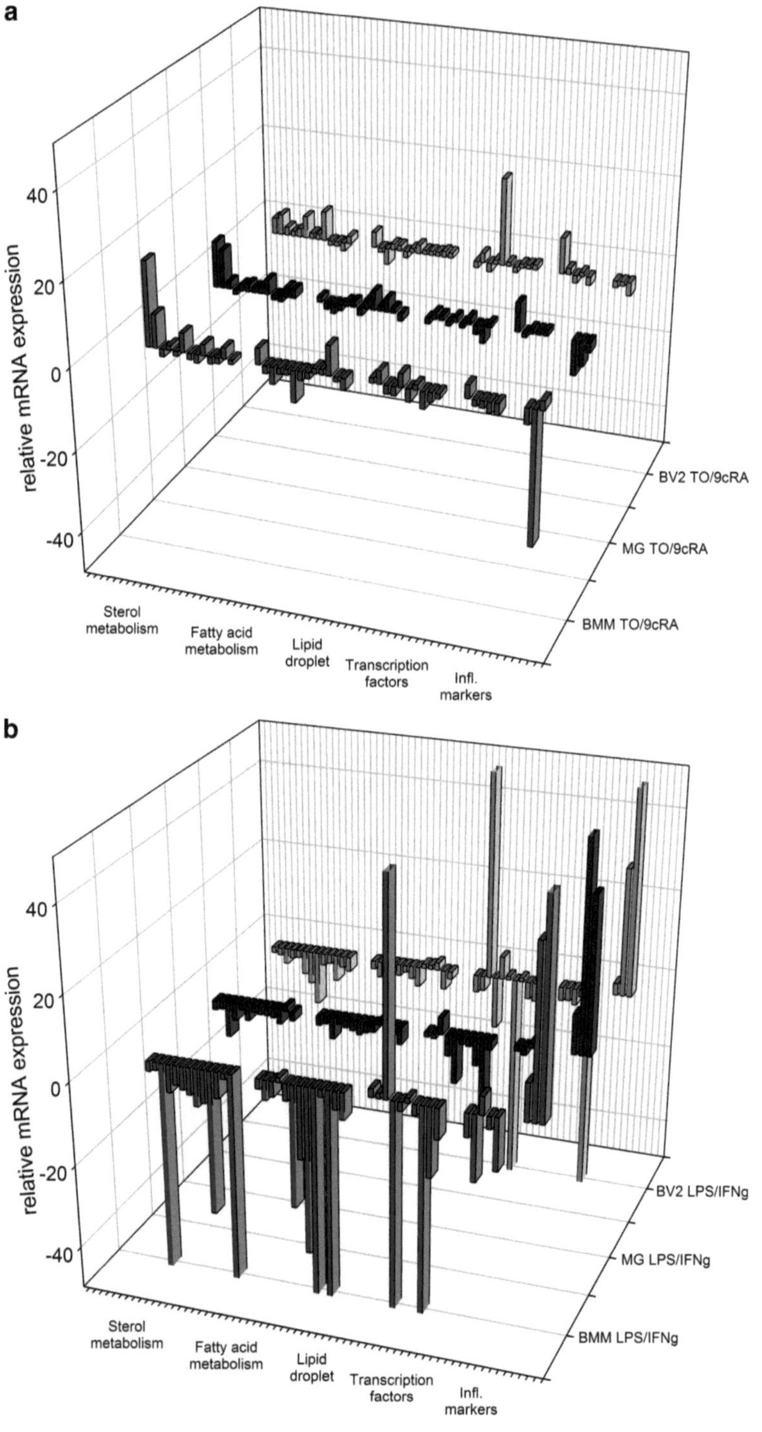

Fig. 2. Differential expression analysis of 41 lipid-related genes and inflammatory markers covered by the Lipidomic Taqman Array. The relative mRNA expression is displayed for bone marrow-derived macrophages (*BMM*), primary brain microglia (*MG*), and BV-2 microglial cells (*BV2*) stimulated with T0901317 + 9-*cis* RA (**a**), LPS + Ifn-γ (**b**), and LPS + DHA (**c**). Relative quantification was performed with the ΔΔCt method and normalization to the reference gene Gapdh.

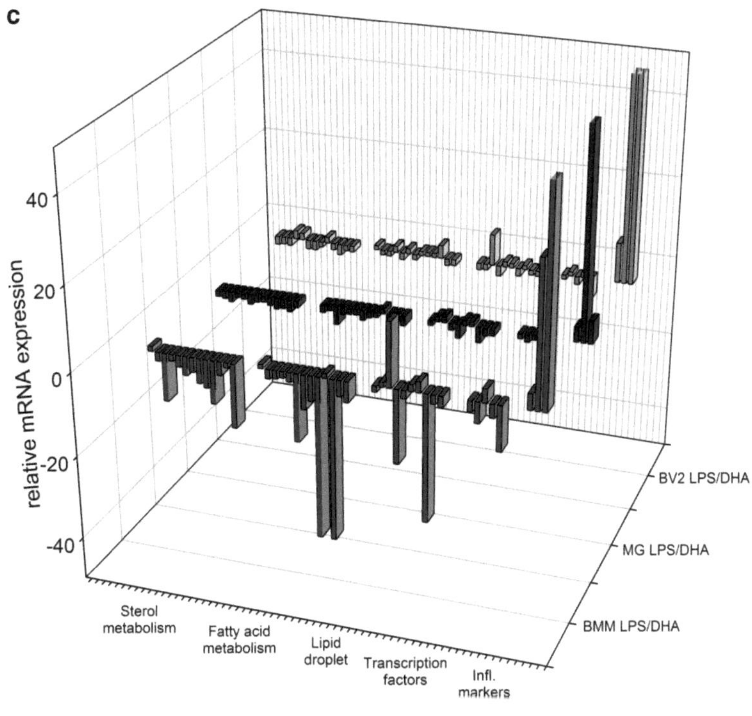

Fig. 2. (continued)

matory stimuli LPS and Ifn-γ strongly repress most lipid-related genes, with the exception of Cox2, and (3) coincubation with DHA attenuates the repressing effect of LPS. These data also implicate that BMM display a highly dynamic lipid-related transcriptome, whereas gene expression in primary brain microglia and the BV-2 cell line is regulated at a lower level. Taken together, the Lipidomic Taqman Array is a suitable tool to study quantitative changes in gene expression of the most dynamic lipid-related genes. For lipidomic studies in macrophage-like cells, primary BMM seem to be the most dynamic model system.

4. Notes

1. We do recommend commercially available RNA purification kits as they provide a good and reproducible quality required for the downstream applications.
2. Any UV/Vis spectrophotometer can be used to measure total RNA concentrations. However, we routinely use the Nanodrop 1000 system, because this method requires only 2 μl RNA sample. The Nanodrop has a broad quantitative range of 2–3,700 ng/μl and directly displays reproducible

concentration plus A260/A280 ratios. The RNA 6000 Nano chip also requires only 1 μl sample and has a quantitative range of 25–500 ng/μl. The 2100 bioanalyzer expert software assigns a specific quality number to the RNA sample based on its electrophoretic profile. The RNA Integrity Number (RIN) ranges from 1 (totally degraded RNA) to 10 (completely intact RNA). Only high-quality RNA, with RIN greater than 8 and A260/280 greater than 1.8 should be considered for Taqman Array analysis.

3. Experimental variation in quantitative RT-PCR is mainly attributable to the reverse transcription step. Optimal reverse transcription efficiency and full-length transcripts are prerequisites for quantitative RT-PCR results. Since primers and probes may be located anywhere in cDNA targets, we recommend to use random hexamer primers for cDNA synthesis of long transcripts. Also, random primers are required to reverse transcribe ribosomal RNAs when using 18S rRNA as endogenous control for normalization. Priming with oligo dT primers may also be suitable after optimization for specific genes. In general, MMLV reverse transcriptases have higher yields than AMV enzymes. For optimal comparison, cDNA synthesis for quantitative RT-PCR should be performed with similar RNA concentrations.

4. AffinityScript multiple temperature reverse transcriptase performs optimally over the full range of 42–55°C. Typically, 42°C is a good starting point. For RNAs containing secondary structure and other challenging targets, a synthesis temperature of 55°C may be used without loss of performance.

5. Pipette the entire 100 μl into the fill reservoir. Do not allow the tip to contact and damage the coated foil beneath the fill port. Be careful when pushing the pipette plunger to its second stop position. No air should be released that could push the reaction mix out of the fill reservoir via the vent port.

6. If there is excess cDNA sample remaining in the fill reservoir, centrifuge the Taqman Array again for 1 additional minute. Do not exceed 331 × *g* or accumulated centrifugation times more than 3 min, since this may deform the Array.

7. We recommend running Taqman Arrays as soon as possible after completing the reaction setup. Prepared Taqman Arrays may be stored for up to 8 h at 4°C in the dark.

8. In case of atypical amplification curves, manual Ct calculation with self-defined baseline and thresholds can be selected. The amplification curve should begin after the maximum baseline and the threshold should be set in the exponential phase of the amplification curve.

9. The ranges of the RQ values can be calculated independently from the RQ manager software by use of the equation RQ = $2^{-\Delta\Delta Ct}$, with $\Delta\Delta$Ct + SD and $\Delta\Delta$Ct – SD of the individual replicate measurements.

Acknowledgments

We thank Dr. Andrea Geiger and Dr. Astrid Ferlinz for helpful comments and discussions. This work was supported in part by Applied Biosystems and the German Research Foundation (LA1203/6-1).

References

1. van Meer, G. (2005) Cellular lipidomics. *EMBO J.* **24**, 3159–3165.
2. Liebisch, G., Binder, M., Schifferer, R., Langmann, T., Schulz, B., Schmitz, G. (2006) High throughput quantification of cholesterol and cholesteryl ester by electrospray ionization tandem mass spectrometry (ESI-MS/MS). *Biochim Biophys Acta.* **1761**, 121–128.
3. Langmann, T., Liebisch, G., Moehle, C., Schifferer, R., Dayoub, R., Heiduczek, S., Grandl, M., Dada, A., Schmitz, G. (2005) Gene expression profiling identifies retinoids as potent inducers of macrophage lipid efflux. *Biochim Biophys Acta.* **1740**, 155–161.
4. Langmann, T., Mauerer, R., Schmitz, G. (2006) Human ATP-binding cassette transporter TaqMan low-density array: analysis of macrophage differentiation and foam cell formation. *Clin Chem.* **52**, 310–313.
5. Ebert, S., Schoeberl, T., Walczak, Y., Stoecker, K., Stempfl, T., Moehle, C., Weber, B.H., Langmann, T. (2008) Chondroitin sulfate disaccharide stimulates microglia to adopt a novel regulatory phenotype. *J Leukoc Biol.* 2008 **84**, 736–740.
6. Weigelt, K., Ernst, W., Walczak, Y., Ebert, S., Loenhardt, T., Klug, M., Rehli, M., Weber, B.H., Langmann, T. (2007) Dap12 expression in activated microglia from retinoschisin-deficient retina and its PU.1-dependent promoter regulation. *J Leukoc Biol.* **82**, 1564–1574.
7. Schmittgen, T.D., Livak, K.J. (2008) Analyzing real-time PCR data by the comparative C(T) method. *Nat Protoc.* **3**, 1101–1108.

Chapter 11

Imaging Lipid Membrane Domains with Lipid-Specific Probes

Françoise Hullin-Matsuda, Reiko Ishitsuka, Miwa Takahashi, and Toshihide Kobayashi

Summary

Imaging membrane lipid domains to characterize their organization and function has been hindered by the lack of reliable lipid-specific probes. In this paper, we provide detailed methods to investigate, mainly by confocal microscopy, the distribution and dynamics of two components of the "lipid rafts," sphingomyelin (SM) and cholesterol, using two specific lipid probes that have been extensively studied in the laboratory: lysenin, a SM-binding toxin and the fluorescent esters of poly(ethylene glycol) cholesteryl ether (PEG-Chol) that label cholesterol-rich domains. The production of nontoxic forms of lysenin as well as its specific binding behavior have allowed monitoring the distribution and the dynamics of SM-rich domains in living cell membranes. Because of its water-solubility and low toxicity, the fluorescent PEG-Chol can be used to follow the reorganization of cell surface cholesterol-rich domains as well as intracellular cholesterol dynamics in living cells. These probes can thus provide important informations on lipid distribution and traffic in living cell membranes.

Key words: Cholesterol, Endocytosis, Lipid probe, Membrane domains, Sphingomyelin, Lipid raft

1. Introduction

Improvement of morphological techniques for analysis of the complex lipid organization in cellular membrane has indicated that lipid molecules are not homogenously distributed but organized into specific domains that can be observed in different cellular localizations (e.g., plasma membrane, Golgi, endosomes). These lipid domains are considered to be important in many cellular processes including signal transduction, membrane trafficking,

Donald Armstrong (ed.), *Lipidomics*, Methods in Molecular Biology, vol. 580,
DOI 10.1007/978-1-60761-325-1_11, © Humana Press, a part of Springer Science+Business Media, LLC 2009

cytoskeletal organization, cell adhesion, cell migration, and pathogen entry *(1–3)*. The existence and function of these domains referred as "lipid rafts" have been first, and are still often, demonstrated by indirect biochemical methods such as using nonionic detergent at low temperature or cholesterol depletion with methyl-β-cyclodextrin (MβCD). However, some problems with these techniques have been pointed out *(4, 5)*. Therefore, the need for reliable methods to visualize the distribution and dynamics of cellular lipids has emerged in the recent years *(6)*. Only a few lipids can be visualized directly because of their autofluorescence and in contrast to the proteins, lipids cannot be modified by molecular biology tools, i.e., by introduction of a fluorescent sequence like the green fluorescent protein (GFP). Thus, two different strategies have been developed both presenting advantages and disadvantages that should be carefully considered according to the purpose of experiments. Because of space limit, we will only present briefly some basic points hoping that these considerations will be useful in designing experiments. To study the distribution and dynamics of cellular lipids, there are two types of methods based first, on labeling the endogenous lipids with lipid-binding molecules and second, on using fluorescent lipid analogs. The use of lipid-binding probes (anti-lipid antibodies, lipid-binding peptides, or toxins) seems the only way to examine the true distribution and dynamics of endogenous lipids based on the specificity of the lipid targeting. However, the binding of the probes is often affected by the organization of the lipids in the membrane, i.e., curvature, lipid clusters, and so on. Furthermore, the binding of the probe itself may alter the distribution and dynamics of the lipids, i.e., shielding lipid accessibility to regulatory proteins, avoiding lipid transbilayer movement. The use of fluorescent lipid analogs such as C6-NBD and C5-BODIPY that carry short fluorescent and relatively polar fatty acids *(7, 8)* is the other alternative. These hydrophilic analogs present the advantage of efficient cellular incorporation and studies of their transbilayer movement suggest that they are useful to examine the role of lipid head-group (but not fatty acids), e.g., to characterize flippase activity *(9)*. However, nobody really knows whether endogenous lipids and exogenously added lipids behave similarly *(10)*. Therefore, the best method will be to image individual lipids without using probes, but unfortunately this is technically quite impossible at the moment. Thus, the best method choice will be dependent on the purpose of the experiment: choosing lipid-binding probes when you want to study cellular distribution of lipids *(11, 12)* and using fluorescent lipid analogs when you want to explore for example, the consequences of a disease on endocytic pathway *(13)*. In addition, the use of more than one probe should be often considered before drawing conclusions.

Visualizing the dynamics of sphingolipid- and cholesterol-rich domains in the membrane with lipid-specific probes has been one of the goals of the authors' laboratory. In this paper, we will explain how to analyze these domains in living and fixed cells using two lipid-specific probes: lysenin, a SM-binding toxin and fluorescent PEG-Chol that labels cholesterol-rich domains *(14, 15)*.

SM is the major sphingolipid in mammalian cells and constitutes with cholesterol the essential component of the lipid rafts. However, compared to other sphingolipids, such as the GM1 ganglioside recognized by the cholera toxin CTx and its nontoxic B subunit CTxB *(16, 17)*, a limited number of probes can specifically recognize SM. Until now, the dynamic of SM has been studied with exogenously added fluorescent SM analogs such as the *N*-((6-(7-nitrobenz-2-oxa-1,3-diazol-4-yl)amino) hexanoyl)sphingosyl phosphorylcholine C6-NBD-SM and the *N*-(4,4-difluoro-5,7-dimethyl-4-bora-3a,4a-diaza-s-indacene-3-pentanoyl)sphingosyl phosphorylcholine C5-BODIPY-SM *(7, 8)*. Recently, a SM-specific cytolysin, pleurotolysin has been described *(18, 19)*. In addition, a fluorescent-tagged peptide derived from the amyloid peptide Ab and containing a sphingolipid binding domain (SBD) known to interact with galactose sphingolipid terminal glycosphingolipids, especially GM1 ganglioside and to a lesser extent SM *(20)*, has been used to study the dynamics of glycosphingolipid-rich detergent-resistant microdomains in living cells *(21)*. In the laboratory, we are currently using lysenin, a pore-forming toxin derived from the earthworm *E. foetida* that binds specifically to SM *(22)*. The binding of lysenin to SM is dependent on the membrane distribution of the lipid, i.e., the presence or not of SM clusters *(11)*. On binding to SM, lysenin oligomerizes and this oligomer forms pores on the target membrane *(22)*. The development of nontoxic forms of lysenin allows using it in living cells to study the distribution of SM-rich domains in the membrane *(23)*.

Cholesterol is considered to be crucial in maintaining raft structure as sphingolipids preferentially form complex with cholesterol *(24)*. The cellular content and distribution of cholesterol is tightly regulated *(25)*. However, the formation and regulation of membrane lipid domains related to cholesterol trafficking are still unclear because of the lack of reliable and nontoxic cholesterol probes that can be used in living cells. The polyene antibiotic, filipin, is commonly used to detect free cholesterol but it is not suitable for living cells because of its cytotoxicity. In addition, its detection is limited by rapid photobleaching *(26, 27)*. Dehydroergosterol (DHE), a fluorescent cholesterol analog, is used to study free cholesterol trafficking in living cells. DHE undergoes a fast transbilayer movement and accumulates in recycling endosomes. However, its detection is requiring a special microscope setting in ultraviolet range *(28)*. Furthermore, it is not known whether DHE is able to partition in lipid domains where endogenous cholesterol is enriched. Recently, perfringolysin O (θ-toxin), a cytolysin produced by *Clostridium perfringens*, and its nontoxic derivatives have been proven useful to selectively bind cholesterol-rich domains in living cells *(12, 29)*. In the laboratory, we are using fluorescent-tagged PEG-Chol to monitor the dynamics of membrane cholesterol-rich domains because of its water solubility and low toxicity *(30, 31)*. Although the physical properties of PEG-Chol are different from that of cholesterol, exogenously added PEG-Chol preferentially partitions to cholesterol-rich membrane domains. Using this probe, it was possible to follow alteration of the cholesterol content and distribution when cells approach confluency. PEG-Chol revealed that the cell-confluency dependent accumulation of cellular cholesterol altered the endocytic pathway of fluorescent sphingolipid analogs as well as that of cholesterol itself by inhibiting the extraction of the small GTPases Rab proteins from endosomal membrane *(31)*.

2. Materials

2.1. Equipment: We Are Currently Using for Cell Observation

1. A Zeiss LSM 510 confocal microscope equipped with a C-Apochromat 63× W korr (1.2 numerical aperture) objective and a Plan-Apochromat 100× oil immersion (1.4 n.a.) objective with processing by LSM Image Browser (Zeiss).
2. A confocal Olympus FV1000 microscope equipped with a 60× oil immersion lens and Plan-Apochromat 100× (1.4 n.a.) objective.
3. Laser lines 488 or 543 nm for excitation and 505–530 band pass filter (fluorescein) or 560–615 band pass filter (rhodamine) for emission, respectively.

2.2. Reagents and Supplies

1. Cells: the lipid-specific probes that we are using can stain either adherent cells or cells grown in suspension from human as well as animal origin. In the experiments described herein (*see also* figures), we used Chinese hamster ovary CHO-Kl and Jurkat cell lines that are available from the American Type Culture Collection (ATCC, Manassas, VA, USA) as well as other cells such as mouse melanoma cell line MEB4 and its glycosphingolipid-deficient mutant GM95 provided by Dr. Yoshio Hirabayashi (Brain Science Institute, RIKEN) *(32)*; cultured skin fibroblasts from control subjects and patients with Niemann-Pick type A (NPA), characterized by a deficiency in acid sphingomyelinase provided by Dr. Hitoshi Sakuraba (Tokyo Metropolitan Institute of Medical Science) *(33)*.

2. Culture medium: Dulbecco's modified Eagle's medium DMEM (for melanoma cells, human skin fibroblasts), Ham's F-10 medium (for NPA cells), and Ham's F-12 medium (for CHO cells) are from GIBCO Invitrogen (Carlsbad, CA, USA). RPMI 1640 medium (for Jurkat cells) and Dulbecco's modified Eagle's medium/nutrient mixture F-12 Ham without phenol red and with 15 mM HEPES (DMEM/F-12) are from Sigma (St Louis, MO, USA). Usually cells are cultured in medium supplemented with 10% fetal calf serum, 100 units/ml penicillin, and 100 μg/ml streptomycin at 37°C with 5% CO_2. DMEM/F-12 medium supplemented with 2.5 mM L-glutamine is used for observation of living cells. As DMEM F-12 contains HEPES, the pH is kept constant during incubation outside the CO_2 incubator.
3. For living cell observation, cells were usually plated on glass-bottom culture dishes (diameter 35-mm dish with diameter 12-mm glass) from Matsunami Glass Ind. Ltd. (Japan) or from Iwaki (Asahi Technoglass, Scitech Division Iwaki brand, Japan). For the confluence effect study, cells are plated at different cell concentrations (*see* **Subheading 3.2.2**) in Lab-Tek chambered coverglass (Nalge Nunc International Corp., Rochester, NY, USA).
4. For fixed cells experiments, cells are grown directly on pre-cleaned sterile glass-coverslips (Matsunami Glass Ind. Ltd.) placed inside 12-well tissue culture plates. Before microscope observation, coverslips are mounted in Mowiol on precleaned NEO glass micro slides.
5. The full-length lysenin peptide is purchased from Peptide Institute, Inc. (Osaka, Japan). Subsequently to the purification and cDNA cloning of lysenin (GenBank accession number D85846) *(34)*, the recombinant protein and its tagged and nontoxic variants have been produced in our laboratory to detect selectively SM-enriched domains in living as well as fixed cells. We will give examples of some of these tagged lysenin forms and their applications.
6. The lysenin cDNA *(34)* is subcloned into pMAL-c2X vector (Amersham Biosciences-GE Healthcare Life Sciences) or pGEX-4T (New England Biolabs). The recombinant proteins fused to MBP (maltose binding protein) or GST (glutathione S-transferase) are produced in *Escherichia coli* and purified on amylose resin or GST-sepharose 4B (New England Biolabs), respectively *(35)*.
7. The nontoxic lysenin (NT-Lys) consisting of amino acids 161–297 of lysenin, the shortest fragment that recognizes SM and its fluorescent derivatives are also produced in the laboratory *(23, 35)*. The GFP-derived fluorescent protein, Venus, and its variant, A206K Venus that does not dimerize *(36)*,

are conjugated at the N-terminus of NT-Lys producing respectively, His-Venus-NT-Lys (HV-NT-Lys) and His-Venus-A206K-NT-Lys (HmV-NT-Lys). The His-tagged monomeric red fluorescent protein (mRFP) NT-Lys is also generated. The fluorescent recombinant proteins are expressed in *E. coli* and purified by affinity chromatography using a nickel column (Amersham BioSciences-GE Healthcare Life Sciences).

8. Fluorescein and tetramethylrhodamine B isothiocyanate poly(ethylene glycol)-derived cholesterol (fPEG-Chol or TRITC-PEG-Chol) are gifts from Dr Satoshi B. Sato (Department of Biophysics, Kyoto university, Japan). They are prepared from a solution of mesylate PEG (50)-Chol (50 is the average number of PEG repeats) conjugated to the fluorochroms as described *(37)*. A 1 mM stock solution is prepared in ethanol, protected from light exposure and kept at −20°C. For cell incorporation, the fluorescent probe is directly dispersed in the aqueous media (0.1% ethanol final concentration) and immediately distributed on the surface of cultured cells avoiding light exposure during the incubation.
9. *N*-(4,4-difluoro-5,7-dimethyl-4-bora-3a,4a-diaza-s-indacene-3pentanoyl) sphingosyl phosphorylcholine (BODIPY-SM) is from Molecular Probes (Eugene, OR).
10. Polyclonal (rabbit) anti-lysenin antiserum is purchased from Peptide Institute, Inc. (Osaka, Japan). Anti-MBP and anti-GST antisera are purchased from New England Biolabs.
11. Alexa Fluor 546-labeled cholera toxin B subunit (CTxB), Alexa Fluor 594-labeled CTxB, Texas Red transferrin Tf, and rhodamine dextran are purchased from Molecular Probes. Anti-CD63 antibody was from Serotec (Oxford, UK). Phycoerythrin-conjugated anti-CD59 (glycosyphosphatidylinositol anchored protein) monoclonal antibody is purchased from Pharmingen (San Diego, CA). Methyl-β-cyclodextrin (mean degree of substitution 10.5–14.7) is from Sigma and is dissolved in water.

3. Methods

3.1. Lysenin for Probing Distribution and Dynamics of SM Domains

Lysenin is a 297 amino acid pore-forming toxin that specifically binds SM when the lipid forms clusters, thus it is a very useful tool to study localization of SM-rich membrane domains. The development of nontoxic forms of lysenin allows using it in living cells. The fact that most mammalian cells are sensitive to lysenin suggests that SM form clusters in most membranes with the exception of the apical membrane of epithelial cells and the plasma membrane of melanoma cells *(11)*. Such binding difference has been linked to the membrane enrichment in glycosphingolipids that decreases the local density of SM. On the other hand, cholesterol facilitates lysenin oligomerization without affecting lysenin binding. This is not related to the fact that cholesterol favors the phase separation, but likely that cholesterol fluidizes SM, increasing the chance of collision of lysenin monomers as observed *in vitro (38)*. We will describe typical procedures to stain SM domains at the plasma membrane as well as intracellularly after permeabilization. Costaining with various lipid or protein markers can also be performed to investigate the dynamics of the SM domains, e.g., CTxB, C6-NBD-SM, CD 63, transferrin.

3.1.1. Labeling of SM with Native and Recombinant Lysenin in Fixed Cells

All manipulations were performed at room temperature unless otherwise noted.

1. Cells are grown on glass coverslips for 2 days to reach 50–80% confluence. They can also be grown on polycarbonate filter (e.g., Transwell insert from Corning Inc. Costar) in case of polarized cells such as MDCK or Caco-2 cells.
2. Wash three times with phosphate buffered saline (PBS) or PBS + containing 1 mM $CaCl_2$ and 0.5 mM $MgCl_2$ in case of filter-grown cells.
3. Fix cells with 3% paraformaldehyde (w/v) in PBS for 20 min or for 30 min at 4°C for filter-grown cells.
4. Wash twice with PBS.
5. Incubate with 50 mM NH_4Cl for 10 min for quenching excess aldehyde.
6. Wash three times with PBS.
7. Permeabilize cells with 50 μg/ml digitonin for 10 min (20 min for filter-grown cells).
8. Wash three times with PBS.
9. Incubate cells with 0.2% gelatin or 1% bovine serum albumin (BSA from Sigma, code A7030, suitable for ELISA assays) in PBS for 30 min.
10. Incubate cells with 1 μg/ml lysenin or 5 μg/ml recombinant lysenin (MBP-lysenin or GST-lysenin) at 4°C for 60 min. Native and recombinant lysenin share the same binding characteristics *(35)*.
11. Then cells are fixed again with paraformaldehyde.
12. Incubate with primary antibodies: anti-lysenin or anti-MBP or anti-GST antibody for 30 min.
13. Wash three times with PBS.
14. Incubate with Alexa 488 or Alexa 546-conjugated anti-rabbit IgG (Molecular Probes) for 30 min.

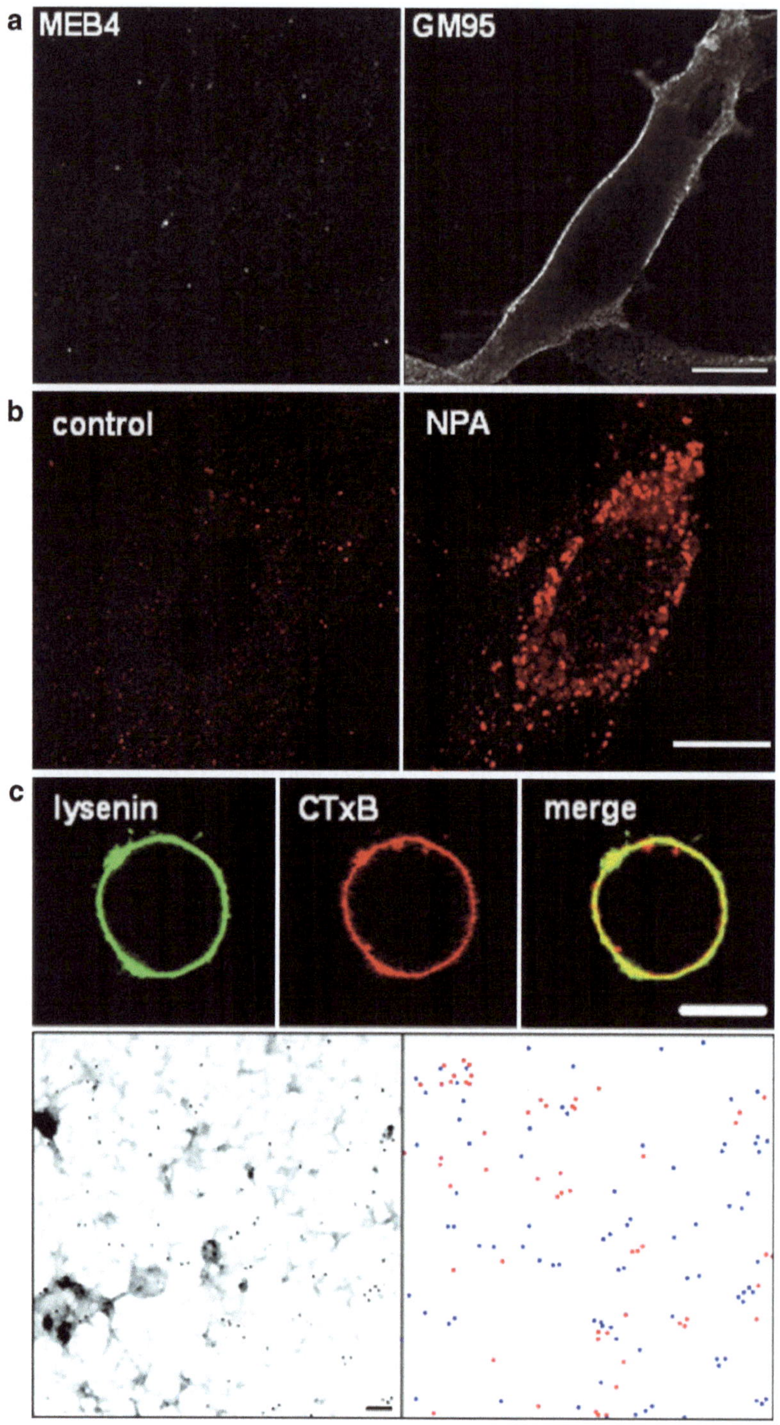

Fig. 1. (**a**) Glycolipid content affects organization of SM domains and lysenin binding. Mouse melanoma cell line MEB4 (*left*) and its glycolipid-deficient mutant GM95 (*right*) are fixed and incubated with native lysenin, then with anti-lysenin antibody and Alexa 488-conjugated anti-rabbit antibody. Scale bar, 10 μm. (**b**) Detection of cellular SM by MBP-lysenin. Normal (*left*) and NPA fibroblasts (*right*, deficient in acid sphingomyelinase) are fixed and permeabilized, then incubated with MBP-lysenin followed by treatment with anti-MBP antibody and with secondary antibody conjugated with Alexa 546. Scale bar, 20 μm. (**c**) Surface distribution of SM and GM1 in Jurkat cells. *Upper panel*, confocal images of doubly labeled

15. Wash three times with PBS.
16. Mount coverslip in Mowiol and after it becomes harder, observe with confocal microscope. If you want to observe immediately, you can secure the coverslip to the slide with nail polish (*see* **Fig. 1a and b**).

3.1.2. Labeling of SM on the Plasma Membrane of Living Cells

1. Cells are grown on glass-bottom culture dishes to reach 50–80% confluence or in suspension for Jurkat cells.
2. Wash cells twice with ice-cold serum-free medium.
3. Incubate cells for 20 min on ice with serum-free medium containing 50 mg/ml of monomeric Venus-NT-Lysenin.
4. Wash cells twice with ice-cold serum-free medium.
5. Place cells onto a glass-bottom culture dish precoated with poly-L-lysine in the case of cells grown in suspension such as Jurkat cells.
6. Observe with confocal microscope (*see* **Fig. 1c**).

3.2. Fluorescent PEG-Cholesterol for Monitoring Distribution and Transport of Cholesterol-Rich Membrane Domains

Fluorescein-tagged or TRITC-tagged PEG-Chol are nonionic amphipatic cholesterol derivatives that contain a fluorescein or a TRITC moiety on the distal end of PEG chain (**Fig. 2a**). The fluorescent PEG-Chol probes are able to partition from an aqueous media into cholesterol-rich membranes both in cells and in model membranes *(30, 37)*. These properties enable to follow the re-organization of cell surface cholesterol as well as intracellular cholesterol dynamics. The bulk PEG moiety of PEG-cholesterol prevents transbilayer movement of the molecule. This inhibits the transport of PEG-cholesterol to the cytoplasmic leaflet and the nonspecific diffusion of the molecule to the cytoplasm when the molecule is inserted to the cell surface. Thus, PEG-Chol can selectively monitor the dynamics of the outer leaflet cholesterol-rich membrane domains. After addition at low temperature to the cells and then, increasing the temperature to 37°C, PEG-Chol is slowly internalized via endocytic pathway. Its colocalization with various lipid or protein markers can be investigated, e.g., dextran, CTxB, and so on. In addition, cholesterol-rich domains in the cytoplasmic leaflet of intracellular organelles can also be analyzed after microinjection *(30)*.

Fig. 1. (continued) cells with monomeric venus-NT-Lysenin (green) and Alexa 594-conjugated CTxB (red); *lower panel*, electron microscopy images of Jurkat cells that were first labeled with HmV-NT-Lysenin and biotinylated CTxB at 4°C and then fixed with 4% paraformaldehyde and 0.02% glutaraldehyde for 10 min at 4°C. Then cells were labeled with anti-GFP rabbit polyclonal antibody at 4°C followed by labeling with goat anti-rabbit IgG 5-nm gold and goat anti-biotin IgG 10-nm gold at 4°C. The distribution of gold particles on the plasma membrane was examined under electron microscope after ripping off the membrane. Scale bar, 100 nm. In the right lower panel, distribution of SM is colored in red, whereas the distribution of GM1 is in blue to visualize more easily that SM and GM1 domains did not cocluster in the plasma membrane *(23)*.

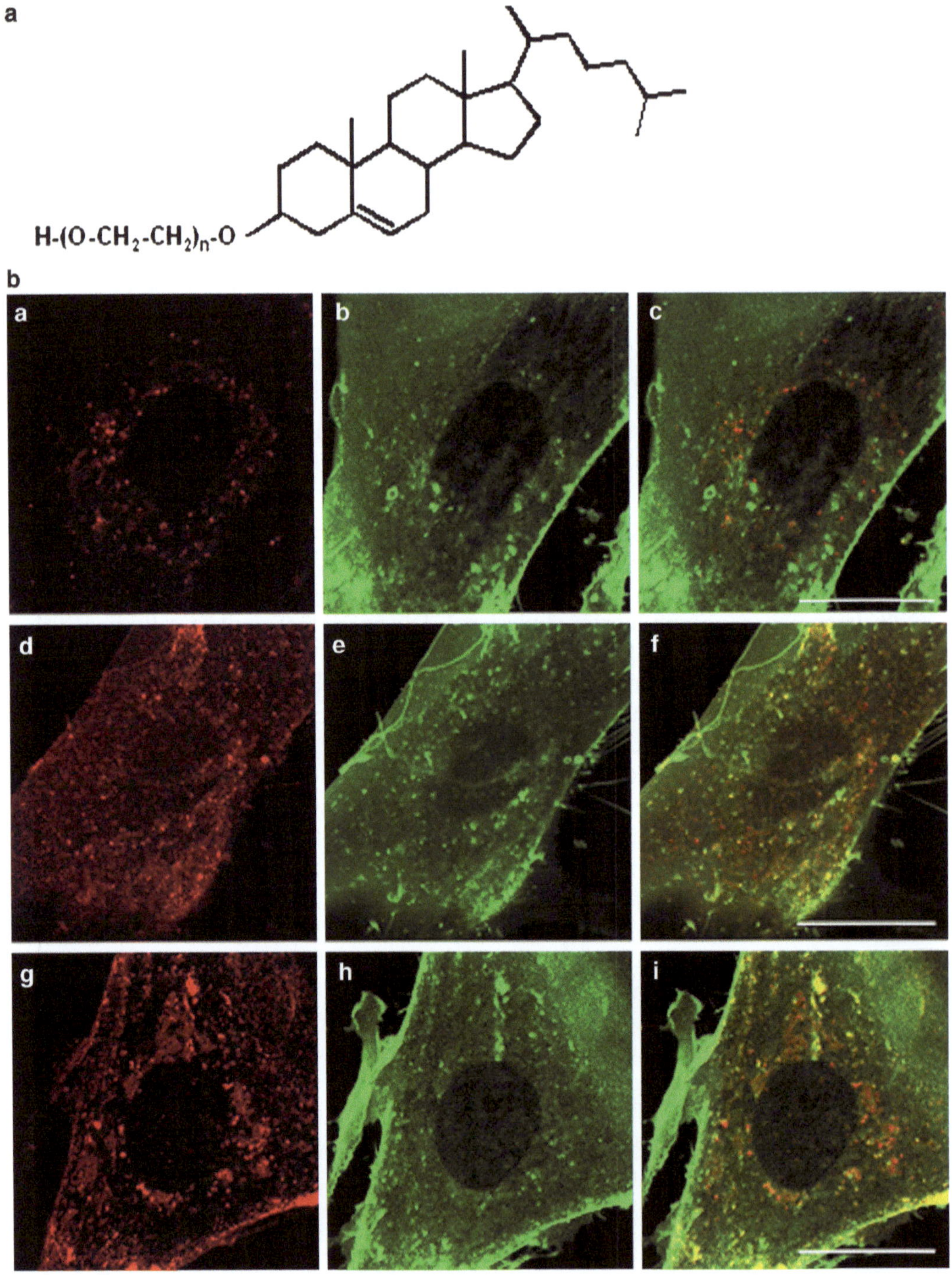

Fig. 2. (**a**) Structure of poly(ethylene glycol)cholesteryl ether (PEG-Chol). *n* is the average of ethylene glycol units repeated number. (**b**) PEG-Chol is internalized together with lipid raft components. Human skin fibroblasts are incubated with 1 mM fPEG-Chol together with Texas Red transferrin Tf (**a–c**), phycoerythrin-conjugated anti-CD59 monoclonal antibody (**d–f**), or AlexaFluor 594-labeled CTxB (**g–i**) for 5 min at room temperature. Cells are then washed and further incubated for 45 min (**a–c, g–i**) or 30 min (**d–f**) at 37°C. Cells are treated with NH_4Cl before taking images. Merged images are in (**c, f**, and **i**). Scale bar, 20 μm. Figure from Ref. (*15*), Oxford University Press.

3.2.1. Characterizing the Internalization of Cell Surface Cholesterol with Fluorescent PEG-Chol in Living Cells

Using low concentrations of PEG-Chol and short incubation time, it is possible to follow the fate of cell surface PEG-Chol in living cells, whereas higher concentration and longer incubation time are known to specifically inhibit clathrin-independent endocytosis *(37, 39)*.

1. Plate cells on glass-bottom culture dishes [about (2.5–4.5) × 10^4 cells/dish] and grow them for 1 day in their regular culture medium to get a 50–60% confluence. Some cells need 2 days to adhere efficiently to the glass bottom. This might be improved by precoating the glass surface with fibronectin (4 mg/cm^2).
2. Wash cells with serum-free medium (two to three times).
3. Various procedures can be used to follow cholesterol internalization:
   (a) incubate cells with 1 mM fPEG-Chol in plain DMEM F-12 without phenol red for 5 min at room temperature. Then, cells are washed with serum-free medium (two to three times) and incubated in DMEM F-12 without phenol red at 37 °C to chase the fluorescent-labeled PEG-Chol. To compare the endocytosis of PEG-Chol with bulk endocytosis, cells are incubated in the presence of 1 mg/ml rhodamine dextran. Then, confocal microscope observation is performed at various times up to 180 min; or
   (b) incubate cells with 1 mM fPEG-Chol in plain DMEM F-12 without phenol red together with Texas Red transferrin (Tf), AlexaFluor 594-labeled CTxB, or phycoerythrin-conjugated anti-CD59 monoclonal antibody for 5 min at room temperature. Then, wash cells with serum-free medium (two to three times) and further incubate cells in plain DMEM F-12 without phenol red for 20 min at 37°C.
   (c) Cells are treated with NH_4Cl 10 mM in plain DMEM F-12 at room temperature before taking images to enhance fPEG-Chol fluorescence in intracellular vesicles by neutralizing the vesicular pH. Then they are immediately observed in plain DMEM F-12 without phenol red with confocal microscope (*see* **Fig. 2b**).

3.2.2. Analyzing Cholesterol-Dependent Endocytosis with Fluorescent PEG-Chol in Living Cells

The content and distribution of cellular cholesterol can be modified by various conditions dependent of cell growth, cell confluence, diseases (e.g., Niemann Pick type C), treatment with reagents such as U18666A, and this will subsequently interfere with lipid traffic *(8, 13, 40)*. We recently demonstrated using Chinese hamster ovary (CHO) cells that internalization of fluorescent PEG-Chol is affected by cell confluency and this also alters internalization of other fluorescent lipid probes, like BODIPY-sphingomyelin

(BODIPY-SM) through a redistribution of the rab GTPase Rab 11 *(31)*.

1. CHO cells (3×10^3) are grown in two chambers (area 4.2 cm^2) to obtain subconfluent cells (Lab-Tek Chambered Coverglass, Nalgen Nunc International Corp., Rochester, NY, USA) or 8.6×10^3 cells are grown in eight chambers (area, 0.8 cm^2) for 3 days to be confluent.
2. Cells are incubated with 0.5 mM fluorescent PEG-cholesterol in DMEM F-12, without phenol red for 1 min (subconfluent) or 5 min (confluent) at 15°C. Five minutes is required for confluent cells because of the low efficiency of insertion of PEG-Chol.
3. Then excess of PEG-cholesterol is washed with plain DMEM F-12.
4. Then cells (transiently transfected with human Tf receptor in the case of CHO cells) are incubated for 15 min at 37°C in plain DMEM F-12 with Alexa 546–conjugated Tf (50 mg/ml).
5. The cells are treated with 10 mM NH_4Cl in DMEM F-12 medium before acquiring image by confocal microscope (*see* **Fig. 3a and b**).
6. To analyze BODIPY-SM internalization, cells are preincubated with 4 mM BODIPY-SM in serum free Ham's F-12 medium supplemented with 0.2% fatty acid-free BSA (Sigma) for 30 min at 15°C. After washing, cells are labeled with TRITC-PEG-Chol as previously and incubated for further 15 min at 37°C before observation by confocal microscope.

3.3. Results

3.3.1. Lysenin to Probe Organization of Membrane SM in Fixed and Living Cells

In **Fig. 1a**, native lysenin brightly stains SM on the cell surface of the glycolipid-deficient mutant GM95, deficient in ceramide glucosyltransferase I catalyzing the first step of glycosphingolipid synthesis (right). In contrast, the mouse melanoma parental cells MEB4 (left) are almost devoid of fluorescence, indicating that high concentration of glycolipids alters the organization of SM domains and thus, the binding of lysenin. This should be always kept in mind as a negative staining does not necessarily indicate a deficiency of SM, but is linked to a lack of SM domains *(11)*.

In **Fig. 1b**, the SM distribution is visualized by MBP-tagged lysenin in Niemann-Pick type A (NPA) human skin fibroblasts, characterized by a deficiency in acid sphingomyelinase (right). Lysenin staining displayed a bright perinuclear labeling consistent with the accumulation of SM (and cholesterol) in the late endosome/lysosome of those cells, whereas small intracellular vesicles are stained in normal (control) cells (left).

In **Fig. 1c**, SM and GM1 on the cell surface of living Jurkat cells are doubly labeled by nontoxic lysenin (green) and CTxB (red), respectively. As both toxins evenly stain the plasma membrane,

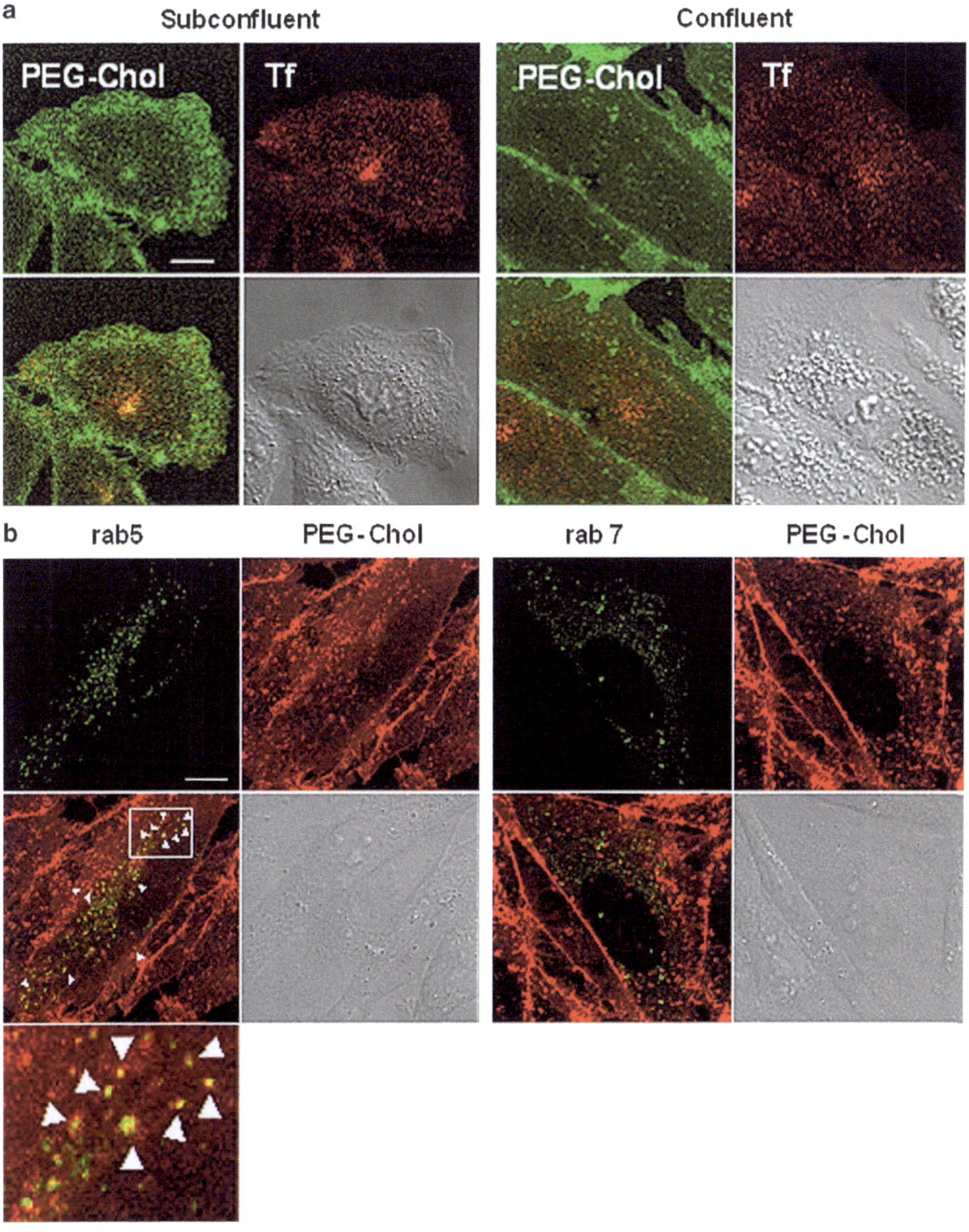

Fig. 3. Cell confluency alters internalization of fluorescent PEG-cholesterol. (**a**) Subconfluent and confluent CHO cells transiently transfected with human transferrin receptor are incubated with Alexa 546-conjugated transferrin (Tf, red) and fluorescein-PEG-cholesterol (PEG-Chol, green) for 15 min at 37°C. Scale bar, 10 μm. (**b**) Confluent CHO cells transfected with GFP-rab5 or GFP-rab7 (green) are incubated with TRICT-PEG-cholesterol (PEG-Chol, red) for 15 min at 37°C. White arrows indicate colocalization of rab5 and TRICT-PEG-Chol (*see* inset for enlarged image). Scale bar, 10 μm.

this supports the idea that domain sizes of SM and GM1 are below the resolution of the fluorescence microscope (**Fig. 1c**, upper panel). However, electron microscopy analysis can demonstrate that

SM- and GM1-rich clusters with a radius of 60–70 nm form spatially distinct domains on the plasma membrane of Jurkat cells [**Fig. 1c** lower panel and *(23)*]. Furthermore, as binding of either of the toxins did not affect binding of the other as observed by confocal microscopy, this emphasizes that SM and GM1 domains are found in different clusters.

3.3.2. Fluorescent PEG-Chol to Probe Distribution and Transport of Cholesterol-Rich Membrane Domains and Cholesterol-Dependent Endocytosis in Living Cells

When human skin fibroblasts are incubated with 1 mM of fPEG-Chol for 5 min, they incorporate 1.8 ± 0.3 pmol of fPEG-Chol/mg of protein *(30)*. After 10 min chase, most of the fPEG-Chol is still on the plasma membrane and then after 60 min, fPEG-Chol is found in intracellular vesicles that do not colocalized with internalized rhodamine dextran *(30)*.

In **Fig. 2b**, fPEG-Chol is internalized together with lipid raft components. Fluorescent PEG-Chol colocalizes with lipid raft markers CD59 (d–f) and GM1 (g–i) endocytosed by clathrin-independent mechanisms, but in contrast does not colocalize with Tf (a–c) internalized via clathrin-dependent endocytosis. During prolonged incubation, fPEG-Chol is further transported to the Golgi apparatus.

In **Fig. 3a**, cell confluency alters internalization of fluorescent PEG-cholesterol. In subconfluent cells (left panels), internalized fPEG-cholesterol is accumulated in recycling endosomes (pericentriolar region), characterized by the accumulation of Tf. In contrast in confluent cells (right panels), fPEG-cholesterol is distributed into early endosomes (small vesicles throughout the cytoplasm) that are identified by their colocalization with rab5, but not with the late endosome marker, rab 7 (*see* **Fig. 3b**, the white arrows indicate colocalization of rab5 and TRICT-PEG-Chol).

We also demonstrated that the cell-confluence dependent accumulation of cholesterol altered the endocytic pathway of BODIPY-SM, but not that of BODIPY-lactosylceramide, indicating that endocytosis of a limited subset of membrane molecules is affected by cell-confluency *(31)*. In subconfluent cells, BODIPY-SM accumulates in recycling endosomes with fluorescent PEG-Chol, whereas both probes are concentrated in early endosomes in confluent cells *(31)*.

4. Notes

1. When labeling living cells, it is better to perform the washes with a serum-free medium instead of PBS to maintain physiological pH and osmotic pressure as well as to provide calcium and magnesium for avoiding cell detachment. Alternatively, Hanks' balanced salt solution (HBSS) containing calcium and magnesium can be used.

2. To reduce background staining and improve probe binding to the cell surface, cells should be incubated with a serum-free medium during probe incorporation.
3. Native lysenin can also be used to label cell surface SM, however this procedure requires carefully controlled conditions to avoid cell permeabilization especially because of the second fixation step mentioned as follows:
   a. Cells are incubated with 0.5 μg/ml lysenin or 1–5 μg/ml recombinant lysenin (MBP-lysenin or GST-lysenin) at 4°C for 60 min.
   b. Then, cells are fixed again with paraformaldehyde.
   c. Wash cells with PBS.
   d. Incubate cells with 0.2% gelatin or 1% BSA in PBS for 30 min.
4. To avoid oligomerization of lysenin, we recommend incubating cells on ice or at 4°C and fixing the cells once again. This second fixation step might also prevent artificial aggregation of secondary antibodies.
5. To avoid internalization of NT-lysenin, living cells (non-fixed) are labelled on ice or at 4°C.
6. To confirm the specific binding of lysenin to SM, we recommend observing whether sphingomyelinase treatment abolishes the lysenin staining. Cells are incubated with recombinant *Bacillus cereus* sphingomyelinase (Sigma) 10 mU/ml for 1 h at 37°C in PBS for permeabilized and fixed cells or in medium without serum for living cells.
7. Alternatively, the content of SM can be decreased by culturing the cells in serum-free medium or in complete medium for 2 days with various concentrations of inhibitors of the sphingolipid biosynthesis, such as the serine palmitoyltransferase inhibitor Myriocin/ISP-1 (usually around 50 nM) or the inhibitor of the ceramide synthase Fumonisin B1 (usually around 40 mM) *(41)*. Both inhibitors are available from Sigma. Growing the cells in serum-free medium improves the SM depletion but also increases cell toxicity. Such effect should be carefully controlled by a cell-toxicity assay (such as MTT assay) in preliminary experiments. In addition, the reduced content of SM can be reversed by culturing the cells in medium containing fatty acid-free BSA-sphingosine complex *(42)* for 3 days (final concentration of d-erythro-sphingosine 2 μM).
8. To observe fluorescent PEG-Chol staining, the cell layer is treated with 10 mM NH_4Cl in phenol red-free DMEM F-12 medium before acquiring image to enhance fPEG-Chol fluorescence in intracellular vesicles by neutralizing the vesicular pH.
9. To confirm specific binding of fluorescent PEG-Chol to cholesterol, we recommend checking that MβCD treatment

abolishes staining. For this purpose, cells can be preincubated with 5 mM MβCD in serum-free medium for 15 min at 37°C, then washed with serum-free medium before fluorescent PEG-Chol staining as described above.

Acknowledgments

The authors thank all the members of the Kobayashi laboratory and acknowledge funding from Ministry of Education, Sciences and Culture of Japan and from RIKEN Frontier Research System to T.K.

References

1. Simons, K., and W.L. Vaz. (2004) Model systems, lipid rafts, and cell membranes. *Annu Rev Biophys Biomol Struct* **33**, 269–295.
2. Mayor, S., and M. Rao. (2004) Rafts: Scale-dependent, active lipid organization at the cell surface. *Traffic* **5**, 231–240.
3. Le Roy, C., and J.L. Wrana. (2005) Clathrin- and non-clathrin-mediated endocytic regulation of cell signalling. *Nat Rev Mol Cell Biol* **6**, 112–126.
4. Shogomori, H., and D.A. Brown. (2003) Use of detergents to study membrane rafts: The good, the bad, and the ugly. *Biol Chem* **384**, 1259–1263.
5. Zidovetzki, R., and I. Levitan. (2007) Use of cyclodextrins to manipulate plasma membrane cholesterol content: Evidence, misconceptions and control strategies. *Biochimica et Biophysica Acta (BBA) - Biomembranes* **1768**, 1311–1324.
6. Hoetzl, S., H. Sprong, and G. van Meer. (2007) The way we view cellular (glyco)sphingolipids. *J Neurochem* **103 Suppl 1**, 3–13.
7. Pagano, R.E., R. Watanabe, C. Wheatley, and M. Dominguez. (2000) Applications of BODIPY-sphingolipid analogs to study lipid traffic and metabolism in cells. *Methods Enzymol* **312**, 523–534.
8. Puri, V., R. Watanabe, R.D. Singh, M. Dominguez, J.C. Brown, C.L. Wheatley, D.L. Marks, and R.E. Pagano. (2001) Clathrin-dependent and -independent internalization of plasma membrane sphingolipids initiates two Golgi targeting pathways. *J Cell Biol* **154**, 535–547.
9. Kol, M.A., A.I. de Kroon, J.A. Killian, and B. de Kruijff. (2004) Transbilayer movement of phospholipids in biogenic membranes. *Biochemistry* **43**, 2673–2681.
10. Wang, T.Y., and J.R. Silvius. (2000) Different sphingolipids show differential partitioning into sphingolipid/cholesterol-rich domains in lipid bilayers. *Biophys J* **79**, 1478–1489.
11. Ishitsuka, R., A. Yamaji-Hasegawa, A. Makino, Y. Hirabayashi, and T. Kobayashi. (2004) A lipid-specific toxin reveals heterogeneity of sphingomyelin-containing membranes. *Biophys J* **86**, 296–307.
12. Sugii, S., P.C. Reid, N. Ohgami, Y. Shimada, R.A. Maue, H. Ninomiya, Y. Ohno-Iwashita, and T.-Y. Chang. (2003) Biotinylated {theta}-toxin derivative as a probe to examine intracellular cholesterol-rich domains in normal and Niemann-Pick type C1 cells. *J Lipid Res* **44**, 1033–1041.
13. Choudhury, A., M. Dominguez, V. Puri, D.K. Sharma, K. Narita, C.L. Wheatley, D.L. Marks, and R.E. Pagano. (2002) Rab proteins mediate Golgi transport of caveola-internalized glycosphingolipids and correct lipid trafficking in Niemann-Pick C cells. *J Clin Invest* **109**, 1541–1550.
14. Shogomori, H., and T. Kobayashi. (2008) Lysenin: A sphingomyelin specific pore-forming toxin. *Biochim Biophys Acta* **1780**, 612–618.
15. Ishitsuka, R., S.B. Sato, and T. Kobayashi. (2005) Imaging lipid rafts. *J Biochem (Tokyo)* **137**, 249–254.
16. Shogomori, H., and A.H. Futerman. (2001) Cholera toxin is found in detergent-insoluble

rafts/domains at the cell surface of hippocampal neurons but is internalized via a raft-independent mechanism. *J Biol Chem* **276**, 9182–9188.

17. Lencer, W.I., and B. Tsai. (2003) The intracellular voyage of cholera toxin: Going retro. *Trends Biochem Sci* **28**, 639–645.
18. Sakurai, N., J. Kaneko, Y. Kamio, and T. Tomita. (2004) Cloning, expression, and pore-forming properties of mature and precursor forms of pleurotolysin, a sphingomyelin-specific two-component cytolysin from the edible mushroom *Pleurotus ostreatus. Biochim Biophys Acta* **1679**, 65–73.
19. Tomita, T., K. Noguchi, H. Mimuro, F. Ukaji, K. Ito, N. Sugawara-Tomita, and Y. Hashimoto. (2004) Pleurotolysin, a novel sphingomyelin-specific two-component cytolysin from the edible mushroom *Pleurotus ostreatus*, assembles into a transmembrane pore complex. *J Biol Chem* **279**, 26975–26982.
20. Mahfoud, R., N. Garmy, M. Maresca, N. Yahi, A. Puigserver, and J. Fantini. (2002) Identification of a common sphingolipid-binding domain in Alzheimer, prion, and HIV-1 proteins. *J Biol Chem* **277**, 11292–11296.
21. Hebbar, S., E. Lee, M. Manna, S. Steinert, G.S. Kumar, M. Wenk, T. Wohland, and R. Kraut. (2008) A fluorescent sphingolipid binding domain peptide probe interacts with sphingolipids and cholesterol-dependent raft domains. *J Lipid Res* **49**, 1077–1089.
22. Yamaji-Hasegawa, A., A. Makino, T. Baba, Y. Senoh, H. Kimura-Suda, S.B. Sato, N. Terada, S. Ohno, E. Kiyokawa, M. Umeda, and T. Kobayashi. (2003) Oligomerization and pore formation of a sphingomyelin-specific toxin, lysenin. *J Biol Chem* **278**, 22762–22770.
23. Kiyokawa, E., T. Baba, N. Otsuka, A. Makino, S. Ohno, and T. Kobayashi. (2005) Spatial and functional heterogeneity of sphingolipid-rich membrane domains. *J Biol Chem* **280**, 24072–24084.
24. London, E. (2002) Insights into lipid raft structure and formation from experiments in model membranes. *Curr Opin Struct Biol* **12**, 480–486.
25. Maxfield, F.R., and I. Tabas. (2005) Role of cholesterol and lipid organization in disease. *Nature* **438**, 612–621.
26. Robinson, J.M., and M.J. Karnovsky. (1980) Evaluation of the polyene antibiotic filipin as a cytochemical probe for membrane cholesterol. *J Histochem Cytochem* **28**, 161–168.
27. Norman, A.W., R.A. Demel, B. de Kruyff, and L.L. van Deenen. (1972) Studies on the biological properties of polyene antibiotics. Evidence for the direct interaction of filipin with cholesterol. *J Biol Chem* **247**, 1918–1929.
28. Mukherjee, S., X. Zha, I. Tabas, and F.R. Maxfield. (1998) Cholesterol distribution in living cells: Fluorescence imaging using dehydroergosterol as a fluorescent cholesterol analog. *Biophys J* **75**, 1915–1925.
29. Hayashi, M., Y. Shimada, M. Inomata, and Y. Ohno-Iwashita. (2006) Detection of cholesterol-rich microdomains in the inner leaflet of the plasma membrane. *Biochem Biophys Res Commun* **351**, 713–718.
30. Sato, S.B., K. Ishii, A. Makino, K. Iwabuchi, A. Yamaji-Hasegawa, Y. Senoh, I. Nagaoka, H. Sakuraba, and T. Kobayashi. (2004) Distribution and transport of cholesterol-rich membrane domains monitored by a membrane-impermeant fluorescent polyethylene glycol-derivatized cholesterol. *J Biol Chem* **279**, 23790–23796.
31. Takahashi, M., M. Murate, M. Fukuda, S.B. Sato, A. Ohta, and T. Kobayashi. (2007) Cholesterol Controls Lipid Endocytosis through Rab11. *Mol Biol Cell* **18**, 2667–2677.
32. Ichikawa, S., N. Nakajo, H. Sakiyama, and Y. Hirabayashi. (1994) A mouse B16 melanoma mutant deficient in glycolipids. *Proc Natl Acad Sci USA* **91**, 2703–2707.
33. Yamaji, A., Y. Sekizawa, K. Emoto, H. Sakuraba, K. Inoue, H. Kobayashi, and M. Umeda. (1998) Lysenin, a novel sphingomyelin-specific binding protein. *J Biol Chem* **273**, 5300–5306.
34. Sekizawa, Y., T. Kubo, H. Kobayashi, T. Nakajima, and S. Natori. (1997) Molecular cloning of cDNA for lysenin, a novel protein in the earthworm Eisenia foetida that causes contraction of rat vascular smooth muscle. *Gene* **191**, 97–102.
35. Kiyokawa, E., A. Makino, K. Ishii, N. Otsuka, A. Yamaji-Hasegawa, and T. Kobayashi. (2004) Recognition of sphingomyelin by lysenin and lysenin-related proteins. *Biochemistry* **43**, 9766–9773.
36. Nagai, T., K. Ibata, E.S. Park, M. Kubota, K. Mikoshiba, and A. Miyawaki. (2002) A variant of yellow fluorescent protein with fast and efficient maturation for cell-biological applications. *Nat Biotechnol* **20**, 87–90.
37. Ishiwata, H., S.B. Sato, A. Vertut-Doi, Y. Hamashima, and K. Miyajima. (1997) Cholesterol derivative of poly(ethylene glycol) inhibits clathrin-independent, but not clathrin-dependent endocytosis. *Biochim Biophys Acta* **1359**, 123–135.
38. Ishitsuka, R., and T. Kobayashi. (2007) Cholesterol and lipid/protein ratio control the oligomerization of a sphingomyelin-specific toxin, lysenin. *Biochemistry* **46**, 1495–1502.

39. Baba, T., C. Rauch, M. Xue, N. Terada, Y. Fujii, H. Ueda, I. Takayama, S. Ohno, E. Farge, and S.B. Sato. (2001) Clathrin-dependent and clathrin-independent endocytosis are differentially sensitive to insertion of poly (ethylene glycol)-derivatized cholesterol in the plasma membrane. *Traffic* **2**, 501–512.

40. Linder, M.D., R.-L. Uronen, M. Holtta-Vuori, P. van der Sluijs, J. Peranen, and E. Ikonen. (2007) Rab8-dependent Recycling Promotes Endosomal Cholesterol Removal in Normal and Sphingolipidosis Cells. *Mol Biol Cell* **18**, 47–56.

41. Yamaji-Hasegawa, A., A. Takahashi, Y. Tetsuka, Y. Senoh, and T. Kobayashi. (2005) Fungal metabolite sulfamisterin suppresses sphingolipid synthesis through inhibition of serine palmitoyltransferase. *Biochemistry* **44**, 268–277.

42. Hanada, K., M. Nishijima, M. Kiso, A. Hasegawa, S. Fujita, T. Ogawa, and Y. Akamatsu. (1992) Sphingolipids are essential for the growth of Chinese hamster ovary cells. Restoration of the growth of a mutant defective in sphingoid base biosynthesis by exogenous sphingolipids. *J Biol Chem* **267**, 23527–23533.

Chapter 12

Monitoring Sterol Uptake, Acetylation, and Export in Yeast

Vineet Choudhary and Roger Schneiter

Summary

Sterols are essential lipid components of eukaryotic membranes. They are synthesized in the endoplasmatic reticulum (ER) from where they are efficiently transported to the plasma membrane, which harbors ~90% of the free sterol pool of the cell. The molecular mechanisms that govern this lipid transport, however, are not well characterized and are challenging to analyze. *Saccharomyces cerevisiae* offers the opportunity to circumvent some of the technical limitations associated with studying this forward transport of sterols from the ER to the plasma membrane, because the organism can also take up sterols from the environment, incorporate them into the plasma membrane and transport them back to the ER, where the free sterol is converted to steryl esters. This reverse sterol transport, however, occurs only under anaerobic conditions, where the cells become sterol auxotroph, or in mutant cells that cannot synthesize heme. The reverse sterol transport pathway, however, is more amenable to experimental studies, because arrival of the sterol in the ER membrane can be monitored unambiguously by following the formation of steryl esters. Apart from sterol acylation, we have recently described a reversible sterol acetylation cycle that is operating in the lumen of the ER. Acetylation occurs on both cholesterol and pregnenolone, a steroid precursor, and serves as a signal for export of the acetylated sterols into the culture media. The time-dependent appearance of acetylated sterols in the culture supernatant thus provides a new means to monitor the forward transport of chemically modified sterols out of the ER.

Key words: Sterol, Cholesterol, Steroids, Pregnenolone, Sterol Acetylation, Lipid, Transport, Membranes, Yeast, *Saccharomyces cerevisiae*

1. Introduction

Sterols are essential lipid components of eukaryotic membranes that determine many membrane properties and are required for polar sorting events during vesicle transport in animal, plant, and fungal cells *(1, 2)*. Changes in sterol levels in membranes can have profound effects on signal transduction pathways, and on

Donald Armstrong (ed.), *Lipidomics*, Methods in Molecular Biology, vol. 580,
DOI 10.1007/978-1-60761-325-1_12, © Humana Press, a part of Springer Science+Business Media, LLC 2009

the trafficking of membrane proteins and lipids. Eukaryotic cells synthesize sterols in the ER from where they need to be efficiently transported to the plasma membrane, which harbors ~90% of the free sterol pool of the cell. This transport process is facilitated by vesicular as well as non-vesicular components *(3)*. This forward transport of sterols, however, is challenging to study because arrival of the newly synthesized sterol at the plasma membrane in not accompanied by any covalent modification of the lipid. This is frequently being circumvented by analyzing radio-labeled cyclodextrin extractable material from the intact plasma membrane *(4–6)*.

Animal cells take up exogenous sterols through receptor-mediated endocytosis of low-density lipoproteins (LDL). Once delivered to late endosome or lysosome, LDL-derived cholesteryl esters are hydrolyzed, and the free cholesterol is rapidly cycled back to the plasma membrane and/or the ER for re-esterification *(7)*. This back transport pathway of sterols from the plasma membrane/endosome to the ER is technically more robust to analyze as arrival of the sterol at the ER is accompanied by a covalent modification of the lipid, i.e. acylation *(8, 9)*.

Fungal cells synthesize ergosterol instead of cholesterol as their main sterol and many aspects of sterol homeostasis are conserved between yeast and human *(10–12)*. Under aerobic conditions, ergosterol is synthesized in the ER membrane and is greatly enriched at the yeast plasma membrane *(13)*. Under these conditions, cells do not take up exogenous sterols, a phenomenon known as "aerobic sterol exclusion." Under anaerobic growth conditions or in mutants that lack heme, however, *S. cerevisiae* becomes auxotrophic for sterols and unsaturated fatty acids, because the synthesis of these lipids requires molecular oxygen *(14)*. Under anaerobic conditions, cells thus induce a sterol uptake pathway to enable growth. The fact that *S. cerevisiae* is a facultative anaerobic organism that displays robust growth when appropriately supplemented indicates that lipid uptake is efficient. Anaerobic or heme-deficient conditions thus allow to investigate the reverse transport of sterols from the plasma membrane to the ER. Arrival of the radio-labeled sterol that is typically supplied to the media of these sterol uptake competent cells is monitored by the time-dependent formation of steryl esters, because the enzymes that convert free sterols into steryl esters, the acyl-CoA:sterol acyltransferases, are located in the ER membrane *(7, 15–17)*.

Apart from this well characterized sterol acylation and deacylation cycle, we have recently described a novel covalent sterol modification: the sterol acetylation and deacetylation cycle *(18)*. Sterol acetylation requires the acetyltransferase *ATF2*, whereas deacetylation requires *SAY1*, a membrane-anchored deacetylase. Both enzymes are located in the ER membrane with their active site facing the ER lumen *(18)*. Exogenous cholesterol is subject to acetylation and deacetylation and cholesterol acetate accumulates only if cells are deleted for the deacetylase, *SAY1*. Lack of *SAY1*, however, results in the secretion of acetylated sterols into the culture media *(18)*. Monitoring the appearance of sterol acetate in the culture medium thus can serve as a novel readout for the forward transport of sterols from the ER to the extracellular space. Similar to cholesterol, the steroid precursor pregnenolone is also subject to Atf2-dependent acetylation, but unlike cholesterol acetate, pregnenolone acetate is not deacetylated by Say1. As a consequence, pregnenolone acetate is rapidly excreted by the cells and accumulates in the culture supernatant. Monitoring the appearance of acetylated pregnenolone thus provides a novel tool to investigate the cellular uptake of steroids as well as the molecular basis for the excretion of acetylated steroids.

This paper discusses the uptake and secretion of radio-labeled sterols by yeast *S. cerevisiae*. Radio-labeled lipids are typically provided to heme-deficient, i.e., uptake competent cells, for some time and then extracted again from whole cell lysates as well as from the culture media. The metabolically modified lipids are then separated by ascending thin layer chromatography (TLC) using plates coated with silica gel (60 Å pore size). This technique of lipid separation is simple, versatile, and highly sensitive with the flexibility to be used both quantitatively and qualitatively *(19)*. After TLC separation, radio-labeled lipids are detected and quantified using either a radio TLC analyzer or by exposing the TLC plate to a phosphorimager screen or an X-ray film.

2. Materials

2.1. Equipment

1. Tabletop centrifuge with appropriate adaptors (Eppendorf, Hamburg, Germany).
2. TLC chamber with cover and Whatman filter paper (GE Healthcare, Piscataway, NJ).
3. Incubator, shaker, and water bath.
4. Cell homogenizer (Precellys 24, Bertin Technologies, France).
5. Liquid scintillation counter (Tri-Carb 2100TR, Perkin-Elmer, Waltham, MA) and scintillation cocktail (Zinsser Analytic, Frankfurt, Germany).
6. Phosphorimager (Bio-Rad, Hercules, CA).
7. Radio TLC-linear analyzer (Berthold Technologies, Bad Wildbad, Germany).

2.2. Reagents and Supplies

1. Yeast growth media. Rich media, YEPD: 1% (w/v) yeast extract, 2% (w/v) bactopeptone, 2% (w/v) glucose. Synthetic media, SC-LEU: 0.67% (w/v) nitrogen base without amino acids, 2% (w/v) glucose, 0.07% (w/v) amino acid drop out mix without leucine *(20)*. Heme-deficient cells must be grown either with lipid supplementation (5 mg/ml Tween 80, which substitutes for unsaturated fatty acids, and 20 μg/ml cholesterol or ergosterol) or in the presence of aminolevulinic acid (ALA, 10 μg/ml), which bypasses the *hem1Δ* deficiency *(17)*.
2. Glass beads: 0.25–0.3 mm diameter (Braun Melsungen, Melsungen, Germany).
3. Silica gel 60 TLC plates (Merck, Darmstadt, Germany).
4. Solvent system for neutral lipid separation by TLC: petroleum ether/diethyl ether/glacial acetic acid (70:30:2, v/v/v). The organic solvents, i.e., chloroform and methanol, are of analytical grade. All solvent manipulations are to be carried out in a ventilated hood while wearing protective clothing, because chloroform is listed as harmful, and methanol is toxic (*see* **Note 1**).
5. Falcon tubes, 50 ml (BD Biosciences, San Jose, CA).
6. Eppendorf safe-lock tubes, 1.5 and 2 ml (Eppendorf, Hamburg, Germany).
7. 0.5% tergitol solution (Sigma-Aldrich, St. Louis, MO).
8. Tween 80 (Carl Roth, Karlsruhe, Germany).
9. Cholesterol 99.9% pure (Sigma-Aldrich, St. Louis, MO).
10. [4-^{14}C]-cholesterol (0.1 mCi/ml; American Radiolabeled Chemicals, Inc. St. Louis, MO).
11. [9,10-^{3}H]-palmitic acid (10 mCi/ml; American Radiolabeled Chemicals, Inc.).
12. [7-^{3}H]-pregnenolone (1 mCi/ml; American Radiolabeled Chemicals, Inc.).

3. Methods

3.1. [^{14}C]-Cholesterol Uptake

1. Grow heme-deficient yeast cells (*hem1Δ*) in a falcon tube containing 20 ml of SC-LEU supplemented with aminolevulinic acid at 24°C overnight.
2. Harvest 5 OD_{600} units of cells by centrifugation at 2,000 rpm for 5 min and wash the cells twice with SC-LEU media.
3. Resuspend the cells in 10 ml SC-LEU media supplemented with cholesterol and Tween 80.

4. Add [^{14}C]-cholesterol (0.025 μCi/ml) to the culture (*see* **Note 2**).
5. Let the culture grow at 24°C overnight or to the time point required.
6. Next morning control the OD_{600} and take the volume necessary to obtain 30 OD units of cells.
7. Centrifuge the cells at 2,000 rpm for 5 min. Pour off the supernatant and discard the radio-labeled waste.
8. Wash the cells twice with 5 ml of 0.5% tergitol solution, centrifuge each time at 2,000 rpm for 7 min. Pour off the supernatant and discard the radio-labeled waste.
9. Prepare a 1:100 dilution of [^{3}H]-palmitic acid (in ethanol) to be added as internal standard to control the efficacy of lipid extraction.
10. Add 600 μl of chloroform/methanol (1: 1, v/v) to the cell pellet and then transfer it to a 2 ml safe-lock tube containing 300 μl of glass beads. Add 1 μl of the diluted [^{3}H]-palmitic acid as internal standard.
11. Freeze the samples at −20°C.

3.1.1. Lipid Extraction from Whole Cells and TLC Analysis

1. Thaw the cell pellet in a water bath set to 24°C. Freeze the samples again in liquid nitrogen for 1 min. Repeat this freeze–thaw cycle once more.
2. Break cells in a Precellys 24, cell homogenizer at 5,000 rpm, 4°C, using three cycles of homogenization for 30 s interrupted by a pause of 30 s each. Alternatively, the cells can be homogenized on a Vortex set at top speed for 30 min at 4°C.
3. Centrifuge samples at 13,000 rpm for 5 min and transfer the supernatant to a new 2 ml Eppendorf safe-lock tube.
4. Wash the glass beads with 300 μl chloroform/methanol (1: l, v/v), centrifuge at 13,000 rpm for 5 min and pool the supernatant. Repeat this wash step twice.
5. Centrifuge the samples at 13,000 rpm for 5 min and transfer the supernatant to a new tube. Avoid carry over of glass beads.
6. Take 5 μl of the lipid extract and mix with 2 ml of scintillation cocktail and vortex briefly for 20 s.
7. Count radioactivity of the lipid samples by liquid scintillation counter and remove a sample volume corresponding to 20,000 cpm. Dry this lipid sample under a stream of nitrogen gas.
8. Resuspend the dried lipids in 20 μl of chloroform/methanol (1:1, v/v) and spot the sample on a TLC plate (*see* **Note 3**). Wash the tube again with 20 μl chloroform/methanol (1:1, v/v) and place onto the first spot on the TLC.

9. Let the TLC plate dry at room temperature for 5–10 min. Saturate the TLC chamber with the solvent system, petroleum ether/diethyl ether/glacial acetic acid (70:30:2, v/v/v) for 40–60 min prior to run (*see* **Note 4**). Separate neutral lipids for 30 min at room temperature (*see* **Note 5**).
10. After separation is complete, the TLC plate is removed from the chamber and left to dry in a ventilated hood for at least 1–2 h. The radio-labeled lipid classes (free cholesterol, cholesterol esters, and triacylglycerols) are then quantified by scanning using a radio TLC analyzer (*see* **Note 6**). Alternatively, the TLC plate can be exposed to phosphorimaging plate for 24 h and then be analyzed, using a phosphorimager (Bio-Rad) (*see* **Notes 7 and 8**).

3.2. [^{14}C]-Cholesterol Export

1. Grow heme-deficient yeast strains in a lipid supplemented medium at 24°C overnight.
2. Next day harvest 10 OD_{600} units of cells in a tabletop centrifuge at 4,000 rpm for 5 min. Pour off the supernatant and discard.
3. Wash the cells twice with SC-LEU media.
4. Resuspend the cell in 10 ml of SC-LEU media supplemented with cholesterol and Tween 80. Add [^{14}C]-cholesterol (0.025 μCi/ml) and grow cells overnight at 24°C to let them take [^{14}C]-cholesterol.
5. Next day pellet the cells at 4,000 rpm for 5 min. Pour off the supernatant and discard the radio-labeled waste.
6. Wash the cells twice with SC-LEU media. Remove supernatant and discard the radio-labeled waste.
7. Resuspend the cell pellet in 10 ml of SC-LEU media supplemented with cold cholesterol (20 μg/ml) and Tween 80 (5 mg/ml).
8. Let the cells grow at 24°C overnight, or for the time required, to allow export of [^{14}C]-cholesterol or acetylated [^{14}C]-cholesterol into the culture medium.
9. Next morning pellet the cells at 4,000 rpm for 5 min and keep aside the culture supernatant in a 50 ml falcon tube.
10. Wash the cells with 5 ml 0.5% tergitol and centrifuge at 2,000 rpm for 5 min. Pour off the supernatant and discard the radio-labeled waste.
11. Resuspend the cells in 400 μl of chloroform/methanol (1:1, v/v) and transfer them into a 2 ml safe-lock tube containing 200 μl glass beads.
12. Freeze the cell pellet (–20°C).
13. Lipid extraction from whole cells and TLC analysis of radio-labeled lipids is performed as described in Section 3.1.1.

3.2.1. Lipid Extraction from The Culture Supernatant

1. Add 30 ml of chloroform/methanol (1: 1, v/v) to the 10 ml culture supernatant in a 50 ml falcon tube.
2. Vortex well for 2–3 min.
3. Centrifuge at 4,000 rpm for 10 min to separate organic and aqueous phases.
4. Carefully transfer the lower organic phase with a 5 ml pipette into a new 50 ml falcon tube (avoid the aqueous phase as well as the interphase).
5. Pour off the aqueous phase and discard the radio-labeled waste.
6. Let the tube containing the organic phase stand for 1 h at room temperature with caps open, and aspirate carefully the aqueous layer formed on the top (if any).
7. Count 5 μl of the organic phase by liquid scintillation counter and aliquot sample volumes corresponding to 10,000 cpm (i.e., same counts for the lipids extracted from the cell pellet) in 2 ml tube and dry under a stream of nitrogen gas.
8. Load samples on TLC, separate and quantify lipids as described in Section 3.1.1.

3.3. [^{3}H]-Pregnenolone Uptake and Secretion

1. Grow yeast in a suitable medium (e.g., YEPD) overnight at 24°C.
2. Control the optical density and harvest 5 OD_{600} units.
3. Pellet the cells at 1,500 rpm for 5 min in a tabletop centrifuge.
4. Resuspend the cells in 5 ml fresh YEPD media, and add 1 μCi/ml [^{3}H]-pregnenolone.
5. Incubate the cells for 2 h with shaking at 24°C.
6. Check the optical density, pellet the cells at 4,000 rpm for 5 min. Keep aside the culture supernatant in a 50 ml falcon tube at 4°C.
7. Wash the cell pellet with 1 ml YEPD, centrifuge at 4,000 rpm for 5 min and transfer the supernatant to the culture supernatant that was kept aside.
8. Resuspend the cells in 400 μl of chloroform/methanol (1:1, v/v) and transfer into a 2 ml safe-lock tube containing 200 μl glass beads.
9. Freeze the cells at –20°C.
10. Continue with lipid extraction from the whole cells as described under Section 3.1.1.
11. Dry down the whole lipid extract and analyze by TLC as described under Section 3.1.1.

12. Extract lipids from the culture supernatant once with 20 ml chloroform/methanol (1: 1, v/v), vortex well for 2–3 min.
13. Centrifuge at 4,000 rpm in a tabletop centrifuge for 5 min to separate the organic and aqueous phases.
14. Carefully transfer the lower organic phase with a 5 ml pipette into a new 50 ml falcon tube (avoid the aqueous phase as well as the interphase).
15. Pour off the supernatant and discard the radio-labeled waste.
16. Dry down the organic phase under a stream of nitrogen gas and analyze and quantify the lipids after TLC separation as described in Section 3.1.1.

3.4. Results

1. Cholesterol uptake under the conditions described in Section 3.1 is time-dependent and the internalized cholesterol is efficiently back transported to the ER membrane where it is converted to steryl esters (**Fig. 1**). This lipid uptake pathway is affected in a number of sterol uptake and transport mutants that were isolated in a genome-wide screen for mutants that fail to grow under anaerobic conditions *(12, 17)*.

Fig. 1. Uptake and esterification of [^{14}C]-cholesterol in heme-deficient cells. *hem1Δ* cells were grown in the presence of [^{14}C]-cholesterol and Tween 80 at 24°C. At the indicated time points, equal OD units of cells were harvested, lipids were extracted and analyzed by TLC. *FC*, free cholesterol; *STE*, steryl esters.

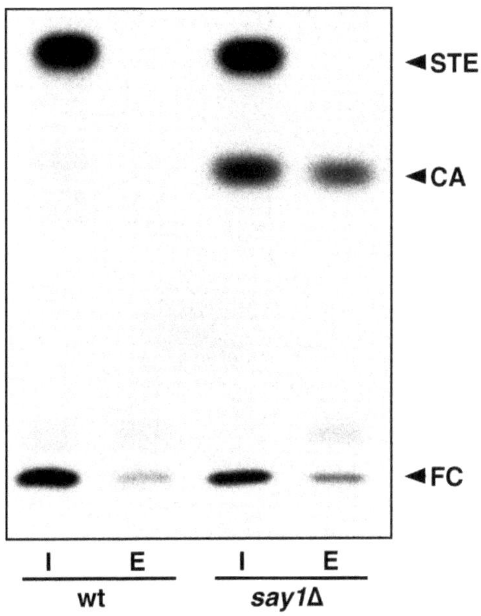

Fig. 2. Acetylated sterols are secreted. Heme-deficient wild-type and *say1Δ*-mutant cells were labeled with [^{14}C]-cholesterol for 16 h, diluted into fresh media containing cold cholesterol, and cultivated for 16 h. Lipids were extracted from the cell pellet (I) and from the culture media (E) and analyzed by TLC. *FC*, free cholesterol; *CA*, cholesterol acetate; *STE*, steryl esters.

2. Formation of cholesterol acetate is barely detectable in wild-type cells, but accumulates in cells lacking the sterol deacetylase (*say1Δ*) (**Fig. 2**). Discrimination between intra- and extracellular lipids reveals that *say1Δ* mutant cells excrete sterol acetate. Excretion of this lipid is selective for acetylated sterols because long-chain steryl esters are not detectable in the culture supernatant of wild-type or *say1Δ* mutant cells. The identity of the modified sterol that accumulates in *say1Δ* mutant cells was independently confirmed by mass spectrometry *(18)*.
3. Radio-labeled pregnenolone is rapidly acetylated and excreted from wild-type and deacetylase (*say1Δ*) mutant cells (**Fig. 3**). The fact that acetylation-deficient (*atf2Δ*) mutant cells accumulate radio-labeled long-chain ester intracellularly indicates that acetylation is required for export of the modified pregnenolone and that a failure to acetylate results in the intracellular accumulation of this steroid precursor in form of long-chain esters. The identity of these long-chain pregnenolone esters is based on the fact that formation of this product is abolished in cells lacking the two acyl-CoA:sterol acyltransferases, *ARE1* and *ARE2 (17)*.

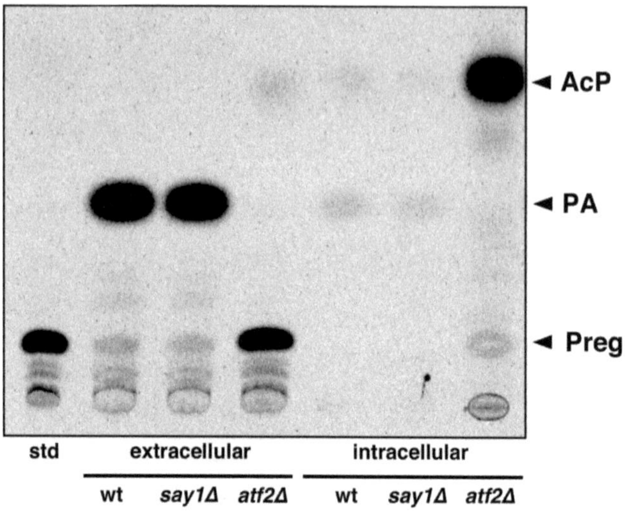

Fig. 3. Export of steroids is controlled by sterol acetylation/deacetylation. Atf2 is required for acetylation and secretion of pregnenolone. Wild-type, *say1*Δ, and *atf2*Δ mutant cells were labeled with [^{3}H]-pregnenolone; lipids were extracted from the culture media (extracellular), and the cell pellet (intracellular), and analyzed by TLC. [^{3}H]-pregnenolone was loaded as standard (std). *Preg*, free pregnenolone; *PA*, pregnenolone acetate; *AcP*, acylated pregnenolone.

4. Notes

1. The organic solvents used are toxic or even carcinogenic. Chloroform/methanol mixture will rapidly leach the skin lipids from hands. Further contact with the solvents will give rise to irritation. It is therefore advisable to handle all the solvents with care. If mixtures of chloroform/methanol spill over a body part, that region should immediately be rinsed with cold running water to minimize the burning sensation, which could last for about 10 min.
2. All radioactive mixes should be handled with utmost care and precautions, always wearing protective gloves. If there is a spill of radioactive mix over any part of the body, that region should be immediately rinsed with plenty of running cold water and then seeking the advice of a physician.
3. At a given humidity, the amount of water adsorbed by the silica gel increases as pore size decreases. The water content of the silica gel increases the polarity of the adsorbent and hence its chromatographic properties. For good separations, the water content of the silica gel on TLC plates therefore must be carefully controlled. To remove the water, the silica gel is "activated" by heating the plates immediately before use for 10 min at temperature above 100°C.
4. Only the purest solvents are used in the development of the plates, and the component solvents should be thoroughly

mixed. When solvent systems containing large proportions of polar solvents such as methanol are employed, the chambers should be lined with filter paper to help saturate the atmosphere. However, with nonpolar solvents such as petroleum ether or diethyl ether, the lining of chambers is not necessary.

5. The time taken for a TLC plate to develop depends on the ambient temperature and the solvent system employed. For example, with a solvent system to separate nonpolar lipids, a standard 20 × 20 cm plate will be developed fully in approximately 30 min at room temperature.
6. Although lipid classes can be identified by reference to published R_f values, the application of commercially available lipid standards, either as mixtures or individually, alongside the lipid being analyzed, greatly aids in the identification of the components present in the lipid sample. Within any laboratory, the R_f values, of lipid classes in a given solvent system are not always constant owing to day-to-day variations in temperature, humidity, and perhaps even the batch of plates used. By routinely analyzing lipid standards alongside samples, such variations can be taken into account.
7. For the TLC plates to be exposed to [^{32}P]- or [^{3}H]-phosphorimager screens, it is important to dry the plates very well, or otherwise the acetic acid used in the solvent system will stick to the screens and contaminate them.
8. To analyze the radioactive lipid bands by phosphorimaging, an exposure time of 24 h is usually adequate. Care should be taken to avoid exposure of the screen to light before scanning.

Acknowledgments

This research was supported by the Swiss National Science Foundation (grants PP00A3-110450 and 3100A0-20650).

References

1. Maxfield, F. R., and Tabas, I. (2005) Role of cholesterol and lipid organization in disease. *Nature* **438**, 612–621.
2. Ikonen, E. (2008) Cellular cholesterol trafficking and compartmentalization. *Nat Rev Mol Cell Biol* **9**, 125–138.
3. Maxfield, F. R., and Wustner, D. (2002) Intracellular cholesterol transport. *J Clin Invest* **110**, 891–898.
4. Kaplan, M. R., and Simoni, R. D. (1985) Transport of cholesterol from the endoplasmic reticulum to the plasma membrane. *J Cell Biol* **101**, 446–453.
5. Heino, S., Lusa, S., Somerharju, P., Ehnholm, C., Olkkonen, V. M., and Ikonen, E. (2000) Dissecting the role of the Golgi complex and lipid rafts in biosynthetic transport of cholesterol to the cell surface. *Proc Natl Acad Sci USA* **97**, 8375–8380.
6. Baumann, N. A., Sullivan, D. P., Ohvo-Rekila, H., Simonet, C., Pottekat, A., Klaassen, Z., Beh, C. T., and Menon, A. K. et al. (2005) Transport

of newly synthesized sterol to the sterol-enriched plasma membrane occurs via nonvesicular equilibration. *Biochemistry* **44**, 5816–5826.

7. Chang, T. Y., Chang, C. C., Ohgami, N., and Yamauchi, Y. (2006) Cholesterol sensing, trafficking, and esterification. *Annu Rev Cell Dev Biol* **22**, 129–157.
8. Lange, Y. (1994) Cholesterol movement from plasma membrane to rough endoplasmic reticulum. Inhibition by progesterone. *J Biol Chem* **269**, 3411–3414.
9. Hao, M., Lin, S. X., Karylowski, O. J., Wustner, D., McGraw, T. E., and Maxfield, F. R. (2002) Vesicular and non-vesicular sterol transport in living cells. The endocytic recycling compartment is a major sterol storage organelle. *J Biol Chem* **277**, 609–617.
10. Sturley, S. L. (2000) Conservation of eukaryotic sterol homeostasis: New insights from studies in budding yeast. *Biochim Biophys Acta* **1529**, 155–163.
11. Schulz, T. A., and Prinz, W. A. (2007) Sterol transport in yeast and the oxysterol binding protein homologue (OSH) family. *Biochim Biophys Acta* **1771**, 769–780.
12. Schneiter, R. (2007) Intracellular sterol transport in eukaryotes, a connection to mitochondrial function? *Biochimie* **89**, 255–259.
13. Schneiter, R., Brugger, B., Sandhoff, R., Zellnig, G., Leber, A., Lampl, M., Athenstaedt, K., Hrastnik, C., Eder, S., Daum, G., Paltauf, F., Wieland, F. T., and Kohlwein, S. D. (1999) Electrospray ionization tandem mass spectrometry (ESI-MS/MS) analysis of the lipid molecular species composition of yeast subcellular membranes reveals acyl chain-based sorting/remodeling of distinct molecular species en route to the plasma membrane. *J Cell Biol* **146**, 741–754.
14. Lewis, T. A., Taylor, F. R., and Parks, L. W. (1985) Involvement of heme biosynthesis in control of sterol uptake by *Saccharomyces cerevisiae*. *J Bacteriol* **163**, 199–207.
15. Yang, H., Bard, M., Bruner, D. A., Gleeson, A., Deckelbaum, R. J., Aljinovic, G., Pohl, T. M., Rothstein, R., and Sturley, SL (1996) Sterol esterification in yeast: A two-gene process. *Science* **272**, 1353–1356.
16. Li, Y., and Prinz, W. A. (2004) ATP-binding cassette (ABC) transporters mediate nonvesicular, raft-modulated sterol movement from the plasma membrane to the endoplasmic reticulum. *J Biol Chem* **279**, 45226–45234.
17. Reiner, S., Micolod, D., Zellnig, G., and Schneiter, R. (2006) A genomewide screen reveals a role of mitochondria in anaerobic uptake of sterols in yeast. *Mol Biol Cell* **17**, 90–103.
18. Tiwari, R., Koffel, R., and Schneiter, R. (2007) An acetylation/deacetylation cycle controls the export of sterols and steroids from *S. cerevisiae*. *EMBO J* **26**, 5109–5119.
19. Henderson, J. R., and Tocher, D. R. (1992) Thin layer chromatography, in *Lipid Analysis: A Practical Approach* (Hamilton, R. J. and Hamilton, S., eds.), Oxford University Press, Oxford, UK, pp. 65–111.
20. Burke, D., Dawson, D., and Stearns, T. (2000) *Methods in yeast genetics*. A cold spring harbor laboratory course manual. 2000 Edition. Cold spring harbor laboratory press.

Chapter 13

Methods to Monitor Fatty Acid Transport Proceeding Through Vectorial Acylation

Elsa Arias-Barrau, Concetta C. DiRusso, and Paul N. Black

Summary

The process of fatty acid transport across the plasma membrane occurs by several mechanisms that involve distinct membrane-bound and membrane-associated proteins and enzymes. Among these are the fatty acid transport proteins (FATP) and long-chain acyl CoA synthetases (Acsl). Previous studies in yeast and adipocytes have shown FATP and Acsl form a physical complex at the plasma membrane and are required for fatty acid transport, which proceeds through a coupled process-linking transport with metabolic activation termed vectorial acylation. At present, six isoforms of FATP and five isoforms of ACSL have been identified in mice and man. In addition, there are a number of splice variants of different FATP and Acsl isoforms. The different FATP and Acsl isoforms have distinct tissue expression profiles and different cellular locations suggesting they function in the channeling of fatty acids into discrete metabolic pools. The concerted activity of these proteins is proposed to allow cells to discriminate different classes of fatty acids and provides the mechanistic basis underpinning the selectivity and specificity of the fatty acid transport process.

Key words: Fatty acid, FATP, Acsl, Transport, Trafficking, Vectorial acylation, Acyl-CoA

1. Introduction

The principle behind vectorial acylation is that exogenous fatty acids are transported across the plasma membrane concomitant with activation to CoA thioesters. This idea was put forth by Overath et al. *(1)* in studies describing the role of Acsl from *Escherichia coli* in fatty acid metabolism showing intracellular free fatty acids could not be detected following incubation with exogenous long-chain fatty acids. Rather acyl CoA derivatives were detected, which were destined for either complex lipid synthesis or ß-oxidation. Bacterial cells lacking Acsl are also unable

Donald Armstrong (ed.), *Lipidomics*, Methods in Molecular Biology, vol. 580,
DOI 10.1007/978-1-60761-325-1_13, © Humana Press, a part of Springer Science+Business Media, LLC 2009

to transport exogenous fatty acids further supporting this concept. The process of vectorial acylation was subsequently proven to be operational in bacteria by our group *(2)*. In yeast, we have shown exogenous fatty acids are converted to acyl CoAs on transport, but with an added layer of complexity. Cells defective in the two major Acsls (Faa1p and Faa4p) are unable to accumulate intracellular oleoyl CoA following incubation with oleate *(3)*. This is also the case for yeast cells defective in FATP (designated Fat1p) *(4)*. Ensuing studies by our laboratory have shown these two proteins interact and function in concert at the plasma membrane. Along with our earlier studies, these data confirm the mechanistic roles of FATP and Acsl in the vectorial acylation of exogenous fatty acids *(5)*. The situation in mammalian cells is more complex, as there are at least six FATP and five Acsl isoforms (**Table 1**). Yet in mammalian cells, there is compelling data suggesting specific FATPs and Acsls mediate vectorial acylation in a manner analogous to that defined in yeast. When the first FATP was identified using expression cloning, an often overlooked result was the identification of Acsl1, which when over expressed in Cos7 cells also increased fatty acid transport *(6)*. When both FATP1 and Acsl1

Table 1
Tissue and subcellular localization of the FATP and Acsl isoforms

	Tissue expression	Subcellular localization
FATP1	Heart, adipose tissue, skeletal muscle, brain, lung, kidney	Plasma membrane, MAM[1], mitochondrial membrane
FATP2	Liver, intestine, kidney	Endoplasmic reticulum, peroxisome
FATP3	Lung, testis, ovary, brain, adrenal gland, kidney	Cytoplasmic vesicles
FATP4	Small intestine, skin, brain, adipose tissue, muscle, heart, liver, kidney	Plasma membrane, endoplasmic reticulum, peroxisome, MAM
FATP5	Liver	Endoplasmic reticulum
FATP6	Heart	Plasma membrane
Acsl1	Liver, skeletal muscle, adipose tissue, heart	Endoplasmic reticulum, MAM, plasma membrane
Acsl3	Brain, testis, liver	Cytoplasmic vesicles
Ascl4	Brain, liver, skeletal muscle, adrenal gland, adipose tissue	MAM, peroxisome
Acsl5	Brain, liver, intestine, adipose tissue	Mitochondrial outer membrane
Acsl6[2]	Brain, heart	Plasma membrane, endoplasmic reticulum, MAM

[1]MAM, mitochondrial-associated membrane
[2]Acsl6 subcellular localization has not been defined experimental, but predicted by its activity

are coexpressed in NIH 3T3 cells, there is a synergistic rise in fatty acid transport *(7)* suggesting these two proteins functionally interact. More recent studies have shown FATP1 and Acsl1 form a complex at the plasma membrane in mouse adipocytes analogous to what we defined in yeast *(5, 8)*. When we express the murine FATP isoforms in yeast defective for transport, only FATP1, FATP2, and FATP4 are functional in this pathway. In the yeast system, each of these proteins must necessarily interact with the endogenous yeast Acsl (Faa1p or Faa4p) to catalyze transport *(9)*. Thus of the studies completed to date, there is compelling data that vectorial acylation of exogenous long-chain fatty acids requires both an FATP and an Acsl. There are, however, other studies suggesting several of the FATP isoforms may function alone in vectorial acylation. Milger et al. *(10)*, for example, have shown FATP4 functions to drive fatty acid transport, even though it is primarily localized in the endoplasmic reticulum. Further, partially purified FATP4 has both long chain ($C_{16:0}$) and very long chain ($C_{24:0}$) acyl CoA synthetase activities *(11)*. On the basis of these studies, three scenarios can be drawn regarding the protein components involved in this process (**Fig. 1**).

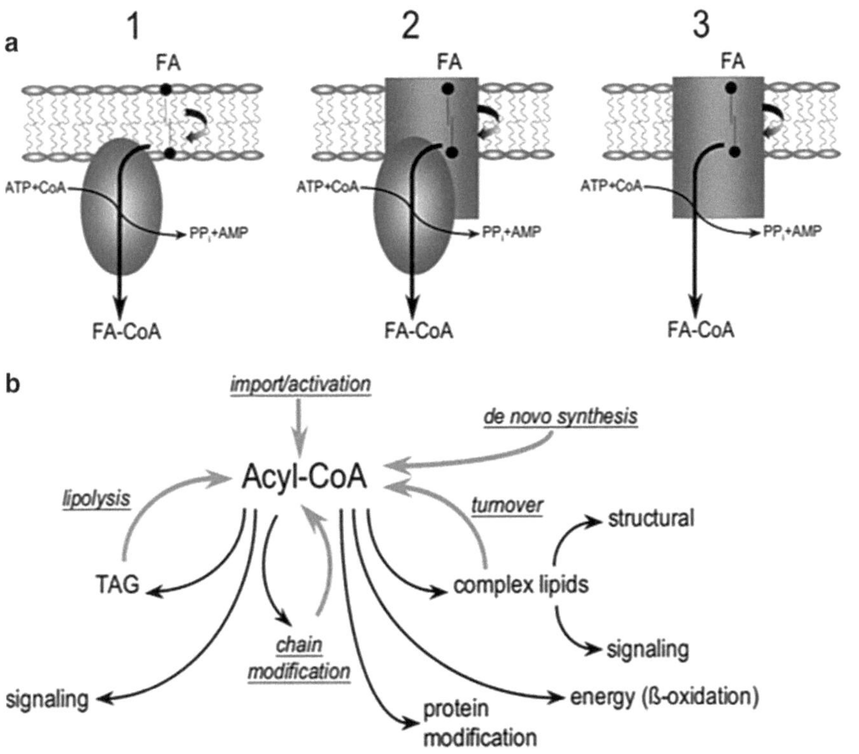

Fig. 1. Process of vectorial acylation. (**a**) Three different scenarios are illustrated. (1) Acsl, when membrane associated, functions to abstract, and activate fatty acids concomitant with transport; (2) A membrane-bound FATP functions cooperatively with Acsl to transport and activate fatty acids; (3) A membrane-bound FATP, with the appropriate acyl CoA synthetase specificity, functions to activate fatty acids concomitant with transport. In all three cases, the final product of transport is acyl CoA, hence vectorial acylation. (**b**) Trafficking of acyl CoA; the acyl CoA pools are likely to be distinct in different cellular compartments as a consequence of specific FATP and Acsl activity. Adapted from Ref. *(22)*.

In the simplest scenario, and as demonstrated in *E. coli*, the association of Acsl with the plasma membrane facilitates both the abstraction and concomitant activation of exogenous fatty acids. The second scenario is more complex and requires both a membrane-bound FATP and a membrane-associated Acsl that function cooperatively to transport and activate exogenous fatty acids. The third scenario is similar to the first, but requires only a membrane-bound FATP that functions in the concomitant transport and activation of exogenous fatty acids. There are experimental data supporting each of these proposals with the commonality residing in the formation of acyl CoA upon transport rendering this process unidirectional. The different FATP and Acsl isoforms are proposed to function in directing distinct classes of fatty acids as acyl CoAs into fairly discrete metabolic pools and thus provide both selectivity and specificity to this process.

This paper details the fundamental methods used to monitor fatty acid transport and gives examples of the experimental outcomes using both yeast and mammalian model systems.

2. Materials

2.1. Equipment

1. 96-well plates for C_1-BODIPY-C_{12} fatty acid transport assay: BIOCOAT Collagen I-coated 96-Well Black/Clear Plate (Becton Dickinson); Microtest 96-Well Plate Optilux™ Black/Clear Bottom (BD Falcon, USA).
2. Fluorescence 96-well plate reader for C_1-BODIPY-C_{12} fatty acid transport assay: BioTek Synergy HT multidetection microplate reader (Bio-Tek Instruments, Inc., USA).
3. Fluorescence microscopy to monitor steady-state accumulation C_1-BODIPY-C_{12} following transport: Zeiss LSM510 confocal laser scanning microscope (Carl Zeiss MicroImaging, Inc., Germany).
4. Radioactive scintillation counting: LS 6500 Multi-Purpose Scintillation Counter (Beckman Coulter, USA).
5. Acyl CoA synthetase assay: Shaking adjustable temperature water bath.
6. Thin layer chromatography of complex lipids: Molecular Image FX scanner (Bio-Rad, USA).
7. Data analysis: Prism software (GraphPad Software, Inc., USA).

2.2. Supplies

1. C_1-BODIPY-C_{12} fatty acid transport assay: C_1-BODIPY-C_{12} (4,4-difluoro-5-methyl-4-bora-3a,4a-diaza-s-indacene-3-dodecanoic acid) (Invitrogen, Molecular Probes, USA); 0.4% solution trypan blue (Sigma, USA); Fatty acid-free BSA (Calbiochem, USA); phosphate saline buffer (PBS)

(Hy-clone, USA); Hank's balanced salt solution (Hy-clone); 3-(4,5-dimethylthiazol-2-yl)-2,5-diphenyltetrazolium bromide (MTT) (Sigma).

2. Thin layer chromatography of complex lipids: methyl acetate; 2-propanol (Fisher, USA); chloroform (Fisher); cyclohexane (Fisher); ethyl acetate (Fisher); Whatman Partisil LK5 TLC plates (250 μm, 150 Å silica); Lipid standards (Sigma).
3. Confocal microscopy to monitor C_1-BODIPY-C_{12} transport: Four-well tissue culture-treated microscope slides (BD Falcon; BD Biosciences); Elvanol mounting media (DuPont, Wilmington, Del., USA).
4. Radio-labeled fatty acid transport and acyl CoA synthetase activity assays: [^{3}H] palmitate ($C_{16:0}$), [^{3}H]oleate ($C_{18:1}$), [^{14}C] arachidonate ($C_{20:4}$) and [^{14}C]lignocerate ($C_{24:0}$) (Perkin-Elmer, USA); Sorbitol (Fisher); KH_2PO_4 (Fisher); Tris Base (Fisher); $MgCl_2$ (Sigma); Triton-X 100 (Acros); dithiothreitol (DTT) (Sigma); ATP (Sigma); Coenzyme A (Sigma); 2-propanol (Fisher); *n*-Heptane (Fisher); H_2SO_4 (Fisher).
5. Mammalian cell culture: DMEM high glucose (Hy-clone); EMEM (Hy-clone); MEM without phenol red (Gibco, USA); Modified DMEM (Gibco); fetal bovine serum (FBS, Hy-clone); EGM-2 Bullet Kit (Cambrex); MCDB 131 (Gibco); Biocoat Intestinal Epithelium Differentiation Media Pack (BD Bioscience); Epidermal growth factor (BD Bioscience); Hydrocortisone (Sigma); Insulin (Sigma); Dexamethasone (Sigma); L-glutamine (Sigma); Methylisobutylxanthine (Sigma); Zeocin (Invitrogen); Hygromycin (Invitrogen); Blasticidin (Invitrogen).
6. Yeast cell culture: YNBD yeast media (yeast minimal medium containing 0.67% yeast nitrogen base, 2% dextrose, 20 mg/L adenine, 20mg/L uracil, and amino acids as required); β-estradiol (Sigma).

2.3. Reagents

1. C_1-BODIPY-C_{12} fatty acid transport assay: Yeast cells: 1.25 μM C_1-BODIPY-C_{12}, 0.75 μM fatty acid-free BSA, 2.1 mM trypan blue. Mammalian cells: 5 μM C_1-BODIPY-C_{12}, 5 μM fatty acid-free BSA, 3.9 mM trypan blue (*see* **Note 1**).
2. Thin layer chromatography of complex lipids: for phospholipids, the solvent system is methyl acetate, 2-propanol, chloroform, 0.25% KCl (0.9:0.9:1.0:0.25). For to neutral lipids, the solvent system is cyclohexane, ethyl acetate (3:2).
3. Acyl CoA synthetase activity: Dole's reagent [isopropyl alcohol:*n*-heptane:1M H_2SO_4 (40:10:1)].

2.4. Cell Lines

1. Yeast strains. *Saccharomyces cerevisiae* stains LS2086 (*faa*1Δ *fat*1Δ) and LS2089 (*faa1Δfaa4Δfat1Δ)* and derivatives harboring expression plasmids derived from pDB121 (covered by United States patent 7,070,944).

2. Mammalian cell lines. 3T3-fibroblast and adipocytes (ATCC, CL-173); b-END3 cells (ATCC, CRL-2299); Caco-2 cells (ATCC, HTB-37); HAEC (Cambrex, CC-2535); HepG2 cells (ATCC, HB-8065); 293 T-REx (Invitrogen) and derivatives harboring expression plasmids derived from pcDNA4 and pcDNA5 (*see* **Note 2**).

2.5. Vectors

1. For expression in yeast: pDB121 (9) and derivatives; pRS-416GAL4-ER-VP16 encoding a synthetic trans-criptional activator, which is a protein fusion between the Gal4 DNA binding domain, a β-estradiol-responsive regulatory domain, and a VP16 RNA polymerase activation domain *(12)*.
2. For expression in mammalian cells: pcDNA4/TO/myc-His-A (pcDNA4) and pcDNA5/TO (pcDNA5) (Invitrogen) and derivatives (*see* **Note 2**).

3. Methods

3.1. Experiment

1. Experimental setup of 96-well plates using yeast expressing a mammalian FATP. Yeast cells transformed with a pDB121-derived expression plasmid and pRS416GAL4-ER-VP16 are pregrown in YNBD media without leucine and uracil (YNBD-leu-ura), subcultured to an initial absorbance 0.02 at 600 nm in the same medium containing 10 nM β-estradiol to induce expression and then grown at 30°C to mid-log phase (0.8–1.2 $Å_{600}$). The cells are harvested by centrifugation (5 min, 5000 × *g*) and resuspended in PBS to a final cell density of 6×10^7/ml.
   In a standard assay, 50 μl of yeast cells (3×10^6 cells) are added into each well of a 96-well plate (black wells and clear bottom). One hundred and fifty microliters of a mixture containing C_1-BODIPY-C_{12} (1.25 μM final concentration), fatty acid-free BSA (0.75 μM final concentration), and trypan blue (2.1 mM final concentration) is added to each well to give a final volume of 200 μl. The 96-well plate is incubated in the dark for varying lengths of time (1–30 min depending on the experimental needs) at 25°C and the fluorescence intensity measured using a Bio-Tek Synergy HT multidetection microplate reader (filter sets of 485 ± 20 nm excitation and 528 ± 20 nm emission). Trypan blue, a vital dye that cannot enter the cell, quenches the extracellular fluorescence of C_1-BODIPY-C_{12} thereby allowing only cell-associated fluorescence of C_1-BODIPY-C_{12} to be measured, which is reflective of transport (*see* **Note 1**).
2. Experimental setup of 96-well plates to measure transport in mammalian cells. Mammalian cells are grown under standard

conditions in Microtest 96-well plate (black wells/clear bottom), except Caco-2 cells, which are seeded in BIOCOAT Collagen I 96-well black/clear plates. The seeding density is routinely 6×10^4 cells/cm^2 for 3T3-L1 fibroblast and adipocytes, b-END3 cells, and HMEC; and 2.5×10^5 cells/cm^2 for 293 T-REx, Caco-2, and HepG2 cells. Cells are grown at 37°C in 5% CO_2 for 3 days; Caco-2 cells require 4 days of growth because of the differentiation process; 293 T-REx cells are seeded the same day of the experiment following induction of the target FATP or Acsl using tetracycline. The final volume of each well is 250 µl. For the assay, the growth media is removed and the cells starved for 1 h in 250 µl of MEM without phenol red. The media is removed and replaced by 50 µl of MEM without phenol red and 50 µl of a C_1-BODIPY-C_{12} reaction mixture prepared in Hank's balanced salt solution to give a final volume of 100 µl per well (final concentrations are 5 µM C_1-BODIPY-C_{12}, 5 µM fatty acid-free BSA and 3.9 mM trypan blue). The plates are maintained in the dark to preserve the C_1-BODIPY-C_{12} and are incubated for varying lengths of time (1–30 min depending on the experiment) at 37°C in 5% CO_2. The plates are read using a BioTek Synergy HT multidetection microplate reader following excitation at 485 nm and emission at 528 nm (*see* **Note 1**).

As noted above, the trypan blue quenches the extracellular fluorescence; the cell-associated fluorescence of C_1-BODIPY-C_{12} is reflective of transport.

3. Mammalian cell growth conditions. Caco-2 cells (ATCC, HTB-37) are maintained in EMEM containing 20% FBS. Biocoat Intestinal Epithelium Differentiation Media Pack and BIOCOAT Collagen I 96-well black/clear plates are used for the differentiation process. HepG2 cells (ATCC, HB-8065) are grown in EMEM supplemented with 10% FBS. 3T3-L1 cells (ATCC, CL-173) are maintained in modified DMEM and 10% FBS. Cells are treated with methylisobutylxanthine (0.5 µM), dexamethasone (1 µM), and insulin (1.75 µM) as detailed by Student et al. *(13)* for the differentiation into 3T3-L1 adipocytes. HMEC-1 (CDC) are maintained in MCDB 131 containing l-glutamine (10 mM), hydrocortisone (1 µg/ml), epidermal growth factor (10 ng/ml) and 10% FBS. b-END3 cells (ATCC, CRL-2299) are grown in DMEM high glucose supplemented with sodium pyruvate and 10% FBS. 293 T-REx cells are maintained in high glucose DMEM containing L-glutamine (2 mM), 10% FBS, blasticidin (5 µg/ml) and zeocin, or hygromycin (150 µg/ml or 200 µg/ml respectively; only for the stable cell lines).

3.2. Results

3.2.1. Kinetics of Long-Chain Fatty Acid Transport

The methods detailed above will define the apparent K_T and maximal rates (V_{max}) of fatty acid transport using C_1-BODIPY-C_{12} transport in different cell types (yeast, mouse, human) *(14)*. The catalytic efficiency (k_{cat}/K_T) of transport proceeding through vectorial acylation can be assessed by assuming the rate-limiting step in the transport process is reflected by the V_{max} and thus k_{cat}. The constants derived using these methods must be given as "apparent" because the total concentration of C_1-BODIPY-C_{12} is used in the calculations as opposed to the free concentration.

An example using this approach is shown in **Fig. 2** for 3T3-L1 fibroblasts and 3T3-L1 adipocytes; transport follows typical Michaelis–Menten kinetics *(14)*. The catalytic efficiency (k_{cat}/K_T) of this process increases over twofold (2.1×10^6–4.5×10^6 $s^{-1}M^{-1}$) upon adipocyte differentiation (**Table 2**) (*see* **Note 3**). This is in agreement with other studies that show that the long-chain fatty acids uptake increases during preadipocytes differentiation *(15, 16)*.

The excess accumulation of fatty acids leads to chronic cellular dysfunction and injury and is broadly termed lipotoxicity *(17, 18)*. Consequently, to assure that the high concentrations of fatty acids used in these experiments (up to 60 μM) are not toxic to the cells,

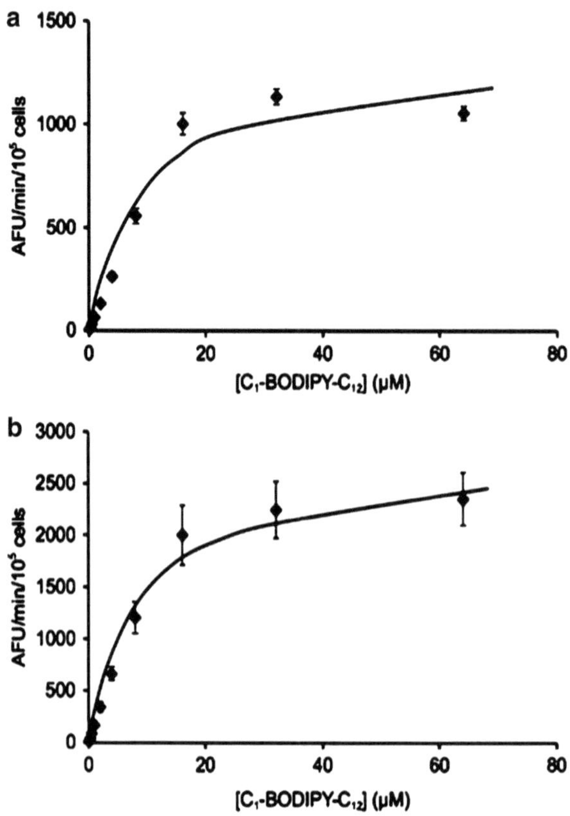

Fig. 2. Kinetics of C_1-BODIPY-C_{12} transport in 3T3-L1 fibroblasts (**a**) and adipocytes (**b**). Adapted from Ref. *(14)* with permission.

Table 2
Kinetic constants derived from transport of C_1-BODIPY-C_{12} in 3T3-L1 fibroblasts and adipocytes. Adapted from Ref. *(14)* with permission

	V_{max} (AFU min^{-1}, ×10^5 cells^{-1}) (SE)[2]	K_T (μM)[3] (SE)	k_{cat}/K_T (sec^{-1} M^{-1})
3T3-L1 fibroblast	1398.0 (68.4)	11.11 (1.55)	2.1×10^6
3T3-L1 adipocyte	2921.0 (203.3)	10.84 (2.17)	4.5×10^6

viability is routinely determined using the MTT assay *(19)*, following incubation with C_1-BODIPY-C_{12} and oleic acid at a final concentration of 100 μM. Caco-2 cells, for example, treated with C_1-BODIPY-C_{12} retained 99.1% (±0.3%) viability, and those treated with oleic acid retained 90.8% (±0.8%) viability. Thus, the fatty acid concentrations (i.e., C_1-BODIPY-C_{12} or oleate) are not toxic at the concentrations used in these experiments.

3.2.2. Real Time Fatty Acid Uptake: Evidence Supports the Concept of Protein-Mediated Fatty Acid Transport

The methods detailed above to monitor the fatty acid transport using C_1-BODIPY-C_{12} can be extended to define this process in real time *(14)*. Cells are grown under standard conditions and the 96-well plate prepared as detailed above; the C_1-BODIPY-$C_{12,}$ fatty acid-free BSA, trypan blue mixture is added; and within readings taken at 12 s intervals using a BioTek Synergy HT multidetection microplate reader. Transport is rapid (between 30 and 120 s the accumulation of C_1-BODIPY-C_{12} is generally in the linear range); after 120 s, accumulation plateaus to a steady-state level (**Fig. 3**). This type of kinetic behavior is reflective of a protein-mediated process (*see* **Note 3**).

3.2.3. Lipid Analysis: C_1-BODIPY-C_{12} is Activated upon Transport via Vectorial Acylation and Incorporated into Complex Lipid Pools

The incorporation of C_1-BODIPY-C_{12} into complex lipids following transport can be defined in yeast expressing a specific FATP isoform *(20)*. The *S. cerevisiae fat*1Δ *faa*1Δ strain transformed with the specific expression plasmid are grown in YNBD (-leu, -ura) containing 10 nM β-estradiol to 1.0 $Å_{600}$ at 30°C; harvested; and resuspended in PBS to a final cell density of 6×10^7/ml. C_1-BODIPY-C_{12} (1.25 μM final concentration) in fatty acid-free BSA (0.75 μM final concentration) is added and incubated for 30 min (*see* **Note 4**). Lipids are extracted using routine methods *(3)* and separated by TLC using Whatman Partisil LK5 plates. The C_1-BODIPY-C_{12} labeled lipids are visualized using a BioRad Molecular Image FX scanner, equipped with an external laser using filter sets of 488 nm excitation and 530 nm emission. Specific classes of lipids are identified by comparisons to internal standards resolved on the same TLC plate. As shown in **Fig. 4**, C_1-BODIPY-C_{12} was

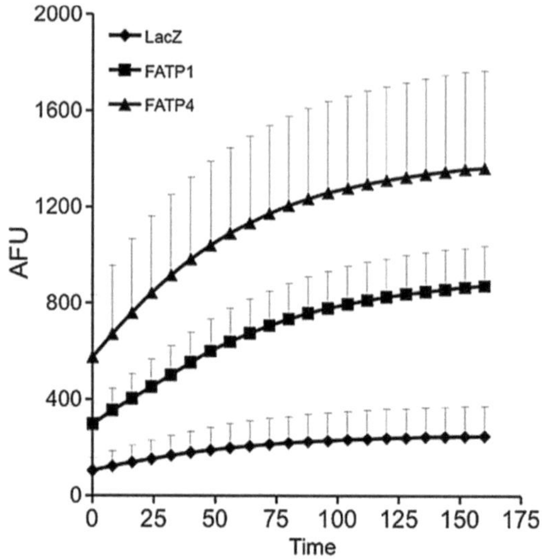

Fig. 3. Transport of C_1-BODIPY-C_{12} (arbitrary fluorescence units, AFU) as a function of time in 293T-REx cells expressing FATP1 or FATP4 relative to control cells expressing LacZ. Note that $t = 0$ is equivalent to 30 s after the initiation of the reaction; readings are taken at 12 s intervals. The bars represent the standard deviation ($n = 4$).

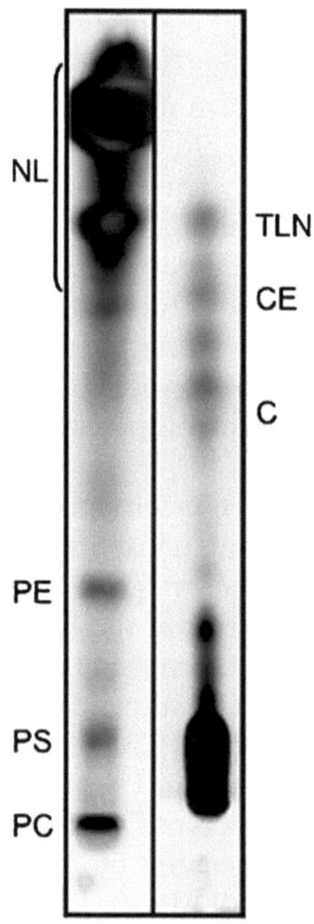

Fig. 4. High performance thin layer chromatography of C_1-BODIPY-C_{12}-labeled lipids (glycerophospholipids, left; neutral lipids, right) from yeast expressing FATP2. Modified from Ref. *(20)* with permission.

incorporated into phospholipids and neutral lipids in yeast expressing human FATP2. The C_1-BODIPY-C_{12} must necessarily be activated to a CoA derivative following transport prior to incorporation into complex lipids in support of the concept of vectorial acylation.

3.2.4. Confocal Microscopy Defines the Patterns of C_1-BODIPY-C_{12} Accumulation in Different Cell Types

C_1-BODIPY-C_{12} accumulation can be visualized using confocal microscopy as first described in yeast *(9, 21)* and later in mammalian cells *(14, 20)*.

Mammalian cells are cultured on four-well tissue culture-treated microscope slides under the standard conditions detailed above. The media is removed and cells washed with MEM. A mixture of 5 μM C_1-BODIPY-C_{12} 5 μM fatty acid-free BSA (final concentrations) is added to the slides for 3 min. Then cells are washed with MEM and Elvanol mounting media is applied to each well. The accumulation of the C_1-BODIPY-C_{12} is monitored using confocal laser microscopy equipped with argon laser (excitation at 488 nm and emission at 505 nm). **Figure 5** illustrates the accumulation of C_1-BODIPY-C_{12} in 3T3-L1 fibroblasts and adipocytes. These results are in agreement with the kinetic parameters obtained for these cell types as noted above.

3.2.5. Validation of the C_1-BODIPY-C_{12} Method Using Radio-labeled Fatty Acid Substrates

The studies detailed above monitor fatty acid transport using a fluorescent fatty acid analog (C_1-BODIPY-C_{12}). As this analog is not a native fatty acid, it is important to validate these methods using natural fatty acid ligands. Stable cell lines (derived from 293T-REx) expressing the different FATP isoforms have been generated. Using the methods detailed above, the 293T-REx cell line expressing FATP1 accumulates C_1-BODIPY-C_{12} at levels significantly greater that cells harboring the empty vector (*see* **Fig. 3**). Fatty acid transport can also be evaluated using radio-labled fatty acid substrates to validate the use of C_1-BODIPY-C_{12}. Cells are grown and starved under the conditions as detailed above in T-75 flasks; the media is removed and the cells are incubated with mixture of 5 μM [^{3}H]oleate 5 μM fatty acid-free BSA for 3 min. The cells are placed on ice; washed three times with 30 μM fatty acid-free BSA; harvested and the radioactivity was counted. Results are expressed as pmol fatty acid/1 × 10^5 cells (*see* **Note 5**). As shown in **Fig. 6**, the expression of the FATP1 results in increased rates of fatty acid transport when compared with cells harboring the empty vector. These results validate the use of C_1-BODIPY-C_{12} as a standard (and nonradioactive) method to monitor fatty acid transport.

3.2.6. Acyl CoA Synthetase Provides an Essential Role in Vectorial Acylation

In addition to FATP, there is considerable evidence showing that long-chain acyl CoA synthetase (Acsl) also provides an important role in vectorial acylation. This is supported by studies showing the yeast FATP (Fat1p) and Acsl (either Faa1p or Faa4p)

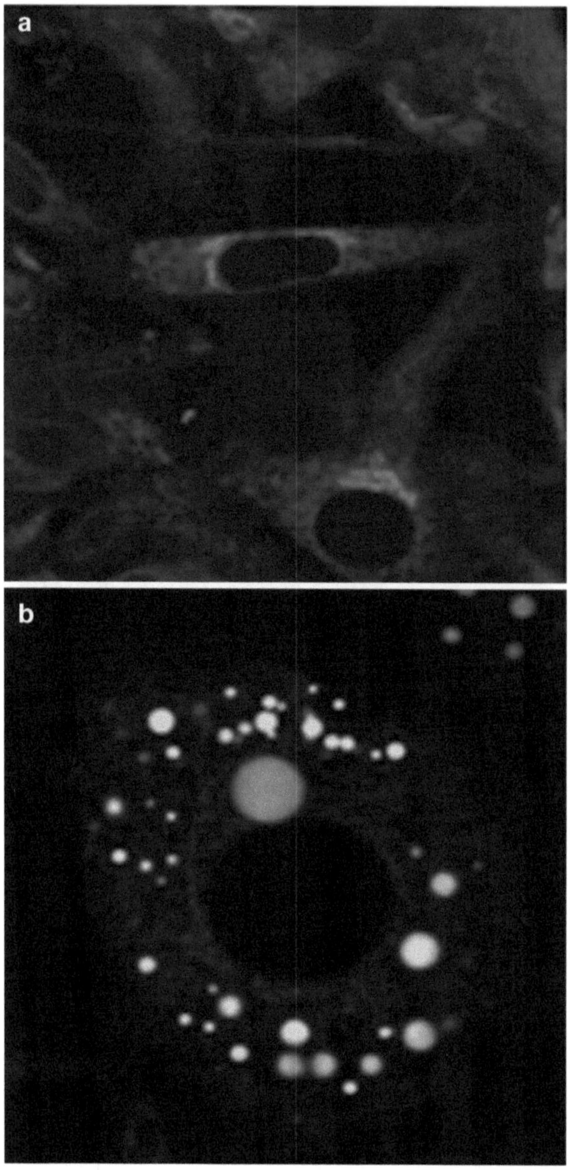

Fig. 5. Transport of C_1-BODIPY-C_{12} in 3T3-L1 fibroblasts and adipocytes visualized using confocal microscopy (upper panel, fibroblasts; lower panel adipocytes). Adapted from Ref. *(14)* with permission.

form a functional complex at the plasma membrane *(5, 22)*. This also appears to be the case for FATP1 and Acsl1 in murine adipocytes *(8)*.

To demonstrate Acsl is essential to the process of vectorial acylation, two yeast strains were used: LS2086 [Δ *fat*1 Δ *faa*1 (Faa4p remains functional)] and LS2089 [Δ *fat*1 Δ *faa*1 Δ *faa*4 (lacking Fat1p (the yeast FATP orthologue (and both of the long-chain acyl-CoA synthetases; these cells are unable to transport or

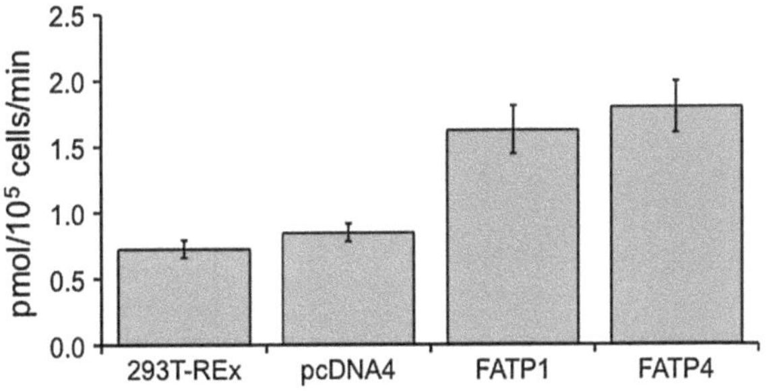

Fig. 6. Transport of [^{3}H] oleate in 293T-REx cells expressing FATP1 or FATP4; 293T-REx are the untransfected cells; pcDNA are 293T-REx cells transfected with vector alone. The error bars represent the standard error of the mean ($n = 4$).

activate exogenous fatty acids)]. These strains are transformed with the vector pDB121 (derived from YEpGALSET983) *(9, 23)* or pDB127 (pDB121 expressing human FATP4) to assess FATP4 function in vectorial acylation.

Acyl-CoA synthetase activities are determined using an adaptation of protocols previously described *(3, 24–26)*. In this modification, 25 ml of yeast cells are grown to mid-log phase (0.8–1.0 $Å_{600nm}$) under standard conditions, harvested by centrifugation (10 min, 3000 × *g*) and washed once with 10 ml PBS. The cell pellet is resuspended in 1 ml of lysis buffer (1.2M sorbitol/20 mM KH_2PO_4, pH 7.4), and cells lysed using three cycles of sonication at 0°C. The extracts are clarified by centrifugation (15 min, 15,000 × *g*) and the resultant supernatant decanted and kept on ice. The reaction mixtures consist of 50 mM Tris-HCl (pH 8), 10 mM $MgCl_2$, 0.01% Triton-X, 300 µM DTT, 10 mM ATP, 50 µM [^{3}H]oleate or 10 µM [^{14}C]arachidonate or [^{14}C]lignocerate, and 100 µl of cell extract in a final volume of 500 µl. The reaction is initiated by the addition of coenzyme A (200 µM final concentration) and incubated at 30°C for 20 min (may be varied for kinetic experiments). The reactions are terminated by the addition of 2.5 ml Dole's reagent [isopropyl alcohol:*n*-heptane:1M H_2SO_4 (40:10:1)]. The radioactive free fatty acid is removed by successive organic extraction with *n*-heptane. Acyl-CoA formed during the reaction remains in the aqueous fraction and is quantified by scintillation counting. Data are routinely expressed as nmol(or pmol) acyl CoA formed/min/mg protein.

The expression of FATP4 in both yeast strains confers modest long-chain ([^{3}H]oleate) and substantial very long-chain ([^{14}C] arachidonate and [^{14}C]lignocerate) acyl-CoA synthetase activities to both yeast strains (**Fig. 7**). In parallel experiments, fatty acid transport is monitored using C_1-BODIPY-C_{12} in both strains of yeast expressing FATP4 (methods detailed above). The *S. cerevisiae fat1Δ faa1Δ* strain expressing FATP4 increases C_1-BODIPY-C_{12} transport; the *S. cerevisiae faa1Δ faa4Δ fat1Δ* strain expressing FATP4 fails to transport C_1-BODIPY-C_{12} (**Fig. 8**). These results

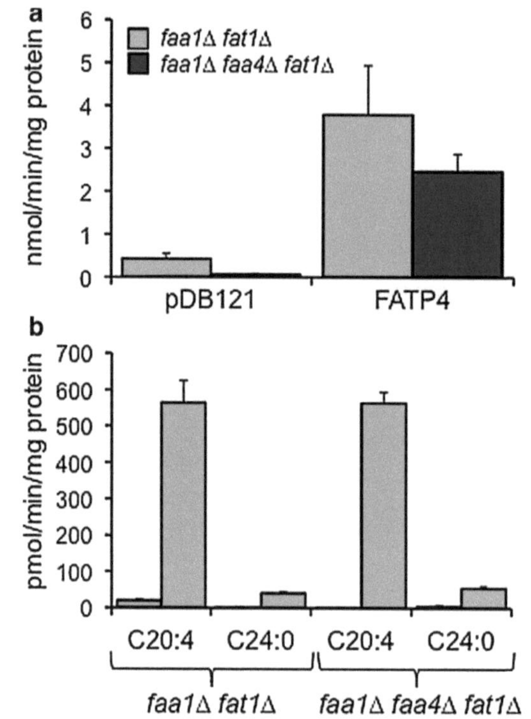

Fig. 7. Acyl CoA synthetase activities in *faa1Δ faa4Δ fat1Δ* and *faa1Δ fat1Δ* strains of yeast expressing FATP4. **a**. Oleoyl CoA synthetase activities and **b**. Arachidonyl and lignoceroyl CoA synthetase activities of cells expressing FATP4 (right bar of the pair) relative to the vector control (pDB121, left bar of the pair). The error bars represent the standard error of the mean ($n = 3$).

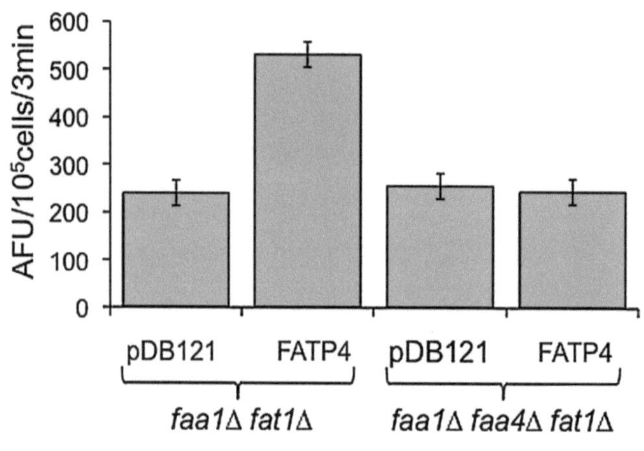

Fig. 8. C_1-BODIPY-C_{12} transport levels in *faa1Δ faa4Δ fat1Δ* and *faa1Δ fat1Δ* strains of yeast expressing FATP4 relative to the vector control (pDB121). The error bars represent the standard error of the mean ($n = 4$).

demonstrate fatty acid transport in this experimental system requires acyl CoA synthetase (either Faa1p or Faa4p); these data support the proposal that fatty acid transport and activation are coupled together in the process of vectorial acylation (*see* **Note 6**).

4. Notes

1. The concentrations of C_1-BODIPY-C_{12} and fatty acid-free BSA can be varied depending on the experimental design. For example, if experiments are directed toward establishing the kinetic constants underpinning the transport of C_1-BODIPY-C_{12}, the concentration of fatty acid-free BSA is held constant and the concentrations of C_1-BODIPY-C_{12} varied. For steady-state transport measurements, a single nonlimiting amount of the C_1-BODIPY-C_{12}/BSA complex should be used.
2. Stable cell lines derived from 293T-REx cells transfected with the pcDNA4- and pcDNA5-derived expression plasmids can be generated using standard methods. The expression of the target protein (e.g., a specific FATP or Acsl) is regulated through the TetR system. As a matter of routine procedure, expression is monitored using both western blots (with antibodies directed against the specific FATP or Acsl) and qPCR.
3. The cell types listed are not all-inclusive, but have been selected to show these methods can be applied to a variety of cell types. These methods will allow investigators to define the kinetic parameters that govern fatty acid transport using C_1-BODIPY-C_{12}. The methods detailed can be extended by holding the concentration of C_1-BODIPY-C_{12} constant and increasing the concentrations of competing fatty acid ligands to assess specificity.
4. In these experiments where the C_1-BODIPY-C_{12} is incorporated into complex lipids, the cells are maintained with darkroom light (red) illumination or in the dark during the incubation period to preserve the integrity of this fluorescent fatty acid. It is best to cover the plates and incubation tubes with foil during the preparation of the final lipid samples.
5. The methods detailed to monitor fatty acid transport using radio-labeled fatty acids are standard; it is essential that the labeling period be short (30–120 s) to maintain the linearity of the reaction, although this needs to be empirically defined for different cell types. Transport is slowed by placing the cells on ice and the use of high concentrations of fatty acid-free BSA in the wash effectively removed fatty acids that are not activated to CoA thioesters via vectorial acylation.

6. The acyl CoA synthetase data described are from yeast cells expressing the different FATP isoforms to demonstrate the importance of Acsl in this process. These methods can be directly applied for the measurement of acyl CoA synthetase activities in the different mammalian cell lines.

Acknowledgments

This work was supported by grants from the National Institutes of Health (RO1-GM56850 and RO1-DK07076) and the Charitable Leadership Foundation to P.N.B. and C.C.D.

References

1. Overath, P., Pauli, G., and Schairer, H. U. (1969) Fatty acid degradation in *Escherichia coli.* An inducible acyl-CoA synthetase, the mapping of old-mutations, and the isolation of regulatory mutants, *Eur J Biochem* 7, 559–574.
2. DiRusso, C. C., and Black, P. N. (2004) Bacterial long chain fatty acid transport: Gateway to a fatty acid-responsive signaling system, *J Biol Chem* **279**, 49563–49566.
3. Faergeman, N. J., Black, P. N., Zhao, X. D., Knudsen, J., and DiRusso, C. C. (2001) The Acyl-CoA synthetases encoded within FAA1 and FAA4 in *Saccharomyces cerevisiae* function as components of the fatty acid transport system linking import, activation, and intracellular Utilization, *J Biol Chem* **276**, 37051–37059.
4. DiRusso, C. C., Connell, E. J., Faergeman, N. J., Knudsen, J., Hansen, J. K., and Black, P. N. (2000) Murine FATP alleviates growth and biochemical deficiencies of yeast fat1Δ strains, *Eur J Biochem* **267**, 4422–4433.
5. Zou,Z.,Tong,F.,Faergeman,N.J.,Borsting,C., Black, P. N., and DiRusso, C. C. (2003) Vectorial acylation in *Saccharomyces cerevisiae.* Fat1p and fatty acyl-CoA synthetase are interacting components of a fatty acid import complex, *J Biol Chem* **278**, 16414–16422.
6. Schaffer, J. E., and Lodish, H. F. (1994) Expression cloning and characterization of a novel adipocyte long chain fatty acid transport protein, *Cell* **79**, 427–436.
7. Gargiulo, C. E., Stuhlsatz-Krouper, S. M., and Schaffer, J. E. (1999) Localization of adipocyte long-chain fatty acyl-CoA synthetase at the plasma membrane, *J Lipid Res* **40**, 881–892.
8. Richards, M. R., Harp, J. D., Ory, D. S., and Schaffer, J. E. (2006) Fatty acid transport protein 1 and long-chain acyl coenzyme A synthetase 1 interact in adipocytes, *J Lipid Res* **47**, 665–672.
9. DiRusso, C. C., Li, H., Darwis, D., Watkins, P. A., Berger, J., and Black, P. N. (2005) Comparative biochemical studies of the murine fatty acid transport proteins (FATP) expressed in yeast, *J Biol Chem* **280**, 16829–16837.
10. Milger, K., Herrmwann, T., Becker, C., Gotthardt, D., Zickwolf, J., Ehehalt, R., Watkins, P. A., Stremmel, W., and Fullekrug, J. (2006) Cellular uptake of fatty acids driven by the ER-localized acyl-CoA synthetase FATP4, *J Cell Sci* **119**, 4678–4688.
11. Hall, A. M., Wiczer, B. M., Herrmann, T., Stremmel, W., and Bernlohr, D. A. (2005) Enzymatic properties of purified murine fatty acid transport protein 4 and analysis of acyl-CoA synthetase activities in tissues from FATP4 null mice, *J Biol Chem* **280**, 11948–11954.
12. Stafford, G. A., and Morse, R. H. (1997) Chromatin remodeling by transcriptional

activation domains in a yeast episome, *J Biol Chem* **272**, 11526–11534.

13. Student, A. K., Hsu, R. Y., and Lane, M. D. (1980) Induction of fatty acid synthetase synthesis in differentiating 3T3-L1 preadipocytes, *J Biol Chem* **255**, 4745–4750.
14. Sandoval, A., Fraisl, P., Arias-Barrau, E., DiRusso, C. C., Singer, D., Sealls, W., and Black, P. N. (2008) Fatty acid transport and activation and the expression patterns of genes involved in fatty acid trafficking, *Arch Biochem Biophys* **477**, 363–371.
15. Abumrad, N. A., Forest, C. C., Regen, D. M., and Sanders, S. (1991) Increase in membrane uptake of long-chain fatty acids early during preadipocyte differentiation, *Proc Natl Acad Sci USA* **88**, 6008–6012.
16. Caserta, F., Tchkonia, T., Civelek, V. N., Prentki, M., Brown, N. F., McGarry, J. D., Forse, R. A., Corkey, B. E., Hamilton, J. A., and Kirkland, J. L. (2001) Fat depot origin affects fatty acid handling in cultured rat and human preadipocytes, *Am J Physiol Endocrinol Metab* **280**, E238–E247.
17. Schaffer, J. E. (2003) Lipotoxicity: When tissues overeat, *Curr Opin Lipidol 14*, 281–287.
18. Weinberg, J. M. (2006) Lipotoxicity, *Kidney Int* **70**, 1560–1566.
19. Mosmann, T. (1983) Rapid colorimetric assay for cellular growth and survival: Application to proliferation and cytotoxicity assays, *J Immunol Methods* **65**, 55–63.
20. Li, H., Black, P. N., Chokshi, A., Sandoval-Alvarez, A., Vatsyayan, R., Sealls, W., and DiRusso, C. C. (2008) High-throughput screening for fatty acid uptake inhibitors in humanized yeast identifies atypical antipsychotic drugs that cause dyslipidemias, *J Lipid Res* **49**, 230–244.
21. Li, H., Black, P. N., and DiRusso, C. C. (2005) A live-cell high-throughput screening assay for identification of fatty acid uptake inhibitors, *Anal Biochem* **336**, 11–19.
22. Black, P. N., and DiRusso, C. C. (2007) Vectorial acylation: Linking fatty acid transport and activation to metabolic trafficking, *Novartis Found Symp* **286**, 127–138; discussion 138–141, 162–123, 196–203.
23. Enomoto, S., Ghen, G., and Berman, J. (1998) Vectors for expressing T7 epitope- and His6 affinity-tagged fusion proteins in *S. cerevisiae*, *Biotechniques* **24**, 782–786.
24. Kameda, K., and Nunn, W. D. (1981) Purification and characterization of acyl coenzyme A synthetase from *Escherichia coli*, *J Biol Chem 256*, 5702–5707.
25. Wanders, R. J., van Roermund, C. W., van Wijland, M. J., Schutgens, R. B., Schram, A. W., van den Bosch, H. M., and Tager, J. M. (1987) Studies on the peroxisomal oxidation of palmitate and lignocerate in rat liver., *Biochim Biophys Acta* **919**, 21–25.
26. Watkins, P. A., Lu, J. F., Steinberg, S. J., Gould, S. J., Smith, K. D., and Braiterman, L. T. (1998) Disruption of the *Saccharomyces cerevisiae* FAT1 gene decreases very long-chain fatty acyl-CoA synthetase activity and elevates intracellular very long-chain fatty acid concentrations, *J Biol Chem* **273**, 18210–18219.

Chapter 14

Activity-Based Profiling of Lipases in Living Cells

Maximilian Schicher, Iris Jesse, and Ruth Birner-Gruenberger

Summary

The ultimate goal of proteomics is to characterize the function of all proteins in parallel and in the most physiologically relevant settings possible. A step toward this goal has been the introduction of activity-based proteomics. The simultaneous detection of individual protein activities can be facilitated directly in the proteome using specific activity-based probes consisting of a recognition site targeting a certain enzyme species, a properly positioned reactive site which forms a covalent bond with the target and a reporter tag for visualization and/or purification of the covalently bound target. As properties like polarity, size, charge, structure, and chemical reactivity of the reporter tag have a large impact on the reactivity of the probes toward the target enzymes probes suitable for reporter tagging after the enzyme-activity probe-binding event were designed. These probes resemble the natural substrates more closely and are small and hydrophobic enough to cross the membrane of living cells. Here the methodology for detection of lipolytic activities in intact living cells, including synthesis of probe and reporter, labeling procedure, and detection of target enzymes is described.

Key words: Activity-based proteomics, Click chemistry, Lipases, In vivo, Functional analysis, Enzymatic activity profiling, Adipocytes, Protein expression

1. Introduction

Lipolytic enzymes hydrolyze acyl esters inside and outside cells thus fulfilling specific functions in lipid metabolism, degradation, mobilization, and signaling. These proteins include triacylglycerol lipases, cholesterol ester hydrolases, retinyl esterases, and (lyso-) phospholipases. We employed activity-based proteomics aiming at analyzing and identifying these enzymes in complex proteomes *(1)*. For this purpose, activity-based probes containing a reactive group which forms a covalent bond with the target enzyme and a tag for visualization and/or purification of the covalently bound

Donald Armstrong (ed.), *Lipidomics*, Methods in Molecular Biology, vol. 580,
DOI 10.1007/978-1-60761-325-1_14,

target were designed *(2, 3)* . However, the reporter tag of such probes has to be carefully selected with respect to its chemical and physical properties. Factors such as polarity, size, charge, structure and chemical reactivity of the reporter tag have a large impact on the reactivity of the probe toward the target enzymes. Especially fluorophores with long excitation wavelengths, which are, in general, preferred over fluorophores absorbing at short wavelengths because of lower background fluorescence, are rather bulky and polar and may affect probe recognition, especially by lipolytic enzymes which prefer hydrophobic substrates. Therefore, we designed directed activity recognition probes suitable for reporter tagging after the enzyme-activity probe-binding event (**Fig. 1**). These probes contain a small functional alkyne group that serves as an anchor for fluorophore tagging after formation of the enzyme–inhibitor complex by click chemistry. The click reaction is a copper (I) catalyzed azide-alkyne [3 + 2] cycloaddition, which was shown to be a very efficient and specific labeling procedure leading to ring formation between an alkyne and azide moiety *(4–6)*. These probes are small and hydrophobic enough to cross the membranes of living cells. Lipolytic activities can thus be tagged under the physiologically most relevant conditions, which are in intact living cells not in cell homogenates, avoiding interfering effects caused by cell homogenization and by the structure of the reporter tag during the reaction of the probe with the active enzyme.

Here detailed protocols for click probe synthesis and the labeling procedure in living cells are described. As two examples, we labeled a transiently expressed lipase in COS7 cells as well as lipolytic enzymes in murine adipocytes.

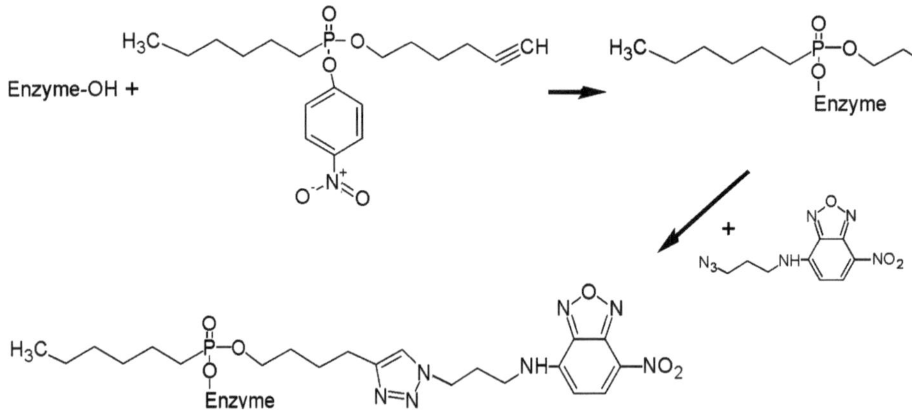

Fig. 1. Specific tagging of lipolytic enzymes by a clickable probe. The probe passes the cell membrane and reacts with the nucleophilic serine in the active site of lipolytic enzymes. The enzyme is irreversibly inactivated and, after cell homogenization, labeled with a fluorophor by click reaction chemistry.

2. Materials

2.1. Equipment

2.1.1. Organic Syntheses

1. Reaction flasks (three-neck, round-bottom).
2. Ground glass stoppers.
3. Dropping funnel.
4. Drying tube filled with calcium chloride.
5. Water-cooled condenser.
6. Filter funnel.
7. Separatory funnel.
8. Rotary evaporator.
9. Oil vacuum pump.
10. Chromatography column.
11. Thin layer chromatography (TLC) chamber.
12. TLC plates: silica gel 60 F_{254} aluminum sheets (0.2 mm; Merck, USA).
13. UV lamp.
14. Hot plate.
15. Sprayer for TLC visualization reagents.
16. Distillation apparatus.

2.1.2. Cell Culture and Activity-Based Labeling of Intact Cells

1. Incubator (37°C, 5% CO_2 atmosphere).
2. Laminar flow hood.
3. Six-well plates.

2.1.2.1. Transient Expression of Lipases in COS-7 Cells

1. 175 cm^2 culture dishes.
2. Centrifuge (maximum force: ~1,000 × *g*)
3. Heating plate (37°C).

2.1.2.2. In vivo Tagging of Lipases

1. Cell scraper.
2. Centrifuge (maximum force: ~1,000 × *g*).
3. Sonicator (Virsonic 457, 10 T; Gardiner, NY, USA).

2.1.3. Gel Electrophoresis

1. Electrophoresis system (Bio-Rad Mini PROTEAN 3 Dodeca cell or PROTEAN Plus Dodeca cell, Bio-Rad, Vienna, Austria).
2. Equipment for gel casting (spacer plates with 1.0 mm integrated spacers, short plates, Mini-PROTEAN combs, 10-well, 1.0 mm, Mini-PROTEAN 3 multi-casting chamber, PROTEAN plus hinged spacer plates 1.5 mm × 200 mm × 205 mm, PROTEAN plus multi-casting chamber; Bio-Rad).

2.1.4. Reporter Tagging

1. Centrifuge (maximum force: ~10,000 × *g*).
2. Thermomixer (Eppendorf Thermomixer compact, Eppendorf, Vienna, Austria).

2.1.5. Visualization

1. Laser scanner (Bio-Rad Molecular Imager™ FX Pro Plus; Bio-Rad).

2.2. Reagents and Supplies

2.2.1. Organic Syntheses

1. Silica gel 60 (230–400 mesh, Merck).
2. Silica sand.
3. 50% sulphuric acid.
4. Phosphomolybdic acid.
5. Sodium sulfate (Na_2SO_4).
6. Dichloromethane (CH_2Cl_2, chromatography grade).
7. Ethyl acetate (EtAc, chromatography grade).
8. Diethyl ether.
9. Chloroform ($CHCl_3$, chromatography grade).
10. Cyclohexane.
11. Methanol (MeOH, chromatography grade).
12. Ammonia (NH_3).

2.2.1.1. Synthesis of Click Probe (hex-5-ynyl 4-nitrophenyl hexylphosphonate, HHPNPP)

1. Hexylphosphonic acid dichloride.
2. Dichloromethane (CH_2Cl_2, absolute).
3. 1-Methylimidazole.
4. Tetrazole.
5. 5-Hexyn-1-ol.
6. 4-Nitrophenol.
7. 1 M aqueous K_2CO_3.
8. Brine (saturated saline solution).

2.2.1.2. Synthesis of Click-Catalyst (tris-(benzyltriazolylmethyl) amine, TBTA)

1. Sodium azide.
2. Dimethylsulfoxide (DMSO, absolute).
3. Benzyl bromide.
4. Ice water.
5. Brine (saturated saline solution).
6. Tripropargylamine.
7. Acetonitrile (absolute).
8. 2,6-Lutidine.
9. Tetrakis(acetonitrile)copper(I)hexafluorophosphate($Cu(MeCN)_4PF_6$).

2.2.1.3. Synthesis of Click Reporter NBD-Azide

1. 1-Bromo-3-aminopropane hydrobromide.
2. Sodium azide.

3. KOH pellets.
4. 4-Chloro-7-nitrobenz-2-oxa-1,3-diazole (NBD).
5. Triethylamine (absolute).
6. N,N-Dimethylformamide (DMF, absolute).
7. Acetic acid p.a. (HAc).

2.2.2. Cell Culture and Activity-Based Labeling of Intact Cells

2.2.2.1. Transient Expression of Lipases in COS-7 Cells

1. COS-7 cells (ATCC # CRL 1651, LGC Standards GmbH, Wesel, Germany).
2. Mammalian expression vectors encoding for lipase and β-galactosidase *(7, 8)*.
3. Cell culture medium (DMEM + 10% FCS + PS): Dulbecco's modified Eagle's medium (DMEM), 10% (v/v) fetal calf serum (FCS), 100 units/ml penicillin and 100 μg/ml streptomycin (PS, antibiotics) (all GIBCO, Invitrogen Life Technologies, Carlsbad, USA).
4. Trypsin-EDTA (GIBCO).
5. Phosphate-buffered saline (PBS) (GIBCO).
6. Cell culture medium without antibiotics (DMEM + 10% FCS).
7. Cell culture medium without antibiotics and without FCS (DMEM).
8. Metafectene™ (Biontex, Martinsried, Germany).

2.2.2.2. Culture and Differentiation of Adipocytes

1. 3T3-L1 preadipocytes (ATCC #CL-173, LGC Standards GmbH, Wesel, Germany).
2. Cell culture medium (DMEM + 10% FCS + PS).
3. Differentiation medium: cell culture medium (DMEM + 10% FCS + PS) containing 1 μM dexamethasone, 500 μM 3-isobutyl-1-methylxanthine (IBMX) and 2 μg/ml insulin.
4. Cell culture medium (DMEM + 10% FCS + PS) containing 2 μg/ml insulin.

2.2.2.3. In Vivo Tagging of Lipases

1. Stock solution of probe HHPNPP, 20× (1 mM) in DMSO.
2. Cell culture medium without antibiotics and without FCS (DMEM).
3. Phosphate-buffered saline (PBS) (GIBCO).
4. 50 mM Tris pH 7.4 supplemented with protease inhibitor mix (20 μg/ml leupeptin, 2 μg/ml antipain, 1 μg/ml pepstatin added from a 1000× stock in DMSO).
5. Bio-Rad protein assay dye reagent concentrate (Bio-Rad).

2.2.2.4. Reporter Tagging

1. Sock solution of Reporter tag NBD-azide, 2.5 mM in DMSO.
2. Stock solution of TCEP, 50 mM in water. Prepare immediately before use.
3. Stock solution of TBTA, 1.7 mM in DMSO:t-butanol 1:4 (v/v).

4. Stock solution of $CuSO_4$, 50 mM in water.
5. 50% (w/v) trichloroacetic acid solution in water.
6. Acetone p.a.
7. SDS-PAGE sample buffer: 20 mM KH_2PO_4, 6 mM EDTA, 60 mg/ml sodium dodecyl sulphate (SDS), 100 mg/ml glycerol, 0.5 mg/ml bromophenol blue, pH 6.8. Add 20 μl/ml mercaptoethanol immediately before use.

2.2.3. Gel Electrophoresis

1. Acrylamide solution 40% (Rotiphorese® Gel 40 (37.5:1 (w/w) acrylamide:bisacrylamide), Lactan, Graz, Austria).
2. 1 M Tris-HCl, pH 6.8.
3. 1 M Tris-HCl, pH 8.8.
4. Sodium dodecyl sulphate (SDS).
5. Ammonium peroxodisulfate (APS).
6. N,N,N′,N′-Tetramethylethylenediamine (TEMED, electrophoresis grade).
7. 1-Butanol saturated with water.
8. Running Buffer: 3 g/l Tris-(hydroxymethyl)-aminomethane (TRIS, buffer grade, Lactan, Austria), 334 mg/l EDTA, 14.2 g/l glycin, and 1 g/l SDS.
9. Fixing solution: 7.5% (v/v) acetic acid, 10% (v/v) ethanol.

3. Methods

3.1. Organic Syntheses (see Notes 1 and 2)

3.1.1. Synthesis of Click Probe (hex-5-ynyl 4-nitrophenyl hexylphosphonate, HHPNPP)

1. Dry a 100 ml three-necked reaction flask, drying tube, dropping funnel, ground glass stoppers and a stir bar at approximately 120°C in a drying oven over night. Immediately assemble the drying tube, dropping funnel, ground glass stoppers, and the reaction flask after taking it out of the drying oven.
2. Dissolve 698 μl (4.075 mmol) hexylphosphonic acid dichloride in 5 ml absolute CH_2Cl_2 in the reaction flask.
3. Add 1-methylimidazole 1.6 ml (20.377 mmol) together with tetrazole (0.45 M in acetonitrile, 35.8 μL (0.016 mmol).
4. Dissolve 442 μl (4.075 mmol) of 5-hexyn-1-ol) in 40 ml absolute CH_2Cl_2 and add it to the stirred solution drop-wise (over a period of 4 h) using the dropping funnel.
5. After complete addition, stir the resulting mixture for at least another 4 h at room temperature (TLC control, mobile phase: CH_2Cl_2/EtAc 1/1 + a few drops acetic acid).

6. Dissolve 4-nitrophenol (2.83 g (20.377 mmol)) together with 1-methylimidazole (1.6 mL (20.377 mmol)) in 8 ml absolute CH_2Cl_2 and add it to the mixture using a dropping funnel.
7. Stir reaction over night at room temperature (TLC control, mobile phase: CH_2Cl_2/EtAc 1/1 + a few drops acetic acid).
8. Evaporate all volatile components under reduced pressure on the rotary evaporator and dissolve the residue in 40 ml CH_2Cl_2.
9. Add 100 ml diethyl ether and extract the mixture, using a separatory funnel, twice with 1 M aqueous K_2CO_3 (120 ml), twice with water (100 ml), and once with brine (100 ml) before drying over sodium sulfate (*see* **Notes 3** and **4**). Remove the sodium sulfate by filtration and discard it.
10. Evaporate volatile components under reduced pressure on the rotary evaporator and dissolve the residue in a small volume of CH_2Cl_2 for purification by column chromatography on silica gel.
11. Fill a chromatography column with silica gel and overlay it with a thin layer of silica sand. Condition the column in CH_2Cl_2 for a few hours, or better over night.
12. Place the dissolved sample on top of the column and start separation. Use at least two column volumes of CH_2Cl_2 as mobile phase, switch to mixtures of CH_2Cl_2 and EtAc with decreasing CH_2Cl_2 ratios and at the end use one column volume pure EtAc (*see* **Note 5**).
13. Combine all fractions containing your desired product and remove the organic solvent on the rotary evaporator (*see* **Note 6**).

3.1.2. Synthesis of Click-Catalyst (tris-(benzyltriazolylmethyl) amine, TBTA)

1. Stir a mixture of sodium azide (7.6 g, 116.9 mmol), absolute dimethylsulfoxide (80 mL), and benzyl bromide (10 g, 5.84 mmol) in a 250 ml reaction flask, containing a drying tube filled with calcium chloride, at room temperature for 5 h (TLC control, mobile phase: cyclohexane/$CHCl_3$ 1/1).
2. Pour the reaction mixture slowly into a reaction flask containing 80 ml ice water and extract three times with 400 ml diethyl ether using a separatory funnel.
3. Wash the combined organic extracts with water (500 ml), brine (500 ml), and dry over sodium sulfate. Remove the sodium sulfate by filtration.
4. Remove volatiles using a rotary evaporator to produce a slightly yellow oil (benzylazide) (TLC control, mobile phase: cyclohexane/$CHCl_3$ 1/1).
5. Dissolve tripropargylamine (1.2 g; 9.2 mmol) in absolute acetonitrile (18 ml) and add sequentially benzylazide

(4.7 g; 35 mmol), 2,6-lutidine (1,070 μl; 9.2 mmol), and $Cu(MeCN)_4PF_6$ (0.167 g; 0.45 mmol, 1.63 mol% with respect to total alkyne units). On addition of the copper salt, the reaction warms and is cooled in an ice bath.

6. Stir for 2 days at room temperature until a white crystalline solid precipitates from the reaction (TLC control, mobile phase: cyclohexane/$CHCl_3$ 1/1).
7. Add absolute acetonitrile (8 ml) to the entire mixture. Heat to 40–50°C for a few minutes and allow cooling to room temperature to induce precipitation of the desired compound.
8. Filtrate through a filter funnel and wash with cold acetonitrile to obtain the product as fine white needle-like crystals. A second crop is obtained by evaporation of the solvent on the rotary evaporator and recrystallization of the residue from hot (40–50°C) acetonitrile (*see* **Note 6**).

3.1.3. Synthesis of Click Reporter NBD-Azide

1. Dissolve 1-bromo-3-aminopropane hydrobromide (10 g; 45.7 mmol) in 140 ml water. Add sodium azide (8.91 g; 137.0 mmol) and stir the solution for 15 h at 80°C (TLC control, mobile phase: $CHCl_3$/MeOH 80/20 + few drops NH_3).
2. Reduce volume to 30 ml under vacuum.
3. Cool in an ice bath. Cautiously add diethyl ether (50 ml) and then KOH pellets (4 g) to the solution keeping the temperature below 10°C (check with thermometer).
4. After separation of the organic phase using a separatory funnel, extract the aqueous phase twice with diethyl ether (30 ml) and combine the organic phases.
5. Dry the combined organic phases over Na_2SO_4, remove the sodium sulfate by filtration and concentrate the product in vacuo on the rotary evaporator.
6. Purify the resulting colorless oil (1-azido-3-aminopropane) by distillation using an oil vacuum pump. Heat the oil to approximately 55°C under reduced pressure (20 mbar) to evaporate it and collect it in a clean reaction flask by cooling it down to room temperature.
7. Dissolve 4-Chloro-7-nitrobenz-2-oxa-1,3-diazole (500 mg; 2.5 mmol) together with 2 equivalents of absolute triethylamine in dry dimethylformamide (5 mL). Add 1-Azido-3-aminopropane (250 mg; 2.5 mmol) and stir the mixture for two hours at room temperature (TLC control, mobile phase: $CHCl_3$/MeOH 80/20 + few drops NH_3).
8. Remove all volatile components under reduced pressure with a rotary evaporator and dissolve the residue in a small volume of CH_2Cl_2.

9. Purify the dissolved sample by silica gel column chromatography as described in steps 10 and 11 in section "Synthesis of Click Probe (hex-5-ynyl 4-nitrophenyl hexylphosphonate, HHPNPP)" with mobile phase composition ranging from CH_2Cl_2/HAc 99/1 to CH_2Cl_2/EtAc/HAc 50/49/1 to obtain pure NBD-azide (*see* **Note 6**).

3.2. Cell Culture and Activity-Based Labeling of Intact Cells

3.2.1. Transient Expression of Lipases in COS-7 Cells

1. Grow COS-7 in monolayers in 175 cm^2 culture dishes in 20 ml cell culture medium (DMEM + 10% FCS + PS) at 37°C in a 5% CO_2 atmosphere to confluence and passage by splitting 1:5 or 1:7 every second or third day by treating with trypsin-EDTA as described in steps 3–5 below and resuspending the appropriate aliquot in fresh medium (*see* **Notes 7–9**).
2. One day before transfection remove the cell culture medium and wash the cells twice with 20 ml of PBS.
3. Treat cells with 12 ml trypsin-EDTA and incubate for 3 min (or until cells separate from the bottom of the culture dish) at 37°C. Stop treatment by adding 12 ml of culture media.
4. Centrifuge cells for 5 min at 2000 rpm and remove the supernatant.
5. Resuspend cells in 10 ml of fresh cell culture medium (DMEM + 10% FCS + PS).
6. Pipette 125 μl of the resuspension into each well of a six-well dish previously filled with 2 ml fresh cell culture medium to obtain 150,000 cells per well.
7. Incubate cells at 37°C in a 5% CO_2 atmosphere over night.
8. One hour before transfection exchange the cell culture medium with 0.6 ml of fresh cell culture media without antibiotics (DMEM + 10% FCS) per well.
9. Mix 1 μg of plasmid DNA encoding the lipase or β-glycosidase as negative control for each well with 50 μl of cell culture medium without antibiotics and without FCS (DMEM) in a sterile Eppendorf tube. In a second tube mix 4.5 μl metafectene™ with 50 μl of cell culture medium without antibiotics and without FCS (DMEM). Mix the contents of both Eppendorf tubes and incubate for 20 min at room temperature.
10. Add the mixture to one well and incubate cells for 4 h at 37°C in a 5% CO_2 atmosphere.
11. Exchange the medium of each well with 2 ml of fresh cell culture medium (DMEM + 10% FCS + PS) and incubate cells for two to three days at 37°C in a 5% CO_2 atmosphere (*see* **Note 10**).

3.2.2. Culture and Differentiation of Adipocytes

1. Grow 3T3-L1 cells in six-well plates in cell culture medium (DMEM + 10% FCS + PS) to confluency.
2. Differentiate cells into adipocytes by the adipogenic cocktail method as described by Student et al. *(9)*. Start the differentiation process by addition of differentiation medium two days after cells have reached confluency.
3. After 3 days exchange cell culture medium with fresh cell culture medium (DMEM + 10% FCS + PS) containing 2 μg/ml insulin and incubate the cells for three days. At that point, maintain cells in growth medium (DMEM + 10% FCS + PS) for another 3–4 days, with medium changes every two days (*see* **Notes 11–12**).

3.2.3. In Vivo Tagging of Lipases

1. Grow cells in six-well plates as described under "Transient Expression of Lipases in COS-7 Cells" or "Culture and Differentiation of Adipocytes".
2. Remove cell culture medium. Incubate cells with 20 μM probe (HHPNPP, added from a 20× stock in DMSO to fresh cell culture medium without antibiotics and without FCS (DMEM), 0.5 ml per well) for 2 h.
3. Wash cells twice with 1 ml PBS. Add 2 ml ice-cold PBS and harvest cells by scraping them off into tubes previously cooled on ice and centrifugation (1000 g, 5 min).
4. Resuspend cells in ice cold 100 μl of 50 mM Tris pH 7.4 supplemented with protease inhibitor mix (*see* **Note 13**) per well.
5. Disrupt cells by sonication on ice for 5 s.
6. Remove cell debris by centrifugation at 1000 × g for 5 min at 4°C.
7. Determine the protein concentration of the cell homogenate (supernatant) using the BIORAD protein assay based on the method of Bradford *(10)*.
8. Adjust protein concentration to 1.0 mg protein per ml with 50 mM Tris-HCl, pH 7.4 supplemented with protease inhibitor mix (*see* **Notes 14–15**).

3.2.4. Reporter Tagging

1. To 30 μl (i.e., 30 μg) of labeled protein samples add 80 μM NBD-azide, followed by 0.8 μM TCEP and 80 μM TBTA. Mix by vortexing gently.
2. Add 0.8 mM $CuSO_4$. Mix by vortexing gently and allow to react at room temperature by shaking at 600 rpm for 1 h *(4, 5)*.
3. Stop the reaction by addition of 10% ice-cold trichloroacetic acid and precipitate protein on ice for 1 h.

4. Harvest protein by centrifugation at 10,000 × g at 4°C for 15 min. Remove the supernatant. Wash pellet with acetone. Resuspend pellet in 20 μl of SDS-PAGE sample buffer and heat to 95°C for 3 min (*see* **Note 16**).

3.3. Gel Electrophoresis

1. Cast a separating gel containing 10% acrylamide. The amounts of reagents required for two gels with dimensions 1 mm × 73 mm × 83 mm are as follows: 8.6 ml 40% acrylamide solution, 12.9 ml 1 M Tris-HCl, pH 8.8, 0.1% SDS, and 12.4 ml bidistilled water. Start polymerization by adding 165 μl of a freshly prepared solution of 10% (w/v) APS in water, and 34 μl TEMED. Mix thoroughly and pour between prepared glass plates. Cover with 1 ml of 1-butanol saturated with water and leave for approximately 45 min to polymerize (*see* **Notes 17–18**).
2. Thoroughly wash off the 1-butanol with water and overlay the separating gel with a 4.5% stacking gel. The amounts of reagents for two gels are: 0.8 ml acrylamide solution 40%, 0.8 ml 1 M Tris-HCl, pH 6.8, 0.1% SDS, and 4.9 ml bidistilled water. Start polymerization by adding 33 μl 10% APS and 7 μl TEMED. Put a comb in the stacking gel and leave it for 45 min to polymerize.
3. Fill cold running buffer in the Bio-Rad Mini PROTEAN 3 dodeca cell and keep it at 15°C throughout the entire electrophoresis run.
4. Load the protein samples in one-dimensional sample buffer in the gel pockets (20 μl per pocket) and put the gels in the electrophoresis cell.
5. Start electrophoresis at 20 mA constant current per gel, and stop electrophoresis when the bromophenol leaves the bottom of the gel.
6. Put the gels in fixing solution and slightly shake them for at least 30 min under protection from daylight.

3.4. Visualization

1. Put the fixed gels on the glass plate of the laser scanner.
2. Scan the gels with the excitation and emission wavelengths appropriate to the optical properties of the fluorescence activity tag used (for NBD, excitation wavelength is 488 nm and emission wavelength is 530 nm) (*see* **Note 19**).

3.5. Results

1. As one example, intact mammalian cells transiently expressing a known lipase, the recently identified adipose triglyceride lipase (ATGL) *(7)*, were incubated with a click probe (**Fig. 2a**). The activity tagged cells were subsequently homogenized and the enzyme--probe complexes were chemically

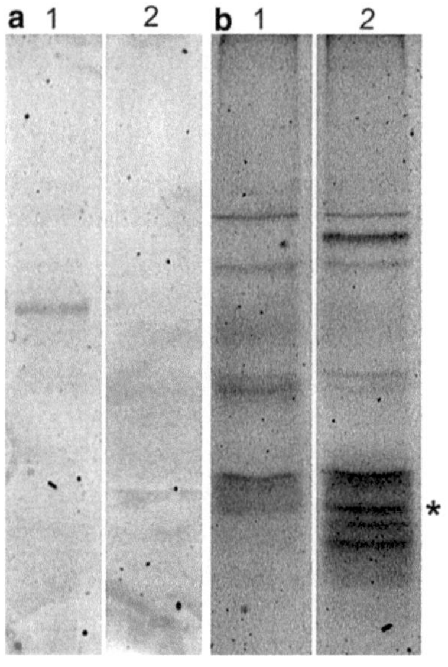

Fig. 2. Lipase detection in intact cells. (**a**) The lipase ATGL was labeled by the click probe in living mammalian (COS7) cells expressing ATGL ***(1)*** as compared with control cells expressing β-galactosidase, which is not a lipase ***(2)***. (**b**) Comparison of murine adipocyte (3T3-L1) cells, incubated with click probe Inh2 after ***(1)*** or before ***(2)*** homogenization and then clicked to NBD-azide. Asterisk (*) indicates the band used for LC–MS/MS analysis in **Table 1**.

coupled to a reporter tag. Labeled proteins were profiled by gel electrophoresis and detected by laser scanning. A band at about 55 kD resembling ATGL was labeled in the cells expressing the lipase (lane 1) but not in control cells (lane 2).

2. As another example, intact murine adipocytes (lane 2) as compared with cell homogenates (lane 1) were labeled with the click probe HHPNPP (**Fig. 2b**). The obtained activity profiles differed with respect to the number of detected bands and to band intensity indicating that cell homogenization has a vast effect on protein activity.

For protein identification, fluorescent bands can be cut out of the gel, proteins can be digested in gel by trypsin and identified by tandem mass spectrometry after reversed phase high-pressure liquid chromatography separation of peptides as detailed elsewhere in this book. For example, we have identified a protein which was stronger labeled in intact murine adipocytes than in their cell homogenates (indicated by * in **Fig. 2b**) as Esterase D (**Table 1**) (*see* **Note 20**).

Table 1
Protein identification by LC–MS/MS analysis and database search

Name	NCBI number	Individual peptides	Sequence coverage (%)	Spectrum Mill score
Esterase 10 (Esterase D)	BAB27115 (gi12846304)	7	25	93.17

Peptide sequences	*m/z*	*z*
(K)AFSGYLGPDESK(W)	636.07	2
(K)AFSGYLGPDESKWK(A)	793.32	2
(K)AYSGSQIDILIDQGK(D)	804.58	2
(K)KAFSGYLGPDESK(W)	699.89	2
(K)SGYQQAASEHGLVVIAPDTSPR(G)	1142.43	2
(K)SGYQQAASEHGLVVIAPDTSPR(G)	762.10	3
(K)VFEHSSVELK(C)	1174.63	1
(K)VFEHSSVELK(C)	587.66	2
(R)SVSAFAPICNPVLCSWGK(K)	997.35	2

The band indicated by * in Fig. 2b was cut out, digested with trypsin and identified by LC–MS/MS analysis and search of the non redundant NCBI protein database using Spectrum Mill (Agilent) Software. Seven individual peptides of Esterase 10 were identified resulting in 25% sequence coverage. Their detected mass/charge *m/z* ratios and charges *z*, as well as the overall protein score are listed

4. Notes

1. Many compounds used in the syntheses are sensitive to humidity, therefore dry glassware must be used and all sensitive chemicals should not be exposed to the air longer then necessary. Also, many compounds are harmful and should thus be handled under a fume hood, as well as disposed of in an appropriate way.
2. Make a TLC control after every reaction step to make sure your starting materials have reacted to the desired (intermediate) product. Therefore, spot a small amount of every reactant and of the reaction mixture near the bottom of the TLC plate and place it in a TLC chamber containing the respective

mobile phase. When the mobile phase reaches the top of the TLC plate, remove the plate and dry it with a common hairdryer. Then check for spots under the UV lamp and mark them with a pencil. Depending on the compounds you spotted on the plate spray the plate either with 50% H_2SO_4 (for all organic compounds) or phosphomolybdic acid (for phosphorous-containing compounds) and heat it on the hot plate until spots get visible.

3. Pour the organic solution in the separatory funnel, then add the aqueous solution and close the funnel. Carefully shake the funnel once and then vent it. Repeat shaking and venting a few times until no more gas exits the funnel on venting. Then shake thoroughly and leave the funnel until the phases are fully separated. Then collect the lower (aqueous) phase and pour the next solution to the organic phase in the funnel.
4. After the last extraction step, add sodium sulfate to the organic phase to eliminate final traces of water. Add a small amount of Na_2SO_4 and swirl the solution. If the Na_2SO_4 is clumped together, then add more. Continue swirling and observing the solution for 5–15 min, adding more drying agent only until a fresh addition no longer forms clumps.
5. After pure CH_2Cl_2 use two column volumes of CH_2Cl_2/EtAc 10/1, then always one volume of CH_2Cl_2/EtAc 5/1, CH_2Cl_2/EtAc 3/1, CH_2Cl_2/EtAc 2/1, CH_2Cl_2/EtAc 1/1, and CH_2Cl_2/EtAc 1/2.
6. To confirm that you synthesized the right product perform NMR and/or MS analysis (Schicher et al., in preparation).
7. COS-7 cells are kidney fibroblasts of African green monkey which show week intrinsic lipolytic and esterolytic activities. Use a low passage (below 15) to ensure maximum expression.
8. cDNA cloning and transient expression of adipose triglyceride lipase (ATGL) and β-galactosidase (LacZ) as a negative control in COS-7 cells was performed as described *(7, 8)*.
9. Prewarm cell culture media and buffers to 37°C before usage.
10. The expression maximum is usually reached after 2 or 3 days and can be assessed as follows: Analysis of transfection efficiency is based on the expression of LacZ and its ability to hydrolyze colorless substrates which results in the release of a blue chromophore *(11)*. Transfect cells with a suitable plasmid (e.g., pcDNA4/HisMax, Invitrogen, Lofer, Austria) containing the coding sequence of LacZ. Two or three days after transfection wash well with

1 ml of PBS for 3 min at room temperature. Fix cells with 0.5% glutardialdehyde in PBS for 5 min at 37°C. Remove the solution, wash the cells with 2 ml PBS and incubate with solution A (10 mM phosphate buffer pH 7.0, 150 mM NaCl, 1 mM $MgCl_2$, 3.3 mM $K_3Fe(CN)_6$ and freshly added 0.2% 5-bromo-4-chloro-3-indolyl-β-galactopyranosid (BCIP)). After 4 h to over-night incubation successfully transfected cells are stained blue.

11. Completion of differentiation can be shown by Nile Red staining as described by Greenspan et al. *(12)* or radioactive assays for measurement of neutral TAG lipase and cholesteryl esterase activity as described by Holm et al. *(13)*.
12. Optionally, after complete differentiation, adipocytes can be incubated with 10 μM isoproterenol and 500 μM IBMX for 2 h at 37°C to stimulate lipolysis by hormone sensitive lipase (HSL).
13. The DMSO stock of the protease inhibitor mix is stable for months when stored at −20°C. Add the mix freshly to buffers before use.
14. Samples are used for reporter tagging or stored at −70°C until use.
15. For *in vitro* tagging of lipases step 2 is omitted. Instead the cell homogenate (1.0 mg protein per ml) is incubated with 20 μM probe and 1 mM Triton X-100 for 2 h at 37°C and 600 rpm. For this purpose, probe and Triton X-100 are added to an empty Eppendorf tube from 20× stocks in chloroform. The solvent is removed by drying with argon or nitrogen before the homogenate is added.
16. Samples can now be used for electrophoresis or stored at −20°C until use.
17. Alternatively, precast gels can be purchased from vendors. In this case, the protocol can be started following step 3.
18. APS and TEMED should be added immediately before pouring the gels, because these compounds start polymerization of the gels.
19. For visualization of the whole protein pattern, stain the gels with SYPRO® Ruby protein gel stain (Invitrogen, Lofer, Austria) for approximately 4 h, then put them in fixing solution for 30 min. Scan the gels at 605 nm using an excitation wavelength of 488 nm.
20. Here protein identification from one gel piece is shown as an example. Detailed description of LC-MS/MS procedures and results is published elsewhere (Schicher et al., in preparation).

Acknowledgments

This work was supported by the "Click it: In vivo functional proteomic screening and cell imaging of lipases" project in the framework of GEN-AU (GENome Research in AUstria; http://www.gen-au.at/), funded by the Austrian Federal Ministry of Science and Research.

References

1. Birner-Gruenberger, R, and Hermetter, A. (2007) Activity-based proteomics of lipolytic enzymes. Curr Drug Discov Technol 4(1): 1–11.
2. Schmidinger, H., Birner-Gruenberger, R., Riesenhuber, G., Saf, R., Susani-Etzerodt, H., Hermetter, A. (2005) Novel fluorescent phosphonic acid esters for discrimination of lipases and esterases. Chembiochem 6(10):1776–1781.
3. Susani-Etzerodt, H., Schmidinger, H., Riesenhuber, G, Birner-Gruenberger, R., Hermetter, A. (2006) A versatile library of activity-based probes for fluorescence detection and/or affinity isolation of lipolytic enzymes. Chem Phys Lipids 144(1):60–68.
4. Speers, A.E., Adam, G.C. and Cravatt, B.F. (2003) Activity-based protein profiling in vivo using a copper(i)-catalyzed azide-alkyne [3 + 2] cycloaddition. J Am Chem Soc 125(16):4686–4687.
5. Speers, A.E., and Cravatt, B.F. (2004) Profiling enzyme activities in vivo using click chemistry methods. Chem Biol 11(4):535–546.
6. Wang, Q., Chan, T.R., Hilgraf, R., Fokin, V.V., Sharpless, K.B. and Finn, M.G. (2003) Bioconjugation by copper(I)-catalyzed azide-alkyne [3 + 2] cycloaddition. J Am Chem Soc 125(11):3192–3193.
7. Zimmermann, R., Strauss, J.G., Haemmerle, G., et al. (2004) Fat mobilization in adipose tissue is promoted by adipose triglyceride lipase. Science 306(5700):1383–1386.
8. Birner-Gruenberger, R., Susani-Etzerodt, H., Waldhuber, M., et al. (2005) The lipolytic proteome of mouse adipose tissue. Mol Cell Proteom 4(11):1710–1717.
9. Student, A.K., Hsu, R.Y. and Lane, M.D. (1980) Induction of fatty acid synthetase synthesis in differentiating 3T3-L1 preadipocytes. J Biol Chem 255(10):4745–4750.
10. Bradford, M.M. (1976) Rapid and sensitive method for quantitation of microgram quantities of protein utilizing principle of protein-dye binding. Anal Biochem 72(1–2):248–254.
11. Lim, K., and Chae, C.B. (1989) A simple assay for DNA transfection by incubation of the cells in culture dishes with substrates for beta-galactosidase. Biotechniques 7(6):576–579.
12. Greenspan, P., Mayer, E.P. and Fowler, S.D. (1985) Nile red: A selective fluorescent stain for intracellular lipid droplets. J Cell Biol 100(3):965–973.
13. Holm, C., Olivecrona, G. and Ottosson, M. (2001) Assays of lipolytic enzymes. Methods in Molecular Biology (Totowa, NJ, United States) 155(Adipose Tissue Protocols): 97–119.

Chapter 15

Histochemistry and Lipid Profiling Combine for Insights into Aging and Age-Related Maculopathy

Christine A. Curcio, Martin Rudolf, and Lan Wang

Summary

Aging is the major risk factor for age-related maculopathy (ARM), the biggest cause of vision loss among the elderly in industrialized societies, and a major change in the affected tissues is the age-related accumulation of neutral lipid in Bruch's membrane (BrM) of the eye throughout adulthood. Here we show that esterified cholesterol (EC) is the major neutral lipid species in this tissue, which has implications for potential sources of this material. The combination of filipin histochemistry and comprehensive lipid profiling made possible this insight on a complex tissue.

Key words: Aging, Age-related maculopathy (ARM), Apolipoprotein B (apoB), Basal deposits, Bruch's membrane (BrM), Cholesterol, Drusen, Esterified cholesterol (EC), Filipin, Lipid profiling, Lipoprotein, Retinal pigment epithelium (RPE), Unesterified cholesterol (UC)

1. Introduction

1.1. Retina and Choroid

Embryologically part of the central nervous system, the retina (**Fig. 1a, b**), converts light energy to an electrochemical signal for transmission to the brain through the optic nerve. The 100 million rod and cone photoreceptors, located at the outer surface of the retinal sheet are supported by the retinal pigment epithelium (RPE). This monolayer serves diverse functions essential for optimal photoreceptor health, including daily phagocytosis of photoreceptor outer segment tips, vitamin A metabolism, maintenance of retinal attachment, and coordination of cytokine-mediated immune protection. The macula is a 6-mm-diameter

Donald Armstrong (ed.), *Lipidomics*, Methods in Molecular Biology, vol. 580,
DOI 10.1007/978-1-60761-325-1_15, © Humana Press, a part of Springer Science+Business Media, LLC 2009

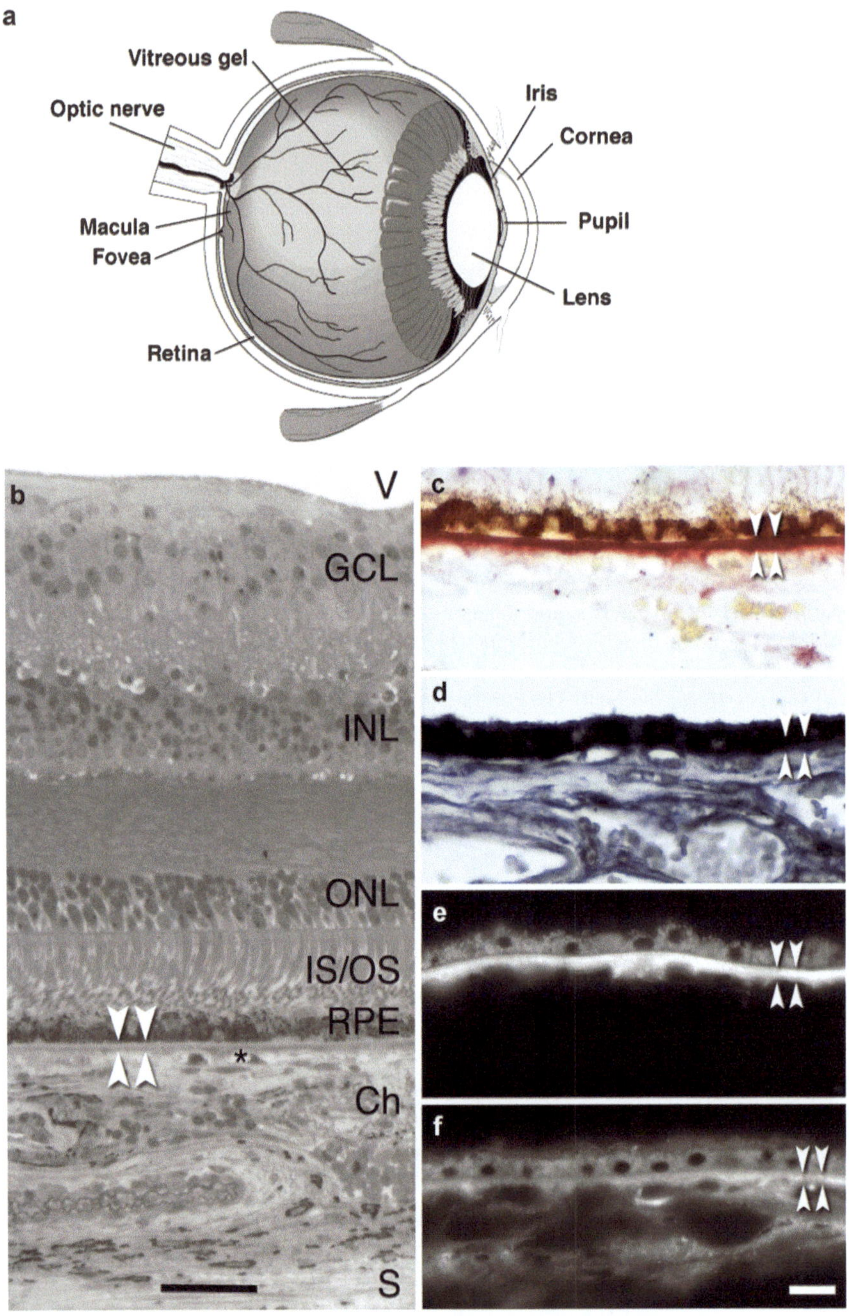

Fig. 1. Human chorioretinal anatomy and histochemical demonstration of lipid in aged Bruch's membrane. (**a**) Cross-section of human eye. (**b**) Histological section of eye wall at the macula, showing neurosensory retina on top (toward center of eye) and choroid and sclera at the bottom (toward exterior of eye). Light shines in from above. *V*, vitreous cavity; *GCL*, ganglion cell layer; *INL*, inner nuclear layer; *ONL*, outer nuclear layer; *IS/OS*, inner and outer segments of photoreceptors; *RPE*, retinal pigment epithelium; *Ch*, choroid; *asterisk*, choriocapillaris; *white arrowheads*, Bruch's membrane; *S*, sclera. Scale bar, 50 μm. (*c–f*) Sections of RPE and choroid from a 79-year-old woman with a normal macula, stained to reveal different lipids. BrM is bracketed by arrowheads. Scale bar in (f), 20 μm. (**c**) Oil red O stains BrM and lipofuscin granules in RPE. (**d**) Sudan Bromine Black B stains BrM, RPE, and cells throughout choroid. (**e**) Filipin for esterified cholesterol stains BrM. RPE lipofuscin autofluorescence is distinguished from filipin by color (golden vs aqua), not shown here. (**f**) Filipin for unesterified cholesterol stains BrM, RPE, and choroidal cells.

(21° visual angle) area centered on the point of highest visual acuity. Two vascular systems supply the human retina, and the photoreceptors and RPE depend on the choroidal vasculature, located external to them. Branches of the ophthalmic artery enter through the sclera and ramify to form the choroid and the choriocapillaris, a dense capillary bed. About 200–300 μm thick, the choroid has the highest blood flow in the body, with sevenfold greater flow in the macula relative to the periphery. The innermost 2–4 μm of the choroid (subjacent to the RPE) is Bruch's membrane (BrM) (**Fig. 1b–f**), a five-layered vessel wall that is laid out flat along one side of the choriocapillaris rather than wrapped around individual lumens. Unhindered translocation across BrM of nutrients to and metabolites from the RPE is essential for normal vision by the photoreceptors *(1)*.

1.2. Retinal Aging and ARM

The leading cause of untreatable vision loss among the elderly in the developed world, age-related maculopathy (ARM) affects four distinct layers: the photoreceptors, the RPE, BrM, and the choriocapillaris. This disorder features characteristic extracellular focal and diffuse lesions on the inner surface of BrM (drusen and basal linear deposits, respectively), dysfunction and loss of RPE, and eventually, dysfunction and loss of the photoreceptors themselves. In ~15% of patients, choriocapillaries break through BrM and ramify under the RPE (choroidal neovascularization), leading to sudden and severe vision loss. Although lifestyle and genetic factors certainly contribute *(2, 3)*, the largest risk factor for ARM is advanced age. Thus, it is important to understand how aging in the affected tissues impels some genetically predisposed individuals toward neovascularization.

A successful strategy for identifying biological pathways perturbed by disease *(4, 5)* is to determine the molecular components of the characteristic lesions. In ARM, this approach involved identifying constituents of drusen *(6–8)*. We with others *(9, 10)* have included BrM and its age changes, which by virtue of sharing the same compartment as drusen and basal linear deposits have great potential for direct involvement in lesion biogenesis. BrM age changes include thickening, accumulation of cross-linked and electron-dense debris, reduced collagen solubility, and the deposition of neutral lipids, advanced glycation end products, and tissue inhibitor of metalloproteinase-3 *(9, 11–14)*. We chose to study lipid deposition because it is a very large effect and because of the recognized role of this process in initiating coronary artery disease *(15)*.

1.3. Neutral Lipids in BrM—Histochemistry, Physical Chemistry

Clinical observations on fluid-filled RPE detachments in older adults led to the hypothesis by Bird and Marshall that a lipophilic barrier in BrM blocked the normal outwardly directed fluid efflux from the RPE *(16)*. Impaired movement of fluids, either derived from RPE or from leaky vessels in choroidal neovascular membranes, was thought to contribute to detachment *(17)*. This hypothesis motivated a seminal laboratory study that *(12)* used three histochemical stains to identify lipids in cryosections of eyes from 30 donors (1 to 95 years) with grossly normal maculas. Bromine Sudan Black B binds phospholipid, triglyceride, Esterified cholesterol (EC), Unesterified cholesterol (UC), and fatty acids, Bromine-Acetone-Sudan Black B binds only acetone-resistant phospholipid after acetone removes other lipids, and oil red O binds triglyceride, EC, and fatty acids *(18, 19)*. All stains gave similar results: no eyes from donors under age 30 exhibited staining, eyes from donors 31 to 60 years exhibited variable staining, and eyes from donors 61 and older exhibited moderate to intense staining (**Fig. 1c, d**). Findings with these classic stains were repeated in two other series *(20, 21)*, with less intense BrM staining noted for peripheral retina *(21)*. Of these methods, only oil red O was localized exclusively to BrM (**Fig. 1c**). The two Sudan dyes also labeled cells throughout the choroid (**Fig. 1d**).

Lipids that bind oil red O increase with age in normal human connective tissues, including the sclera *(22)*, cornea *(23)*, and intima of large arteries *(24)*. The stained material comprises small (60–200 nm) extracellular droplets highly enriched in EC relative to UC *(25–29)*. The source of EC in sclera, cornea, and arterial intima is low-density lipids (LDL) translocated from plasma into connective tissues. EC-enriched particles are thought to arise from smaller LDL particles by extracellular matrix-mediated trapping of LDL, degradation of LDL protein and/or phospholipid components, and fusion of the remaining lipid components *(30)*. It is thus important to determine whether lipid deposition in BrM is an ocular manifestation of this ongoing systemic process or a phenomenon unique to the eye.

Hot stage polarizing microscopy, a physicochemical technique *(31, 32)*, was used to examine the birefringence and melting characteristics of lipids in BrM and sclera *(21)*. EC in tissue sections appears as liquid crystals ("Maltese crosses") when examined through a polarizing filter. When sections are heated and cooled slowly, liquid crystals melt and reform at characteristic temperatures dictated by the saturation level of the major ester. BrM contained Maltese crosses and a linear band (presumably irresolvable small crosses) that melted at a higher temperature than those in sclera. The temperature difference later turned out to be an artifact of fixation and storage rather than a true difference in saturation levels *(33)*, but this study did establish the presence of EC in BrM and also notably observed few birefringent crystals signifying triglyceride.

We approached this question by examining BrM EC and UC content using the fluorescent, polyene antibiotic filipin *(34, 35)*. This compound binds specifically to the 3-b-hydroxy group on cholesterol and other sterols, and it can be used to identify cholesterol in tissue sections *(25, 36)* and cultured cells *(37)*. Importantly, EC can also be localized using filipin, by extracting native UC with ethanol, hydrolyzing EC with the enzyme cholesterol esterase, then binding the newly released unesterified cholesterol with filipin. This method is an improvement over oil red O, because it is specific for one class of neutral lipids, and because the fluorescence, while labile, is readily quantifiable with digital microscopy. Our protocol is modified from that provided by the Kruth Lab *(25, 36)* and is intended for 10-μm-thick cryosections of retina-choroid preserved in 4% paraformaldehyde. Cryosections are produced as described *(38)*. Tissues preserved in fixatives containing glutaraldehyde can also be stained with filipin but have higher levels of background autofluorescence.

2. Materials

2.1. Equipment

1. Incubator at 37°C for cholesterol esterase.
2. Water bath at 50°C for Dako Glycergel.
3. Fluorescence microscope (Nikon Eclipse 80i, Nikon Instruments Inc., Melville NY); 20×, 40×, and 60× (oil immersion) planapochromat objectives; filter set UV-2A—excitation 350/50, emission 420 long-pass.

2.2. Reagents

1. 0.1M potassium phosphate buffer pH 7.4 (0.02M KH_2PO_4, 0.08M K_2HPO_4).
2. Cholesterol esterase solution.
   1.575 ml cholesterol esterase.
   105 ml potassium phosphate buffer.
3. Filipin
   1. Dissolve 5 mg Filipin in 1 ml *N,N*-dimethyl formamide (a carcinogen) in the fume hood.
   2. Add to 100 ml Dulbecco's PBS.
4. Other preparations
   1. Filter Mayer's hematoxylin before use.
   2. Place Dako Glycergel in 50°C water bath.

2.3. Supplies

1. H_2O: double-distilled H_2O.
2. Whatman filters: Catalog# 1202–150 Qualitative 100 150 mm.
3. Aluminum foil.
4. PAP Pen: ImmEdge™.
5. 70% ethanol: Steven's Scientific #7503.

6. Cholesterol esterase: Roche #161772.
7. Mayer's hematoxylin: Poly Scientific R&D #216.
8. Scott's Bluing: Fisher #CS410–4D, store a 4°C.
9. Dulbecco PBS: Gibco #14040–133.
10. Filipin: Sigma #F-9765, store at 4°C.
11. N,N-dimethyl formamide: Sigma #D-8554.

3. Methods

3.1. Rationale/Flow Chart

Slides denoted as EC in the protocol and the flow chart (**Fig. 2**) are stained to reveal esterified cholesterol, meaning that they are filipin-stained after ethanol extraction and cholesterol esterase incubation. Slides denoted as UC are stained to reveal unesterified cholesterol, meaning that they are filipin-stained without prior ethanol or enzyme treatment. Slides denoted as EC-control are negative controls for the enzymatic reaction, meaning that they are filipin-stained after extraction with ethanol but do not go through cholesterol esterase treatment.

3.2. Filipin Staining Protocol

Slides removed from freezer should be warmed on a hot plate (50–55°C) for 30 min.

1. Wash
   a. 70% ethanol (EC, EC-control): 5–10 min.
   b. 70% ethanol (EC, EC-control): dip 2×, 2 s per dip.
   c. H_2O (EC, EC-control): dip 2×, 2 s per dip.
2. Dry
3. Prepare the following:
   a. Circle tissue samples with PAP pen and let dry.
   b. Prepare wet box with filter paper, water, with smaller dish on top.
   c. Mix esterase solution and return enzyme to refrigerator immediately.
4. Enzyme incubation (EC, EC-control only):
   a. Place cholesterol esterase in PAP circle (EC slide).
   b. Place potassium phosphate buffer in PAP circle (EC-control slide).
   c. Put EC and EC-control slides in wet box in incubator 37°C for 3 h.
5. Wash in potassium phosphate buffer: 1 min.

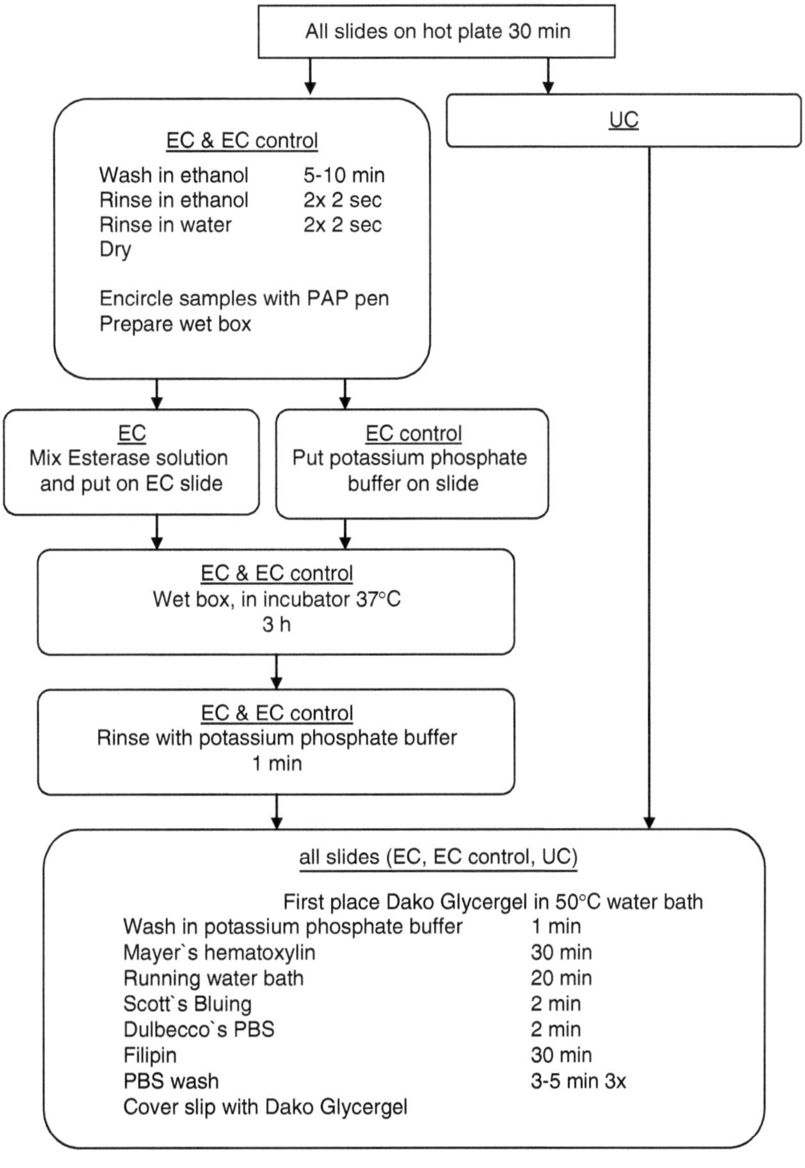

Fig. 2. Flow chart shows how one compound (filipin) is used to label both unesterified and esterified cholesterols (UC and EC).

6. Stain in filtered Mayer's hematoxylin: 30 min.
7. Rinse in gently running water: 20 min.
8. Scott's Bluing in staining dish: 2 min.
9. Dulbecco's PBS on slide: 2 min.
10. Dry slides.
11. Filipin staining (all slides):
    a. Mix Filipin solution and put into PAP circles.
    b. Place slides in closed box: 30 min.

12. Wash with PBS: 3–5 min, 3×.
13. Immediately cover slip with Dako Glycergel.
14. Remove stained slides from light to prevent quenching.
15. Examine slides using fluorescence microscope, viewing EC-control slide first to assess background autofluorescence.

3.3. Results

3.3.1. Histochemical Staining

In sections processed to reveal EC, aged BrM exhibits an intense, diffusely distributed fluorescence **(Fig. 1E)**, the specificity of which can be verified in control sections lacking esterase treatment (not shown). The RPE and retina are negative for EC, with the exception of occasional retinal and choroidal arteries (not shown). Cholesterol esterase can abolish oil red O staining of BrM *(34)*. Thus, we conclude that BrM contains EC as the predominant oil red O-binding lipid. In sections processed with filipin alone, cellular membranes of the neural retina, RPE, and choroid are fluorescent *(8)*, as is BrM **(Fig. 1f)**. UC is thus more widely distributed, in a pattern resembling that revealed by the Sudan dyes **(Fig. 1d)**.

Using filipin fluorescence corrected for BrM autofluorescence *(39)*, we localized EC in macula and temporal periphery of 20 normal eyes from age 17 to 92 years *(40)*. In the macula, EC was undetectable before 22 years and then rose linearly throughout adulthood to reach high and variable levels in aged donors. EC was detectable in periphery at roughly one-seventh the level of macula, but still increased significantly with age. In the same eyes, UC in macular BrM also increased throughout adulthood, although not as steeply as EC, and it did not increase significantly with age in peripheral retina. Like previous work *(12)*, this study implicated a large, constitutive, and completely unknown process in the normal posterior eye that produced a profound age-related change. Extending the original findings, this study identified a specific class of molecules (EC) and quantified their increase with age. More recently *(41)*, we revisited the retina and choroid using a variety of lipid stains (BODIPY 493/503, Nile Red). Of stains employed, only oil red O and filipin for EC stain BrM exclusively.

3.3.2. Validation of Histochemical Results with Direct Assays

Determining BrM lipid composition to validate the histochemical results is challenging because of the size and marked heterogeneity of the choroid *(42, 43)* **(Fig. 1b)**. The choroid contains numerous blood vessels (arteries, veins, and capillaries) with plasma lipoproteins and blood cells, as well as resident cells (vascular endothelium, fibroblasts, melanocytes, macrophages, and neurons) *(44)*. A source of contaminating lipids is therefore phospholipids and UC in cell membranes in extraneous choroidal vessels and stroma left attached to BrM **(Fig. 1d, F)**. Seven studies assayed BrM lipids from tissue extracts (as opposed to lipoproteins released from BrM) *(33, 40, 45–51)*. From these studies, we draw these conclusions (**Table 1**): (1) RPE/BrM/choroid is much more enriched in neutral lipid/ EC than neurosensory retina *(40, 47, 50)*. (2) Within RPE/BrM/choroid, EC is the prominent neutral lipid, exceeding triglyceride by 4–10-fold *(33, 50, 51)*, despite an early report to the contrary *(46)*. **3)** EC represents 50–60% of the cholesterol detected *(40, 50, 51)*.

These compositional analyses of BrM/choroid unavoidably combine neutral lipids of BrM with membrane-associated lipids within choroidal cells and plasma lipoproteins retained in vessels. These concerns would be obviated if particles with lipoprotein characteristics could be isolated from BrM/Ch. Using a double, high-salt buffer extraction from tissue homogenates *(52)*, we showed that such particles could indeed be released for subsequent lipid assays (enzymatic colorimetric methods and electrospray ionization/mass spectrometry *(33)*; gas chromatography and flame ionization detection *(51)*). Some of the salient characteristics of BrM lipoproteins are as follows (**Table 1**): (1) Particles are spherical electron-lucent by negative stain electron microscopy, consistent with a neutral lipid core. (2) Particles form a mono-disperse population with a mean diameter of 67 nm (range, 11–211 nm), similar to plasma VLDL. (3) EC is the predominant cholesterol form (58% of total cholesterol). (4) EC is the predominant neutral lipid. Conspicuous by it paucity is triglyceride, the major core lipid of plasma apoB lipoproteins in the same size class (VLDL, small chylomicrons). EC is 11.6-fold more abundant than triglyceride (moles). (5) Relative to plasma lipoproteins containing apoB, BrM lipoproteins have a distinct distribution of major lipid classes. Within each lipid class, however, BrM lipoprotein fatty acids are remarkably similar to plasma lipoproteins (for EC: linoleate > oleate > palmitate > arachidonate > myristate; for phospholipids: palmitate > oleate > stearate > linoleate > arachidonate). Linoleate levels among nonesterified fatty acids, triglyceride, phosphatidylcholine, and phosphatidylethanolamine are somewhat lower in BrM lipoproteins than in plasma lipoproteins.

3.3.3. Significance of EC Enrichment of BrM

Several authors attempted to deduce the source of BrM lipids from compositional studies, under the accepted and reasonable assumption that composition should reflect the source(s). Converging and repeatable evidence from light microscopic histochemistry, physical chemistry, ultrastructure *(40, 53–56)*, and comprehensive lipid profiling of tissues and isolated lipoproteins have now established EC as the primary lipid in aged BrM. Further, of the major lipid classes, only EC is ***exclusively localized*** to BrM, i.e., not also distributed throughout the choroid. What does the primacy of EC within aged BrM signify about its source? In the eye, various insults can lead to engorgement of RPE cells with large oil red O-binding intracellular droplets *(57–60)*. Such

Table 1
Lipid profile of BrM/choroid and isolated BrM lipoprotein particles

	EC	DG	FA	TG	CL	LYPC	PC	PE	PS	UC	SPM	RE
BrM/ choroid												
mol%, mean[a]	29.8	0.7	3.6	3.0	1.5	0.6	15.4	12.6	6.0	26.7	na	na
BrM lipoproteins, isolated												
mol%, mean[a]	32.4	1.7	6.3	3.3	3.2	1.8	14.2	9.5	4.6	22.9	na	na
nmol/eye, mean[b]	270.6	11.7	44.5	23.6	21.4	11.4	109.7	79.4	35.2	184.2	79.5	0.117

[a]Calculated relative to the sum of all lipids except SPM and RE
[b]Corrected for yield
Data from Wang et al. (2009); BrM/choroid, $n = 3$ (consistent with $n = 10$ in Li et al. (2005), with enzymatic assays); BrM lipoproteins, $n = 7$
EC, esterified cholesterol; *DG*, diglyceride; *FA*, nonesterified fatty acid; *TG*, triglyceride; *CL*, cardiolipin; *LYPC*, lysophosphatidylcholine; *PC*, phosphatidylcholine; *PE*, phosphatidylethanolamine; *PS*, phosphatidylserine; *UC*, unesterified cholesterol; *SPM*, sphingomyelin; *RE*, retinyl ester; *na*, not available
Assays used: SPM, TLC for polar lipids; RE, reversed phase HPLC; all others, LC–GC

cells could conceivably contribute to BrM neutral lipid if they die and release their droplets. However, the intracellular droplets are much larger (1–2 μm) than BrM particles (66 nm), they apparently have little EC *(60)*, few RPE cells exhibit droplets in any one eye, lipoidal degeneration is not widely prevalent across eyes, and the rate of age-related RPE cell death *(61)* is probably too slow to account for the normal, large, and universal age-related accumulation of BrM EC. Excluding this possibility leaves the best-documented way to release EC from a healthy cell, i.e., within the core of a lipoprotein containing apoB.

Elsewhere, we present gene expression, immunohistochemistry, and metabolic-labeling studies that together demonstrate secretion of apoB-containing lipoproteins by RPE *(38, 51, 62)*. These studies are buttressed by indirect evidence against a plasma origin, most notably the disassociation between ARM and elevated plasma cholesterol levels *(63)* Thus, this work helps expand apoB's role in mammalian physiology beyond the classic pathways of hepatic VLDL and intestinal chylomicrons *(64)*. But it also raises pressing new questions about the function of an EC-rich lipoprotein in eye physiology. Previous literature speculated that the age-related accumulation of BrM neutral lipid is related to the RPE's signature activity—phagocytosis of photoreceptor outer segment tips—perhaps by direct basolateral deposition of debris *(65)*. We originally postulated that an RPE apoB-lipoprotein could be an excellent mechanism to dispose of fatty acids released by lysosomal phospholipases after ingestion of outer segments *(62)* but for reasons explained elsewhere *(51)*, little evidence now supports this initially attractive hypothesis. Indeed the close resemblance of BrM EC composition to plasma lipoproteins suggests that they dominate BrM lipoproteins instead. However, which plasma lipoproteins predominate at the RPE LDL-receptor and scavenger receptors *(66–69)* and for what purpose they are taken up is currently unknown. Delivery of cholesterol, xanthophylls, and vitamin E ultimately destined for the neurosensory retina are some plausible reasons for a major plasma lipoprotein input to RPE. Determining the answers to these questions should be critical steps taken on the road to developing treatments for ARM based on manipulating age-related deposition of EC in BrM.

4. Notes

1. Cholesterol esterase solution should be freshly made, as it expires in 10 days.
2. Filipin should be protected from light exposure by covering container in aluminum foil to prevent quenching.

Acknowledgments

Supported by NIH grants EY06109, International Retinal Research Foundation, EyeSight Foundation of Alabama, Research to Prevent Blindness, Inc., Macula Vision Research Foundation, Roger Johnson Prize in Macular Degeneration Research.

References

1. Ethier, C.R., M. Johnson, and J. Ruberti. 2004. Ocular biomechanics and biotransport. *Ann Rev Biomed Eng* **6**: 249–273.
2. Klein, R., T. Peto, A. Bird, and M.R. Vannewkirk. 2004. The epidemiology of age-related macular degeneration. *Am J Ophthalmol* **137**: 486–495.
3. Montezuma, S.R., L. Sobrin, and J.M. Seddon. 2007. Review of genetics in age related macular degeneration. *Semin Ophthalmol.* **22**: 229–240.
4. Smith, E.B., and R.S. Slater. 1972. The microdissection of large atherosclerotic plaques to give morphologically and topographically defined fractions for analysis. *Atherosclerosis* **15**: 37–56.
5. Selkoe, D.J., Y. Ihara, and F.J. Salazar. 1982. Alzheimer's disease: Insolubility of partially purified paired helical filaments in sodium dodecyl sulfate and urea. *Science* **215**: 1243–1245.
6. Mullins, R.F., S.R. Russell, D.H. Anderson, and G.S. Hageman. 2000. Drusen associated with aging and age-related macular degeneration contain proteins common to extracellular deposits associated with atherosclerosis, elastosis, amyloidosis, and dense deposit disease. *FASEB J.* **14**: 835–846.
7. Crabb, J.W., M. Miyagi, X. Gu, K. Shadrach, K.A. West, H. Sakaguchi, M. Kamei, A. Hasan, L. Yan, M.E. Rayborn, R.G. Salomon, and J.G. Hollyfield. 2002. Drusen proteome analysis: An approach to the etiology of age-related macular degeneration. *Proc Natl Acad Sci USA* **99**: 14682–14687.
8. Curcio, C.A., J.B. Presley, N.E. Medeiros, G. Malek, D.V. Avery, and H.S. Kruth. 2005. Esterified and unesterified cholesterol in drusen and basal deposits of eyes with age-related maculopathy. *Exp Eye Res* 81(6): 731–741.
9. Handa, J.T., N. Verzijl, H. Matsunaga, A. Aotaki-Keen, G.A. Lutty, J.M. te Koppele, T. Miyata, and L.M. Hjelmeland. 1999. Increase in the advanced glycation end product pentosidine in Bruch's membrane with age. *Invest Ophthalmol Vis Sci* **40**: 775–779.
10. Kamei, M., and J.G. Hollyfield. 1999. TIMP-3 in Bruch's membrane: Changes during aging and in age-related macular degeneration. *Invest Ophthalmol Vis Sci* **40**: 2367–2375.
11. Feeney-Burns, L., and M.R. Ellersieck. 1985. Age-related changes in the ultrastructure of Bruch's membrane. *Am J Ophthalmol* **100**: 686–697.
12. Pauleikhoff, D., C.A. Harper, J. Marshall, and A.C. Bird. 1990. Aging changes in Bruch's membrane: A histochemical and morphological study. *Ophthalmology* **97**: 171–178.
13. Karwatowski, W.S.S., T.E. Jeffried, V.C. Duance, J. Albon, A.J. Bailey, and D.L. Easty. 1995. Preparation of Bruch's membrane and analysis of the age-related changes in the structural collagens. *Br J Ophthalmol* **79**: 944–952.
14. Fariss, R.N., S.S. Apte, B.R. Olsen, K. Iwata, and A.H. Milam. 1997. Tissue inhibitor of metalloproteinases-3 is a component of Bruch's membrane of the eye. *Am J Patho.* **150**: 323–328.
15. Tabas, I., K.J. Williams, and J. Boren. 2007. Subendothelial lipoprotein retention as the initiating process in atherosclerosis: update and therapeutic implications. *Circulation* **116**: 1832–1844.
16. Bird, A.C., and J. Marshall. 1986. Retinal pigment epithelial detachments in the elderly. *Trans Ophthalmol Soc UK* **105**: 674–682.
17. Marshall, J., A.A. Hussain, C. Starita, D.J. Moore, and A.L. Patmore. 1998. Aging and Bruch's membrane. *In* The Retinal Pigment Epithelium: Function and Disease. M.F. Marmor and T.J. Wolfensberger, editors. Oxford University Press, New York. 669–692.
18. Luna, L.G. 1968. Manual of the Histologic Staining Methods of the Armed Forces Institute of Pathology McGraw-Hill, New York.
19. Adams, C.W.M., and O.B. Bayliss. 1975. Lipid histochemistry. *In* Techniques of Biochemical and Biophysical Morphology. D. Glick and R.M. Rosenbaum, editors. John Wiley and Sons, New York. 100–156.

20. Pauleikhoff, D., S. Wojteki, D. Muller, N. Bornfeld, and A. Heiligenhaus. 2000. [Adhesive properties of basal membranes of Bruch's membrane. Immunohistochemical studies of age-dependent changes in adhesive molecules and lipid deposits]. *Ophthalmologe* 97: 243–250.
21. Haimovici, R., D.L. Gantz, S. Rumelt, T.F. Freddo, and D.M. Small. 2001. The lipid composition of drusen, Bruch's membrane, and sclera by hot stage polarizing microscopy. *Invest Ophthalmol Vis Sci* **42**: 1592–1599.
22. Broekhuyse, R.M. 1972. Lipids in tissues of the eye. VII. Changes in concentration and composition of sphingomyelins, cholesterol esters and other lipids in aging sclera. *Biochim Biophys Acta* **280**: 637–645.
23. Gaynor, P.M., W.Y. Zhang, B. Salehizadeh, B. Pettiford, and H.S. Kruth. 1996. Cholesterol accumulation in human cornea: Evidence that extracellular cholesteryl ester-rich lipid particles deposit independently of foam cells. *J Lipid Res* 37: 1849–1861.
24. Smith, E.B. 1974. The relationship between plasma and tissue lipids in human atherosclerosis. *Adv Lipid Res* **12**: 1–49.
25. Kruth, H.S. 1984. Histochemical detection of esterified cholesterol within human atherosclerotic lesions using the fluorescent probe filipin. *Atherosclerosis* **51**: 281–292.
26. Bocan, T.M., T.A. Schifani, and J.R. Guyton. 1986. Ultrastructure of the human aortic fibrolipid lesion. Formation of the atherosclerotic lipid-rich core. *Am J Pathol* **123**: 413–424.
27. Guyton, J.R., and K.F. Klemp. 1988. Ultrastructural discrimination of lipid droplets and vesicles in atherosclerosis: Value of osmium-thiocarbohydrazide-osmium and tannic acid-paraphenylenediamine techniques. *J Histochem Cytochem* **36**: 1319–1328.
28. Guyton, J.R., and K.F. Klemp. 1989. The lipid-rich core region of human atherosclerotic fibrous plaques. Prevalence of small lipid droplets and vesicles by electron microscopy. *Am J Pathol* **134**: 705–717.
29. Chao, F.F., E. Blanchette-Mackie, Y.-J. Chen, B.F. Dickens, E. Berlin, L.M. Amende, S.I. Skarlatos, W. Gamble, J.H. Resau, W.T. Mergner, and H.S. Kruth. 1990. Characterization of two unique cholesterol-rich lipid particles isolated from human atherosclerotic lesions. *Am J Pathol* **136**: 169–179.
30. Kruth, H.S. 1997. The fate of lipoprotein cholesterol entering the arterial wall. *Curr Opin Lipidol* **8**: 246–252.
31. Waugh, D.A., and D.M. Small. 1984. Identification and detection of in situ cellular and regional differences of lipid composition and class in lipid-rich tissue using hot stage polarizing light microscopy. *Lab Invest* **51**: 702–714.
32. Small, D.M. 1988. George Lyman Duff memorial lecture. Progression and regression of atherosclerotic lesions. Insights from lipid physical biochemistry. *Arteriosclerosis* **8**: 103–129.
33. Li, C.-M., B.H. Chung, J.B. Presley, G. Malek, X. Zhang, N. Dashti, L. Li, J. Chen, K. Bradley, H.S. Kruth, and C.A. Curcio. 2005. Lipoprotein-like particles and cholesteryl esters in human Bruch's membrane: Initial characterization. *Invest Ophthalmol Vis Sci* **46**: 2576–2586.
34. Normal, A., R. Demel, B. de Kruyff, and L. van Deenen. 1972. Studies on the biological properties of polyene antibiotics - evidence for the direct interaction of filipin with cholesterol. *J Biol Chem* **247**: 1918.
35. Norman, A.W., R.A. Demel, B. de Kruyff, and L.L.M. van Deenen. 1972. Studies on the biological properties of polyene antibiotics. *J Biol Chem* **247**: 1918–1929.
36. Kruth, H.S. 1984. Localization of unesterified cholesterol in human atherosclerotic lesions. Demonstration of filipin-positive, oil red O-negative particles. *Am J Pathol* **114**: 201–208.
37. Lakkaraju, A., S.C. Finnemann, and E. Rodriguez-Boulan. 2007. The lipofuscin fluorophore A2E perturbs cholesterol metabolism in retinal pigment epithelial cells. *Proc Natl Acad Sci USA* **104**: 11026–11031.
38. Malek, G., C.-M. Li, C. Guidry, N.E. Medeiros, and C.A. Curcio. 2003. Apolipoprotein B in cholesterol-containing drusen and basal deposits in eyes with age-related maculopathy. *Am J Pathol* **162**: 413–425.
39. Marmorstein, A.D., L.Y. Marmorstein, H. Sakaguchi, and J.G. Hollyfield. 2002. Spectral profiling of autofluorescence associated with lipofuscin, Bruch's Membrane, and sub-RPE deposits in normal and AMD eyes. *Invest Ophthalmol Vis Sci* **43**: 2435–2441.
40. Curcio, C.A., C.L. Millican, T. Bailey, and H.S. Kruth. 2001. Accumulation of cholesterol with age in human Bruch's membrane. *Invest Ophthalmol Vis Sci* **42**: 265–274.
41. Rudolf, M., and C.A. Curcio. 2009. Esterified cholesterol is highly localized to Bruch's membrane, as revealed by lipid histochemistry in wholemounts of human choroid. *J Histochem Cytochem* **in revision, 2/3/09**.
42. Ramrattan, R.S., T.L. van der Schaft, C.M. Mooy, W.C. de Bruijn, P.G.H. Mulder, and P.T.V.M. de Jong. 1994. Morphometric analysis of Bruch's membrane, the choriocapillaris, and the choroid in aging. *Invest Ophthalmol Vis Sci* **35**: 2857–2864.

43. Spraul, C. W., G.E. Lang, and H.E. Grossniklaus. 1996. Morphometric analysis of the choroid, Bruch's membrane, and retinal pigment epithelium in eyes with age-related macular degeneration. *Invest Ophthalmol Vis Sci* **37**: 2724–2735.
44. Lutty, G.A., C. Merges, A.B. Threlkeld, S. Crone, and D.S. McLeod. 1993. Heterogeneity in localization of isoforms of TGF-beta in human retina, vitreous, and choroid. *Invest Ophthalmol Vis Sci* **34**: 477–487.
45. Sheraidah, G., R. Steinmetz, J. Maguire, D. Pauleikhoff, J. Marshall, and A.C. Bird. 1993. Correlation between lipids extracted from Bruch's membrane and age. *Ophthalmology* **100**: 47–51.
46. Holz, F.G., G. Sheraidah, D. Pauleikhoff, and A.C. Bird. 1994. Analysis of lipid deposits extracted from human macular and peripheral Bruch's membrane. *Arch Ophthalmol* **112**: 402–406.
47. Gülcan, H., R. Alvarez, M. Maude, and R. Anderson. 1993. Lipids of human retina, retinal pigment epithelium, and Bruch's membrane/choroid: comparison of macular and peripheral regions. *Invest Ophthalmol Vis Sci* **34**: 3187–3193.
48. Pauleikhoff, D., G. Sheraidah, J. Marshall, A.C. Bird, and A. Wessing. 1994. [Biochemical and histochemical analysis of age related lipid deposits in Bruch's membrane]. *Ophthalmologe* **91**: 730–734.
49. Spaide, R., W. Ho-Spaide, R. Browne, and D. Armstrong. 1999. Characterization of peroxidized lipids in Bruch's membrane. *Retina* **19**: 141–147.
50. Bretillon, L., G. Thuret, S. Gregoire, N. Acar, C. Joffre, A.M. Bron, P. Gain, and C.P. Creuzot-Garcher. 2008. Lipid and fatty acid profile of the retina, retinal pigment epithelium/choroid, and the lacrimal gland, and associations with adipose tissue fatty acids in human subjects. *Exp Eye Res* **87**: 521–528.
51. Wang, L., C.-M. Li, M. Rudolf, O.V. Belyaeva, N.Y. Kedishvili, B.H. Chung, and C.A. Curcio. 2009. Lipoprotein particles of intraocular origin in human Bruch membrane: an unusual lipid profile. *Invest Ophthalmol Vis Sci* **50**: 870–877.
52. Chung, B.H., G. Tallis, V. Yalamoori, G.M. Anantharamaiah, and J.P. Segrest. 1994. Liposome-like particles isolated from human atherosclerotic plaques are structurally and compositionally similar to surface remnants of triglyceride-rich lipoproteins. *Arterioscler Thromb* **14**: 622–635.
53. Ruberti, J.W., C.A. Curcio, C.L. Millican, B.P. Menco, J.D. Huang, and M. Johnson. 2003. Quick-freeze/deep-etch visualization of age-related lipid accumulation in Bruch's membrane. *Invest Ophthalmol Vis Sci* **44**: 1753–1759.
54. Huang, J.-D., J.B. Presley, M.F. Chimento, C.A. Curcio, and M. Johnson. 2007. Age-related changes in human macular Bruch's membrane as seen by quick-freeze/deep-etch. *Exp Eye Res* **85**: 202–218.
55. Johnson, M., J.-D. Huang, J.B. Presley, M.F. Chimento, and C.A. Curcio. 2007. Comparison of morphology of human macular and peripheral Bruch's membrane in older eyes. *Curr Eye Res* **32**: 791–799.
56. Huang, J.-D., C.A. Curcio, and M. Johnson. 2008. Morphometric analysis of lipoprotein-like particle accumulation in aging human macular Bruch's membrane. *Invest Ophthalmol Vis Sci* **49**: 2721–2727.
57. Fine, B.S., and R.P. Kwapien. 1978. Pigment epithelial windows and drusen: An animal model. *Invest Ophthalmol Vis Sci* **17**: 1059–1068.
58. Fine, B.S. 1981. Lipoidal degeneration of the retinal pigment epithelium. *Am J Ophthalmol* **91**: 469–473.
59. Feeney-Burns, L., R. Malinow, M.L. Klein, and M. Neuringer. 1981. Maculopathy in cynomologus monkeys. A correlated fluorescein angiographic and ultrastructural study. *Arch Ophthalmol* **99**: 664–672.
60. Anderson, M.D., W.W. Dawson, J. Martinez-Gonzalez, and C.A. Curcio. 2006. Drusenoid lesions and lipid-filled retinal pigment epithelium cells in a rhesus macula. *Vet Ophthalmol* **9**: 201–207.
61. Del Priore, L.V., Y.-H. Kuo, and T.H. Tezel. 2002. Age-related changes in human RPE cell density and apoptosis proportion in situ. *Invest Ophthalmol Vis Sci* **43**: 3312–3318.
62. Li, C.-M., J.B. Presley, X. Zhang, N. Dashti, B.H. Chung, N.E. Medeiros, C. Guidry, and C.A. Curcio. 2005. Retina expresses microsomal triglyceride transfer protein: Implications for age-related maculopathy. *J Lipid Res* **46**: 628–640.
63. Dashti, N., G. McGwin, Jr, C. Owsley, and C.A. Curcio. 2006. Plasma apolipoproteins and risk for age-related maculopathy. *Br J Ophthalmol* **90**: 1028–1033.
64. Veniant, M.M., E. Kim, S. McCormick, J. Borén, L.B. Nielsen, M. Raabe, and S.G. Young. 1999. Insights into apolipoprotein B biology from transgenic and gene-targeted mice. *J Nutr* **129**: 451S–455S.
65. Grindle, C.F.J., and J. Marshall. 1978. Ageing changes in Bruch's membrane and their functional implications. *Trans Ophthalmol Soc UK* **98**: 172–175.

66. Elner, V.M. 2002. Retinal pigment epithelial acid lipase activity and lipoprotein receptors: effects of dietary omega-3 fatty acids. *Trans Am Ophthalmol Soc* **100**: 301–338.
67. Gordiyenko, N., M. Campos, J.W. Lee, R.N. Fariss, J. Sztein, and I.R. Rodriguez. 2004. RPE cells internalize low-density lipoprotein (LDL) and oxidized LDL (oxLDL) in large quantities in vitro and in vivo. *Invest Ophthalmol Vis Sci* **45**: 2822–2829.
68. Tserentsoodol, N., N.V. Gordiyenko, I. Pascual, J.W. Lee, S.J. Fliesler, and I.R. Rodriguez. 2006. Intraretinal lipid transport is dependent on high density lipoprotein-like particles and class B scavenger receptors. *Mol Vis* **12**: 1319–1333.
69. Duncan, K.G., K.R. Bailey, J.P. Kane, and D.M. Schwartz. 2002. Human retinal pigment epithelial cells express scavenger receptors BI and BII. *Biochem Biophys Res Commun* **292**: 1017–1022.

Part III

Software and Bioinformatics

Chapter 16

Instrument-Independent Software Tools for the Analysis of MS–MS and LC–MS Lipidomics Data

Perttu Haimi, Krishna Chaithanya, Ville Kainu, Martin Hermansson, and Pentti Somerharju

Summary

Mass spectrometry (MS), particularly electrospray-MS, is the key tool in modern lipidomics. However, as even a modest scale experiment produces a great amount of data, data processing often becomes limiting. Notably, the software provided with MS instruments are not well suited for quantitative analysis of lipidomes because of the great variety of species present and complexities in response calibration. Here we describe the use of two recently introduced software tools: lipid mass spectrum analysis (LIMSA) and spectrum extraction from chromatographic data (SECD), which significantly increase the speed and reliability of mass spectrometric analysis of complex lipidomes. LIMSA is a Microsoft Excel add-on that (1) finds and integrates the peaks in an imported spectrum, (2) identifies the peaks, (3) corrects the peak areas for overlap by isotopic peaks of other species and (4) quantifies the identified species using included internal standards. LIMSA is instrument-independent because it processes text-format MS spectra. Typically, the analysis of one spectrum takes only a few seconds.

The SECD software allows one to display MS chromatograms as two-dimensional maps, which is useful for visual inspection of the data. More importantly, however, SECD allows one to extract mass spectra from user-defined regions of the map for further analysis with, e.g., LIMSA. The use of select regions rather than simple time-range averaging significantly improves the signal-to-noise ratio as signals outside the region of interest are more efficiently excluded. LIMSA and SECD have proven to be robust and convenient tools and are available free of charge from the authors.

Key words: HPLC, Mass spectrometry, Software, Lipidomics, Metabolomics, Quantification

1. Introduction

Advances in mass spectrometry (MS), especially electrospray ionization (ESI), have provided a major boost to the development of high-throughput lipidomics. This is largely because of the high sensitivity and resolution of ESI-MS, which far exceed those of traditional

Donald Armstrong (ed.), *Lipidomics*, Methods in Molecular Biology, vol. 580,
DOI 10.1007/978-1-60761-325-1_16, © Humana Press, a part of Springer Science+Business Media, LLC 2009

methods of lipid analysis, such as thin-layer or column chromatography. The selectivity of ESI-MS-based lipid analysis can be greatly enhanced by employing lipid class-specific scanning modes *(1, 2)* or by using online HPLC separation of the sample *(3–6)* ESI-MS together with stable isotope-labeled lipid precursors is also a powerful tool for elucidation of metabolic pathways of lipids *(7–10)*.

An additional benefit of MS is that it can be easily automated, which is not the case with most previously used methods. However, the possibility for automated sample processing accompanied with the vast amount of lipid species readily analyzed by ESI-MS easily results in "data overflow", i.e., data processing becomes the bottle-neck for meaningful lipid analyses. To overcome this, several groups have developed software tools that significantly simplify and speed up the analysis of lipid MS or MS–MS data *(11–15)* or LC-MS datasets (**Note 2**.) *(16)*. Some of these softwares have been designed to work with instruments from a specific vendor, while others are instrument-independent. These softwares have been reviewed recently *(17)*.

Lipid species are often structurally very similar to each other (e.g., only one double bond difference), which causes significant overlap of their isotopic peak patterns. This makes it essential to implement an accurate method of isotope correction (deisotoping). In addition, instrument response varies markedly with lipid structure and (uncontrollable) variations in analytical conditions, thus necessitating inclusion of multiple internal standards and implementation of calibration protocols different from those found in standard MS software *(18)*.

We have recently developed two new software tools, lipid mass spectrum analysis (LIMSA) and spectrum extraction from chromatographic data (SECD), which greatly simplify and speed up quantitative analysis of lipid compositions by MS–MS or LC–MS *(13)*. These tools are easy to learn and available free of charge from the authors. Here we indicate the advantages and disadvantages of these software tools and provide detailed advice for their use.

1.1. Advantages and Disadvantages

1.1.1. The Advantages of LIMSA

1. Instrument-independent; i.e., compatible with all MS software providing a mass spectrum in a text format.
2. Easy to install and use (Excel add-on).
3. Internal lipid data base, which can be easily edited and expanded.
4. Batch mode allows automatic processing of multiple datasets.
5. Compatible with isotope-labeled compounds.
6. Additional data processing and reporting functions and displays are easily added in Excel.

7. Robust and fast.
8. Freeware; the code is available under the GPL license and can be thus modified at will.

1.1.2. The Disadvantages of LIMSA

1. Does not support data-dependent acquisition.

1.1.3. The Advantages of SECD

1. Allows free selection of regions (trapezoids) of the LC–MS map to be chosen for processing. This increases the signal-to-noise ratio as compared with simple time-range averaging, and also allows one to exclude coeluting, overlapping compounds from the spectra.
2. Trapezoids can be easily adjusted with a mouse.
3. Trapezoids can be saved and reloaded for subsequent samples.

1.1.4. The Disadvantages of SECD

1. Trapezoids have to be manually readjusted if retention times vary significantly between runs, which is not uncommon in gradient runs.
2. It is not easy to determine the optimal location of the trapezoids when the signal is weak.

2. Materials

2.1. Software

1. Microsoft Excel.
2. Microsoft.NET Framework 1.1 or 2.0.

3. Methods

3.1. Programming

1. The data processing "core" of LIMSA is written in portable C++. The graphical user interface was implemented using *Visual Basic for Applications* as a Microsoft Excel Addin. SECD was written using managed Visual Basic.NET and C++ for processing NetCDF files.

3.2. Analytical Methods

1. Methods for extraction of lipids, preparation of lipid standards, parameters for analysis of lipid with MS–MS or LC–MS and details of software development can be found elsewhere *(4, 13, 18)*.

3.3. Results

3.3.1. Processing MS or MS–MS Spectra with LIMSA

1. LIMSA is an Excel™ add-on for quantitative analysis of mass spectra of complex lipid samples. LIMSA carries out peak finding, integration, assignment, isotopic overlap correction, and quantification.
2. LIMSA can be downloaded from our website (www.helsinki.fi/science/lipids /software.html). Installation of *LIMSA* is simple (Windows installer) and introduces a new Excel menu "*Limsa*," under which three new submenus "*Search lipids*", "*Batch analyze*" and "*Summarize results*" appear.
3. To start the analysis, paste to the first two columns of an Excel worksheet a mass spectrum, which can be either a line spectrum, i.e., a peak list created by the data acquisition software or a continuous spectrum with equal or nonequal mass spacing.
4. Then click on "LIMSA" in the Excel menu bar and choose "*Search lipids.*" This brings up the LIMSA user interface (**Fig.** 1).
5. Choose in the *Lipid list*-panel the lipid classes to be analyzed. LIMSA has a default list of molecular species for each main phospholipid class. It is possible to remove or add lipids from the internal library to an existing list, or construct completely new lists (e.g., for triacylglycerols, glycosphingolipids, etc.) as required.

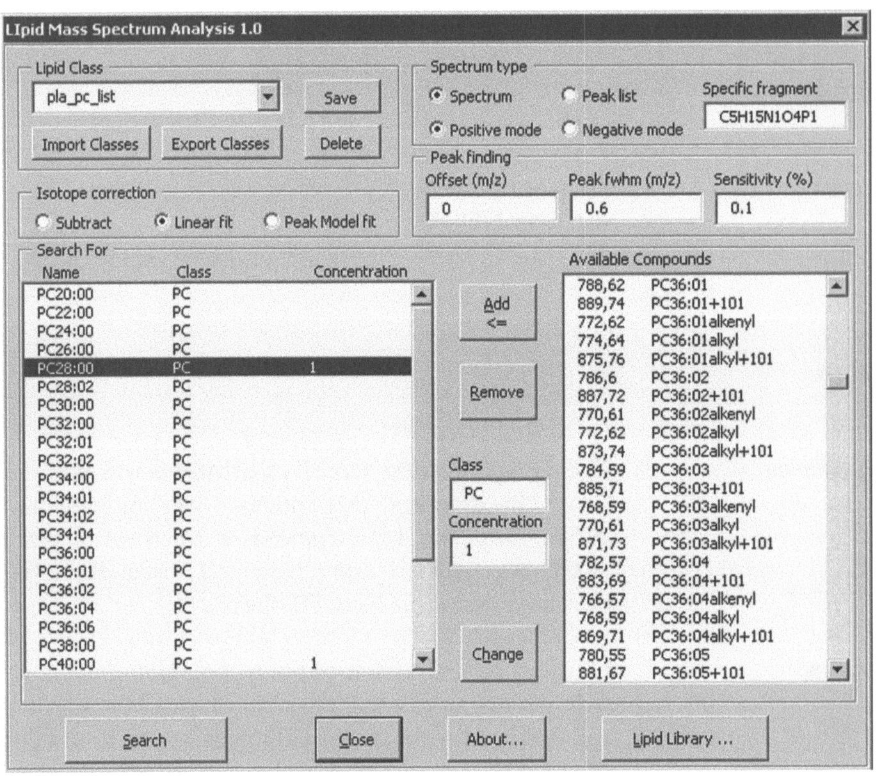

Fig. 1. Input window of LIMSA. This window opens in Excel by clicking on "*Search lipids*" in the *Limsa*-menu. Here one (i) chooses the lipid class to be analyzed, (ii) provides information on the type of spectrum used, and (iii) sets the parameters controlling the analysis. *See* text for further details.

6. To set an internal standard, choose a species in a list, specify the lipid class (e.g., "PC") which it is used for, indicate its amount (e.g., 100 nmol) and click on "*Update.*"
7. In the *Spectrum type*-panel choose: (i) *MS scan polarity* (positive or negative), (ii) the type of MS data, i.e., *Peak list* or *Spectrum*, and (iii) *Specific fragment* for precursor and neutral loss scans (e.g., choline, +184). If this field is left empty, the isotope patterns valid for MS scans are used (the isotopic patterns are not identical for MS vs. MS–MS scans).
8. In the *Peak finding*-panel one can set: (i) *Mass shift*; (corrects for imperfect instrument calibration); (ii) *Peak width* (at half-height); controls peak finding, integration and assignment, and (iii) *Sensitivity*; sets the intensity of the (smallest) relevant peak as the percentage of the intensity of the largest one, thus allowing one to optimize exclusion of noise.
9. In the *Isotope correction*-panel one can choose among three different methods, i.e., a *subtraction method*, a *linear fit method* and a *peak-model method*. For additional information, *see* **Note 1**. We recommend: (i) the subtraction method for well-resolved profile or centroid data, particularly if the spectrum contains unknown, possibly interfering peaks; (ii) the linear fit method for well-resolved profile data and centroid data with no interfering peaks, and (iii) the peak model method for less-resolved profile data.
10. After processing, which typically takes a few seconds, the program outputs: (cf. **Fig. 2**) (i) in panel 1 a spectrum with found peaks labeled with " ♦ " and those assigned and fitted labeled with " Δ ", (ii) a list of found peaks in panel 2, (iii) the fitting residual in panel 3, (iv) the concentration calibration curve drawn based on the peak areas of the internal standards in panel 5, and (v) a table of peak areas and concentrations of the assigned lipids ("6" in the worksheet).

3.3.2. Editing of the Compound Library

When adding a new lipid, one needs to provide a name, sum formula, and the net charge (**Fig. 3**). One can use shorthands when writing the formulas, which can greatly speed up the process, particularly when adding multiple homologous species. For example, (CH2)10 equals C10H20. Negative numbers are allowed, provided that the total number of no element in the formula becomes negative. Thus, (H2)-1 could be used to indicate one double bond, (H2)-2 two double bonds, and so on. Isotope-labeled species are indicated, e.g., by symbols D, Cx, and Nx for a deuterium, ^{13}C and ^{15}N, respectively. Custom isotopic patterns based on measured spectra can be defined for, e.g., partially labeled compounds. The name of the labeled compound must be enclosed in square brackets when used as a part of a molecular formula. Thus, e.g., [Serine-3d] would indicate a serine molecule with three deuteriums.

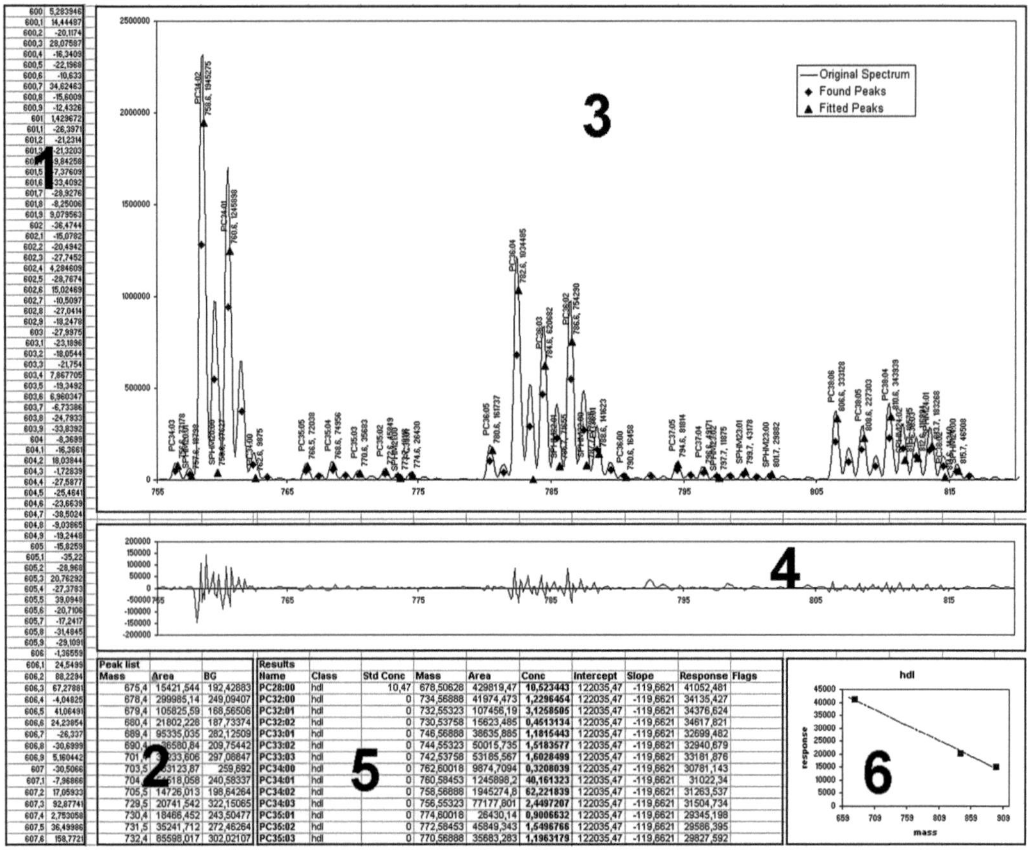

Fig. 2. Data output of LIMSA. (1) Original data with mass (*m/z*) in the first and the corresponding intensity in the second column. (2) List of found peaks with their mass, area, and background values; (3) Graph displaying part of the original spectrum, found peaks (diamonds), and the assigned lipid species (triangles). (4) Graph displaying the residuals of the isotope overlap correction fit. (5) List of found lipids and their concentrations. (6) Calibration curve fit to internal standard values.

3.3.3. Importing and Exporting

Using *Import* and *Export* menu item one can copy the current parameter groups, compounds, or patterns to an empty worksheet. This is useful for making backups or when sharing such parameters with others.

3.3.4. Batch Analysis

One can analyze several data sets in a batch using the same or different sets of parameters. Pressing "*Summarize Results*" collects the results to a single worksheet.

3.4. SECD

1. *SECD* is an independent Windows software designed to (i) visualize LC–MS dataset and (ii) extract MS spectra from select regions of LC–MS maps (to improve the signal-to-noise ratio).
2. *Data import.* SECD can only open MS files stored in the NetCDF-format. This format is supported by most MS instrument software.

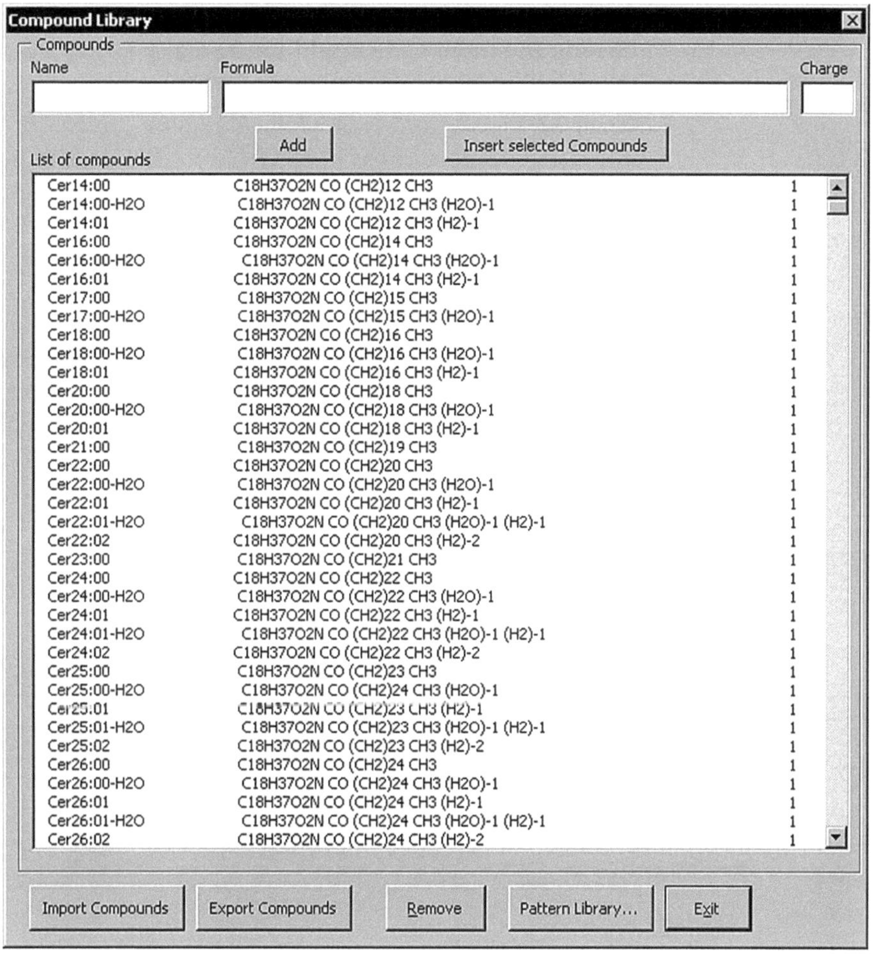

Fig. 3. Lipid library window of LIMSA. This window is used to add new lipids to the internal lipid library (database) of LIMSA. *See* text for details.

3. *Thresholding*. After opening the file (*File* > *Load data*) one can (i) adjust the display by thresholding to exclude noise (*View settings* > *Threshold*), or (ii) smooth the data by averaging select number of data points in the mass and/or time direction. *High Limit* and *Low Limit* values determine the grey scale assigned to specific data values. Settings can be saved for further use.

4. *Zooming and panning*. The map *x*-axis indicates the retention time dimension, while the *y*-axis is the *m/z* dimension. The text boxes below and on the right-hand side of the map panel indicate the limiting time and *m/z* values, respectively, of current view. These can be changed with the mouse or by typing in a new value. To zoom into the map, press the left mouse button and drag across the region of interest. Alternatively, one can zoom in and out by turning the mouse wheel while

holding the *Ctrl*-button. Vertical panning is accomplished by turning the mouse wheel and horizontal panning by pressing the *Shift*-button while turning the wheel.

5. *Creating regions and trapezoids.* A region consists of one or more trapezoids and has a name (e.g., "PC"). A trapezoid is created by pressing the right mouse button and dragging over the region of interest. When the mouse button is released, a rectangular trapezoid appears (**Fig. 4**). Trapezoids can be reshaped by clicking, with the left button pressed, on a corner point or a boundary followed by dragging to a desired position. One can add one or more trapezoids to a (named) region by repeating the procedure elsewhere on the map. All trapezoids of a *Region* have the same color, and the data corresponding to the individual trapezoids of a region are summed up to a single

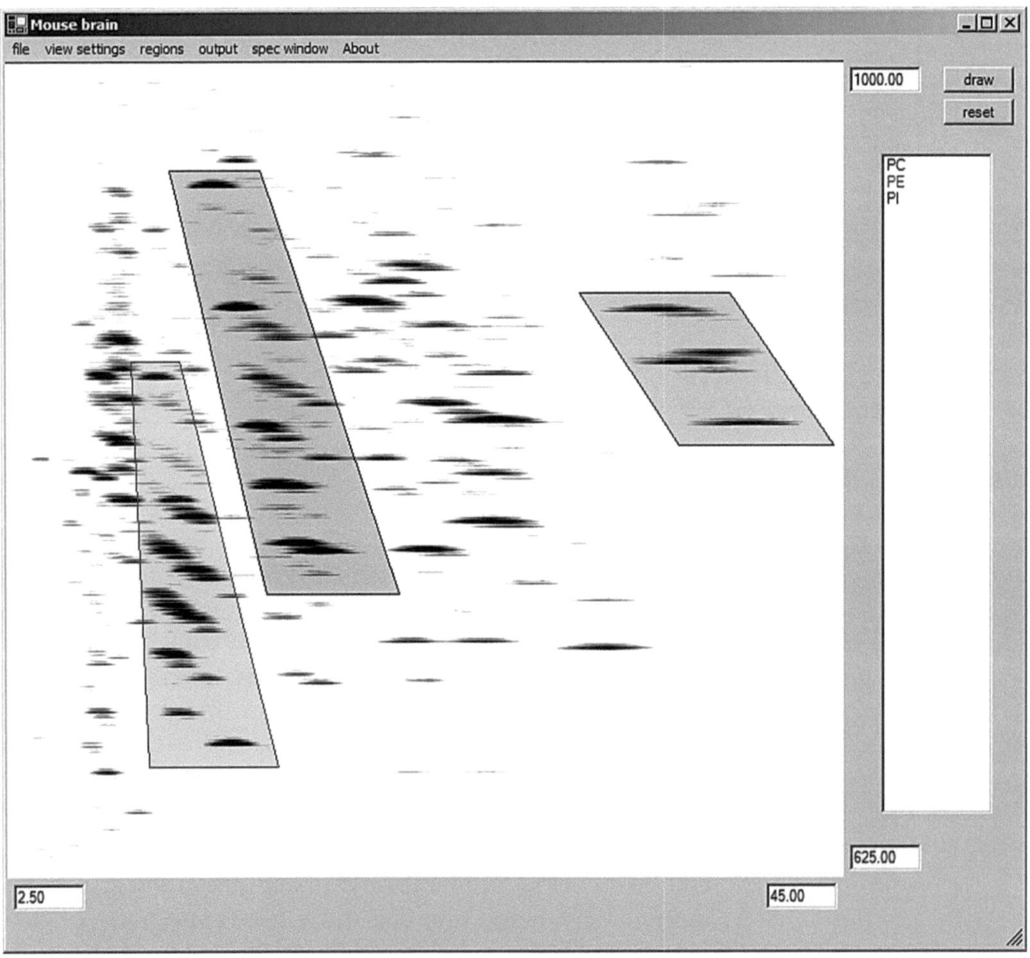

Fig. 4. SECD user interface. The window consists of the menu bar, the map panel, and a list region names on the right-most panel. Trapezoid covering diacyl-PEs (left); diacyl-PCs (middle) and diacyl-PIs (right) are shown as examples. The program displays the trapezoids and corresponding region-names in different colors rather than in gray as here.

spectrum. The regions thus created will appear in the *Region list*. Right-clicking on a name in this panel allows one to rename or delete a region, change its overlay color, or display the spectrum or an ion chromatogram corresponding to the region. Regions can be saved via the *Regions*-menu for future use. The spectra corresponding to different regions can be exported to Excel for analysis in LIMSA, or saved in a text format.

6. The *Output*-menu has the following functions: *Print*, *Copy Picture* (copies the bitmap), *Copy Data* (copies the intensity values to the clipboard), *Save Picture* (saves the unscaled bitmap to a jpg-file). Choosing *Assignment suggestions* followed by left-clicking on a peak in the map opens a new window that indicates the m/z of the peak and the lipid species with that *m/z*.

4. Notes

1. The *subtraction method* calculates the isotope pattern for a lipid based on the integrated area of the first isotope peak and then subtracts the contribution of its isotopic peaks from those they overlap with. The *linear fit method* fits theoretical isotope patterns to all of the integrated isotopic peaks, thus providing slightly more accurate results, but can give erroneous results if a peak of an unassigned compound overlaps an analyte peak. The *peak-model method* fits Gaussian peak patterns to the measured spectrum. It is slowest of the three, but provides most accurate data, particularly with low resolution spectra *(13)*. Note that none of these methods can reliably quantify lipid species with overlapping unknown compounds.
2. We have also developed software allowing fully automated quantitative analysis of LC-MD datasets *(4)*. The use of this software is not, however, as straightforward as SECD-LIMSA because of its command line-driven user interface. It is also more sensitive to the quality of the primary data, including retention time stability, than SECD-LIMSA.

Acknowledgments

Authors thank Dr. Andreas Uphoff for his invaluable contribution in developing SECD and other MS tools. This research was supported by Finnish Academy and University of Helsinki Funds to P.S.

References

1. Han X. and Gross R.W. (2005) Shotgun lipidomics: Electrospray ionization mass spectrometric analysis and quantitation of cellular lipidomes directly from crude extracts of biological samples. Mass Spectrom Rev 24:367–412.
2. Pulfer M. and Murphy R.C. (2003) Electrospray mass spectrometry of phospholipids. Mass Spectrom Rev 22:332–364.
3. Kakela R., Somerharju P, and Tyynela J. (2003) Analysis of phospholipid molecular species in brains from patients with infantile and juvenile neuronal-ceroid lipofuscinosis using liquid chromatography-electrospray ionization mass spectrometry. J Neurochem 84:1051–1065.
4. Hermansson M., Uphoff A, Kakela R, and Somerharju P. (2005) Automated quantitative analysis of complex lipidomes by liquid chromatography/mass spectrometry. Anal Chem 77:2166–2175.
5. Houjou T., Yamatani K, Imagawa M, Shimizu T, and Taguchi R. (2005) A shotgun tandem mass spectrometric analysis of phospholipids with normal-phase and/or reverse-phase liquid chromatography/electrospray ionization mass spectrometry. Rapid Commun Mass Spectrom 19:654–666.
6. Merrill A.H., Jr., Sullards M.C, Allegood J.C, Kelly S, and Wang E. (2005) Sphingolipidomics: High-throughput, structure-specific, and quantitative analysis of sphingolipids by liquid chromatography tandem mass spectrometry. Methods 36:207–224.
7. DeLong C.J., Shen Y.J, Thomas M.J, and Cui Z. (1999) Molecular distinction of phosphatidylcholine synthesis between the CDP-choline pathway and phosphatidylethanolamine methylation pathway. J Biol Chem 274:29683–29688.
8. Boumann H.A., Damen M.J, Versluis C, Heck A.J, de Kruijff B, and de Kroon A.I. (2003) The two biosynthetic routes leading to phosphatidylcholine in yeast produce different sets of molecular species. Evidence for lipid remodeling. Biochemistry 42:3054–3059.
9. Hunt A.N., Clark G.T, Attard G.S, and Postle A.D. (2001) Highly saturated endonuclear phosphatidylcholine is synthesized in situ and colocated with CDP-choline pathway enzymes. J Biol Chem 276:8492–8499.
10. Kainu V., Hermansson M, and Somerharju P. (2008) Electrospray ionization mass spectrometry and exogenous heavy isotope-labeled lipid species provide detailed information on aminophospholipid acyl chain remodeling. J Biol Chem 283:3676–3687.
11. Ivanova P.T., Milne S.B, Forrester J.S, and Brown H.A. (2004) LIPID arrays: New tools in the understanding of membrane dynamics and lipid signaling. Mol Interv 4:86–96.
12. Ejsing CS, Duchoslav E, Sampaio J, et al. (2006) Automated identification and quantification of glycerophospholipid molecular species by multiple precursor ion scanning. Anal Chem 78: 6202–6214.
13. Haimi P., Uphoff A, Hermansson M, and Somerharju P. (2006) Software tools for analysis of mass spectrometric lipidome data. Anal Chem 78:8324–8331.
14. Kurvinen J.P., Aaltonen J, Kuksis A, and Kallio H. (2002) Software algorithm for automatic interpretation of mass spectra of glycerolipids. Rapid Commun Mass Spectrom 16:1812–1820.
15. Song H., Hsu F.F, Ladenson J, and Turk J. (2007) Algorithm for processing raw mass spectrometric data to identify and quantitate complex lipid molecular species in mixtures by data-dependent scanning and fragment ion database searching. J Am Soc Mass Spectrom 18:1848–1858.
16. Katajamaa M., and Oresic M. (2007) Data processing for mass spectrometry-based metabolomics. J Chromatogr A 1158:318–328.
17. Yetukuri L., Ekroos K, Vidal-Puig A, and Oresic M. (2008) Informatics and computational strategies for the study of lipids. Mol Biosyst 4:121–127.
18. Koivusalo M., Haimi P, Heikinheimo L, Kostiainen R, and Somerharju P. (2001) Quantitative determination of phospholipid compositions by ESI-MS: Effects of acyl chain length, unsaturation, and lipid concentration on instrument response. J Lipid Res 42:663–672.

Chapter 17

Computer-Assisted Interpretation of Triacylglycerols Mass Spectra

Josef Cvačka and Edita Kofroňová

Summary

Triacylglycerols (TGs) are principal components of vegetable oils and animal fats. Natural TGs form extremely complex mixtures composed of tens or hundreds of molecular species. HPLC/MS suits well for their analyses, but manual data processing is laborious and time-consuming. Specialized software algorithms are needed to accelerate the interpretation process. Here we present software named *TriglyAPCI* for interpreting APCI, APPI, or ESI MS/MS spectra of TGs. The chapter shows how to build and use the software, what are its advantages and limitations. The algorithm uses diacylglycerol fragments and molecular adducts for determining TG structure. Each ion in a spectrum is tested whether it might be a fragment or a molecular adduct. If so, the number of carbons and double bonds is assigned to it. The relations among the ions are searched and possible structures are suggested. *TriglyAPCI* allows interpreting spectra of single compounds, mixtures, or incomplete spectra lacking one of the diagnostic ions. The fragment intensities are used to distinguish regioisomers. An efficient chromatographic separation of TGs is shown to be crucial for correct spectra interpretation and avoiding false positive results. *TriglyAPCI* performance is demonstrated for HPLC/APCI-MS data of TGs isolated from bumblebee *Bombus rupestris*.

Key words: HPLC, Mass spectrometry, Programming, Software development, Spectra interpreting, Triglycerides

1. Introduction

Triacylglycerols (TGs) are the most abundant lipids in nature known as the main components of vegetable oils and animal fats. They serve as energy depots and play important roles in metabolism *(1)*. TGs consist of three fatty acids (FAs) esterified to glycerol. Each of the three positions in molecule may be occupied by a different FA **(Fig. 1)**. Not considering enantiomers, $(N^3 + N^2)/2$ molecular species theoretically exist for N FAs. Natural

Donald Armstrong (ed.), *Lipidomics*, Methods in Molecular Biology, vol. 580,
DOI 10.1007/978-1-60761-325-1_17, © Humana Press, a part of Springer Science+Business Media, LLC 2009

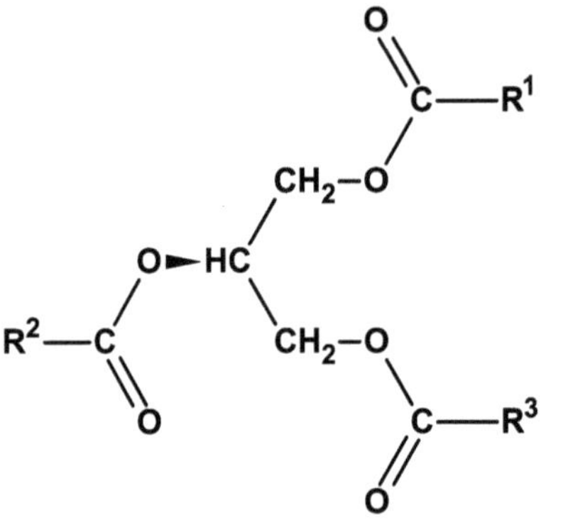

Fig. 1. Structure of TGs.

TGs are complex mixtures with tens to hundreds of species. A new term "triacylglycerolomics" was recently suggested *(2)* for lipidomics focused on complex TG mixtures. High-performance liquid chromatography (HPLC) is a convenient method for analyzing TGs, particularly when coupled to mass spectrometry. Because of the extreme complexity of natural TG mixtures, an in-depth insight into their composition requires extremely powerful chromatography. Two HPLC systems were shown to separate TGs quite efficiently: (i) nonaqueous reversed-phase systems with octadecyl modified silicagel and (ii) argentation chromatography using stationary phases with silver ions *(3)*. TG ionization is achieved using common ion sources for electrospray ionization (ESI), atmospheric pressure chemical ionization (APCI), or atmospheric pressure photoionization (APPI) *(4, 5)*. All techniques yield protonated molecules $[M + H]^+$ or other molecular adducts (e.g., $[M + NH_4]^+$). APCI and APPI spectra already contain important fragments for structure assignment, whereas ESI usually does not. Fragmentation in ESI is achieved by collision-induced dissociation (CID). The main fragment ions are of three types: diacylglycerol ions $[M + H\text{-}R_iCOOH]^+$, monoacylglycerol ions $[M + H\text{-}R_iCOO\text{-}R_jCO]^+$, and acyl ions $[R_iCO]^+$. The most intense fragments are diacylglycerol ions, and they are mostly used for structure assignment. Interpretation of TG spectra is usually straightforward, consisting in the revelation of FA neutral losses. It can be, however, quite time-consuming, when large amount of spectra has to be processed. Data interpretation usually requires incomparably more time than the sample analysis. Specialized software algorithms extremely accelerate data interpretation. In contrast to proteomics, programs for lipid spectra interpretation are not available commercially. Software tools being developed in research laboratories allow to process data from high-throughput analyses of complex lipid mixtures *(6–8)*. The spectra are mostly generated by ESI and the interpretation strategies are based on accurate masses or data-dependent scanning and database searching. When dealing with a particular lipid class, dedicated separation methods and software tools allows more thorough characterization of molecular species. Several tools for interpreting spectra of TGs have been developed. Program MSPECTRA developed by a group of Prof. Kallio allows interpreting ammonia negative ion chemical ionization CID spectra of TGs. It identifies molecular species and calculates their relative proportions *(9)*. The same group later published an algorithm for identification and quantification of glycerolipids based on ESI mass spectra *(10)*. Our contribution to the field is *TriglyAPCI*, an algorithm for interpreting spectra of TGs *(11)* originally developed for APCI spectra.

This chapter describes *TriglyAPCI*, which is being developed in our laboratory. It shows how the software is built, how works and how can be used for interpreting spectra of TGs. We believe that the ideas behind *TriglyAPCI* might be useful for those working on their own interpretation computer programs. This chapter also demonstrates that efficient chromatographic separations are essential for correct characterization of TG molecular species. Mass spectra used as examples in this work originate from our research on insect lipids.

2. Materials

2.1. Equipment

1. Computer with Microsoft Visual Basic 6.0.
2. TSP liquid chromatograph (Thermo Separation Product, USA): SCM 1000 vacuum membrane degasser, P 4000 quaternary gradient pump, AS 3000 autosampler. Additional HPLC pump (Waters 515) for postcolumn addition of cationization agent.
3. An ion-trap mass spectrometric detector LCQ classic (LCQ Fleet) for low-resolution mass detection or LTQ Orbitrap XL hybrid mass spectrometer for high-resolution work (both Thermo Fisher Scientific, USA) equipped with an APCI source.
4. Two stainless steel columns Nova-Pak C18 (300 × 3.9 mm and 150 × 3.9 mm, 4 μm, Waters) connected in series.
5. Laboratory balances.
6. Sonicator.
7. Laboratory freezer.
8. Fume hood.

2.2. Reagents and Supplies

1. Acetonitrile.
2. Isopropanol.

3. Chloroform.
4. Methanol.
5. Hexane.
6. Diethylether.
7. Formic acid.
8. Rhodamine 6G (0.05% solution in ethanol).
9. Ammonium acetate.
10. 2,6-Di-*tert*-butyl-4-methylphenol (BHT).
11. Glass TLC plates (36 × 76 mm) coated with Adsorbosil-Plus [Applied Science Labs; layer thickness 0.2 mm with gypsum (12%)].
12. Argon cylinder and gas regulator.
13. 1.5-ml Glass autosampler vial and caps.
14. Glass capillaries.
15. Small beaker (250 ml).
16. Nebulizer spray bottle.
17. Gas-tight Hamilton syringes (250 μl).
18. Glass Pasteur pipettes.
19. Stainless steel tweezers and dissector.

3. Methods

3.1. Programming

The algorithm was developed using Microsoft Visual Basic 6.0 (Microsoft Corporation, Redmond, WA, USA) under the Microsoft Windows XP operating system. Data export is optimized for Xcalibur Qual Browser version 2.0.7 (Thermo Fisher Scientific, MA, USA).

3.2. Sample Processing and Storage

Insects (bumblebee males or queens) are immobilized in a freezer and their peripheral fat bodies are dissected. The fat body tissue is transferred into a glass vial with 100 ml of $CHCl_3/CH_3OH$ (1:1, v/v) containing 25 mg/ml BHT, and the sample is sonicated for 15 min. The tissue is further crushed using a glass stick and extracted three times with 100 ml of $CHCl_3/CH_3OH$ (1:1, v/v). The crude extract is separated on precleaned TLC glass plates using hexane/diethyl ether/formic acid (80:20:1) mobile phase. TLC zones are made visible by spraying Rhodamine 6G solution. TGs are the most intense zone in a TLC plate ($R_F = 0.6$). The band of silica containing the TGs is scraped off the plate and extracted with 7 ml of freshly distilled diethyl ether. The solvent is evaporated to dryness under argon stream. The residues are reconstituted in chloroform to a concentration of 1% and stored in sealed glass ampoules in argon at −18°C until analyzed.

3.3. Standards Preparation and Storage

Stock solutions of synthetic TG standards are prepared in chloroform at a concentration of 2.0 mg/ml and stored in sealed glass ampoules at −18°C. Diluted solutions are prepared before using.

3.4. HPLC Separation

The HPLC gradient method is adapted from **Ref.** *(12)*. TGs are sequentially eluted over 150 min as the proportion of 2-propanol (B) in acetonitrile (A) is increased. Simultaneously, mobile phase flow rate decreases because of high column backpressure. The last step of the gradient is restoration of the initial column conditions. The gradient program: 0 min—100% of A (1 ml/min); 108 min—30% A, 70% of B (1 ml/min); 150 min—100% of B (0.7 ml/min); 160 min—100% of A (0.4 ml/min); 162 min—100% of A (1 ml/min). The columns are kept at 30°C during analyses. The mobile phase is mixed postcolumn in a low-dead volume T-piece with 100 mM ammonium acetate prepared in 2-propanol/water 1:1 (v/v), flow rate 10 µl/min. Samples dissolved in $CHCl_3$ are injected using autosampler or manual injector.

3.5. MS Detection

Mass spectrometer equipped with an APCI probe is coupled to the HPLC instrument. The full scan MS or CID MS/MS centroided spectra are recorded in the range of 75–1300 *m/z*. The temperature of vaporizer and heated capillary are set to 400°C and 200°C, respectively. Corona discharge current is 4.5 µA. Other ion source parameters are optimized using a standard procedure recommended by the manufacturer. Caution! A large amount of toxic solvents are evaporated in the ion source; the vapors must be drain out to the hoods. Surfaces of the APCI source are hot!

3.6. Data Analysis and Spectra Interpretation

The chromatogram is best viewed in base peak projection. Mass spectra of individual compounds or mixtures are exported in ASCII format to Windows Clipboard. Relative proportions of TGs are determined after integration of peaks in reconstructed chromatograms calculated for $[M + H]^+$ and $[M + NH_4]^+$ adducts.

3.7. Results

3.7.1. The Algorithm Overview

The algorithm facilitates interpretation of atmospheric pressure ionization mass spectra of TGs. It uses diacylglycerole fragments ("fragments" in the following text) and molecular adduct ions for determining TG structure. The algorithm evaluates each ion in the spectrum and determines whether the ion may be a fragment or a molecular adduct. If so, the number of carbons and double bonds is assigned to it. The relations among the ions are searched and possible TG structures are suggested. The algorithm processes MS or MS/MS spectra of a single compound,

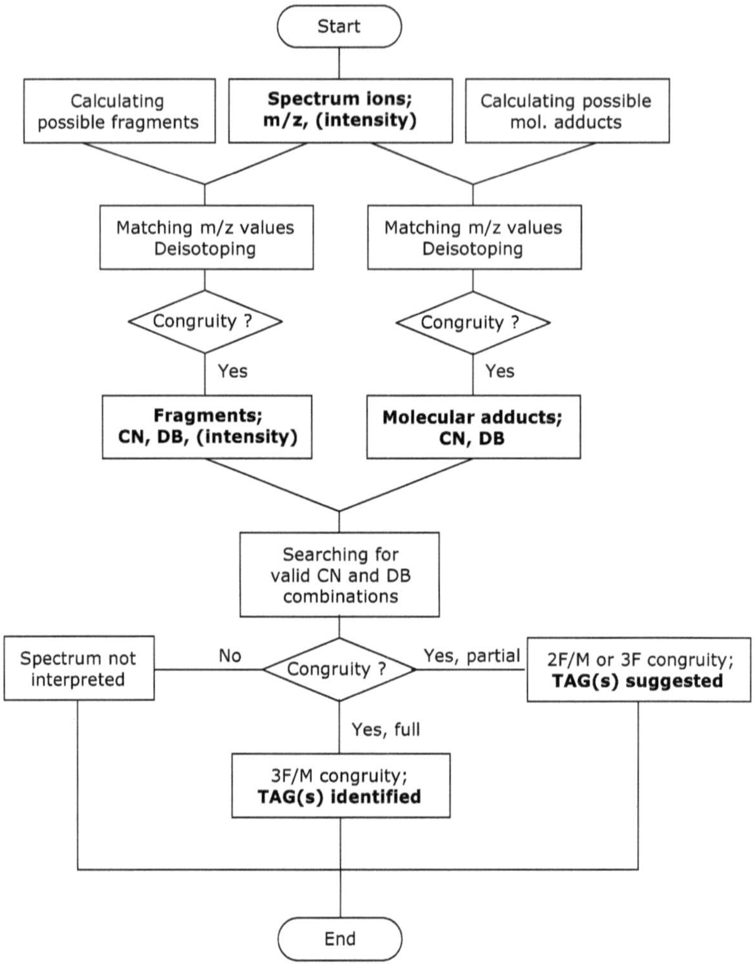

Fig. 2. Flowchart of the *TriglyAPCI* algorithm.

mixtures, or incomplete spectra lacking one of the diagnostic ions. The intensities of the fragments are used to estimate TG stereochemistry. A simplified flowchart of *TriglyAPCI* is shown in **Fig. 2**.

3.7.2. Data Input and Initial Spectra Processing

Mass spectra are imported from a data system using Windows Clipboard; the current version of *TriglyAPCI* supports import from Xcalibur™ (Thermo Fisher Scientific, USA). The data in ASCII format contain *m/z* values and the corresponding ion intensities. The low-intensity ions removal (threshold cutting) is applied to reduce the spectrum noise. The threshold values can be set independently for *m/z* ranges of fragments and molecular adducts. Optionally, spectra (*m/z* values) can be entered manually, one after another.

3.7.3. Selecting Relevant Ions

After data input, the algorithm evaluates each ion in the spectrum and determines whether the ion may be a fragment or a molecular adduct. TGs, as well as many other lipid classes differ from one another only by FAs. The mass of any TG molecule ($m_{(M)}$) can be expressed using several number of carbon atoms (CN) and double bonds (DB) in acyls,

$$m_{(M)} = m_{C_3H_2O_6} + (m_{CH_2}) \times CN - (m_{H_2}) \times DB, \qquad (1)$$

where CN ≥ 3 and DB ≤ (CN-4). Similar equations apply for ions. For protonated molecular adduct one can write

$$\begin{aligned} m_{(M+H^+)} &= m_{C_3H_3O_6} + (m_{CH_2}) \times CN - (m_{H_2}) \times DB \\ &= 134.9924 + 14.0151 \times CN - 2.0151 \times DB, \end{aligned} \qquad (2)$$

and similar equations exist for any other molecular adduct, e.g.,

$$\begin{aligned} m_{(M+NH_4^+)} &= m_{C_3H_6NO_6} + (m_{CH_2}) \times CN - (m_{H_2}) \times DB \\ &= 152.0190 + 14.0151 \times CN - 2.0151 \times DB. \end{aligned} \qquad (3)$$

Analogically, masses of diacylglycerol fragment ions can be calculated according to the Eq. 4:

$$\begin{aligned} m_{(M+H-RiCOOH^+)} &- m_{C_3H_3O_4} + (m_{CH_2}) \times CN - (m_{H_2}) \times DB \\ &= 103.0026 + 14.0151 \times CN - 2.0151 \times DB, \end{aligned} \qquad (4)$$

where CN and DB are number of carbons and double bonds in a diacylglycerole fragment, respectively. It is obvious that ion masses in TG spectra attain only certain values. All possible ion masses are calculated within defined CN and DB ranges. Comparing the measured *m/z* values of the ions with the calculated set enables to identify fragment and molecular adduct ions. *TriglyAPCI* characterizes all relevant ions by their CN and DB values and stores them in two independent data arrays (fragments and molecular adducts). As calculating with the isotopic peaks generate false positive results, it is necessary to remove them. Common deisotoping procedures consist in calculating theoretical isotope distribution and subtracting it from the measured pattern, but the procedure can be substantially simplified for TGs. The appearance of isotopic clusters is mostly determined by the number of carbon atoms. The isotopic peaks occur because of ^{13}C (1.108% in natural mixture of carbon isotopes). Interestingly, the ratio between intensities of the third (two ^{13}C) and the second (one ^{13}C) isotopic peak ($R_{3/2}$) for C_1–C_{100} depends linearly on CN:

$$R_{3/2} = 0.0054 \times CN + 0.0054 \qquad (5)$$

In analogous equations calculated for TG fragments and common molecular adducts the values are slightly different, but the slope is always very close to 0.005 and the intercept to zero.

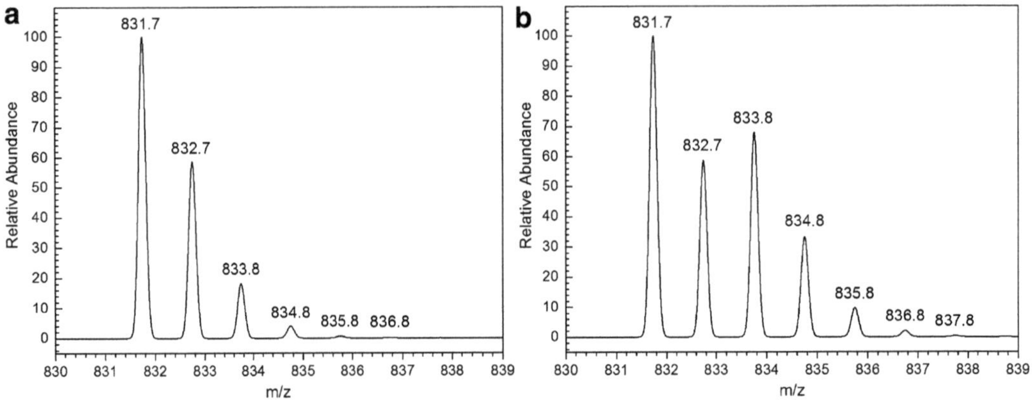

Fig. 3. Simulated isotopic clusters for protonated TG 50:2 ($C_{53}H_{99}O_6$; $[M + H]^+$) (**a**), and for a mixture of two protonated TGs (50:2, $C_{53}H_{99}O_6$ and 50:1, $C_{53}H_{101}O_6$) in the ratio of 2:1 (**b**).

The Eq. 5 can be used for simplified calculation of isotopic peak intensities and thus for simple deisotoping. **Figure 3a** shows an isotopic cluster simulated for $C_{53}H_{99}O_6$ (TG 50:2; $[M + H]^+$). To get the correct interpretation, the algorithm has to work only with the first peak at *m/z* 831.7. All others have to be removed. The ions with even integer masses (*m/z* 832.7 and 834.8) are eliminated during selecting ions based on the Eq. 2. However, all ions with odd integer masses (*m/z* 833.8 and 835.8) fulfill the Eq. 2 and have to be removed before the next calculations. Coeluting of two TGs is common even in high efficient separations. **Figure 3b** shows an isotopic cluster simulated for two $[M + H]^+$ ions differing by one double bond; TG 50:2 ($C_{53}H_{99}O_6$) and TG 50:1 ($C_{53}H_{101}O_6$). Their intensities are in the ratio 2:1. Only two *m/z* values have to be extracted for correct structures assignment (*m/z* 831.7 and 833.8); all other isotopic peaks would generate false positive results. Taken together, a deisotoping routine has to inspect the third (fifth, seventh, ...) isotopic peak in a cluster to determine, whether it should be completely removed (single compound) or its intensity reduced (coeluting compounds differing by two mass units). The deisotoping routine in *TriglyAPCI* compares intensity of the tested ion with the intensity of the previous ion (at *m/z* –1) in an isotopic cluster. If there is no peak at *m/z*-1 (e.g., when testing *m/z* 831.7 in **Fig. 3**), the tested peak is considered to contain only ^{12}C and remains for further calculations. If there is a peak at *m/z* –1 (e.g., when testing *m/z* 833.8 in **Fig. 3**), the algorithm calculates intensity ratio between the tested ion and the ion one unit mass lower (*m/z* –1). In our example (**Fig. 3**, testing *m/z* 833.8), the algorithm calculates intensity ratio between *m/z* 833.8 and 832.7. The ratio is compared with a critical value $R_{3/2}$ obtained from Eq. 5. When the calculated value is lower than the critical value, the algorithm does not consider the tested peak for further calculations.

If the calculated value is higher, the peak remains for further calculations. Its intensity can be recalculated by subtracting intensity of the isotopic peak calculated from Eq. 5. In our example, $R_{3/2}$ is 0.28 (for CN = 50). Therefore, the deisotoping routine removes *m/z* 833.8 in the example shown in **Fig. 3a**, but not in **b**. TG fragments are deisotoped in an analogous way.

It is important to take into account that LC/MS instruments do not always show spectra with isotopic clusters exactly matching the theory. The spectrum noise affects low-intensity signals and the limited dynamic range might be also a problem. Fast chromatography usually does not provide enough time to get sufficient number of scans ensuring good ion statistics. Therefore, the criterion for deisotoping is set higher (0.5–1) than the calculated values of $R_{3/2}$.

3.7.4. Mass Accuracy

Mass accuracy of the spectrometer (analyzer) must be taken into account when comparing spectrum masses with the calculated ones. *TriglyAPCI* considers error window around a calculated value (calculated mass ± mass error). Wide mass error window ensures that even poorly calibrated spectra can be processed, but the total number of ions for further calculations considerably increases. When narrowing the error window, the number of ions decreases and the calculation is faster. On the basis of our experience, mass error of 0.5 u works well for unit mass resolution analyzers and 0.01 u or less works well for high-resolution spectra. It is important to mention that the nominal mass of methylene group is the same as the nominal mass of 14 hydrogens. The difference (CH_2–14H) is 0.094 u, which is too low to be resolved by the low-resolution mass analyzers (ion traps or quadrupoles). Therefore, TGs having seven more double bonds and being one methylene group larger have almost the same mass and cannot be distinguished. To measure spectra with better mass accuracy, high-resolution instruments, e.g., LTQ Orbitrap XL, are needed. Fortunately, TGs containing FAs with a very high number of double bonds are extremely rare in nature. FAs with the highest number of double bonds ever found in nature contain 28 carbons and seven or eight double bonds. They were detected in marine dinoflagellates *(13, 14)*. Therefore, unless working with quite unusual samples, high-resolution MS detectors are not essential. *TriglyAPCI* algorithm can be set to report only TGs containing FAs with up to six double bonds. The use of the Orbitrap is yet advantageous, as MS and MS/MS spectra are cleaner *(15)* than spectra recorded in the LCQ.

3.7.5. Calculating TG Structures

In the next step *TriglyAPCI* looks for fragments, which are related to the molecular adducts. The characterization of the ions by just two parameters, CN and DB, is convenient, as very simple relations exist between the ions. The number of carbons (or double bonds) in a molecular adduct equals the sum of the carbon atoms (or double bonds) in FA moieties:

$$CN_{(M+H)^+} = CN_{R_1COOH} + CN_{R_2COOH} + CN_{R_3COOH} \quad (6)$$

$$DB_{(M+H)^+} = DB_{R_1COOH} + DB_{R_2COOH} + DB_{R_3COOH} \quad (7)$$

Simultaneously, the sum of carbons (or DB) in a fragment and corresponding FA gives CNs (or DB) in the molecular adduct:

$$CN_{(M+H)^+} = CN_{(M+H-R_1COOH)^+} + CN_{R_1COOH} \quad (8)$$

$$CN_{(M+H)^+} = CN_{(M+H-R_2COOH)^+} + CN_{R_2COOH} \quad (9)$$

$$CN_{(M+H)^+} = CN_{(M+H-R_3COOH)^+} + CN_{R_3COOH} \quad (10)$$

$$DB_{(M+H)^+} = DB_{(M+H-R_1COOH)^+} + DB_{R_1COOH} \quad (11)$$

$$DB_{(M+H)^+} = DB_{(M+H-R_2COOH)^+} + DB_{R_2COOH} \quad (12)$$

$$DB_{(M+H)^+} = DB_{(M+H-R_3COOH)^+} + DB_{R_3COOH} \quad (13)$$

A combination of the previous equations gives Eqs. 14 and 15:

$$CN_{(M+H)^+} = \frac{CN_{(M+H-R_1COOH)^+} + CN_{(M+H-R_2COOH)^+} + CN_{(M+H-R_3COOH)^+}}{2} \quad (14)$$

$$DB_{(M+H)^+} = \frac{DB_{(M+H-R_1COOH)^+} + DB_{(M+H-R_2COOH)^+} + DB_{(M+H-R_3COOH)^+}}{2} \quad (15)$$

These equations show the relationship between CNs and DBs of fragments and molecular adducts. If the ions satisfy Eqs. 14 and 15 simultaneously, the individual acyls are calculated based on the Eqs. 8–13. Results are reported and the interpretation process is finished. Note that Eqs.6–15 are valid for any molecular adduct ion, as CNs (DBs) for the same TG equal:

$$CN_{(M+H)^+} = CN_{(M+NH_4)^+} + CN_{(M+Na)^+} = \cdots \quad (16)$$

$$DB_{(M+H)^+} = DB_{(M+NH_4)^+} + DB_{(M+Na)^+} = \cdots \quad (17)$$

The spectra interpretation could be alternatively performed using ion masses, without prior transformation into CNs and DBs. In this case, one can write:

$$m_{(M+H)^+} = \frac{m_{(M+H-R_1COOH)^+} + m_{(M+H-R_2COOH)^+} + m_{(M+H-R_3COOH)^+} - m_{C_3H_5}}{2} \quad (18)$$

This approach, however, can have drawbacks. First, mass errors of all the fragment ions are combined, so that poorly calibrated spectra may not be interpreted correctly. Second, any calculation with fractional numbers requires more system resources and slows down the interpretation process considerably.

3.7.6. Interpreting Single Compound Spectra

Figure 4 shows APCI mass spectra of OOP (dioleoyl–palmitoyl–glycerol) averaged across a chromatographic peak recorded using various instruments and techniques. The first spectrum **(Fig. 4a)** was measured using an ion trap instrument. It contains all expected ions at *m/z* 577.7, 603.7 (fragments), 859.7 (protonated molecular adduct), and 876.5 (adduct with NH_4^+). There is also low-intensity ion at *m/z* 897.5 (adduct with K^+). *TriglyAPCI* interprets the spectrum as [18:1, 18:1, 16:0]. When deisotoping is not used, one incorrect result ([18:0, 16:0, 18:1]) is reported as well, obviously because of calculating with isotopic peaks at *m/z* 579.7, 605.7, 861.7, and 878.4. When high number of double bonds is allowed (up to 8 for a FA), two more suggestions are reported ([19:8, 19:8, 16:0] and [18:1, 18:1, 17:7]). In principle, even these two hits are correct and they cannot be excluded unless other information about the sample are known [e.g., data from GC analysis of FA methyl esters (FAMEs) after transesterification]. However, this uncertainty can be eliminated when a high-resolution analyzer is used. **Figure 4b** gives APCI spectrum of the same compound recorded by the Orbitrap at a resolution of 30.000. If the *TriglyAPCI* mass error window is narrowed to 0.01 u, one single correct result ([18:1, 18:1, 16:0]) is reported. Next example **(Fig. 4c)** is a MS/MS spectrum of NH_4^+ adduct of OOP obtained by CID and recorded by the Orbitrap. As the precursor isolation width was set to 1, no isotopic peaks are present. Obviously, this clean spectrum does not need deisotoping and the correct results are obtained using narrow mass error window.

3.7.7. Interpreting Spectra of Mixtures

Nonseparated or partially separated peaks of TGs are common even in highly efficient separation techniques. Mass spectra of unresolved peaks contain signals of two or more compounds making the spectra interpreting troublesome. As one cannot tell which fragments belong to a particular molecular adduct, the results are burdened by some uncertainty. It can be worked out by successive selecting molecular adducts and performing MS/MS.

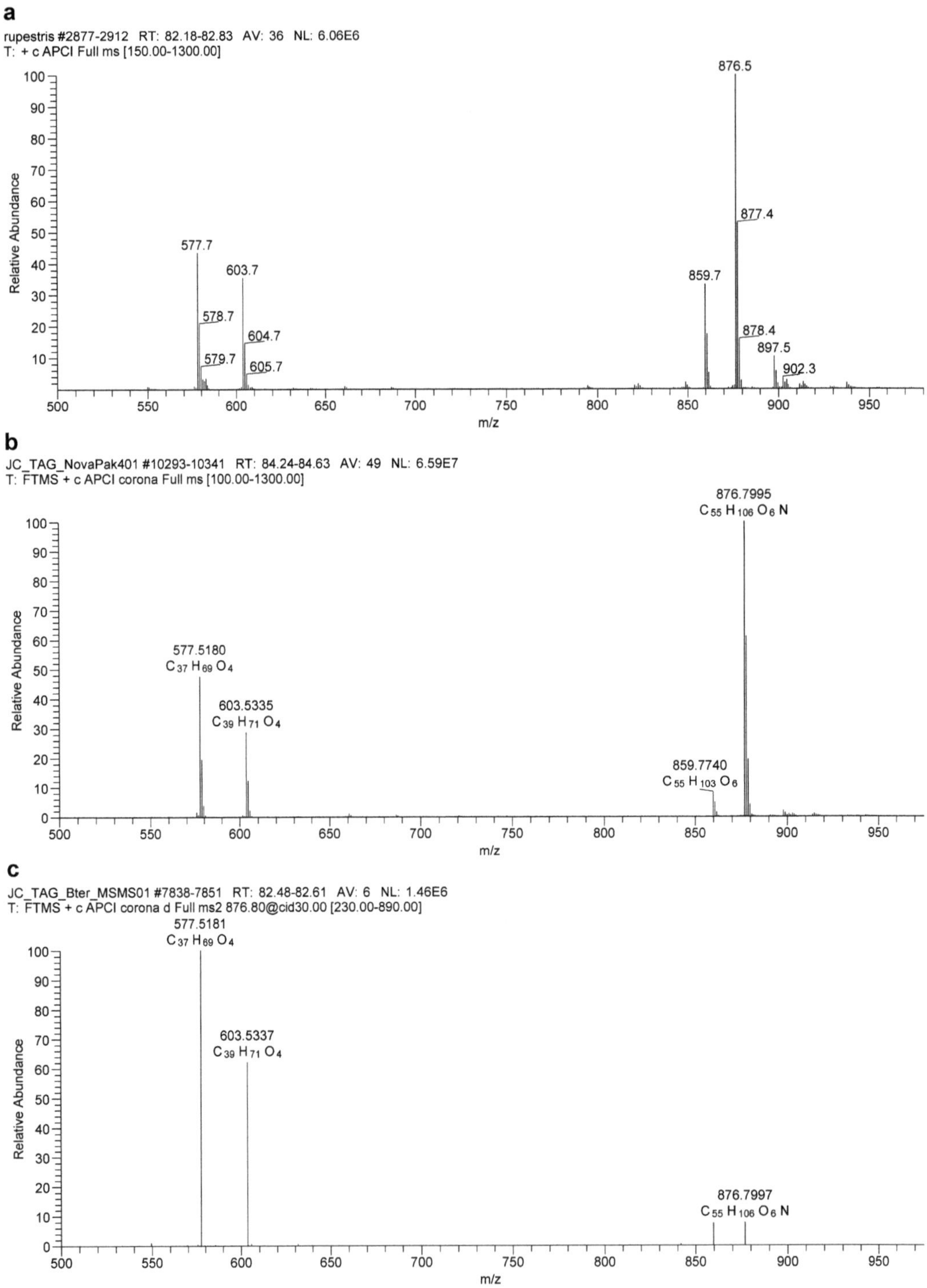

Fig. 4. APCI mass spectrum of OOP recorded by an ion trap (**a**) and the Orbitrap (**b**) instrument. APCI MS/MS spectrum of ammonium adduct of OOP recorded by the Orbitrap (**c**). Spectra were obtained by HPLC/MS analysis of TGs isolated from bumblebees *Bombus rupestris* (**a**) *and Bombus terrestris* (**b** and **c**).

However, when TGs with the same molecular weight elute at the same time, the MS/MS has no particular advantage over the first stage APCI (APPI) MS. *TriglyAPCI* allows obtaining the correct structural information even from spectra of mixtures. As mentioned above, the algorithm searches all combinations of the fragments and molecular adducts to obtain a valid result. It is obvious that calculation with incorrect subset of the ions (i.e., ions, which originated from several different TGs) gives false positive results. With the more complicated TG mixtures the more false positive results are obtained. The incorrect results showing nonexisting FA(s) can be eliminated if the overall composition of FA is known. **Figure 5a** depicts APCI MS spectrum of coeluting TGs, two of them being the most abundant. When interpreting with *TriglyAPCI*, the following results are obtained: [16:1, 16:1, 18:1] and [18:1, 14:1, 18:1] for the molecular adducts at *m/z* 829.7 and 846.8 ($[M + H]^+$ and $[M + NH_4]^+$, respectively) and [18:1, 18:1, 18:3] for the molecular adducts at *m/z* 881.8 and 898.8 ($[M + H]^+$ and $[M + NH_4]^+$, respectively). MS/MS experiments **(Fig. 5b, c)** performed with the ammoniated molecular adducts clearly revealed, that the correct results are [16:1, 16:1, 18:1] and [18:1, 18:1, 18:3] only. The suggestion [18:1, 14:1, 18:1] is a false positive result because of the simultaneous occurrence of fragments *m/z* 547.5 and 603.5. **Figure 6a** shows MS/MS spectrum of an ammonium adduct at *m/z* 848.8. The number of fragments indicates that the spectrum represents a mixture of compounds with the same precursor mass. *TriglyAPCI* reports five hits ([18:2, 18:0, 14:0], [16:1, 18:0, 16:1], [18:1, 14:0, 18:1], [18:1, 16:1, 16:0] and [16:0, 18:2, 16:0]), some of them are false positives. This particular example demonstrates that MS/MS does not bring any advantage in case of co eluting species with the same molecular weights. In conclusion, *TriglyAPCI* is able to interpret spectra of mixtures, however, false positive results cannot be entirely avoided.

3.7.8. Interpreting Low-Intensity Spectra

Low abundant TGs provide spectra showing weak signals buried in noise. Such spectra are difficult to interpret, as distinguishing diagnostic ions from noise is not straightforward. As *TriglyAPCI* can evaluate each ion regardless the noise level, even such spectra can be correctly interpreted. An example of low-intensity spectrum is in **Fig. 6b**. When threshold is properly adjusted, the spectrum can be quickly interpreted as [18:0, 20:1, 18:1]. Manual interpretation is in this case almost impossible.

3.7.9. Interpreting Incomplete Spectra

Quite low abundant TGs provide spectra that might lack important diagnostic ion(s). Molecular adducts may be missing for several reasons. Too high voltage settings on various parts of the ion source or high collision energy in MS/MS result in complete fragmentation of the precursor. Fully saturated TGs in APCI show

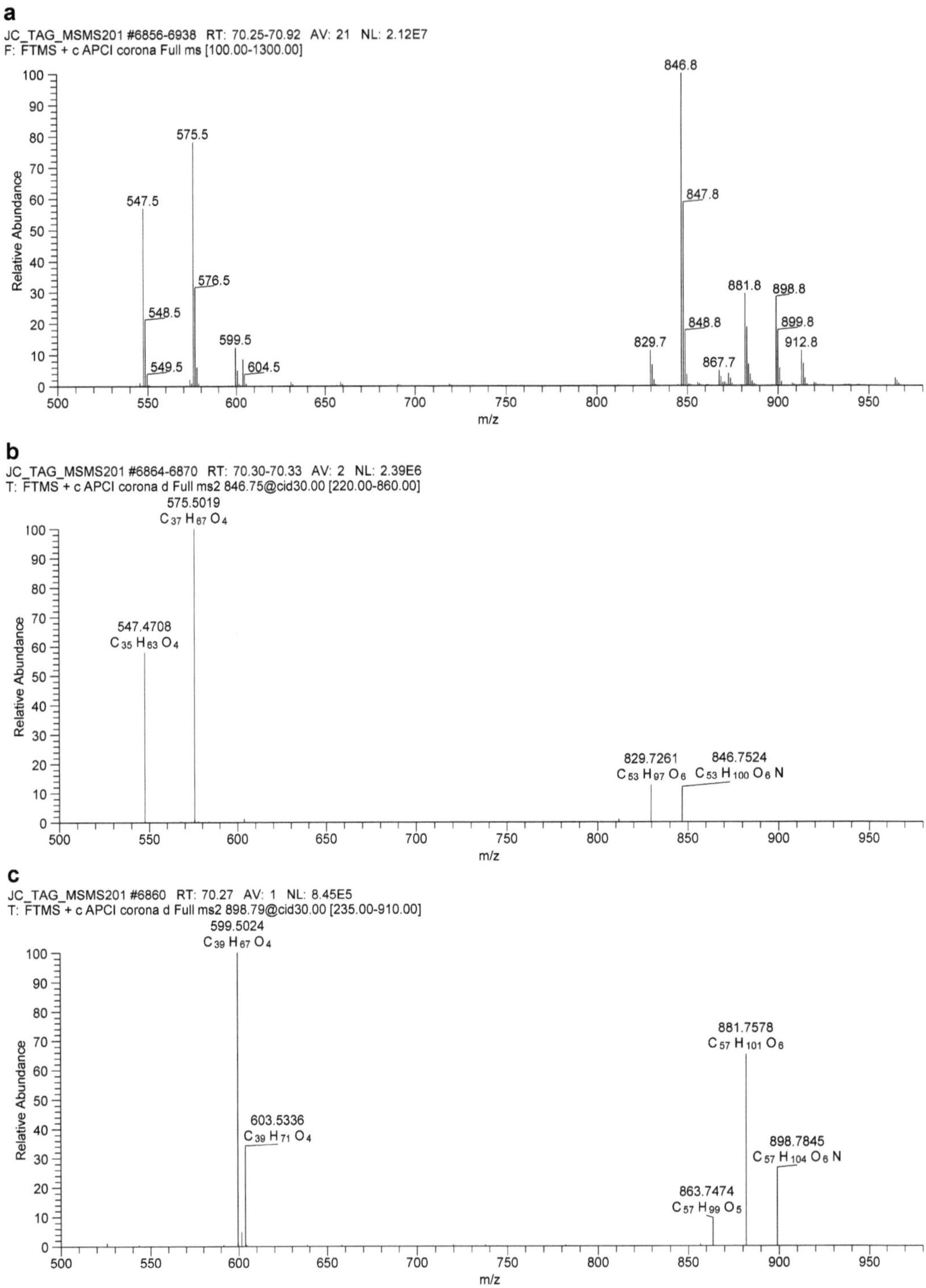

Fig. 5. APCI full scan mass spectrum of TGs eluting as one chromatographic peak (t_R = 70.6 min) in HPLC/APCI analysis of TGs isolated from *Bombus terrestris* (**a**) and MS/MS spectra of two most abundant molecular adducts at *m/z* 846.8 (**b**) and *m/z* 898.8 (**c**).

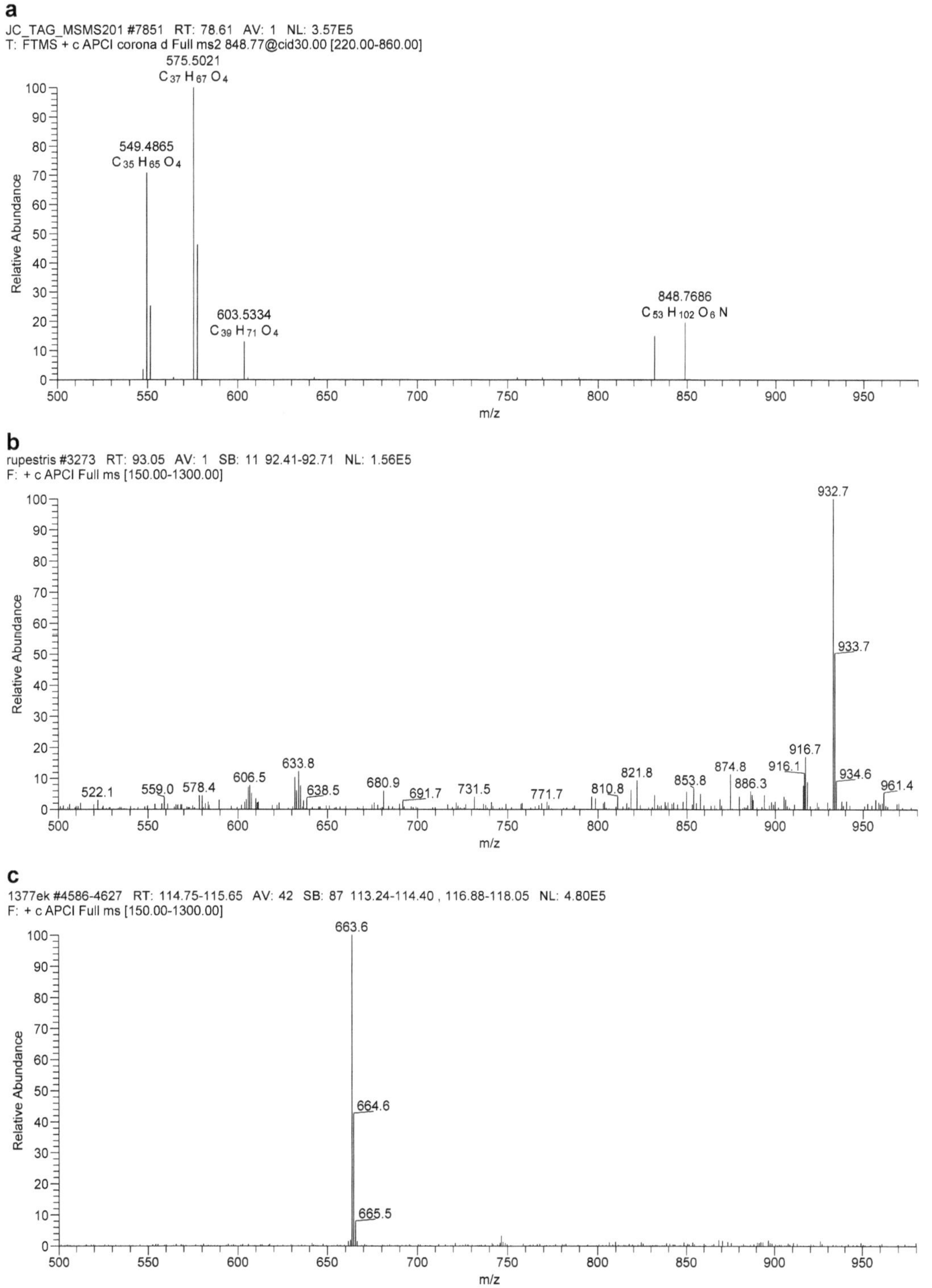

Fig. 6. APCI MS/MS spectrum of a TG ammonium adduct at *m/z* 848.8 from *Bombus terrestris* (**a**), low-intensity APCI MS spectrum of [18:0, 20:1, 18:1] from *Bombus rupestris* (**b**), and APCI MS spectrum of standard of triarachidin [20:0, 20:0, 20:0] (**c**).

protonated molecules at extremely low intensity, or do not show $[M + H]^+$ at all *(16)*. The algorithm can process incomplete spectra, providing that only one diagnostic ion is missing. In the first example, let us suppose that one fragment is missing in a spectrum of LnLnP (linolenoyl–linolenoyl–palmitoyl–glycerol). Two ions at *m/z* 573.3 (fragment) and 851.7 (protonated molecule) are used for calculation. Despite the missing ion the correct TG [18:3, 18:3, 16:0] is reported. As two FAs in LnLnP are the same, also two terms in Eqs. 14 and 15 equal, one explicit solution exists and only one hit is reported. The interpretation is more complicated when a TG contains three different FA. Interpretation of incomplete LLnO (linoleyl–linolenoyl–oleoyl–glycerol) spectrum containing ions at *m/z* 597.5, 601.5 (fragments), and 879.7 (molecular adduct) gives three hits, [18:2, 18:3, 18:1] (correct), and [18:1, 18:1, 18:4], [18:3, 18:3, 18:0] (false positives). APCI spectrum of fully saturated AAA (triarachidoyl–glycerol) shows only one ion, a fragment at *m/z* 663.6 **(Fig. 6c)**. This ion suffices for correct interpretation without any false positives. APCI spectrum of saturated PPS (dipalmitoyl–stearoyl–glycerol) shows fragments at *m/z* 551.5 and 579.5 and no $[M + H]^+$. The correct hit [16:0, 16:0, 18:0] is reported together with three other false positives ([16:0, 16:0, 16:0], [17:0, 17:0, 15:0], [17:0, 17:0, 17:0]). Some of the false positives can be excluded considering the chromatographic behavior of the particular peak [gradual increase of equivalent carbon number (ECN; ECN = CN-2DB) in case of reversed-phase chromatography, or analogous increase of DB in silver ion chromatography with the retention time]. The incorrect hits with nonexisting FA(s) can be eliminated based on the known overall composition of FAs.

3.7.10. Regioisomerism

Relative intensities of TG fragments depend on several factors. First, FA losses from the *sn*-1 and *sn*-3 positions are more likely and equally favored. The lowest probability of the loss is in position *sn-2*, which results in the least intense diacylglycerol fragment. Second, the ion intensities are influenced by the nature of FAs, e.g., by the number of double bonds and chain lengths. Finally, ion source design and ion optics settings should be also taken into account. Therefore, mass spectra allow distinguishing regioisomers, but *sn*-position assignment is not completely confident without authentic standards. Nevertheless, fragment intensities can be used as a “first guess” to distinguish FAs bonded to the *sn*-2 position from those attached to the *sn*-1/*sn*-3 positions. If TG has all three FAs different, the *sn-2* position assignment to the FA related to the least intense fragment is straightforward. If two FAs of TG are the same, only two fragment ions are obtained. Their relative intensity ratio (less intense ion to more intense ion) enables us to assign FA in the *sn*-2 position. If the ratio is lower than 0.5, then the TG is of ABA type. If the ratio is between

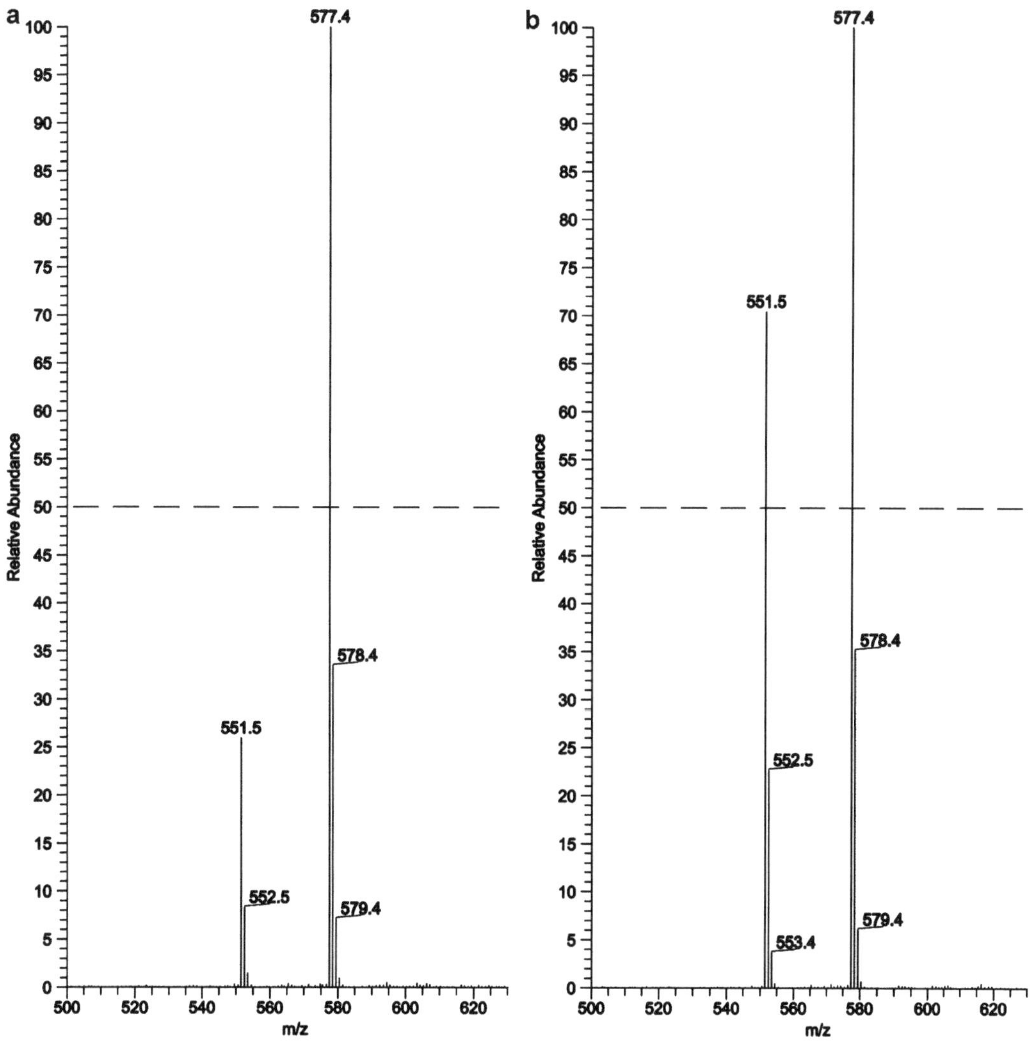

Fig. 7. Sections of APCI mass spectra of 1,2-dipalmitoyl-3-oleoyl glycerol, PPO (**a**) and 1,3-dipalmitoyl-2-oleoyl glycerol, POP and (**b**) showing abundances of the diacylglycerole fragment ions. Dash line is a theoretical boundary line for TGs of AAB type (ion with lower intensity is below the boundary) and ABA type (ion with lower intensity is above the boundary).

0.5 and 1.0, the TG is of AAB type. **Figure 7** shows an example of fragment ions of PPO (palmitoyl–palmitoyl–oleoyl–glycerol) and POP (palmitoyl–oleoyl–palmitoyl–glycerol). The intensity of fragment at *m/z* 551.5 is less then 50% of the second one at *m/z* 577.4 for PPO and more than 50% for POP. The intensity ratios are in a good agreement with the theoretical expectations.

3.7.11. Interpreting HPLC/MS Data with TriglyAPCI

An example of LC/APCI-MS data that have been processed by *TriglyAPCI* is given in **Fig. 8**. It shows a base peak chromatogram of TGs isolated from a male of cuckoo bumblebee *Bombus (Psithyrus) rupestris.* The separation was carried out in reversed-phase system using 45-cm-long column, which allowed

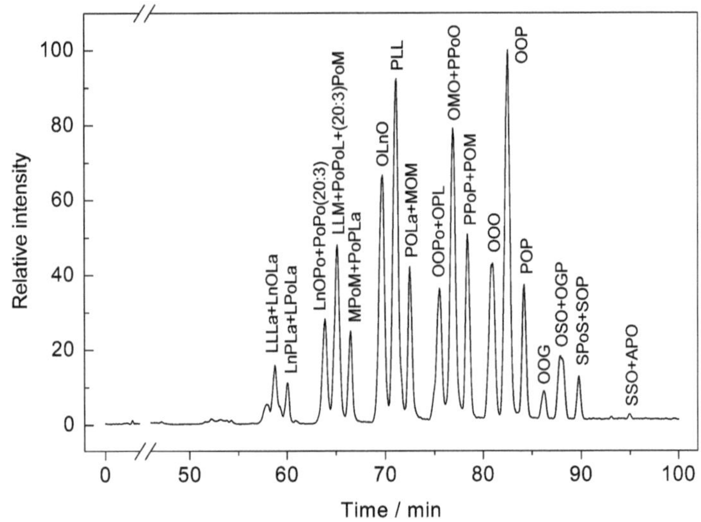

Fig. 8. Base peak chromatogram of a TG mixture isolated from *Bombus rupestris*. Separation conditions: nonaqueous reversed phase system with two stainless steel columns Nova-Pak C18 (300 × 3.9 mm and 150 × 3.9 mm, 4 μm) connected in series and acetonitrile/2-propanol gradient. Detection: APCI in positive ion mode. Abbreviations: *G*, gondoic acid; *L*, linoleic acid; *La*, lauric acid; *Ln*, linolenic acid; *M*, myristic acid; *O*, oleic acid; *P*, palmitic acid; *Po*, palmitoleic acid; S, stearic acid.

us to achieve a good separation. TGs that differ in ECN value are well separated and also separation within the ECN groups is more than acceptable. Unfortunately, even carefully optimized methods do not ensure full separation of all isomers with the same CN and DB values. Our method uses postcolumn addition of ammonium acetate to promote formation of $[M + NH_4]^+$ adducts. It is advantageous for saturated TGs, which usually do not provide protonated molecular adduct. Both $[M + H]^+$ and $[M + NH_4]^+$ molecular adducts are observed for unsaturated TGs, which predominate in the sample. More than 70 TGs was identified using *TriglyAPCI*. They are listed in **Table 1**. Their structures were critically evaluated taking into account both the retention behavior and FAs known from transesterification and subsequent GC/MS.

4. Notes

1. *TriglyAPCI* algorithm for interpreting mass spectra of TGs has been developed. It speeds up data interpretation process to a great extent.
2. *TriglyAPCI* process APCI, APPI MS spectra, or MS/MS spectra from molecular adducts generated by APCI, APPI, or ESI.

Table 1
TGs of *Bombus rupestris* identified by *TriglyAPCI* algorithm

ECN	CN:DB	t_R (min)	Triacylglycerol(s)	Peak Area (%)
36	48:6	46.02	18:3, 18:3[a], 12:0	0.0
	44:4	45.57	18:4, 14:0[a], 12:0; 18:4, 14:0[a], 12:0	0.0
38	52:7	51.79	18:3, 16:1[a], 18:3	0.1
	50:6	52.78	18:3, 18:3[a], 14:0	0.1
	48:5	51.57	18:3, 16:1[a], 14:1	0.1
	46:4	52.15	18:3, 16:1[a], 12:0	0.2
	44:3	53.3	18:3, 14:0[a], 12:0	0.1
	42:2	52.06	12:0, 18:2[a], 12:0; 16:1, 12:0[a], 14:1	0.0
40	54:7	57.99	18:3, 18:1[a], 18:3	0.5
	52:6	57.38	18:3, 18:3[a], 16:0; 18:2, 16:1[a], 18:3;	0.4
	50:5	57.93	16:1, 16:1[a], 18:3	0.8
	48:4	58.78	18:2, 18:2[a], 12:0; 18:3, 18:1[a], 12:0	1.7
	46:3	60.07	18:3, 16:0[a], 12:0; 18:2, 16:1[a], 12:0	1.0
	44:2	58.78	16:1, 16:1[a], 12:0	0.3
	42:1	59.86	14:0, 16:1[a], 12:0	0.2
42	54:6	63.57	18:2, 18:2[a], 18:2	0.6
	52:5	63.9	18:3, 18:1[a], 16:1; 16:1, 16:1[a], 20:3	3.6
	50:4	65.14	**18:2, 18:2[a], 14:0; 16:1, 16:1[a], 18:2; 20:3, 16:1[a], 14:0**	5.2
	48:3	64.14	16:0, 18:3[a], 14:0; 16:1, 16:1[a], 16:1; 14:0, 14:0[a], 20:3	2.9
	46:2	65.04	16:1, 16:1[a], 14:0	1.4
	44:1	66.39	14:0, 16:1[a], 14:0; 16:1, 16:0[a], 12:0	1.1
44	56:6	69.9	18:2, 18:2[a], 20:2	0.2
	54:5	69.73	**18:1, 18:3[a], 18:1**	7.8
	52:4	71.15	**18:2, 18:2[a], 16:0**	10.2
	50:3	72.62	16:0, 16:0[a], 18:3; 16:1, 16:1[a], 18:1; 18:1, 18:1[a], 14:1	4.6
	48:2	71.14	16:1, 18:1[a], 14:0; 18:1, 18:1[a], 12:0	4.3
	46:1	72.5	16:0, 18:1[a], 12:0; 14:0, 18:1[a], 14:0	3.1
	44:0	74.3	16:0, 16:0[a], 12:0; 14:0, 14:0[a], 16:0	0.1

(continued)

Table 1 (continued)

ECN	CN:DB	t_R (min)	Triacylglycerol(s)	Peak Area (%)
46	56:5	75.13	18:1, 18:1[a], 20:3; 20:1, 18:3[a], 18:1	1.3
	54:4	76.8	18:1, 18:1[a], 18:2;	3.6
	52:3	75.53	**18:1, 18:1[a], 16:1; 18:1, 16:0[a], 18:2; 18:2, 18:1[a], 16:0**	6.8
	50:2	76.91	**18:1, 14:0[a], 18:1; 18:1, 16:1[a], 16:0**	8.8
	48:1	78.39	16:0, 16:1[a], 16:0; 16:0, 18:1[a], 14:0	3.9
48	56:4	80.25	20:1, 18:1[a], 18:2; 18:1, 18:1[a], 20:2	0.1
	54:3	80.89	**18:1, 18:1[a], 18:1**	6.2
	52:2	82.53	**18:1, 18:1[a], 16:0**	10.8
	50:1	84.2	16:0, 18:1[a], 16:0	2.8
	48:0	86.18	16:0, 16:0[a], 16:0	0.1
50	56:3	86.2	18:1, 18:1[a], 20:1	1.0
	54:2	87.83	18:1, 18:0[a], 18:1; 18:1, 20:1[a], 16:0	2.4
	52:1	89.75	18:0, 16:1[a], 18:0; 18:0, 18:1[a], 16:0	0.9
52	58:3	91.07	20:1, 18:1[a], 20:1; 18:1, 18:1[a], 22:1	0.1
	56:2	92.92	18:1, 18:1[a], 20:0; 18:0, 20:1[a], 18:1; 18:0, 18:0[a], 20:2	0.2
	54:1	94.86	18:0, 18:0[a], 18:1; 20:0, 16:0[a], 18:1	0.2
54	58:2	97.88	18:1, 18:1[a], 22:0; 20:2, 26:0[a], 12:0	0.0
	56:1	99.68	18:0, 18:0[a], 20:1; 20:0, 16:0[a], 20:1	0.0

The most abundant TGs are shown in bold
[a]FA in the *sn*-2 position

3. Unambiguous results are obtained for single compound spectra. Spectra of mixtures, incomplete or noisy spectra can be processed as well, but the results must be critically evaluated. The worse is the spectrum quality the higher probability of false positive results exists. Clearly, good chromatography is very important to obtain correct results. The number of false positives can be efficiently reduced when overall composition of FA is known (e.g., from GC/MS of FAMEs) and when considering chromatographic behavior of TGs (ECN values in reversed phase systems or number of double bonds in argentation chromatography). Direct infusion MS cannot provide as detailed results as LC/MS.

4. Limited information about stereochemistry can be mined from the relative intensities of fragments, but drawing rigorous conclusions without authentic standards cannot be recommended.
5. MS/MS detection is fundamental in ESI, but it has only limited importance for APCI or APPI. MS/MS allows distinguishing coeluting TGs with different molecular weights, but has no advantage for coeluting TGs with the same molecular weight.
6. High-resolution MS providing accurate masses is useful for distinguishing TGs having seven more double bonds from those being one methylene group larger. They appear at the same nominal masses and cannot be resolved by low-resolution analyzers. Fortunately, TGs with more than six double bonds are very rare in nature. Therefore, low-resolution instruments suffice for most samples.

Acknowledgments

The authors thank the Academy of Sciences of the Czech Republic (Z40550506 and A4055403) and Ministry of Education, Youth and Sports of the Czech Republic (MSMT 2B06007) for financial support, and to Dr. Oldřich Hovorka and Dr. Jiří Kindl for insect specimens.

References

1. Gunstone, F. D., Harwood, J. L. and Dijkstra, A. J. (2007) *The Lipid Handbook*, CRC Press, Boca Raton.
2. Holčapek, M. and Lísa, M. (2008) *Triacylglycerolomics—Characterization of Complex Triacylglycerol Mixtures in Plant Oils and Animal Fats*, Proceedings of the 56th ASMS Conference on Mass Spectrometry and Allied Topics, Denver, USA, TPHH-205.
3. Christie, W. W. (2003) *Lipid Analysis, Isolation, Separation, Identification and Structural Analysis of Lipids*. The Oily Press, Bridgwater.
4. Byrdwell, W. C. (2005) *Modern Methods for Lipid Analysis by Liquid Chromatography/Mass Spectrometry and Related Techniques*. AOCS Press, Champaign.
5. Cai, S. S. and Syage, J. A. (2006) Atmospheric pressure photoionization mass spectrometry for analysis of fatty acid and acylglycerol lipids. *J Chromatogr A* **1110**, 15–26.
6. Leavell, M. D. and Leary, J. A. (2006) Fatty acid analysis tool (FAAT): An FT-ICR MS lipid analysis algorithm. *Anal Chem* **78**, 5497–5503.
7. Haimi, P., Uphoff, A., Hermansson, M. and Somerharju, P. (2006) Software tools for analysis of mass spectrometric lipidome data. *Anal Chem* **78**, 8324–8331.
8. Song, H., Hsu, F. F., Ladenson, J. and Turk, J. (2007) Algorithm for processing raw mass spectrometric data to identify and quantitate complex lipid molecular species in mixtures by data-dependent scanning and fragment ion database searching. *J Am Soc Mass Spectrom* **18**, 1848–1858.
9. Kurvinen, J. P., Rua, P., Sjovall, O. and Kallio, H. (2001) Software (MSPECTRA) for automatic interpretation of triacylglycerol molecular mass distribution spectra and collision induced dissociation product ion spectra

obtained by ammonia negative ion chemical ionization mass spectrometry. *Rapid Commun Mass Spectrom* **15**, 1084–1091.

10. Kurvinen, J. P., Aaltonen, J., Kuksis, A. and Kallio, H. (2002) Software algorithm for automatic interpretation of mass spectra of glycerolipids. *Rapid Commun Mass Spectrom* **16**, 1812–1820.
11. Cvačka, J., Krafková, E., Jiroš, P. and Valterová, I. (2006) Computer-assisted interpretation of atmospheric pressure chemical ionization mass spectra of triacylglycerols. *Rapid Commun Mass Spectrom* **20**, 3586–3594.
12. Holčapek, M., Lísa, M., Jandera, P. and Kabátová, N. (2005) Quantitation of triacylglycerols in plant oils using HPLC with APCI-MS, evaporative light-scattering, and UV detection. *J Sep Sci* **28**, 1315–1333.
13. Mansour, M. P., Volkman, J. K., Holdsworth, D. G., Jackson, A. E. and Blackburn, S. I. (1999) Very-long-chain (C-28) highly unsaturated fatty acids in marine dinoflagellates. *Phytochemistry* **50**, 541–548.
14. Van Pelt, C. K., Huang, M. C., Tschanz, C. L. and Brenna, J. T. (1999) An octaene fatty acid, 4,7,10,13,16,19,22,25-octacosaoctaenoic acid (28:8n-3), found in marine oils. *J Lipid Res* **40**, 1501–1505.
15. Olsen, J. V., de Godoy, L. M. F., Li, G., Macek, B., Mortensen, P., Pesch, R., Makarov, A., Lange, O., Horning, S. and Mann, M. (2005) Parts per million mass accuracy on an orbitrap mass spectrometer via lock mass injection into a C-trap. *Mol Cell Proteom* **4**, 2010–2021.
16. Holčapek, M., Jandera, P., Zderadička, P. and Hrubá, L. (2003) Characterization of triacylglycerol and diacylglycerol composition of plant oils using high-performance liquid chromatography-atmospheric pressure chemical ionization mass spectrometry. *J Chromatogr A* **1010**, 195–215.

Chapter 18

Visualization of Complex Processes in Lipid Systems Using Computer Simulations and Molecular Graphics

Jelena Telenius, Ilpo Vattulainen, and Luca Monticelli

Summary

Computer simulation has become an increasingly popular tool in the study of lipid membranes, complementing experimental techniques by providing information on structure and dynamics at high spatial and temporal resolution. Molecular visualization is the most powerful way to represent the results of molecular simulations, and can be used to illustrate complex transformations of lipid aggregates more easily and more effectively than written text. In this chapter, we review some basic aspects of simulation methodologies commonly employed in the study of lipid membranes and we describe a few examples of complex phenomena that have been recently investigated using molecular simulations. We then explain how molecular visualization provides added value to computational work in the field of biological membranes, and we conclude by listing a few molecular graphics packages widely used in scientific publications.

Key words: Biological membrane, Lipid, Molecular visualization, Computer simulation, Molecular mechanics, Molecular dynamics, Force field, Coarse-grain

1. Introduction

Seeing is believing. Many of us have friends who like fishing, claiming that once they almost captured a fish so huge that "you would not believe how large it was." Correct, we do not believe that. Not without a photo or other piece of evidence. In the same spirit, it is much easier to be convinced of something spectacular holding true if one has *seen* that to really happen, regardless of its context. Here, we discuss how that could be rationalized in fascinatingly small and complex systems dealing with nano-sized biomolecular engines and molecular complexes.

Donald Armstrong (ed.), *Lipidomics*, Methods in Molecular Biology, vol. 580,
DOI 10.1007/978-1-60761-325-1_18, © Humana Press, a part of Springer Science+Business Media, LLC 2009

Biological systems are intrinsically complex in terms of their molecular composition, time and length scales, interaction patterns and the consequent biochemical networks. Even seemingly simple entities such as biological membranes are fascinatingly complex, exemplified by the interplay of thousands of different membrane proteins and lipids which jointly are responsible for a major fraction of cellular functions taking place over a multitude of time and length scales.

Biological membranes are an excellent example to highlight the complexity of phenomena taking place in biological matter *(1)*. Elucidating and visualizing the numerous time-dependent processes in membranes is far from simple. Many of the cellular functions are due to membrane proteins acting as nano-sized engines, and characterizing their action in time through experiments is exceedingly difficult due to the extremely short time and length scales involved in those functions. Structural studies of membrane proteins are largely based on crystallizing them, thus obtaining insight into their transient structures with little understanding of their dynamics. As for membranes themselves, techniques such as atomic force microscopy provide an almost molecular resolution view into the structure of membranes. However, these images are also static. The dynamics can be explored, for example, using fluorescent probes bound to a molecule of interest, thus detecting positions of individual molecules in time *(2)*. However, the drawback in this case is the spatial resolution of about 200 nm.

While the above examples are far from exhaustive, they do illustrate how difficult it is to characterize nano-scale processes involving individual molecules with space and time-dependent information at the same time. Visualizing biological processes in time, with full atomic or molecular detail, would enormously increase the chances of understanding them.

To complement experiments, computer simulations have become an appealing tool to gauge complex processes in biological matter with a level of detail not within reach through experiments. In the context of membranes, simulations can provide information on, e.g., the structure of lipid aggregates with atomic resolution, the dynamics of lipids and proteins over time scales from picoseconds to microseconds, and on transformations of lipid aggregates including formation of lipid domains, vesicle fusion and changes in phase behavior *(3)*. Essentially, the present level of simulation methodology allows one to simulate a significant fraction of systems and processes dealing with biological membranes.

Despite this highly positive view, there is reason to keep in mind that doing simulations is challenged by numerous obstacles. Starting from model development, which is perhaps the most central step in simulations of soft condensed matter, any mistake at this stage may be critical regarding the validity of the model. The second step following model development is perhaps less inspiring but equally important: collecting data through extensive simulations. Having done that, one faces the third and perhaps the most exciting part of a research project: the stage for analysis where one has to answer the questions on the phenomena that has been simulated, possibly in a quantitative manner. In many cases, and for a number of reasons, the last part of the work is the most difficult. Having at hand a large amount of numerical descriptors for the simulated process (e.g., coordinates, distances, velocities, forces), it still might prove difficult to understand what the driving forces are, and how the system evolves. Even if one has an intuitive view of the results, the process might be intrinsically so complex that there is no way to see through the data and describe the process under study in terms of simple physical or biological concepts. So, what to do? Visualize. Simulations yield trajectories describing the time dependence of all particles included in a system, and visualizing the trajectories shows how the process evolves in time. Once the mechanism of the process has been found, or the simulation has revealed how the interplay of the molecules involved in a process gives rise to the function studied, it is far easier to invent means to describe quantitatively the process. The keyword is indeed visualization.

In this contribution, we briefly discuss means to perform simulations of lipids and the added value they yield, thus complementing experiments. Most important, although, we focus on the added value of visualization on the basis of simulation data that can be generated through different models. A few examples will highlight some of the benefits of visualization, hence closing this chapter.

2. Simulation Methodology

2.1. Molecular Modeling of Lipids

Computer simulations can provide detailed information on membrane structure and dynamics on a broad range of time scales, from femtoseconds to microseconds. Different methodological approaches are employed to model the dynamics of biological macromolecules depending on the length and time scale of the process of interest. These approaches can be classified in two groups: molecular mechanical (MM) and quantum mechanical (QM) methods. In QM simulations, electrons are considered explicitly while in MM simulations they are not. QM calculations have a very high computational cost and are nowadays limited to systems consisting of tens to hundreds of atoms. Therefore, they are currently used mainly to provide information about the reorganization of electrons around nuclei and creating or breaking chemical bonds. Molecular mechanics instead is more commonly used to describe the collective motion of lipid aggregates.

In molecular mechanics, simple potential energy functions (e.g., harmonic oscillators, Coulomb law) are used to model molecular systems. Different representations of molecular systems exist, ranging from all-atoms models to less detailed ones, in which several atoms are grouped together in one interaction site. The last ones include so-called united-atoms models (in which nonpolar hydrogen atoms are "included" in the heavy atoms to which they are chemically attached) and coarse-grained models (in which several atoms are represented by one particle). Atoms (or particles) are then connected using harmonic springs, and all the interactions within the molecule are described in the molecular topology (for details, *see* **Subheading 2.3** on force fields).

Software is available to easily "build" molecular structures (i.e., sets of atomic coordinates and the connectivity) for any individual lipid molecule. The description of the interactions among all atoms in the system is usually referred to as "force field" and is normally not done automatically. Some details of force fields commonly used for lipid simulations are explained below.

2.2. Simulation System Setup for Lipid Bilayers and Monolayers

Before starting a simulation, an initial guess for the structure of the system (i.e., the atomic coordinates of each individual molecule) has to be provided. Two routes are common: to start from a random distribution of lipid molecules in water and simulate their self-assembly; or else, to start from a structure we know to be realistic for the particular lipid composition of interest.

Numerous simulations have been reported about the self-assembly of lipid systems into micelles *(4)*, bilayers *(5)*, hexagonal phases *(6)*, and vesicles *(7)*. Yet, when one wants to investigate the properties of a certain lipid aggregate, it is far more common to start the simulation from a preassembled system.

To "build" the starting configuration models for any lamellar phase it is generally sufficient to generate multiple copies of an individual lipid molecule and translate them in one plane. While translating molecules, one should try to maintain the "correct" surface density of molecules (if this is known). This method based on simple translation generates extremely ordered structures, which is often not what one wants to simulate. In some cases one can decrease ordering by manipulating the geometry of the individual lipid molecules (adding random rotation and translation along the molecule main axis, for example). In most cases, long "equilibration" simulations need to be performed to achieve complete relaxation of the starting structure.

Usually model cell membranes are simulated as infinite, flat lipid bilayers, piled on top of each other to form an infinite stack of bilayers. This is achieved by using so-called "periodic boundary conditions": a relatively small molecular system is replicated in all three spatial dimensions infinitely. This effectively mimics experimental conditions in multilamellar stacks. The bilayers are separated from each other by a certain amount of water molecules, which depends on the hydration level that one wants to simulate. In the case of full hydration, at least 30 or 40 water molecules per lipid are used, and each bilayer does not have a significant interaction with its periodic image. This setup enables macroscopic properties to be calculated from simulations using relatively small number of particles.

For model lung membranes, the setup usually includes one slab of water bounded by vacuum on both sides with two symmetric monolayers at the two water–vacuum interfaces *(8, 9)*. Few water molecules usually escape the bulk water phase, leading to a vapor phase. For more complex, nonlamellar geometries (hexagonal phases, cubic phases, etc.) generating an initial structure might prove more challenging.

2.3. Force Fields: Atomistic and Coarse-Grained Representations

In atomistic simulations, all the atoms of the system are represented. Forces between the atoms are calculated based on simple models, which constitute the so-called force field. A force field consists of a set of equations chosen to model the potential energy and their associated parameters. The typical form of an atomistic force field is similar to:

$$
\begin{aligned}
V(r) = & \sum_{\text{bonds}} k_b (b - b_0)^2 + \sum_{\text{angles}} k_\theta (\theta - \theta_0)^2 \\
& + \sum_{\text{dihedrals}} k_\phi \left(1 + \cos(n\phi - \phi_0)\right) \\
& + \sum_{\text{impropers}} k_\psi (\psi - \psi_0)^2 \\
& + \sum_{\substack{\text{nonbonded} \\ \text{pairs}(i,j)}} 4\varepsilon_{ij} \left[\left(\frac{\sigma_{ij}}{r_{ij}} \right)^{12} - \left(\frac{\sigma_{ij}}{r_{ij}} \right)^{6} \right] \\
& + \sum_{\substack{\text{nonbonded} \\ \text{pairs}(i,j)}} \frac{q_i q_j}{\varepsilon_D r_{ij}}
\end{aligned}
$$

The potential function $V(r)$ depends only on the position of particles in the system. Despite occasional extra terms in some force fields, the functional form of the potential function is essentially the same in all common force fields. Bonds and angles between atoms are modeled by harmonic springs, dihedral angles often by periodic functions. In most cases, a harmonic potential is used to preserve planarity of certain functional groups and the chirality. Interactions between atoms that are not directly connected through bonds (nonbonded interactions) consist of dispersion interactions, a hard core repulsion and electrostatics. Dispersion interactions (van der Waals) and the repulsive term at short interatomic distances are represented using a Lennard–Jones potential, while electrostatic interactions are calculated using the Coulomb law.

Generally, different Lennard–Jones parameters (ε, σ) are used for each "atom type." Atom types may vary depending on the chemical features and the position of a certain atom in a molecule. Specific bonds, angles, and dihedral parameters depend on the specific nature of two, three, and four atoms, respectively. Among the most commonly used force fields for biological molecules are AMBER *(10)*, CHARMM *(11)*, GROMOS *(12, 13)*, and OPLS *(14–16)*.

The first force fields for biological macromolecules were developed starting from over 25 years ago. As computer power increased enormously in these years, the parameterization procedure changed over time and became more and more sophisticated. For example, initially most force fields were developed using the so-called "united-atom" scheme, in which hydrogen atoms were not explicitly considered, while most of the recent force fields describe explicitly the interactions involving hydrogen atoms.

Two force fields are most commonly used today for simulations of lipids: the all-atoms CHARMM force field *(17)* and the united-atoms force field by Berger et al. *(18)*. Both models offer advantages and disadvantages, and reproduce a number of experimental observations on the structure and dynamics of lipid bilayers reasonably well. The most extensively studied lipids are those including a phosphatidylcholine head group and two acyl tails, which are very common in most living organisms.

In coarse-grained models, a certain number of atoms are grouped together into one effective interaction site (CG particle or bead) and the description of the forces acting among the particles is somewhat simplified. For example, usually long-range nonbonded interactions are disregarded in the coarse-grained approach. Typically, the speed-up reached in CG simulations compared to atomistic ones is between 2 and 4 orders of magnitude. Among coarse-grained force fields, we mention here the pioneering work of Smit et al. *(19)*, numerous works by the groups of Voth *(20)*, Klein *(21, 22)*, Vattulainen *(23)*, and the recent development of the MARTINI force field by Marrink et al. *(24–26)* (*see* **Fig. 1**).

2.4. Molecular Mechanics

The term "molecular mechanics" (MM) refers to a large number of simulation methods in which molecular system are modeled using simple, empirical potential energy functions. MM simulation methods can be classified in two groups: the first group of methods does not rely on equations of motion, while in the

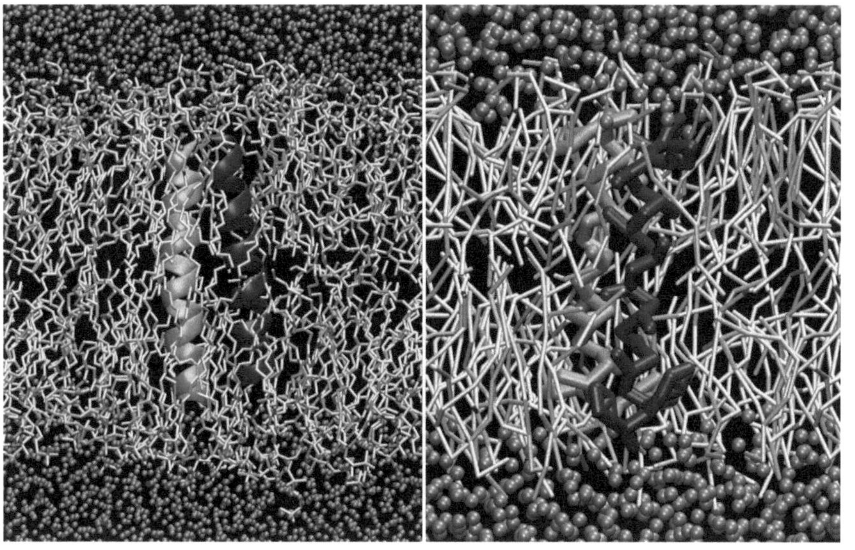

Fig. 1. Atomistic (left panel) and coarse-grained (right panel) representations for transmembrane peptides in a DOPC lipid bilayer. Adapted from reference *(26)*.

second group equations of motion are used. Methods of the first group include energy minimization, normal mode analysis, and Monte Carlo simulations. They do not provide direct information on the dynamics of the system. In methods of the second group, the positions of all the atoms (or particles) are calculated as a function of time by integrating the equations of motion in discrete time steps. The most common techniques are molecular dynamics (MD), Langevin dynamics (LD), Brownian dynamics (BD), and dissipative particle dynamics (DPD). When an all-atoms representation is used, the integration time step is up to a few femtoseconds and simulations typically consist of millions of steps. When coarse-grained representations are used, the integration time step is usually much longer and depends on the degree of "coarseness" of the model.

Molecular dynamics is by far the most common simulation technique for biomolecular systems. It has been used in conjunction with both all-atom and CG representations. In the following we will describe briefly the basic principles of the MD simulation technique.

2.5. The MD Simulation Technique

Molecular dynamics simulates the motion of atoms in a molecular system, from which average properties of the system can be calculated using the tools of statistical mechanics. Sets of atomic positions are derived as a function of time by solving Newton's equation of motion:

$$m_i \frac{\partial^2 r}{\partial t^2} = F_i$$

The equation of motion for all atoms is integrated by breaking the calculation in a series of time steps, typically between 0.5 and 2 femtoseconds. At each step, the forces on all atoms are computed. Using the current position and velocity, the force is used to calculate the displacement of each atom. The force is assumed to be constant during this short time interval. After the atoms are moved to the new positions, an updated set of forces is calculated and so on. The result of the molecular dynamics calculation is a set of positions and velocities for all the atoms of the system as a function of time, which is normally referred to as a "trajectory."

The equations of motion can be modified to take into account the temperature and the pressure of interest for a certain phenomenon. This is particularly interesting for self-assembled lipid systems that undergo phase transitions by changing the temperature or the pressure.

The typical length scale of the systems simulated using atomistic molecular dynamics is a few nanometers. These systems include typically hundreds or a few thousands of lipid molecules. For these systems, phenomena occurring on a time scale of tens or hundreds of nanoseconds are currently accessible with present-day computer facilities, while longer length and time scales require simplification in the description of the systems. CG simulations allow the description of complex transformations and collective motions of larger lipid systems on the microsecond time scale. For example, the elastic properties and the permeability of lipid bilayers to fullerene aggregates has recently been characterized using microsecond time scale coarse-grained MD simulations on systems including thousands of lipids *(27)*.

3. What Kind of Information Can We Obtain from Simulation of Lipids?

Computer simulation of lipid aggregates has made very significant progress over the past decade. In the following section we will summarize some of the most remarkable achievements in this area. Because of the very rapid development of the field, it is impossible to give a broad overview that will do justice to any of these simulations, thus we focus here on some suggestive example of recent work.

3.1. Membrane Phenomena Studied Recently by Molecular Simulations

The first computer simulations of lipid bilayers were performed on single components lipid bilayers, and only recently computer simulations have been used to investigate lipid mixtures. One of the first simulations of many-component lipid mixtures was a model of a mixture of cholesterol with phosphatidylcholine (PC) and sphingomyelin (SM) lipids by Niemelä et al. *(28)*. This simulation of a lipid raft like mixture increases the biological relevance of the simulation models and is particularly interesting from the point of view of the formation of biologically relevant lipid domains rich in cholesterol. Lipid rafts *(29–31)* are predicted to form spontaneously in lipid mixtures containing sphingomyelin and cholesterol, in addition to phospholipids. They are characterized by distinct structural, dynamic, and elastic properties, e.g., tighter packing, slower diffusion, and a distinct lateral pressure profile. The formation of rafts has been proposed to have important effects on cellular functions, like hindering the diffusion of lipid signal transduction molecules within some distinct parts of the cellular membranes, or affecting membrane protein activity.

In recent years, transformations between different phases of lipids aggregates have been investigated with computational tools. More than 5 years ago the first simulation of a lipid cubic phase was published *(32)*. The selected phase was diamond subtype of the cubic phase, and the lipid of choice was GMO (mono-oleyl-glycerol). In the simulations, the starting structure was built based on ideal cubic phase geometries. The model was then used to simulate a phase transition from cubic phase to inverted hexagonal phase *(33)*. The simulation revealed many details of the mechanism of the phase transition that were not accessible experimentally.

Subsequent studies have revealed many phase transitions in other lipids. For example, a simulation of dipalmitoyl-phosphatidylcholine (DPPC)/palmitic acid two-component lipid system revealed a phase transition from a gel phase to inverted hexagonal phase *(34)*. The temperature of the transition was in good agreement with previous experimental results. Another recently published survey *(35)* illustrates the effect of LPA (lyso-phosphatidate, a phospholipid with only one hydrocarbon chain in *sn*-1 position) on phase transition in DOPA/DOPE (dioleyl-phosphatidate/dioleyl-phosphatidyl-ethanolamine) mixtures. It was shown that the addition of LPA lipids (which have a large head group) to a bilayer consisting mostly of PE lipids (having small head groups) hinders the transition from planar bilayer to hexagonal phase.

Computer simulations have been recently used to characterize the effect of peroxidation on the properties of lipid membranes as well. A recent molecular dynamics study focused on the effect of peroxidation on PLPC (palmitoyl-linoleyl-phosphatidylcholine) *(36)*. Four common oxidation products were selected and for each oxidation product a number of simulations were performed with different concentrations of oxidized lipids. The presence of oxidized products altered significantly structural and dynamic properties of the lipid membrane, and a significant increase in water permeability was observed, in agreement with experimental results. What experiments could not tell is the molecular origin of these changes. Through the simulations it was possible to show that all the changes in macroscopic properties of the membrane are related to a conformational change in the oxidized lipid tails: the oxidized part of the acyl chain turns toward the water molecules and polar lipid head groups. This is due to the presence of oxygen atoms in the oxidized acyl chain. It was hypothesized that the high degree of oxidation can lead to cellular damage through an increase in membrane permeability.

Recently, computer simulations of the formation of lipid vesicles have been reported. Vesicles are crucial vehicles in cellular transport of proteins and other molecules within living cells, as well as in exporting molecules from cells (exocytosis) and taking molecules into cells (endocytosis). Knecht et al. modeled the initial stages of vesicle fusion: the starting point of the simulation was the moment when the fusing vesicles first get in contact with each other *(37)*. The simulation led to a hemifused state, where the outer monolayers of the vesicle membranes were already fused but the inner monolayers were still separate. Information on the mechanism of vesicle fusion and on the structure of the fusion intermediates is extremely difficult to obtain experimentally.

One of the most important topics in the computer simulation field at the present time is the modeling of membrane-bound proteins. Membrane proteins are difficult to crystallize because most of them need the lipid environment to preserve their native, functional conformation, and lipids (in their physiologically relevant state) are highly flexible. Even when crystallization is possible, the resolution of the X-ray data is usually quite low. NMR experiments on membrane proteins are also very difficult to the relatively large size of lipid–protein complexes. Structural information on the membrane proteins often derives from experiments involving paramagnetic or fluorescent labels, or from crosslinking, and is usually fragmentary and largely incomplete. Simulation of membrane proteins *(38)* can provide a significant aid in the interpretation of experimental data regarding protein structure, dynamics, and functioning.

One area in which computer simulations are an invaluable tool is the prediction of the membrane-spanning part of a membrane protein. This prediction can be assessed using a range of experimental techniques (e.g., site-directed spin labeling, cysteine-scanning mutagenesis, tryptophan-scanning mutagenesis with fluorescence spectroscopy and two-dimensional infrared spectroscopy). The main drawback in this case is the high cost of the experimental procedures, which at the moment cannot be considered "high-throughput." In a recent study, Sansom et al. *(39)* proposed a very simple procedure to predict the position of the protein in the membrane, using coarse-grained molecular dynamics simulations. The procedure consists in running simulations starting from the crystal structure of the protein in the presence of a random distribution of lipids in water. During the simulations, spontaneous self-assembly of a lipid bilayer around the protein is observed.

Another area in which computer simulations are proving to be useful is the simulation of the mechanism of functioning of ion and water channel proteins. In many cases, crystal structures for these proteins are available only in the open or in the closed state. In other cases, both the open and closed state structures are available, but still the mechanism of conformational change is not known. Molecular dynamics can be used to explore possible mechanisms that can be tested in experiments. Marrink et al. simulated the opening of a mechanosensitive channel *(40)* using coarse-grained MD simulations. To trigger the opening of the channel, a high negative pressure was applied in the membrane plane. The process, which is thought to take place in vivo on the microsecond time scale, was simulated using relatively modest computational resources.

3.2. Limitations of Computer Simulations of Lipids

In principle, all the properties of a system that depend on positions and velocities of the atoms can be calculated from a trajectory, using statistical mechanics tools. In practice, a number of limitations are present in all MD simulations. The limitations are related to two aspects: the accuracy in the calculation of the forces and the time and length scale of the trajectory.

Forces in molecular systems are calculated using a so-called "force field." Force fields are simplified descriptions of the interactions between atoms and contain a number of approximations. There are many ways to check that the approximations contained in a force field yield a realistic description of the system we want to characterize. In the case of lipid simulations, a number of quantities can be calculated from computer simulations and directly compared to experimental results: for example, the structure factors derived from X-ray scattering, the electron density profile, the order parameter of the acyl chains, and the average area per lipid molecule. Besides structural features, also dynamic properties can be compared to experiments: for example, the diffusion coefficient of the lipids and the permeation rate of a certain molecule.

Comparison of simulations with experiments is used not only to validate a simulation force field, but sometimes also in the development of the force field itself. Once a certain force field has been shown to reproduce realistically some quantities experimentally measured, we can expect it to perform equally well in predicting properties that have not been measured, provided that simulation conditions are similar. For example, if a certain model reproduces structural and dynamic properties for a given lipid X, for which experimental data is available, then we can expect that it will predict reasonably well the properties of another lipid Y, especially when lipids X and Y are chemically similar.

One problem with calibrating a force field based on experimental data is that many experimental data require some interpretation. For example, in lipid bilayers the area per molecule is not a direct experimental measure but require multiple measures and a theoretical model for their interpretation. It is worth noticing that the area per lipid of a DPPC bilayer, probably the best characterized of all lipid bilayers, has changed many times during the last 25 years, ranging from 0.56 to 0.72 nm^2 *(41)*.

The limited size and time scale of the simulations is often referred to as the "sampling problem": to sample correctly motions that take place on a certain time scale, the simulations are required to be at least as long as the time scale of the phenomena, and typically much longer. Moreover, the size of the modeled systems should be larger than the length scale of the motion one wants to characterize. As a consequence, it is nowadays still difficult and sometimes impossible to use molecular simulations to study transformations and motions that take place on time scales beyond the microsecond and lengths beyond tens of nanometers. Unfortunately this involves many biologically interesting phenomena. Techniques are available to overcome some of these problems by accelerating molecular motions and transformations. The development of these techniques is a very active area in the field of molecular simulations.

4. Added Value of Visualization of Lipids

Everybody who has taught a course or given lectures in a conference knows that the more senses are attracted, the better the audience will understand the core issues of the lecture. Plain speech may contain all the relevant information, but its pedagogical value is limited. If complemented by pictures, the talk will surely attract more attention and facilitate understanding. Further, if the talk contains animations instead of still pictures, the contact between the speaker and the audience will be even better. Nowadays, animations of complex biomolecular systems such as cells including music or narration [http://multimedia.mcb.harvard.edu/media.html] are also becoming more and more common, and it is likely that in the near future flavor and sense of touch will also be part of teaching and conference presentations. In amusement parks showing three-dimensional movies of funny rides along the river, this is common practice already.

In the spirit of the above, it is evident that visualization of complex biological processes has added value. In experiments, this is usually done by marking individual molecules by fluorescent probes, which after excitement by laser will rapidly decay and emit light at some characteristic wavelength, which is visible and can correspond to any of the colors that our eyes can recognize. In studies of biological and model membranes, this is one of the most common techniques to follow in real time the dynamics of lipids and proteins in the membrane plane *(2)*, including both nano-scale phenomena such as protein complex formation and large-scale processes such as membrane domain formation. In studies of cells, fluorescent probes are constantly used to follow the trafficking of individual molecules such as cholesterol *(42)*, or a number of molecules involved in diseases such as cancer *(43)*.

In molecular simulations, the system of interest can be described by a number of ways, the most extensive one being the trajectory, that is, the coordinates of all atoms as a function of time. Visualization of the trajectory provides one with a fascinating possibility to look at and better understand how individual molecules are involved in complex many-particle processes. Important for appreciating this seemingly simple task is the fact that experiments can not yield similar insight. The nano-scales associated with molecular phenomena are beyond the resolution of most experimental techniques. Crystal structures usually offer atomistic detail, but they describe the state of a system during crystallization and provide little information on the dynamics. Visualization of simulation data, on the other hand, allows one to see all processes that occur in the first place.

In practice, the atomic coordinates resulting from a simulation can be used to produce a video clip of the migration of the atoms and molecules in the simulated system. The velocities of each molecule can be visualized as well to show molecular flows and velocity fields. Equally easily one can visualize the strength of forces, pressure, and tension exerted on any part of the system. One can also focus on a small set of molecules involved in a complex reaction and forward the structure of this complex to quantum mechanical simulations to characterize the reaction mechanism. Any part of the system can be highlighted by coloring the atoms and molecules based on any of the molecular properties.

Essentially, molecular graphics tools help the computational scientist to find the most interesting processes taking place in a system, prior to any quantitative analysis. Also, once the analysis has been completed, visualization provides a means to demonstrate the key processes in a manner that is as clear and concise as possible, allowing one to focus on results that are most central to understanding the phenomenon at hand.

In the following, we provide a few examples of how visualization of molecular simulations can be used to illustrate important concepts about the structure and dynamics in lipid systems.

4.1. Morphology of the Ripple Phase Through Schematic Head-and-Tail Representations

The spatial arrangement of lipid molecules within lipid bilayer is highly dependent on the environmental factors such as temperature and salt concentration. For example, while cooling a lecithin bilayer, a fluid lipid bilayer transforms into a rippled form, where the lipids no longer reside on a plane but the bilayer forms a curvy structure (*see* **Fig. 2**). Molecular dynamics simulations of this system *(44)* yielded a major body of atomistic data describing the assembly of lipids in the ripple phase, complementing experiments which provide insight of similar systems over larger scales.

The simulation approach not only recreated the rippled bilayer structure seen in experiments but also revealed that the lipids in the different parts of the ripples differ distinctly in their phase behavior: in some parts of the ripple the lipids were in a disordered phase while other parts had highly ordered gel-like structures (*see* **Fig. 2**). By visualizing the lipids, one can easily see the change in the orientation of the molecules within the ripple structure. Domains of interdigitated, highly ordered acyl chains are seen alternating with domains of non-interdigitated, disordered chains. The simulations predict the ripple phase to consist of separate, alternating domains of ordered and disordered lipids, highlighted by visualization that suggests a number of ways to analyze this nano-scale phase separation mechanism in atomic detail.

4.2. Visualization of Conformational Flexibility of Lipids Through Multiple Overlapping Snapshots

The degree of ordering and the flexibility of lipid acyl chains have a profound influence on the fluidity of lipid aggregates such as membranes. They also play a role in entanglement of lipid hydrocarbon chains, which in turn affects the diffusion of lipids in membranes and even more so in polymer melts such as the core of lipid droplets. Suggestive information on ordering and flexibility can be readily gathered by visualization of computer simulation data, to be complemented by more detailed quantitative analyses once the overall picture has emerged through visualization.

One example of this is provided in a recent manuscript by Wong-Ekkabut et al. *(36)*. In this work, molecular dynamics simulations were used to investigate the effect of lipid peroxidation on

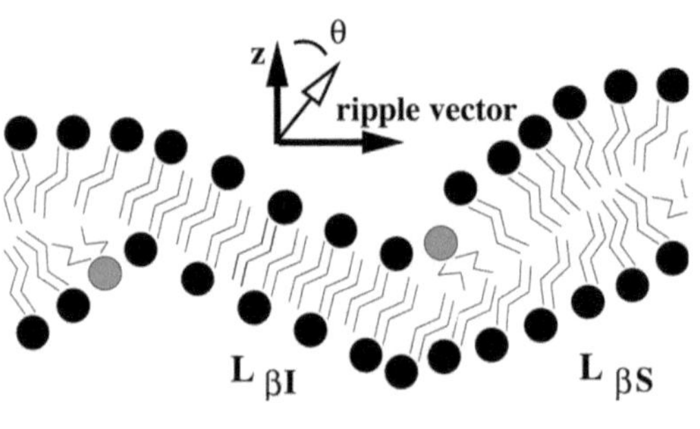

Fig. 2. Schematic representation of the organization of the lipids in the ripple structure. Adapted from reference *(44)*.

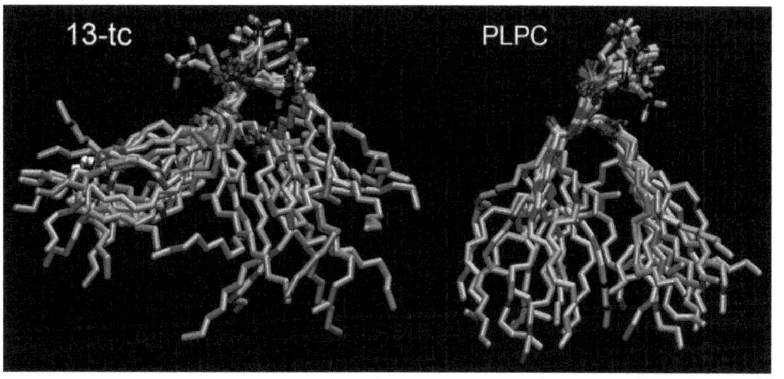

Fig. 3. Snapshots of a single PLPC and a peroxidized analogue (13-tc), taken at 5 ns intervals. Molecules are oriented along the membrane normal and superimposition was done on the phosphorus and oxygen atoms. Adapted from reference *(36)*.

the properties of model membranes with different concentrations of peroxidized lipids. It was found that the oxidized lipids form a thinner bilayer with increased permeability of water. This effect could let the small polar, or even ionized, molecules to "leak" through the membranes compromising the vitally important ion permeability barriers.

Visualization of the simulation data shows that the oxidized acyl chains are much less ordered than the unoxidized ones, and the oxidized phospholipids tend to occupy a larger cross-sectional area in the plane of the membrane. This effect is clearly illustrated in **Fig. 3**, showing how the high degree of flexibility of the oxidized lipids is determined by the presence of oxygen atoms in the acyl chain. These oxygen atoms show a preference for the polar–apolar interface, knocking the lipids aside in such a way that their area in the bilayer plane is bound to increase. The figure also demonstrates that the conformations thermally accessible to the oxidized lipids are more numerous and more diverse compared to nonoxidized lipids. Thus, without any quantitative analysis, using insight gained through visualization only, one can conclude that the enhanced flexibility and increasingly disordered nature of membrane regions containing oxidized lipids leads to an increased water permeability of the lipid membrane, a decreased thickness and increased area per lipid.

4.3. Visualization of Lipid Domains Through Density-Dependent Coloring

Lipid membranes are characterized by membrane heterogeneity and domain formation. The biological relevance of membrane domains is beautifully captured by the so-called lipid raft model *(29)*, which essentially concludes that proteins embedded in lipid systems are not working alone but in a close coupling with the lipids surrounding them. Consequently, a great deal of effort has been directed to elucidate the principles of domain formation and the dependence of domain properties on lipid composition.

One of the main obstacles to achieve these objectives has, although, been the scales associated with lipid domains. For instance, lipid raft domains have been proposed to be as small as a few nanometers in size, which implies that traditional experimental techniques employing fluorescent probes to characterize domain structures are not sufficient.

Visualization of simulation data can make a difference in this regard. Recent coarse-grained simulation studies by Marrink et al. *(45)* have brought about beautiful images of rafts, allowing better understanding of phenomena that need to be analyzed quantitatively to characterize domain properties in detail. Marrink et al. used this insight in their study with success.

Meanwhile, an equally interesting topic of research is lipid monolayers often studied through experiments. In addition, monolayers of lipids constitute the lipid entity found in the lung surfactant, providing us with a nice excuse to present some appealing related computational work.

In a recent study, Baoukina et al. *(46)* calculated the pressure–area isotherm of a DPPC monolayer using molecular dynamics simulations, finding intriguing support for domain formation in monolayer systems. The authors reported that, in addition to the different phases in different regions of the pressure–area isotherm, in one region of the isotherm two different phases coexisted in the monolayer: a liquid condensed and a liquid expanded phase. In the simulations, multiple small domains were formed and subsequently merged, to finally yield two large regions with different properties. Since one of the main differences between the liquid condensed and the liquid expanded phase is the surface density of the lipids, the authors represented the formation and time evolution of the coexisting phases by using different colors for different surface densities in the monolayer, as represented in **Fig. 4**.

4.4. Visualization of Phase Transformations via Multiple Snapshots

In addition to the most common fluid–gel (main) phase transition, lipid aggregates are characterized by a number of other phase transformations involving rather complex morphological changes. To describe the complexity of these structural changes would be extremely difficult without the aid of molecular visualization tools.

In a recent study, Marrink et al. investigated the phase behavior of a binary mixture of DOPC (dioleoylphosphatidylcholine) and DOPE (dioleoylphosphatidyl-ethanolamine) *(6)*. The mixture underwent transformations from a lamellar phase to either an inverted hexagonal or a rombohedral phase, depending on the temperature, lipid composition, and lipid hydration state. **Figure 5a** shows a multilamellar stack of pure DOPE bilayers, which is stable at low temperature (273 K). When the temperature is raised to 308 K, the lamellar structure becomes unstable **(Fig. 5b)** and transforms into an inverted hexagonal phase **(Fig. 5c)** within a

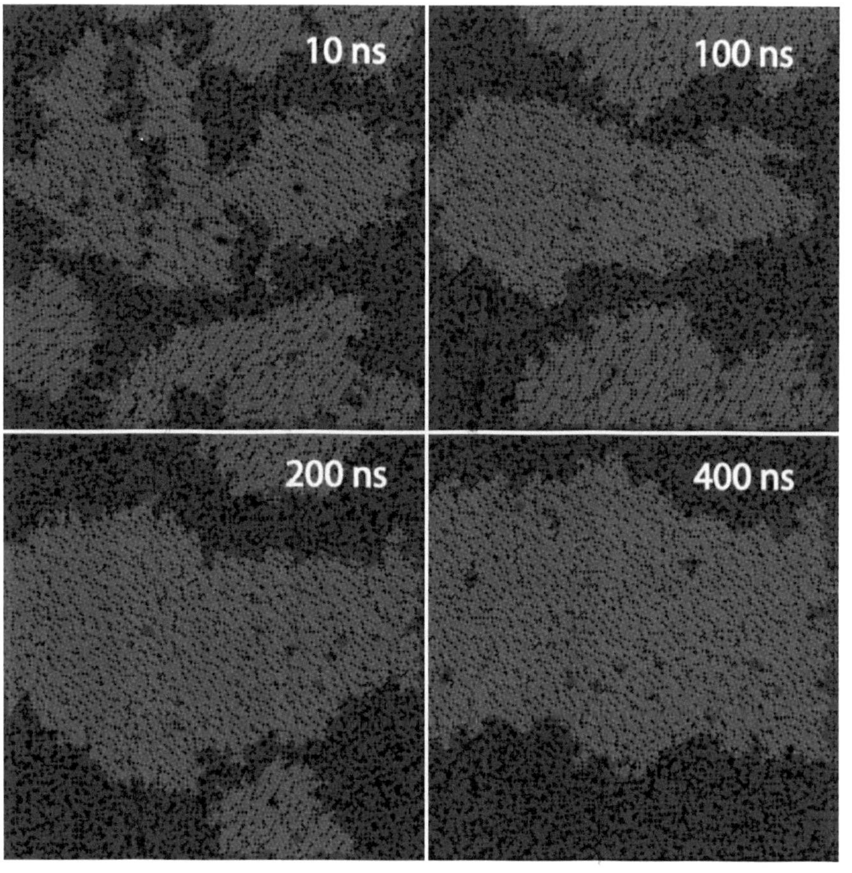

Fig. 4. Coexistence of liquid-expanded (darker color) and liquid-condensed (lighter color) phases in a DPPC monolayer at an area per lipid of 0.51 nm^2. Distribution of lipids between domains is obtained using Voronoi analysis. The simulation time is reported in the top right corner of each snapshot. Snapshots taken at different simulations times show the evolution of the domains. Adapted from reference ***(46)***.

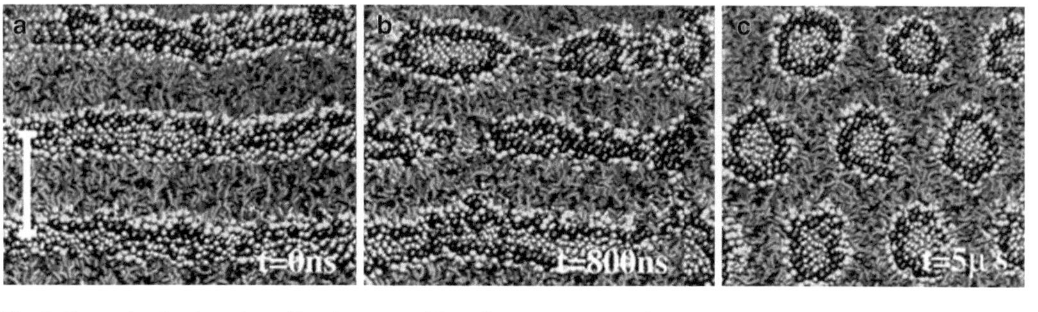

Fig. 5. Thermotropic phase transition from a multilamellar to an inverted hexagonal phase for DOPE at low hydration (nine water molecules per lipid). Closeups of cross sections perpendicular to a system of three independent bilayers of 512 lipids each. The multilamellar stack of pure DOPE bilayers (**a**) is stable at T = 273 K. On increasing the temperature to $T = 308$ K, stalks form (**b**), rearranging themselves into a hexagonal lattice and forming an inverted hexagonal phase (**c**). Adapted from reference ***(6)***.

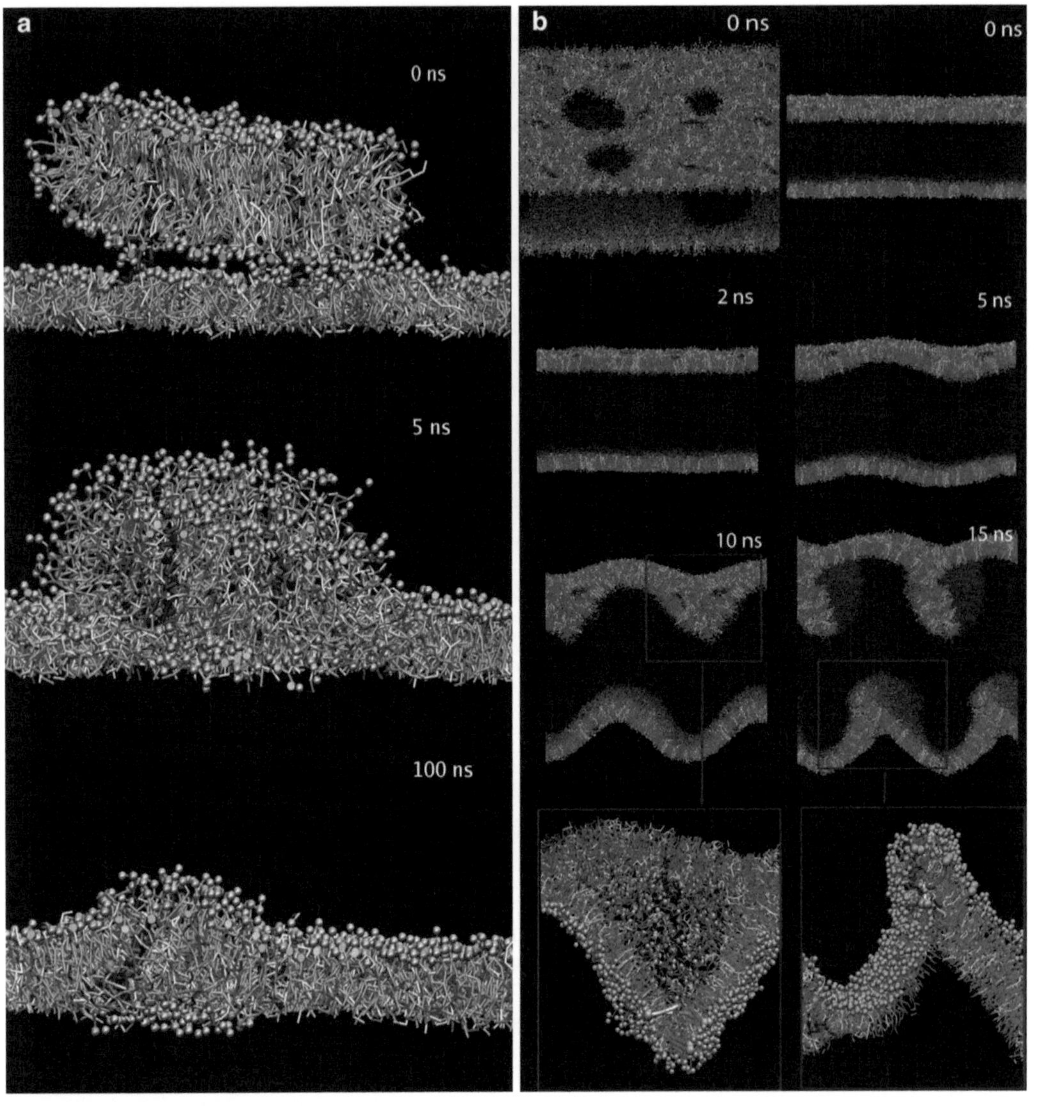

Fig. 6. (**a**) Adsorption of a bilayer patch in the pore region of an expanded monolayer (water not shown) after 0 ns, 5 ns, and 100 ns of MD simulation. (**b**) Collapse of the lung surfactant monolayer and formation of bilayer folds in the water subphase. On compression, an expanded monolayer with pores (0 ns, top and side views) forms a liquid-condensed phase (2 ns); the surface of the monolayer buckles (5 ns); the buckles grow in amplitude (10 ns); and the monolayer folds to form a bilayer (15 ns). Adapted from reference *(8)*.

few microseconds. In the inverted hexagonal phase, water forms "tubes" surrounded by lipids with a hexagonal arrangement. A similar phase transformation was also observed at low temperature (273 K) when the hydration level of DOPE was reduced to four water molecules per lipid. The figure shows in a nutshell how visualization can easily capture the essence of a very complex process, in this case the transformation from one phase to another including substantial change in morphology, which is difficult to describe with words.

Another point of view to highlight the added value of visualization in understanding complex lipid phase transition processes is provided by the study by Baoukina et al. *(8)*. In **Fig. 6a**, the authors illustrate the process of adsorption of a bilayer patch onto a lipid monolayer at the air–liquid interface. At very high surface tensions, the bilayer patch undergoes complete fusion with the monolayer at the air–water interface during a 100 ns simulation. **Figure 6b** further illustrates the transformation of a lung surfactant model membrane on lateral compression. As the lateral pressure increases, the surface density of the lipids increases and the surface tension drops. When the surface tension at the interface becomes negative, the monolayer is not stable and collapses, forming a bilayer in the water phase. The process starts with buckling of the monolayer, as illustrated in the figure.

While the final result of monolayer compression is observable through experiments, the mechanism of fusion of the bilayer patch and the mechanism of monolayer folding are not accessible with current experimental techniques. This is largely due to the combination of time and space resolution needed to characterize processes and patterns involved in these complex processes.

5. Visualization Software Packages

A large variety of molecular visualization programs is available. Some of these programs are freeware, and some of them are free for academic users. We hereby report a very brief description of the basic features of a few molecular graphics software packages that are appropriate for particle-based simulations: VMD, Maestro, Pymol, RasMol, USCF Chimera, gOpenMol, and Molden. This list of programs is far from complete, and our choice for the selection was guided by two criteria: the widespread use in scientific publications and their availability for Linux, Mac OS X and Microsoft Windows operating systems.

RasMol (http://rasmol.org/), USCF Chimera (www.cgl.ucsf.edu/chimera), gOpenMol (www.csc.fi/english/pages/g0penMol), and Molden (www.cmbi.ru.nl/molden/molden.html) are handy in the daily work of the computer simulation scientist: they are freeware or free for academic users and they can be used to perform small tasks regarding visualization and structural analysis. They provide basic representations of atoms and molecules (e.g., ball and stick or van der Waals representations), can display atom names and numbers and atomic coordinates, and can calculate interatomic distances and bond angles. The requirements for hardware (in terms of graphic processing unit and memory) are minimal, so they can easily be used on any personal computer or laptops.

Pymol, Maestro, and VMD are more sophisticated programs, including a broad range of graphical representations for most biological macromolecules and capable of performing simple analyses on the structures displayed.

Pymol (www.pymol.org) is free for academic users. It is intended for molecular representation of biological macromolecules and small organic molecules, and it supports a large variety of molecular representations, including different kinds of molecular surfaces and molecular interactions, as well as volumetric data. It can be used both for the representation of still images and molecular animations.

Maestro (www.schrodinger.com) is also free for academic users. This software is particularly useful in building simulation models of complex organic molecules and biological macromolecules (DNA, proteins, polysaccharides, etc.) from simple building blocks (organic functional groups, amino acids, etc.). Moreover, it includes numerous functionalities allowing for complex molecular representations. Maestro has a handy GUI-based interface and can be used to produce high quality graphics.

VMD *(47)* (www.ks.uiuc.edu/Research/vmd) is freely available to all users. The visualization procedure is very intuitive and flexible. The program uses graphical interface through which the visualization commands are written, but the command line can be used as well. This is particularly convenient when complex instructions need to be repeated on a large set of molecular structures. VMD can be used to visualize both individual structure files and trajectories, and can produce illustrative movies directly from simulation trajectory files. It can also be used for structural analysis: for example, to calculate distances, angles and dihedrals within molecules, calculate and represent the secondary structure of proteins, calculate the deviation of a series of structures from each other, and so on.

6. Concluding Remarks

Computer simulations of lipid membranes have become a very common tool in the study of the structure and dynamics of biological systems. They complement experimental techniques by providing information with high spatial and temporal resolution, which is often not easily accessible to experiments. Their use is normally coupled with molecular visualization software, yielding an easy and powerful way to describe a broad range of complex phenomena: lipid self-assembly, vesicle fusion, phase transformations, and permeation through biological membranes, just to mention a few examples. In the present work we have reviewed aspects of computer simulation methodology and a few examples of problems that can presently be tackled using molecular simulations. We also have showed examples of how molecular graphics software can be used to simplify the description of structural properties and the dynamic transformations of lipid systems. Considering the continuous progress in computer technology, both simulations and visualization are expected to play an increasingly important role in the characterization of complex lipid systems.

References

1. Mouritsen OG. Life - as a matter of fat: Springer; 2004.
2. Bagatolli LA. (2006) To see or not to see: Lateral organization of biological membranes and fluorescence microscopy *Biochim Biophys Acta* 1758, 1541–56.
3. Marrink SJ, de Vries AH, Tieleman DP. (2009) Lipids on the move: Simulations of membrane pores, domains, stalks and curves *Biochimica et Biophysica Acta-Biomembranes* 1788(1), 149–168.
4. Marrink SJ, Tieleman DP, Mark AE. (2000) Molecular dynamics simulation of the kinetics of spontaneous micelle formation *J Phys Chem B* 104, 12165–73.
5. Marrink SJ, Lindahl E, Edholm O, Mark AE. (2001) Simulation of the spontaneous aggregation of phospholipids into bilayers *J Am Chem Soc* 123, 8638–9.
6. Marrink SJ, Mark AE. (2004) Molecular view of hexagonal phase formation in phospholipid membranes *Biophys J* 87, 3894–900.
7. Marrink SJ, Mark AE. (2003) Molecular dynamics simulation of the formation, structure, and dynamics of small phospholipid vesicles *J Am Chem Soc* 125, 15233–42.
8. Baoukina S, Monticelli L, Amrein M, Tieleman DP. (2007) The molecular mechanism of monolayer-bilayer transformations of lung surfactant from molecular dynamics simulations *Biophys J* 93, 3775–82.
9. Duncan SL, Larson RG. (2008) Comparing experimental and simulated pressure-area isotherms for DPPC *Biophys J* 94, 2965–86.
10. Duan Y, Wu C, Chowdhury S, et al. (2003) A point-charge force field for molecular mechanics simulations of proteins based on condensed-phase quantum mechanical calculations *J Comp Chem* 24, 1999–2012.
11. MacKerell AD, Bashford D, Bellott M, et al. (1998) All-atom empirical potential for molecular modeling and dynamics studies of proteins *J Phys Chem B* 102, 3586–616.
12. Daura X, Mark aE, Van Gunsteren WF. (1998) Parametrization of aliphatic CHn united atoms of GROMOS96 force field *J Comput Chem* 19, 535–47.
13. Oostenbrink C, Villa A, Mark AE, Van Gunsteren WF. (2004) A biomolecular force field based on the free enthalpy of hydration and solvation: The GROMOS force-field parameter sets 53A5 and 53A6 *J Comput Chem* 25, 1656–76.
14. Jorgensen WL, Tirado Rives J. (1988) The OPLS potential functions for proteins - energy minimizations for crystals of cyclic peptides and crambin *J Am Chem Soc* 110, 1657–66.
15. Jorgensen WL, Maxwell DS, TiradoRives J. (1996) Development and testing of the OPLS all-atom force field on conformational energetics and properties of organic liquids *J AmChem Soc* 118, 11225–36.
16. Kaminski GA, Friesner RA, Tirado-Rives J, Jorgensen WL. (2001) Evaluation and reparametrization of the OPLS-AA force field for proteins via comparison with accurate quantum chemical calculations on peptides *J Phys Chem B* 105, 6474–87.
17. Feller SE, MacKerell AD. (2000) An improved empirical potential energy function for molecular simulations of phospholipids *J Phys Chem B* 104, 7510–5.
18. Berger O, Edholm O, Jahnig F. (1997) Molecular dynamics simulations of a fluid bilayer of dipalmitoylphosphatidylcholine at full hydration, constant pressure, and constant temperature *Biophys J* 72, 2002–13.
19. Smit B, Hilbers PAJ, Esselink K, Rupert LAM, Vanos NM, Schlijper AG. (1990) Computer simulations of a water oil interface in the presence of micelles *Nature* 348, 624–5.
20. Ayton GS, Noid WG, Voth GA. (2007) Multiscale modeling of biomolecular systems: in serial and in parallel *Curr Opin Struct Biol* 17, 192–8.
21. Shelley JC, Shelley MY, Reeder RC, Bandyopadhyay S, Klein ML. (2001) A coarse grain

model for phospholipid simulations *J Phys Chem B* 105, 4464–70.

22. Shelley JC, Shelley MY, Reeder RC, Bandyopadhyay S, Moore PB, Klein ML. (2001) Simulations of phospholipids using a coarse grain model *J Phys Chem B* 105, 9785–92.
23. Murtola T, Falck E, Patra M, Karttunen M, Vattulainen I. (2004) Coarse-grained model for phospholipid/cholesterol bilayer *J Chem Phys* 121, 9156–65.
24. Marrink SJ, de Vries AH, Mark AE. (2004) Coarse grained model for semiquantitative lipid simulations *J Phys Chem B* 108, 750–60.
25. Marrink SJ, Risselada HJ, Yefimov S, Tieleman DP, de Vries AH. (2007) The MARTINI forcefield: coarse grained model for biomolecular simulations *J Phys Chem B* 111, 7812–24.
26. Monticelli L, Kandasamy SK, Periole X, Larson RG, Tieleman DP, Marrink SJ. (2008) The MARTINI coarse-grained force field: extension to proteins *J Chem Theory Comput* 4, 819–34.
27. Wong-Ekkabut J, Baoukina S, Triampo W, Tang IM, Tieleman DP, Monticelli L. (2008) Computer simulation study of fullerene translocation through lipid membranes *Nature Nanotech* 3, 363–8.
28. Niemela PS, Ollila S, Hyvonen MT, Karttunen M, Vattulainen I. (2007) Assessing the nature of lipid raft membranes *PLoS Comput Biol* 3, 304–12.
29. Simons K, Ikonen E. (1997) Functional rafts in cell membranes *Nature* 387, 569–72.
30. Ikonen E, Simons K. (1998) Protein and lipid sorting from the trans-Golgi network to the plasma membrane in polarized cells *Seminars In Cell & Developmental Biology* 9, 503–9.
31. Simons K, Ikonen E. (2000) Cell biology - How cells handle cholesterol *Science* 290, 1721–6.
32. Marrink SJ, Tieleman DP. (2001) Molecular dynamics simulation of a lipid diamond cubic phase *J Am Chem Soc* 123, 12383–91.
33. Marrink SJ, Tieleman DP. (2002) Molecular dynamics simulation of spontaneous membrane fusion during a cubic-hexagonal phase transition *Biophys J* 83, 2386–92.
34. Knecht V, Mark AE, Marrink SJ. (2006) Phase behavior of a phospholipid/fatty acid/water mixture studied in atomic detail *J Am Chem Soc* 128, 2030–4.
35. May ER, Kopelevich DI, Narang A. (2008) Coarse-grained molecular dynamics simulations of phase transitions in mixed lipid systems containing LPA, DOPA, and DOPE lipids *Biophys J* 94, 878–90.
36. Wong-ekkabut J, Xu Z, Triampo W, Tang IM, Tieleman DP, Monticelli L. (2007) Effect of lipid peroxidation on the properties of lipid bilayers: a molecular dynamics study *Biophys J* 93, 4225–36.
37. Knecht V, Marrink SJ. (2007) Molecular Dynamics Simulations of Lipid Vesicle Fusion in Atomic Detail *Biophys J* 92, 4254–61.
38. Ash WL, Zlomislic MR, Oloo EO, Tieleman DP. (2004) Computer simulations of membrane proteins *Biochim Biophys Acta - Biomembranes* 1666, 158–89.
39. Scott KA, Bond PJ, Ivetac A, Chetwynd AP, Khalid S, Sansom MSP. (2008) Coarse-grained MD simulations of membrane protein-bilayer self-assembly *Structure* 16, 621–30.
40. Yefimov S, van der Giessen E, Onck PR, Marrink SJ. (2008) Mechanosensitive membrane channels in action *Biophys J* 94, 2994–3002.
41. Nagle JF, Tristram-Nagle S. (2000) Structure of lipid bilayers *Biochim Biophys Acta* 1469, 159–95.
42. Ikonen E. (2008) Cellular cholesterol trafficking and compartmentalization *Nature Rev Mol Cell Biol* 9, 125–38.
43. Hoffman RM. (2005) The multiple uses of fluorescent proteins to visualize cancer in vivo *Nature Rev Cancer* 5, 796–806.
44. de Vries AH, Yefimov S, Mark AE, Marrink SJ. (2005) Molecular structure of the lecithin ripple phase *Proc Natl Acad Sci USA* 102, 5392–6.
45. Risselada HJ, Marrink SJ. (2008) The molecular face of lipid rafts in model membranes *Proc Natl Acad Sci USA* 105, 17367–72.
46. Baoukina S, Monticelli L, Marrink SJ, Tieleman DP. (2007) Pressure-area isotherm of a lipid monolayer from molecular dynamics simulations *Langmuir* 23, 12617–23.
47. Humphrey W, Dalke A, Schulten K. (1996) VMD: visual molecular dynamics *J Mol Graph* 14, 33–8.

Chapter 19

Bioinformatics Strategies for the Analysis of Lipids

Craig E. Wheelock, Susumu Goto, Laxman Yetukuri, Fabio Luiz D'Alexandri, Christian Klukas, Falk Schreiber, and Matej Orešič

Summary

Owing to their importance in cellular physiology and pathology as well as to recent technological advances, the study of lipids has reemerged as a major research target. However, the structural diversity of lipids presents a number of analytical and informatics challenges. The field of lipidomics is a new post-genome discipline that aims to develop comprehensive methods for lipid analysis, necessitating concomitant developments in bioinformatics. The evolving research paradigm requires that new bioinformatics approaches accommodate genomic as well as high-level perspectives, integrating genome, protein, chemical and network information. The incorporation of lipidomics information into these data structures will provide mechanistic understanding of lipid functions and interactions in the context of cellular and organismal physiology. Accordingly, it is vital that specific bioinformatics methods be developed to analyze the wealth of lipid data being acquired. Herein, we present an overview of the Kyoto Encyclopedia of Genes and Genomes (KEGG) database and application of its tools to the analysis of lipid data. We also describe a series of software tools and databases (KGML-ED, VANTED, MZmine, and LipidDB) that can be used for the processing of lipidomics data and biochemical pathway reconstruction, an important next step in the development of the lipidomics field.

Key words: Bioinformatics, Lipid, Lipidomics, Pathway reconstruction, KEGG, KGML-ED, VANTED, MZmine, LipidDB

1. Introduction

The complexity and challenges involved in the study of lipids is adequately demonstrated by the fact that there are numerous definitions of the term lipid. In its broadest sense, "lipid" defines substances as oils, fats, and waxes that can only be characterized by a large array of properties (http://www.cyberlipid.org/). They have

Donald Armstrong (ed.), *Lipidomics*, Methods in Molecular Biology, vol. 580,
DOI 10.1007/978-1-60761-325-1_19, © Humana Press, a part of Springer Science+Business Media, LLC 2009

been loosely defined as biological substances that are in general insoluble in water, but soluble in organic solvents. An alternative definition states that lipids are hydrophobic or amphipathic small molecules that may originate entirely or in part by carbanion-based condensations of thioesters and/or by carbocation-based condensations of isoprene units *(1)*. Lipids possess a number of vital functions including serving as structural components of membranes, energy storage, and signal transduction molecules *(2)*. The study of lipids is necessary for the understanding of cellular physiology and metabolism as well as the etiology of numerous pathologies *(3–6)*. Recent advances in analytical technologies have made it possible to study lipids at different levels using targeted and global screening strategies *(7, 8)*. The rapid increases in acquisition of lipid data have resulted in a concomitant need for bioinformatics resources capable of analyzing this wealth of data *(9, 10)*. In particular, the structural diversity of lipids creates a number of unique obstacles in performing bioinformatics analyses and biochemical pathway reconstruction. Accordingly, it is necessary that developments in bioinformatics match the analytical developments in a number of areas including: (1) raw data processing, (2) efficient storage and data management, (3) linking of metadata with experimental data, and (4) overall integration with other omics level information in biochemical pathways and networks *(11, 12)*.

The complexity of lipid nomenclature represents another unique challenge for both the lipid bioinformatician as well as the experimentalist. For example, the lack of systemic names for many lipid species renders it difficult to perform comprehensive literature searches. Efforts have been made to develop a new ontology to cope with the complications of the lipid bibliosphere, but multiple challenges remain *(1, 12)*. A number of online resources are currently available for lipid classification and databasing as well as lipid-specific biochemical pathway analysis (**Table** 1). However, the field of "lipidomics" has not received the same level of attention as that of many other omics sciences. A major gap in the field is the lack of tools to relate fluctuations in lipids to phenotypic changes and integrate this information into comprehensive biological networks with genomics and proteomics data. The recent emergence of community-wide efforts such as the LIPID MAPS consortium (http://www.lipidmaps.org/) *(13)*, European Lipidomics Initiative (http://www.lipidomics.net/) *(14)*, and similar developments in Japan (http://www.lipidbank.jp/) *(15)* clearly show the present need and interest for lipid research at the molecular level. Other lipid-centric databases such as LIPIDAT (http://www.lipidat.ul.ie/) focus on thermodynamic data and associated information on lipid phase transitions.

LipidBank is the official database of the Japanese Conference on the Biochemistry of Lipids (JCBL) and provides information

Table 1
Online resources for lipid classification and pathway analysis[a]

Website	Description
Lipid classification and references	
http://www.lipidmaps.org/	Classification scheme and extensive tools
http://www.cyberlipid.org/	Description of lipid classes, with references
http://www.lipidlibrary.co.uk/	Extensive highly referenced information
http://lipidbank.jp/	Official database of Japanese Conference on the Biochemistry of Lipids
http://lipidsearch.jp/	Identification tool for phospholipid batch processing
http://www.massbank.jp/	High resolution mass spectral database
http://www.lipidat.ul.ie/	Lipid thermodynamic and phase transition information
http://www.chem.qmul.ac.uk/iupac/lipid/	IUPAC lipid nomenclature
Pathway analysis tools	
http://www.genome.jp/kegg/	KEGG biochemical pathways
http://vanted.ipk-gatersleben.de/	VANTED network and pathway visualization
http://kgml-ed.ipk-gatersleben.de/	Graphical KGML pathway editor
http://www.biocarta.com/	Molecular and cellular pathways
http://sphingolab.biology.gatech.edu/	Pathway map for sphingolipid biosynthesis
http://www.ingenuity.com/	Commercial pathway program
http://www.genego.com/	Commercial pathway program

[a]There are many additional pathway analysis tools besides those listed here. An overview of tools for visualizing biological networks is provided by Suderman and Hallett *(29)*

on identified natural lipids including fatty acids, glycerolipids, sphingolipids, steroids, and various vitamins. Currently, LipidBank contains over 6,000 molecules that are classified into 26 groups *(16)*. The database contains molecular structures (ChemDraw and MOL format), lipid names (common and IUPAC names), spectral information (molecular mass, UV, IR, NMR, and other if available), and most importantly, literature information that reports lipid identification. All molecular information has been manually curated and approved by experts in lipid research (recorded as *Informant* of data in each entry). Recent advances in LipidBank include an active connection to MassBank (http://www.massbank.jp/), the Institute for Bioinformatics Research and Development (BIRD, http://www-bird.jst.go.jp/index_e.html), and Lipid Search (http://lipidsearch.jp/) *(15)*.

The LIPID MAPS consortium has created a number of useful resources for the bioinformatics analysis of lipids. Specifically, the new naming system that assigns a unique 12-character signature for biologically relevant lipids affords automated processing of lipidomics data *(1)*. The LIPID MAPS online suite of tools enables the drawing of lipid structures and prediction of possible structures from mass spectrometry data *(17)*. These tools are useful for providing conformity in the field regarding the generation of lipid structures and nomenclature. The LIPID MAPS structure database is a relational database that encompasses structures and annotations of biologically relevant lipids curated from a number of different sources *(18)*. The LIPID MAPS Proteome Database is an object-relational database of lipid-associated protein sequences and annotations *(19)*. Recently, the list of lipid-specific keywords was expanded, and in addition to GO and Kyoto Encyclopedia of Genes and Genomes (KEGG) term description, the description field of the UniProt records and the EntrezGene names were also scanned for human and mouse. Taken together, the LIPID MAPS suite of tools provides vital conformity in lipid nomenclature and structural analysis, addressing some of the major lipid-specific obstacles in the advancement of lipidomics research.

The logical next step in the field is the development of tools capable of integrating lipidomics data with gene and enzyme data to perform biochemical pathway reconstruction and flux analyses. The reconstruction of lipid pathways will require three main basic building blocks: (1) advanced automatic data processing software, (2) richly annotated lipid databases, and (3) strategies for pathway mapping of lipid data in a context-dependent manner. Given the structural diversity of lipid species, these tasks will be challenging and require a combination of novel and existing bioinformatics resources. In this chapter, we present a brief description of the KEGG database and focus on its suite of tools for applications in lipid-based analyses. In addition, the application of KGML-ED (KEGG Markup Language Editor) and VANTED (Visualization and Analysis of Networks containing Experimental Data), two specific tools for expanding the functionality of KEGG pathway diagrams, are presented. We then describe MZmine, a software package that enables the differential analysis of liquid chromatography mass spectrometry (LC/MS)-based lipidomics data, and the in silico spectra database LipidDB (a custom database of lipid structures). Finally, we present a strategy for combining MZmine with LipidDB and the KEGG PATHWAY database to perform lipid pathway reconstruction.

2. Using KEGG as a Resource for Lipid-Based Research

2.1. The Database Overview

The KEGG is a database of biological systems that integrates genomic, chemical, and network information (http://www.genome.jp/kegg/) *(20)*. KEGG was initiated in 1995 under the auspices of the Human Genome Program of the Ministry of Education, Science, Sports, and Culture of Japan *(21)*. The initial objective of KEGG was to develop computer-based methods for storing and analyzing the genomic information that was just beginning to be produced in large quantities [the first full genome of *Haemphilius influenzae* was produced by the TIGR center in 1995 *(22)*]. Since then, the suite of KEGG databases has greatly expanded and includes comprehensive and organism-specific metabolic and regulatory pathways as well as chemical compound data, disease-specific information and "omics" data analysis tools. As of August 2008, KEGG consists of 19 different component databases that are broadly categorized into three distinct areas: systems, genomic, and chemical information (**Table 2**). Systems information is the core of the KEGG databases, which link genomic and chemical information via pathway diagrams in the PATHWAY database and functional hierarchies of proteins and chemical compounds in the BRITE database. KEGG ATLAS is a new graphical interface for the PATHWAY and BRITE databases that consists of a global metabolism map with newly developed viewers. The MODULE and DISEASE databases are relatively recent additions to the KEGG resources that store functional modules and complexes in pathways, and disease-related information linking genes, pathways, and drugs, respectively.

The genomic information comprises functional annotation of genes from complete and draft genomes. EST contigs have also been added for some eukaryotic species whose genomes are not yet available. All these annotations are stored in GENES (complete genomes and manual functional annotation), DGENES (draft genomes with automatic annotation), and EGENES (EST contigs with automatic annotation). KO (KEGG Orthology) is a unified description of gene/protein functions that is used for linking genomic and pathway information. SSDB (Sequence Similarity Database) stores all-versus-all sequence similarities (Smith-Waterman scores) of genes in the GENES database, which enables the visualization of homologue relationships. KO and SSDB are the basis for functional annotation in KEGG.

For chemical information, we have created a composite database called LIGAND *(23, 24)* that consists of six individual databases: COMPOUND for chemical compounds and their 2-dimensional structures, DRUG for the medically related components of COMPOUND, GLYCAN for carbohydrate

Table 2
Databases in KEGG and lipid-related content[a]

Category	Database	Content (Lipid-related resources)	No. of Entries
Systems information	PATHWAY	Pathway maps (16 reference pathway maps, including a lipid structure map for unsaturated fatty acids; classified in "Lipid Metabolism")	93,644
	BRITE	Functional hierarchies ("Lipid Metabolism" in pathway and KEGG Orthology classification, and "Lipids" classification based on LIPID-MAPS)	15,687
	MODULE	Pathway modules (80 modules are defined for pathways in "Lipid Metabolism")	669
	DISEASE	Diseases	78
Genomic information	ORTHOL-OGY	KEGG Orthology (KO) groups (orthology groups for enzymes in "Lipid Metabolism")	11,094
	GENOME	Organisms in KEGG	876
	GENES	Genes in high-quality genomes with manual annotation (genes annotated via KO groups for lipid metabolism)	3,445,423
	DGENES	Genes in draft genomes with automatic annotation	454,577
	EGENES	Genes as EST contigs with automatic annotation	2,222,350
	VGENOME	Viral genomes in KEGG	3,167
	VGENES	Genes in viral genomes	65,289
	OGENES	Genes in organelle genomes	74,252
	SSDB	Sequence similarities with best hit relations	
Chemical information	COMPOUND	Metabolites and other chemical compounds ("Lipids" classified in BRITE hierarchy)[b]	15,179
	DRUG	Drugs	7,516
	GLYCAN	Glycans	10,966
	ENZYME	Enzymes	5,010
	REACTION	Enzymatic and other reactions in metabolisms	7,647
	RPAIR	Reactant pairs and chemical transformation	9,906

[a]Entry information is current as of August 14th, 2008
[b]Lipids in KEGG COMPOUND are cross-linked to both LIPID MAPS *(1)* and LipidBank *(16)*

structures represented by monosugars as building blocks rather than chemical atoms, ENZYME for enzyme nomenclature from IUBMB (International Union of Biochemistry and Molecular Biology), REACTION for metabolic reactions in PATHWAY, and RPAIR for reactant pairs in REACTION and their chemical transformation patterns. These individual databases constitute the main components of the PATHWAY database, which is discussed in greater detail below. Each component of the KEGG database suite is given a unique object identifier, enabling the nonredundant categorizing of all entries (**Table 3**).

KGML is an exchange format for KEGG graph objects, especially the KEGG pathway maps that are manually drawn and updated. KGML enables automatic drawing of KEGG pathways and provides facilities for computational analysis and modeling of protein networks and chemical networks. The KGML files for KEGG metabolic pathways contain two types of graph objects: boxes (enzymes) that are linked by a relation and circles (compounds) that are linked by a reaction in the KEGG pathway diagrams. In contrast, the KGML files for KEGG regulatory pathways contain only the aspect of boxes (proteins) that are linked by a relation. In the current KEGG system, boxes are identified by KO identifiers, but for historical reasons boxes in the metabolic pathways are marked with EC numbers in the actual pathway diagrams.

KEGG has been expanding its resources for lipid analysis and currently includes >1,300 individual lipid species (with dedicated

Table 3
KEGG Object Identifiers

Prefix	Content	Database
K	Gene/protein orthologue group	ORTHOLOGY
C	Chemical compound	COMPOUND
D	Drug	DRUG
G	Glycan	GLYCAN
R	Reaction	REACTION
(code)	Pathway map	PATHWAY
(code)	Hierarchical text file	BRITE
M	Pathway module	MODULE
H	Human disease	DISEASE
(code)	Organism	GENOME

KEGG object identifiers) and several sources of information useful for lipid-based bioinformatics analyses (**Table 2**). These lipids are organized according to 16 different biosynthetic pathways under the "Lipid Metabolism" menu of KEGG PATHWAY (**Table 4**). An overview of lipid metabolism in KEGG displaying the interactions between the different pathways is shown in **Fig. 1**. Individual lipid species in KEGG are organized into five distinct levels based on LIPID MAPS classification. The first level consist of eight different lipid groups: fatty acyls (FA), glycerolipids (GL), glycerophospholipids (GP), sphingolipids (SP), sterol lipids (ST), prenol lipids (PR), saccharolipids (SL), and polyketides (PK). The second level divides the lipid groups into distinct classes based on structural similarity; for example, the FAs include FA01 fatty acids and conjugates, FA02 octadecanoids, FA03 eicosanoids, FA04 docosanoids, FA05 fatty alcohols, FA06 fatty aldehydes, FA07 fatty esters, FA08 fatty amides, FA09 fatty nitriles, FA10 fatty ethers, FA11 hydrocarbons, FA12 oxygenated hydrocarbons, and FA00 other. The third level divides the classes into smaller groups, for example, FA03 eicosanoids

Table 4
Lipid metabolism pathways in KEGG

Fatty acid biosynthesis
Fatty acid elongation in mitochondria
Fatty acid metabolism
Synthesis and degradation of ketone bodies
Biosynthesis of steroids
Bile acid biosynthesis
C21-Steroid hormone metabolism
Androgen and estrogen metabolism
Glycerolipid metabolism
Glycerophospholipid metabolism
Ether lipid metabolism
Sphingolipid metabolism
Arachidonic acid metabolism
Linoleic acid metabolism
α-Linolenic acid metabolism
Biosynthesis of unsaturated fatty acids

being further divided into: FA0301 prostaglandins, FA0302 leukotrienes, FA0303 thromboxanes, FA0304 lipoxins, etc. The fourth and fifth levels break the smaller groups into individual compounds (e.g., C00427 Prostaglandin H2). Each individual lipid entity has a unique KEGG object identifier, which enables it to be analyzed using the range of KEGG tools.

2.2. Systems Information in KEGG

2.2.1. Browsing Pathways and Functional Hierarchies

The KEGG PATHWAY database provides biochemical pathway diagrams representing molecular wiring diagrams of interaction and reaction networks. It can be browsed from the KEGG PATHWAY entry point that is available by clicking the "KEGG PATHWAY" link on the KEGG homepage. In addition to metabolic pathways, PATHWAY provides protein interaction networks stored in the sections "Genetic Information Processing," "Environmental Information Processing," and "Cellular Processes." The "Human Diseases" section is also useful for examining protein interactions where proteins reported to be involved in the disease are highlighted. "Drug Development" stores another type of pathway diagram where the individual history of a drug's design is represented with the structural changes in the drug during the development process.

In the "Metabolism" menu of PATHWAY, each pathway diagram can be browsed by clicking the pathway name. In addition, the pathways shown in **Fig. 1** are interactive and can be expanded on by clicking on an individual lipid pathway. For example, **Fig. 2** shows the pathway diagram for arachidonic acid metabolism, where the boxes (arrows) and the circles represent enzymatic reactions and chemical compounds, respectively. The pathway diagram shown in **Fig. 2a** is a reference pathway that is created by combining information from the peer-reviewed literature for a range of organisms. Organism-specific pathways are automatically produced by coloring the corresponding enzymes based on the annotation of genes, as shown for human arachidonic acid metabolism in **Fig. 2b**. The help button at the upper right provides a detailed description of objects in the pathway diagram including protein interaction networks.

In the case of unsaturated fatty acid biosynthesis, a combination of elongases and desaturases are utilized for producing a wide-range of lipids with varying degrees of unsaturation and overall length. Because of this complexity, these structural combinations are represented differently from other KEGG metabolic pathway diagrams. The elongases and desaturases are drawn from top to bottom and from left to right, respectively, in a grid-like form. This network of enzymes can therefore account for the various synthetic processes that produce the range of unsaturated fatty acids. The enzymes involved in unsaturated fatty acid biosynthesis have been analyzed based on protein sequences and classified to show organism specificities *(25)*.

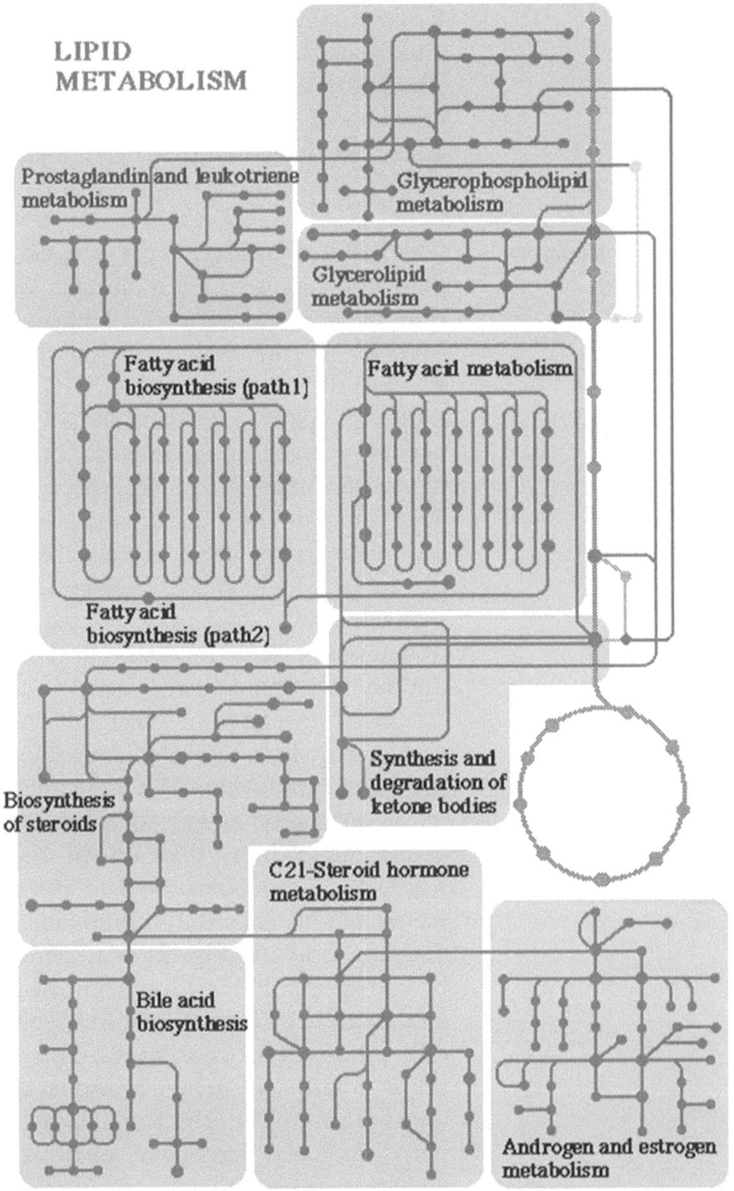

Fig. 1. KEGG reference pathway for lipid metabolism.

The pathway diagrams in the PATHWAY database are classified in hierarchies. Enzymes can be classified according to the pathway classification if they are drawn in pathway diagrams. However, some enzymes and proteins are also assigned specific functions and can be classified according to function. The BRITE database is a collection of such functional hierarchies for proteins and other molecules, and works as a supplement to the PATHWAY database. In addition to the network hierarchies, protein

a

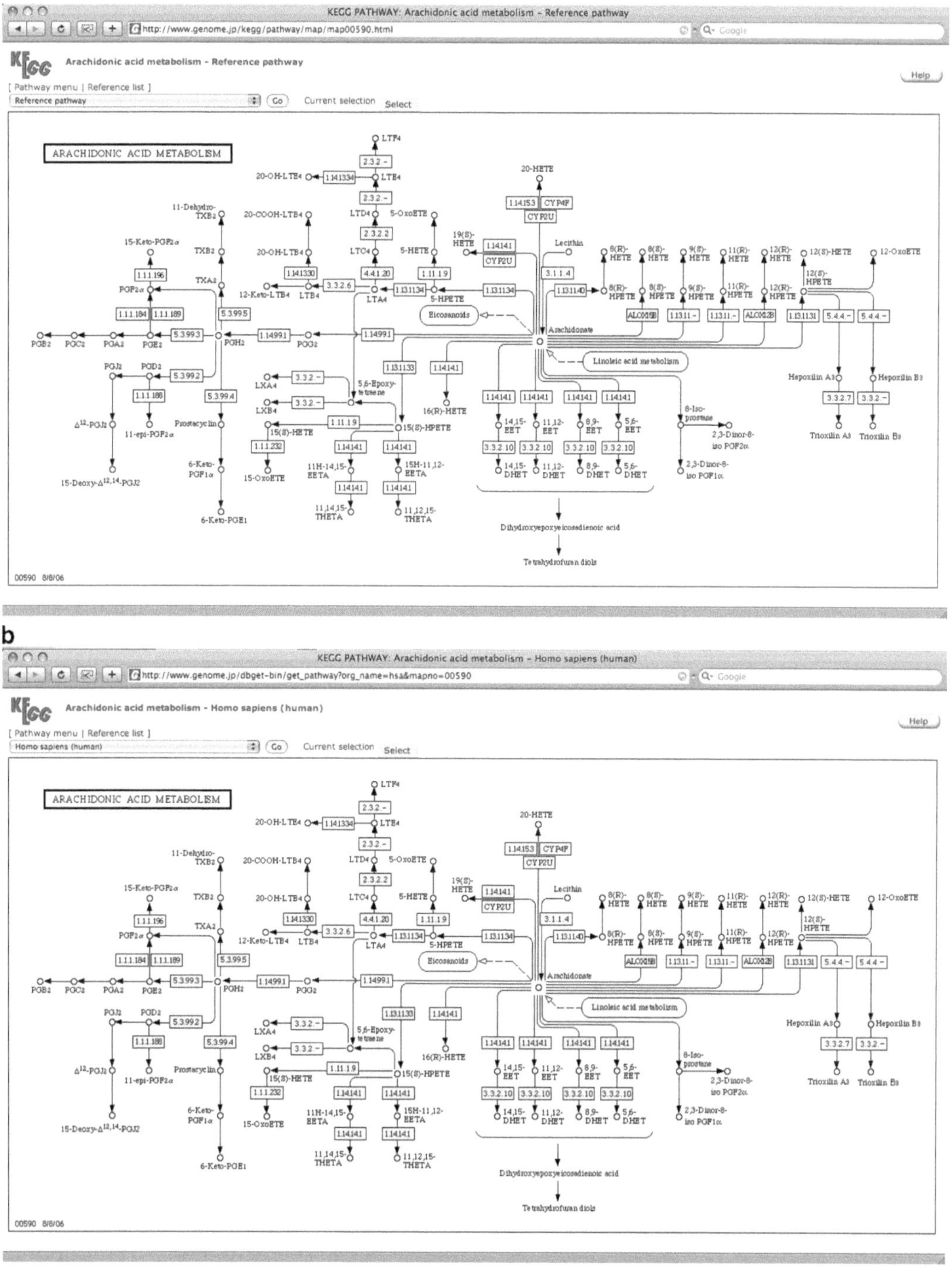

Fig. 2. Pathway diagram for arachidonic acid metabolism, the *boxes* and *circles* represent enzymatic reactions and chemical compounds, respectively. (**a**) KEGG reference pathway. (**b**) KEGG pathway for *Homo sapiens.*

families, and hierarchies for compounds, reactions, drugs, diseases, and organisms are available. The lipid classification system employed is based on that of LIPID MAPS *(1)*.

2.2.2. Searching Pathway Information

Besides a standard keyword search for the definitions such as "arachidonic acid metabolism" in the PATHWAY database, searching and custom coloring objects in the pathway diagrams can be performed (**Fig. 3a**). By selecting target organisms from pull-down menus and inputting a list of object ID and color pair, pathway diagrams including the specified objects (genes and/or chemical compounds) are listed with the hit objects. Clicking the pathway name on the list displays a colored pathway as shown for arachidonic acid metabolism in **Fig. 3b**. This functionality is useful for mapping so-called "omics" data onto pathways. Because pathway diagrams contain genes and proteins in addition to chemical compounds, KEGG can accept omics data input from a range of platforms including metabolomics, proteomics and transcriptomics data. A stand-alone Java application called KegArray has been developed to map omics data to pathway diagrams (Wheelock, et al., 2009). For transcriptomics or proteomics data, KegArray can use two different tab-delimited text data formats: precalculated ratios between a treated and control sample (ratios) or four raw data values representing total signal and background noise for treated and control samples, respectively (KEGG EXPRESSION format; see http://www.genome.jp/kegg/expression/). For metabolomics results, the data format is similar to that of gene/protein data; however, only ratio values can be used. KegArray also provides a mapping tool to genome maps and BRITE functional hierarchies that covers protein families and complexes whose pathway information is not available. KegArray is freely downloadable from http://www.genome.jp/download/

2.3. Genomic Information in KEGG

2.3.1. Browsing Orthology, Genes, and Genomes

Each box in the KEGG pathway diagrams represents several aspects of the objects (i.e., enzymes, orthologies, reactions, and genes). Among them, orthologies are the most important regarding the reconstruction of organism-specific pathway diagrams from the reference pathways. KO is a database for defining the function of each orthologue group and is used to annotate genes in the GENES database. For example, in the case of the box with "1.14.99.1" in **Fig. 2a**, the function is defined as "prostaglandin-endoperoxide synthase" with the orthologue ID K00509, E1.14.99.1, or PTGS (**Fig. 4**). The annotated genes with this function are listed in the "Genes" field with a three-letter organism code, hsa for human, ptr for chimpanzee, etc. (see below for detailed information regarding the three-letter codes).

Each gene ID in the "Genes" field is a link to the KEGG GENES database. **Figure 5** is the human PTGS1 entry.

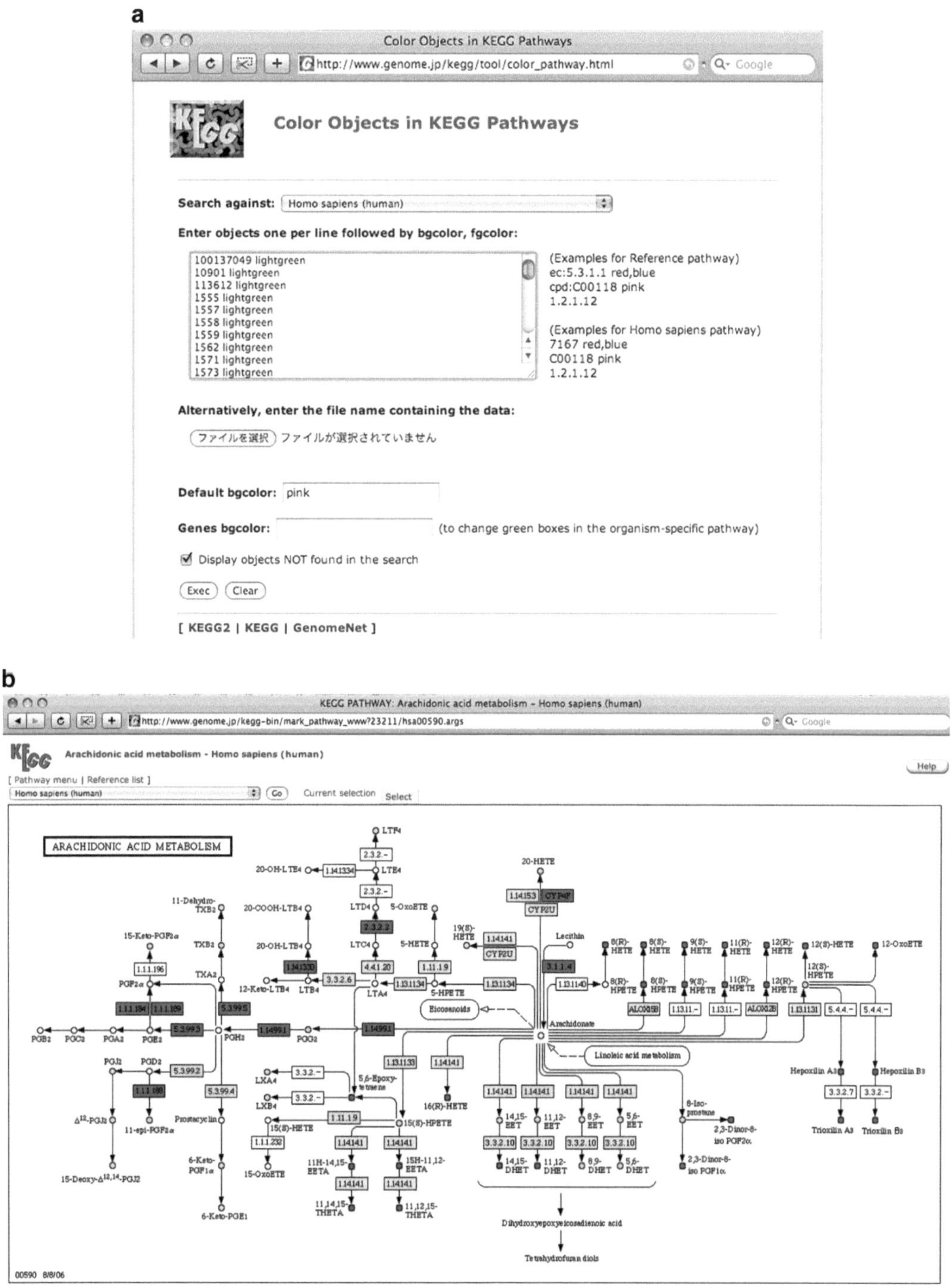

Fig. 3. Custom coloring of objects in KEGG PATHWAY diagrams can be performed. (**a**) Object ID and color pair, pathway diagrams including the specified objects (genes and/or chemical compounds) are listed with the hit objects. (**b**) Colored pathway for arachidonic acid metabolism.

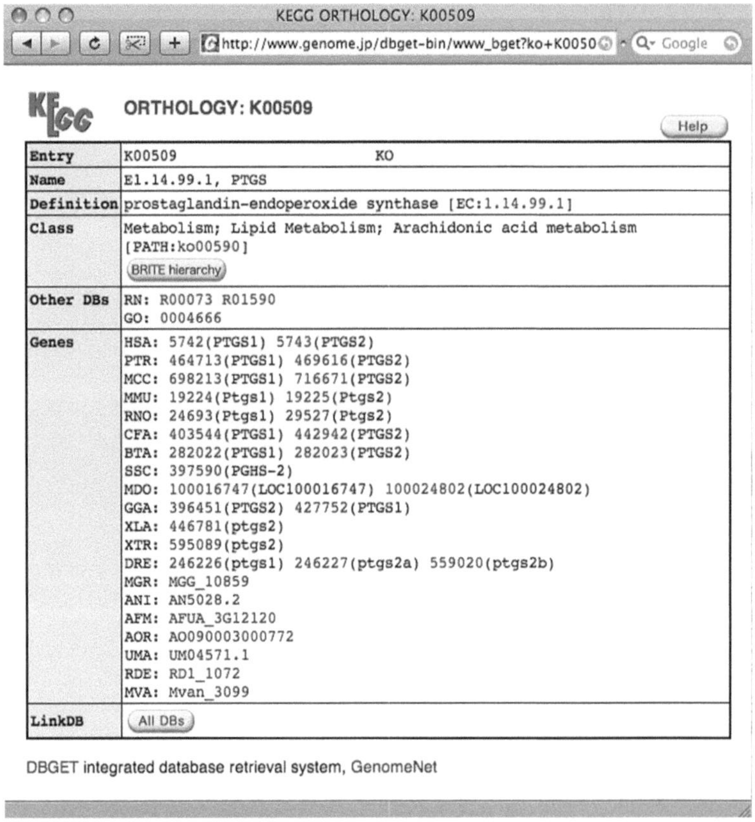

Fig. 4. KEGG orthology for "prostaglandin-endoperoxide synthase" with the orthologue ID K00509, E1.14.99.1, or PTGS. The annotated genes with this function are listed in the "Genes" field with a three-letter organism code, hsa for human, ptr for chimpanzee, etc.

Each entry in GENES contains ID, name, definition, links to other databases, chromosomal position, and amino acid and nucleotide sequences. The amino acid sequences in the GENES database are subject to all-versus-all sequence similarity search using a search program that implements the Smith-Waterman algorithm *(26)* and the results are stored in SSDB.

Similar sequences from other organisms are available from the "Ortholog" button in the SSDB field of a GENES entry (**Fig. 6**). In the same way, similar sequences from the organism in question can be obtained from the "Paralog" button. Relative to a simple homology search, the SSDB also provides a best-hit relations from a genome level comparison as shown in the "best" column in **Fig. 6**. The symbol "<->" indicates bidirectional best hit where the query is the most similar to the sequence in the target organism (cfa for dog as the top hit in **Fig. 6**) and vice versa. The number after the symbol indicates the total number of similar sequences in the target organism.

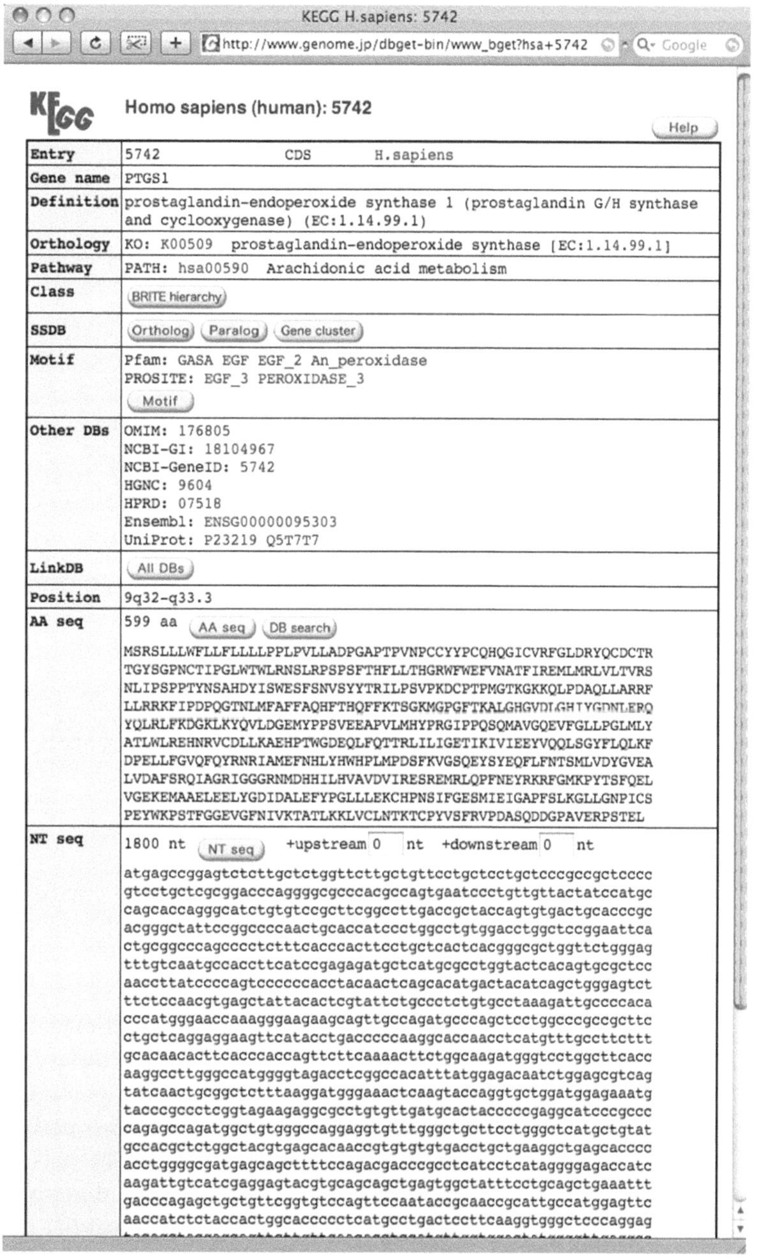

KEGG H.sapiens: 5742

http://www.genome.jp/dbget-bin/www_bget?hsa+5742

Homo sapiens (human): 5742

Entry	5742 CDS H.sapiens
Gene name	PTGS1
Definition	prostaglandin-endoperoxide synthase 1 (prostaglandin G/H synthase and cyclooxygenase) (EC:1.14.99.1)
Orthology	KO: K00509 prostaglandin-endoperoxide synthase [EC:1.14.99.1]
Pathway	PATH: hsa00590 Arachidonic acid metabolism
Class	BRITE hierarchy
SSDB	Ortholog Paralog Gene cluster
Motif	Pfam: GASA EGF EGF_2 An_peroxidase PROSITE: EGF_3 PEROXIDASE_3 Motif
Other DBs	OMIM: 176805 NCBI-GI: 18104967 NCBI-GeneID: 5742 HGNC: 9604 HPRD: 07518 Ensembl: ENSG00000095303 UniProt: P23219 Q5T7T7
LinkDB	All DBs
Position	9q32-q33.3
AA seq	599 aa AA seq DB search MSRSLLLWFLLFLLLLPPLPVLLADPGAPTPVNPCCYYPCQHQGICVRFGLDRYQCDCTR TGYSGPNCTIPGLWTWLRNSLRPSPSFTHFLLTHGRWFWEFVNATFIREMLMRLVLTVRS NLIPSPPTYNSAHDYISWESFSNVSYYTRILPSVPKDCPTPMGTKGKKQLPDAQLLARRF LLRRKFIPDPQGTNLMFAFFAQHFTHQFFKTSGKMGPGFTKALGHGVDLGHIYGDNLERQ YQLRLFKDGKLKYQVLDGEMYPPSVEEAPVLMHYPRGIPPQSQMAVGQEVFGLLPGLMLY ATLWLREHNRVCDLLKAEHPTWGDEQLFQTTRLILIGETIKIVIEEYVQQLSGYFLQLKF DPELLFGVQFQYRNRIAMEFNHLYHWHPLMPDSFKVGSQEYSYEQFLFNTSMLVDYGVEA LVDAFSRQIAGRIGGGRNMDHHILHVAVDVIRESREMRLQPFNEYRKRFGMKPYTSFQEL VGEKEMAAELEELYGDIDALEFYPGLLLEKCHPNSIFGESMIEIGAPFSLKGLLGNPICS PEYWKPSTFGGEVGFNIVKTATLKKLVCLNTKTCPYVSFRVPDASQDDGPAVERPSTEL
NT seq	1800 nt NT seq +upstream 0 nt +downstream 0 nt atgagccggagtctcttgctctggttcttgctgttcctgctcctgctcccgccgctcccc gtcctgctcgcggacccaggggcgcccacgccagtgaatccctgttgttactatccatgc cagcaccagggcatctgtgtccgcttcggccttgaccgctaccagtgtgactgcacccgc acgggctattccggccccaactgcaccatccctggcctgtggacctggctccggaattca ctgcggcccagcccctctttcacccacttcctgctcactcacgggcgctggttctgggag tttgtcaatgccaccttcatccgagagatgctcatgcgcctggtactcacagtgcgctcc aaccttatccccagtcccccacctacaactcagcacatgactacatcagctgggagtct ttctccaacgtgagctattacactcgtattctgccctctgtgcctaaagattgccccaca cccatgggaaccaaagggaagaagcagttgccagatgcccagctcctggcccgccgcttc ctgctcaggaggaagttcatacctgacccccaaggcaccaacctcatgtttgccttcttt gcacaacacttcacccaccagttcttcaaaacttctggcaagatgggtcctggcttcacc aaggccttgggccatggggtagacctcggccacatttatggagacaatctggagcgtcag tatcaactgcggctctttaaggatgggaaactcaagtaccaggtgctggatggagaaatg tacccgccctcggtagaagaggcgcctgtgttgatgcactacccccgaggcatcccgccc cagagccagatggctgtgggccaggaggtgtttgggctgcttcctgggctcatgctgtat gccacgctctggctacgtgagcacaaccgtgtgtgtgacctgctgaaggctgagcacccc acctggggcgatgagcagcttttccagacgacccgcctcatcctcataggggagaccatc aagattgtcatcgaggagtacgtgcagcagctgagtggctatttcctgcagctgaaattt gacccagagctgctgttcggtgtccagttccaataccgcaaccgcattgccatggagttc aaccatctctaccactggcacccctcatgcctgactccttcaaggtgggctcccaggag

Fig. 5. KEGG GENES entry for human PTGS1. Each entry in GENES contains ID, name, definition, links to other databases, chromosomal position, and amino acid and nucleotide sequences.

The organism name in the Entry field of a GENES entry is a link to the GENOME database (*Homo sapiens* in **Fig. 5**). As well as the literature, taxonomy, and statistics such as the number of genes, it provides an entry point for various information on the organism including pathway maps, gene catalogs in BRITE hierarchy,

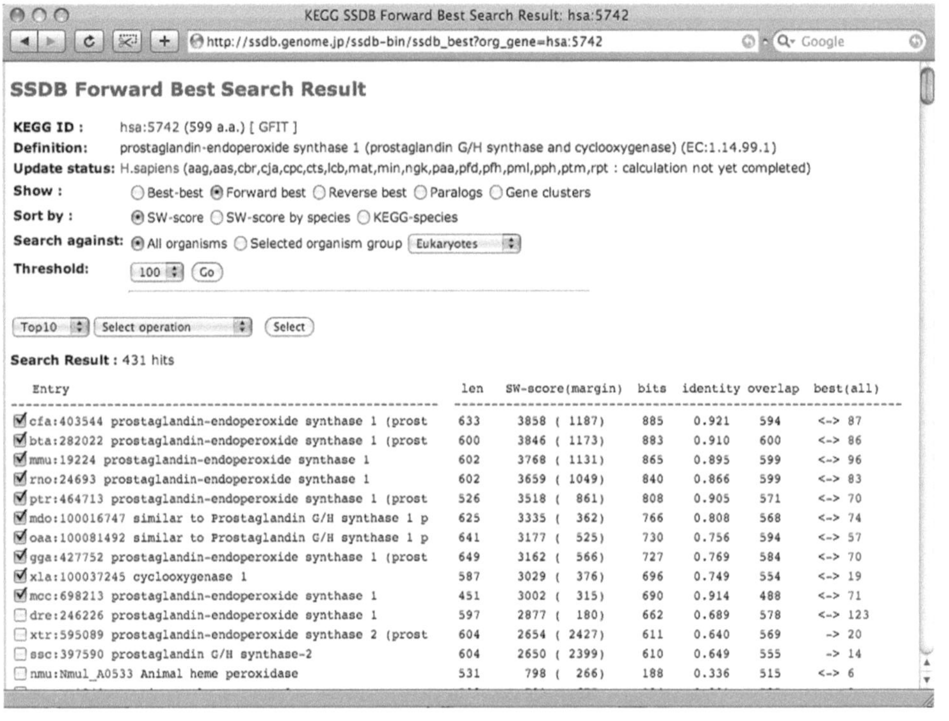

Fig. 6. KEGG SSDB (Sequence Similarity DataBase) search result for human PTGS1.

and genome maps if available. The full list of the organisms with their three-letter code is available by following the link "Organism list."

2.4. Chemical Information in KEGG

2.4.1. Browsing Chemical Information

Small circles in each KEGG pathway diagram represent chemical compounds and are clickable to show chemical structures. **Figure 7** is the COMPOUND entry for prostaglandin H2 that is linked to the arachidonic acid metabolism pathway in **Fig. 2**. Each compound entry contains the fields ID, Name, Formula, Mass (exact molecular mass), Structure (2-dimensional picture), and MDL/Mol file, as well as links to external databases including LIPID MAPS *(1)* and LipidBank *(16)*.

As shown in the previous section, (**Subheading 2.3**) each box in the pathway diagram represents a chemical reaction as well as the gene and orthology. The pathway diagram with the reaction links is available by selecting the "Reference pathway (Reaction)" menu from the pull-down menu and pressing the "Go" button. **Figure 8a** shows the REACTION entry obtained after clicking "1.14.99.1" of the pathway in **Fig. 2**. The reaction scheme is displayed with the structure diagram where the structural changes in the reaction can be seen.

Usually, an enzyme reaction transforms substrates to products in a local part of a chemical compound, and several reactions

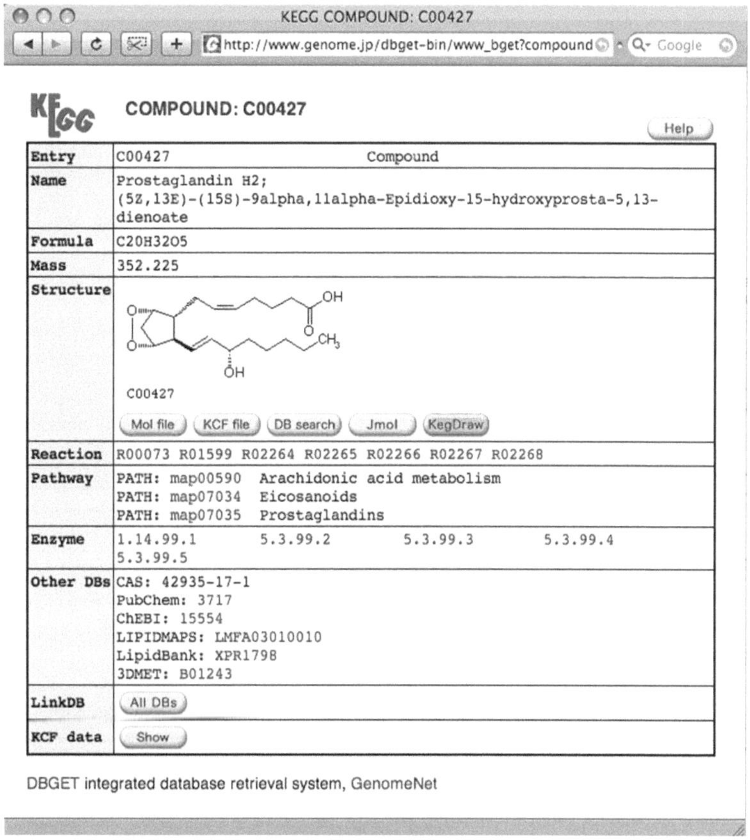

KEGG COMPOUND: C00427

http://www.genome.jp/dbget-bin/www_bget?compound

COMPOUND: C00427

Entry	C00427 Compound
Name	Prostaglandin H2; (5Z,13E)-(15S)-9alpha,11alpha-Epidioxy-15-hydroxyprosta-5,13-dienoate
Formula	C20H32O5
Mass	352.225
Structure	C00427 Mol file KCF file DB search Jmol KegDraw
Reaction	R00073 R01599 R02264 R02265 R02266 R02267 R02268
Pathway	PATH: map00590 Arachidonic acid metabolism PATH: map07034 Eicosanoids PATH: map07035 Prostaglandins
Enzyme	1.14.99.1 5.3.99.2 5.3.99.3 5.3.99.4 5.3.99.5
Other DBs	CAS: 42935-17-1 PubChem: 3717 ChEBI: 15554 LIPIDMAPS: LMFA03010010 LipidBank: XPR1798 3DMET: B01243
LinkDB	All DBs
KCF data	Show

DBGET integrated database retrieval system, GenomeNet

Fig. 7. KEGG COMPOUND entry for prostaglandin H2.

employ the same transformation rules in terms of the local changes between substrates and products. The transformation rules can be extracted automatically by aligning atoms between substrates and products, and KEGG provides such rules in the reactant pair (RPAIR) database, which has been manually curated following automatic extraction from the literature.

Figure 8b shows the RPAIR entry for the reaction in **Fig. 8a**. The structures of substrate and product are shown in KCF (KEGG Chemical Function) format, which is designed to represent atom types with environmental information. In the alignment diagram, the region aligned is colored either in green, orange, or red, and the one not aligned is in blue. We define the red atom as the reaction center because it is placed at the border of aligned and nonaligned regions, and those next to the reaction center as different atom and matched atom for nonaligned and aligned regions, respectively. This information is listed in the RDM field as the format "reaction center:different atoms:matched atoms." Related pair shows the other RPAIR entries with the same RDM, so that similar reactions can be easily identified.

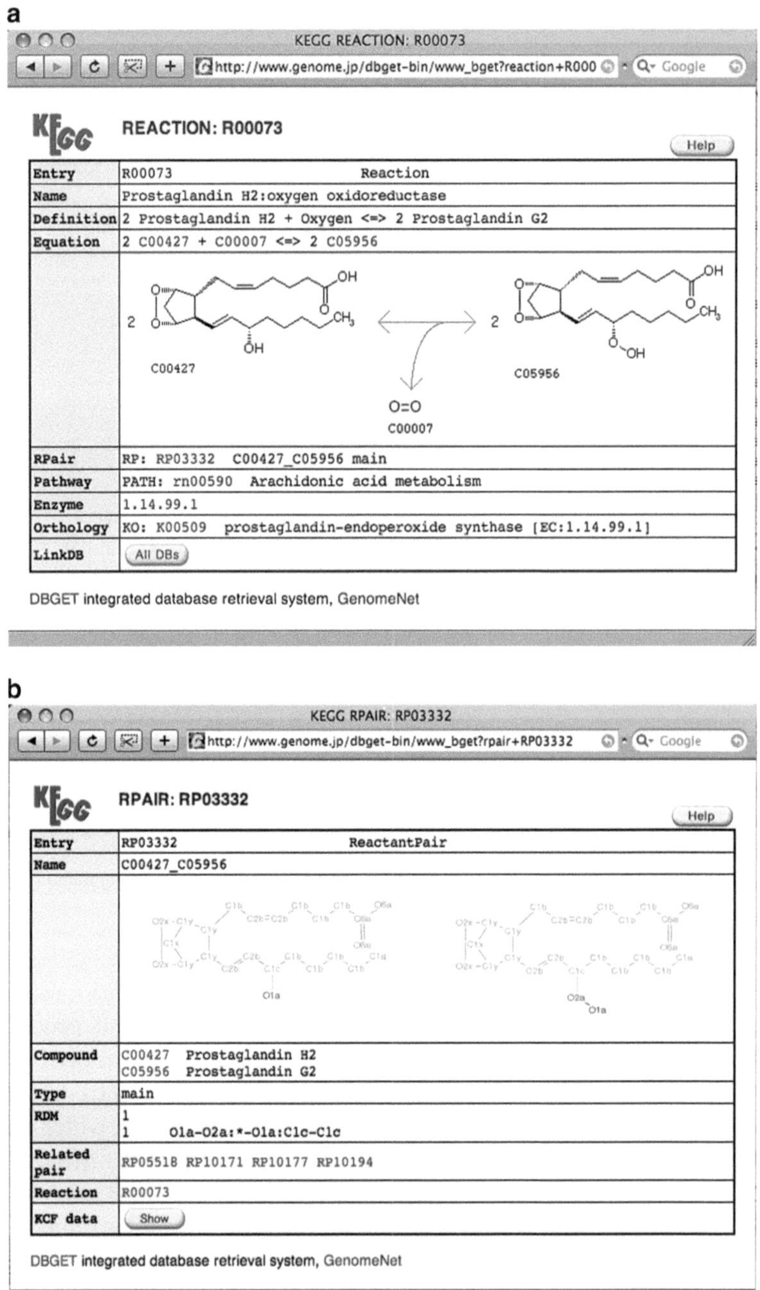

Fig. 8. (**a**) KEGG REACTION entry obtained for prostaglandin H2 synthase. The reaction scheme is displayed with the structure diagram where the structural changes in the reaction can be seen. (**b**) KEGG RPAIR entry for the reaction in (**a**). The structures of substrate and product are shown in KEGG Chemical Function (**KCF**) format, which is designed to represent atom types with environmental information.

2.4.2. Searching Chemical Information

The chemical information in KEGG is stored in the LIGAND composite database, which is maintained in a relational database as well as flat file database for DBGET search *(27)*. The query

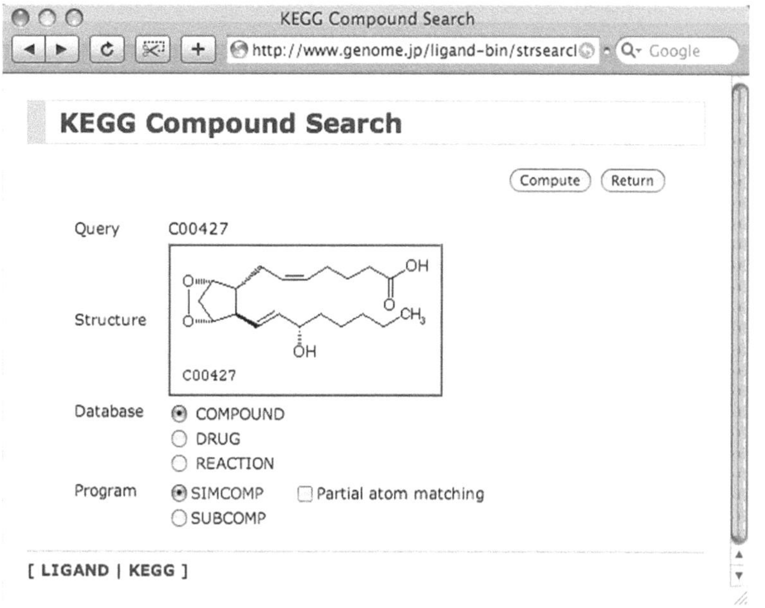

Fig. 9. Results of a structure search, where the programs SIMCOMP and SUBCOMP can be used for similar structure and substructure searches, respectively.

interface for the relational database is available for searching via keyword, formula, and molecular weight (exact mass). Another search facility to LIGAND is to search similar structures or substructures for the query chemical compound structure. By clicking the "DB search" button in a compound entry (e.g., **Fig.** 7), the query interface for the structure search is shown (**Fig. 9**), where SIMCOMP and SUBCOMP are programs for similar structure and substructure searches, respectively. **Figure 10** shows a SIMCOMP search result ordered according to the similarity score where the aligned part is colored. The main page for the compound search allows the input structure in the form of either MDL/MOL, SMILES (simplified molecular input line entry system), or KEGG COMPOUND ID. By using the stand-alone Java application KegDraw, the compound structure can be drawn and submitted to the main page.

3. KGML-ED and VANTED

3.1. Introduction

In addition to the tools directly available from the KEGG website, there are several software applications that interact with the KEGG database and provide additional analysis and visualization capabilities. In the following section, we introduce two tools: (1) KGML-ED, which allows the dynamic exploration and editing of KEGG

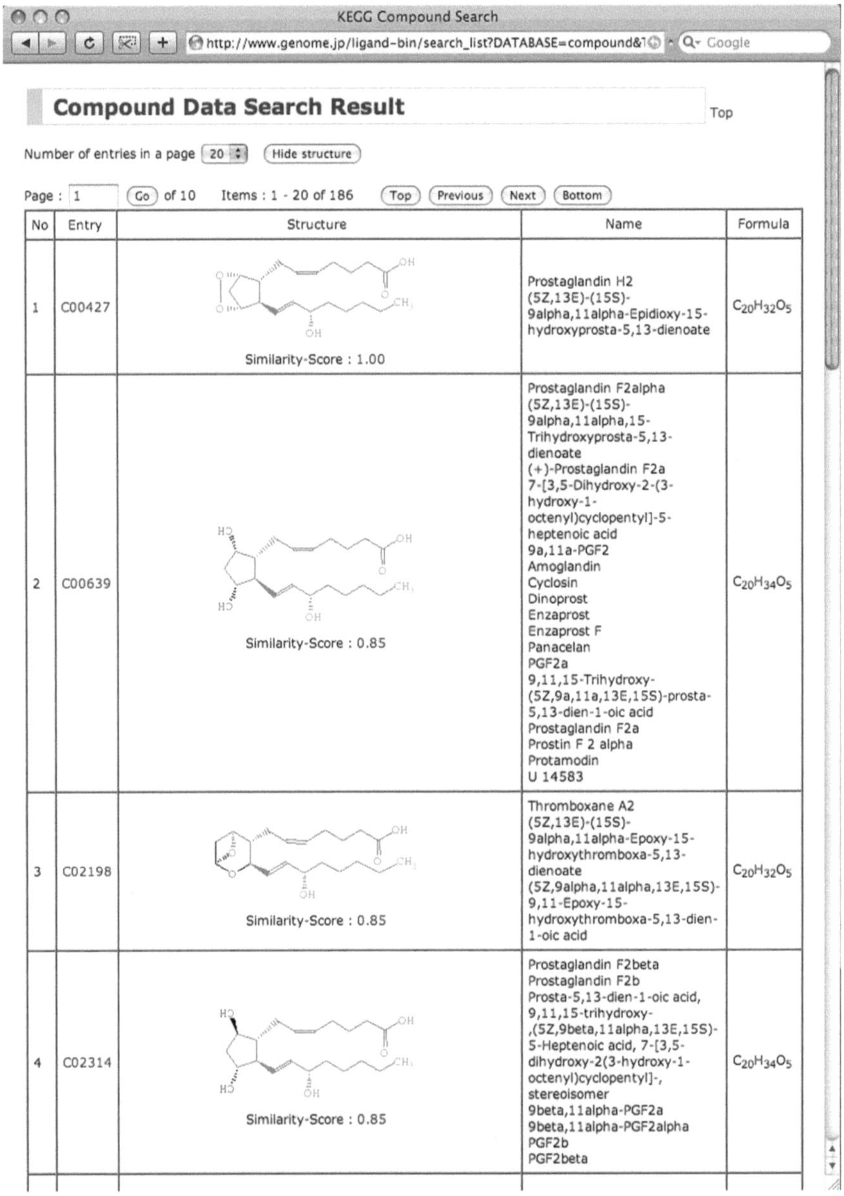

Fig. 10. Results of a SIMCOMP search (similar structure) ordered according to the similarity score.

diagrams and (2) VANTED, which supports a dynamic mapping of experimental high-throughput data onto KEGG pathways and the subsequent analysis of this context enriched data.

3.2. KGML-ED

KGML-ED is a graphical network editor that provides read- and write support for the KGML file format. The KEGG diagrams provide a huge amount of information. However, there are some restrictions that arise from the use of predefined diagrams to present the different pathways as shown in **Fig. 2**. KEGG PATHWAY diagrams only provide a static view of the data, which cannot be edited by the user. In addition, navigating between pathways is often restricted to links to other diagrams where the result of the navigation (the new diagram) replaces the current image. These limitations are addressed by KGML-ED (http://kgml-ed.ipk-gatersleben.de), which provides both editing and dynamic exploration of KEGG pathway diagrams *(28)*.

KGML-ED takes advantage of the KGML and can import, edit, and export files in this format. The structure as well as the attributes of pathways given as KGML files can be modified, deleted, or even newly created. For example, a reaction including its substrates, products, and connections to enzymes, as well as its visual attributes (e.g., position, coloring, and node sizes) can be changed in a consistent fashion. During the export process, the tool confirms that the new pathway is a correct KGML file.

KGML-ED supports advanced visualization and exploration methods for KEGG pathways. An overview network can be generated for either a given set of pathways or all pathways, where each node of the network represents a KEGG pathway and each edge represents the connection between pathways. For example, for all pathways, this is done by clicking on the "Create Pathway-Map-Overview" button in the side panel "Process" and subsequent selection of the species. This overview network can be extended by automatic replacement of a node (representing a particular pathway) with its complete pathway by double-clicking onto the node. The pathways are arranged automatically taking into account the well-known KEGG layout of single pathways. Pathway overlaps can be easily removed by using the menu command "Layout → Separate Pathway-Subgraphs." A set of pathways can be loaded by choosing a pathway category (e.g., "Lipid Metabolism") or by marking all relevant pathways in the panel "Load." All loaded pathways can be either combined into one pathway (in one window) or opened as separate pathways in different windows. **Figure 11** shows the overview network for all lipid metabolism pathways, with two extended pathways. It is also possible to extend pathways stepwise in KGML-ED. Starting with a given pathway, each link to another pathway can be replaced by this pathway, thereby building a pathway map of increasing complexity. Also the reverse function of collapsing pathways after a detailed investigation into an overview node is supported. As KGML-ED provides a complete dynamic access to pathways, it allows the user to custom-build specific pathways that can be exported as KGML-files or pictures. In conclusion, KGML-ED provides novel methods for the editing and dynamic exploration of KEGG pathway diagrams.

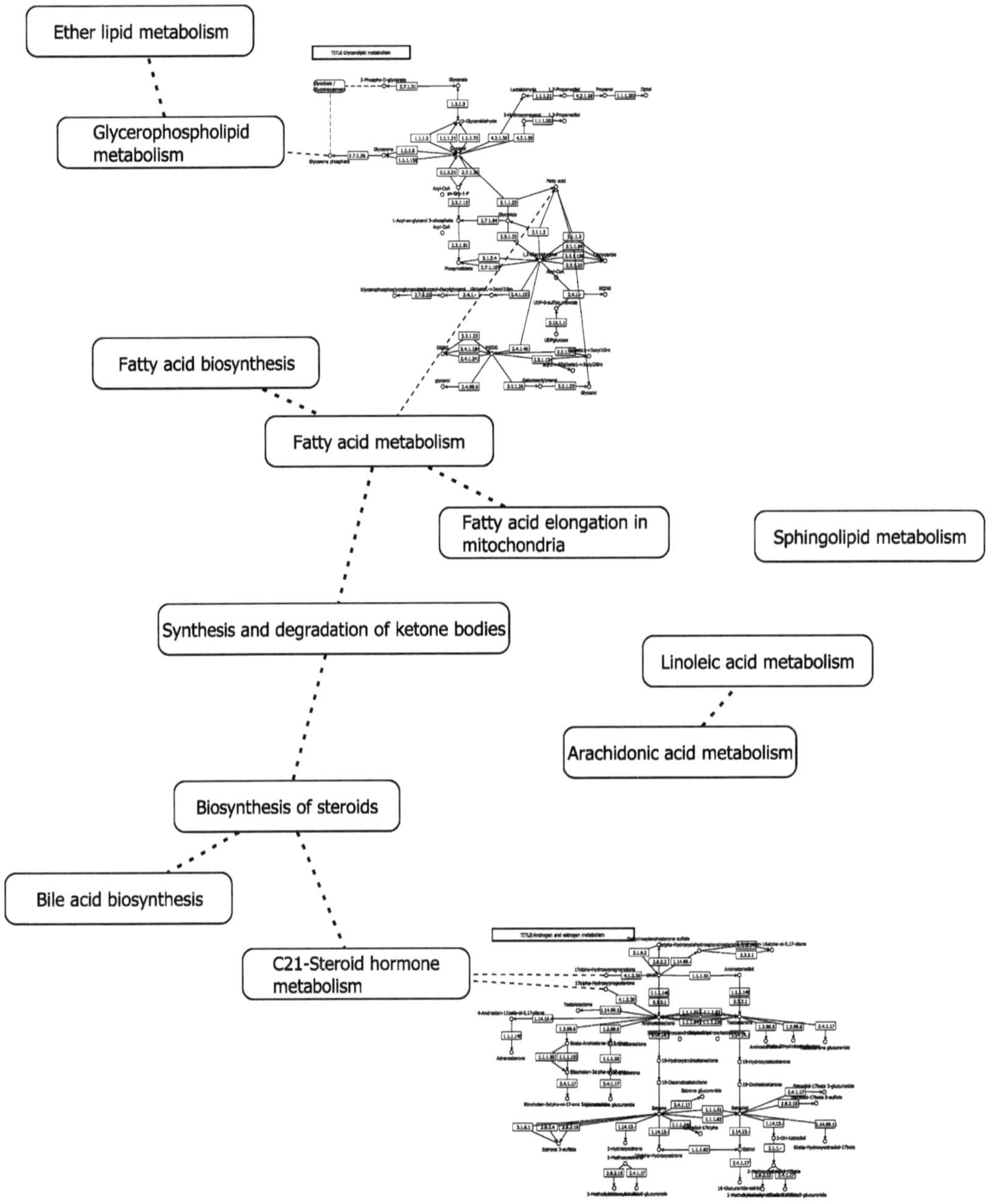

Fig. 11. Application of KGML-ED to KEGG lipid metabolism pathways with two extended KEGG pathways displayed ("glycerolipid metabolism" and "androgene and estrogen metabolism").

3.3. VANTED

The increase in systems-orientated research approaches is leading to the development of numerous pathway analysis and visualization software *(29)*. VANTED (http://vanted.ipk-gatersleben.de) is a tool for the visualization and analysis of networks with related experimental data that maps complex data sets (e.g., transcriptomics, proteomics, metabolomics, and fluxomics data) onto

KEGG pathways and other relevant biological networks *(30, 31)*. The application also provides a variety of functions for network editing, data processing, statistical analysis, and visualization.

Data from experiments can be uploaded into the software using different file formats such as KEGG EXPRESSION files, Microsoft Excel-based forms, and CSV (comma-separated values) files. For example, to access data from an Excel file the panel "Experiments" is used where the "Load Dataset" command allows the upload of data. The data can then be mapped onto pathways, for example, accessed from the KEGG PATHWAY database and analyzed in the pathway context. The complete KEGG hierarchy is easily accessible through VANTED by the panel "Pathways" and the subpanel "KEGG." Marking one or several pathways and clicking "Load Selected Pathways" loads the selected pathways. To allow multiple analyses of the data, the same large data set can be mapped onto different pathways. An example is shown in **Fig. 12**, where gene expression levels from a study on the influence of glucose repression *(32)* are mapped to a lipid pathway and shown inside the nodes in the form of a bar-chart. Red bars inside the diagrams show levels under glucose repression situation in comparison to the control (gray color). As all graphical

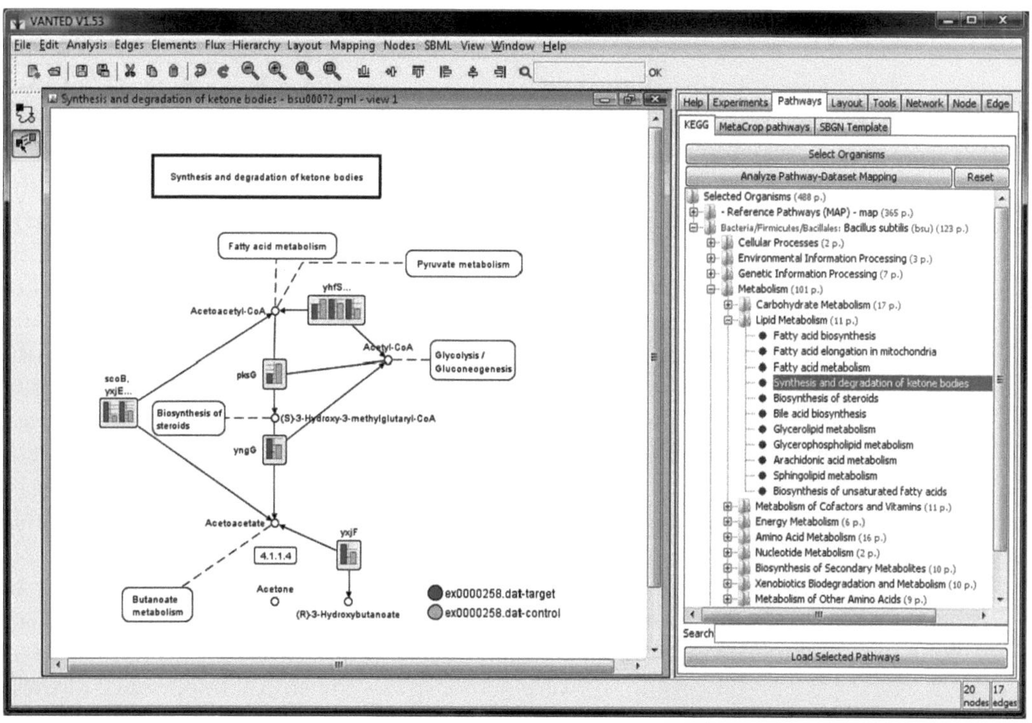

Fig. 12. VANTED screenshot, showing the KEGG pathway "synthesis and degradation of ketone bodies" for *Bacillus subtilis*. Gene expression levels have been mapped onto the pathway and shown inside the nodes in the form of a bar-chart (*dark gray bars*: level under glucose repression situation, *light gray bars*: level in control).

elements are dynamically changeable, any type of diagram can be incorporated into KEGG pathway nodes enabling the representation of multimodal and multidimensional data. Other available visualization options include color-coding (heatmap), pie-charts, and line-charts for time series data. VANTED has been used by the LIPID MAPS consortium to construct custom lipid pathways and to map experimental data *(33)*. In conclusion, VANTED supports the analysis and visualization of high-throughput data in the context of KEGG pathway diagrams.

4. MZmine, LipidDB, and Pathway Reconstruction Efforts

There are numerous ongoing efforts toward achieving the necessary lipid bioinformatics goals, some of which were discussed in detail in earlier sections. Emerging community-wide efforts such as LIPID MAPS provides several lipid research tools including universal nomenclature and a classification system for lipids and various mass spectrometry-based tools for the prediction of lipid mass spectra. Lu et al. *(34)* introduced a novel cognoscitive-contrast-angle algorithm and database to increase the accuracy of identifying lipid-mediators. Software tools such as LipidProfiler *(35)* enable automatic identification and multiple internal standard-based quantification for glycerophospholipids. Similarly, Kurvinen et al. *(36)* developed software that can process glycerolipid data from LC/MS analysis using prespecified parameters. Other open source software tools such as SECD and LIMSA enable the display of chromatograms and processing of mass spectrometry-based lipid data *(37)*. This suite of tools has been augmented by the recently introduced MZmine toolbox, which was designed to perform differential analysis of LC/MS data *(38)*. The following subsections describe the data processing tool, MZmine (**Subheading 4.1**), an in silico database (LipidDB) in (**Subheading 4.2**), and propose a strategy for biochemical pathway reconstruction of lipids (**Subheading 4.3**) *(39)*.

4.1. MZmine-Based Data Processing

The analytical technologies employed for lipidomics studies enable the simultaneous screening of hundreds of lipids and are capable of rapidly generating large data sets. Accordingly, the quality of automated data processing is a crucial step in the bioinformatics pipeline for lipidomics. The recent increase in interest in data processing software demonstrates the critical importance of these tools. We have developed an open source Java-based data processing tool called MZmine that implements all the key processing steps for metabolomics data, with a particular focus on managing lipidomics data obtained from LC/MS-based experiments *(40)*.

MZmine software offers several key methods for data processing stages that allow differential analysis of LC/MS lipidomics data across multiple samples. The data processing stages include spectral filtering, peak detection, alignment and normalization. Additionally, recursive peak search algorithm and peak picking methods are implemented for improving already aligned results. Different visualization methods are implemented for comparative display of data across multiple samples. The latest version, MZmine2 beta (http://mzmine.sourceforge.net/) provides additional features including an improved toolbox framework for better expandability, methods for processing high-resolution instrument data, a new 3D visualiser, and new implementations of chromatogram, spectra, and 2D visualiser, as well as storage and project parameters for defining sample properties. **Figure 13** shows a screen shot of the MZmine2 toolbox. The modular framework of the MZmine2 toolbox offers greater flexibility for the incorporation of new algorithms, which is an important characteristic for customization.

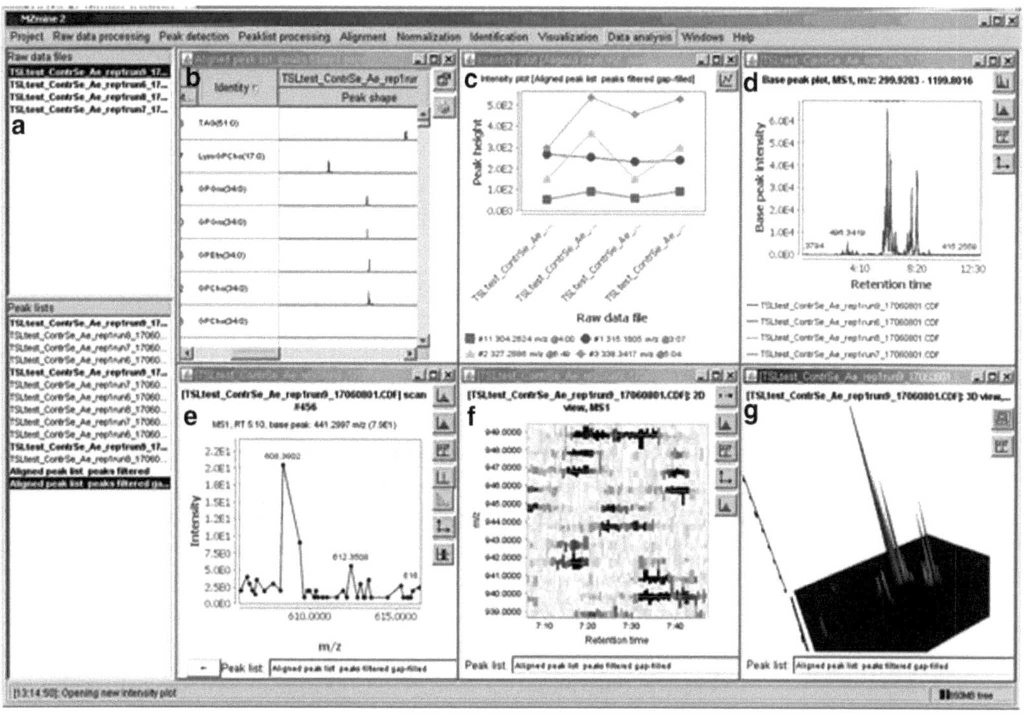

Fig. 13. Screen shot of MZmine2 toolbox (http://mzmine.sourceforge.net/), with the different functionalities shown: (**a**) sample list, (**b**) extracted ion chromatograms, (**c**) selected ion peak heights or areas across multiple samples, (**d**) base peak intensity across multiple samples, (**e**) individual data points for a selected scan, (**f**) 2D plot of LC/MS data, with retention time on **x**-axis and **m/z** ratio on **y**-axis, and (**g**) 3D plot of LC/MS data, with peak heights on the **z**-axis.

4.2. LipidDB: In Silico Spectral Libraries

Improved analytical methods and matured data processing software tools require the construction of reference spectral libraries to facilitate identification at the systems level. Currently available lipid resources such as LIPID MAPS provide the necessary information on lipids and serve as guidelines for lipid databases. In silico LipidDB comprises the main lipid classes including glycerolipids, glycerophospholipids, sphingolipids, and sterol esters. These classes may vary structurally in terms of one or more fatty acid moieties and the head group. In order to facilitate automatic identification of lipids from lipidomics experiments, the most likely occurring fatty acids are utilized as "seeds" to computationally generate the theoretical spectra of lipids. The choice of seeds can be modified to generate representative lipids depending on the experimental design and area of interest.

A SMILES representation is used to store fatty acid seeds. SMILES is an easily understandable linear indexing system in terms of atoms and bonds that follows specific grammar rules. It allows the construction of lipid classes and their substructures by exploiting the class-specific structure of each class. A canonical version of SMILES representation is generated for each lipid using the Daylight SMILES toolkit (http://www.daylight.com). Isotopic distributions are computed for each lipid and lipid nomenclature complies with the recommendations of the LIPID MAPS consortium.

4.3. Reconstruction Efforts of Lipid Pathways

Biochemical pathway mapping is a fundamental requirement to gain mechanistic understanding of the underlying phenotype as well as achieving in-depth insights by pathway modeling and simulation studies. This is feasible with the construction of databases that serve as knowledge bases for biochemical pathways, development of data processing software tools, and strategies for pathway mapping mechanisms. Properly organized databases enable the viewing and/or construction of biochemical pathways. At present, several databases containing information on different pathway levels are available. The KEGG PATHWAY database hosts information on the majority of well-known metabolic pathways, including lipid pathways for several organisms as described in **Subheading 2.1**. The information depicted as biochemical pathway maps includes metabolic reactions, underlying enzymes, and metabolites as discussed above. The KEGG database also provides organism-independent biochemical pathways that serve as reference pathways for constructing organism-specific pathways allowing users to select those enzymes on reference pathways that are relevant to the organism of interest. Other notable databases such as MetaCyc *(41)* provide similar useful information. The available databases therefore provide the necessary basic building blocks for pathway research, which can be combined with newly developed lipid-specific databases.

Modern biochemical research consists of mature analytical techniques that have evolved to provide substantial quantities of information. Accordingly, existing databases need to reflect these changes and be able to comply with this new influx of information. This necessity is particularly acute in the case of lipids. While lipid analytical techniques are capable of providing detailed information at the lipid species level, current lipid pathway information is mostly restricted to the general lipid class level. At this level, the structure of many lipids on biochemical pathways contains one or more fatty acids and head groups. Given the possible number of fatty acids and head groups that are substitutable for that particular entry, lipid pathway reconstruction can easily lead to combinatorial explosion. The complexity varies from pathway to pathway. The bioinformatics approaches should therefore meet these challenges to enable context-specific studies of lipid pathways. We propose here a strategy for pathway reconstruction using MZmine, LipidDB, and other pathway mapping strategies utilizing biochemical pathway templates from the KEGG PATHWAY database.

As a starting point towards such a context-specific pathway reconstruction studies, the proposed strategy bridges the gap between modern analytical knowledge and existing lipid pathways in connection with lipidomics data *(39)*. Every lipid entry can be linked to information on reference lipid pathways and experimental information such as retention time that is necessary for peak assignment from lipidomics experiments. As a practical step toward avoiding this combinatorial problem in pathway mapping, generic pathway templates are utilized to create molecular instance pathways for molecular species selected based on multivariate and coregulation analyses.

Figure 14 shows the essential building blocks involved in the molecular pathway instantiation. We use the following steps for lipid pathway reconstruction: (1) select the appropriate biochemical pathway for the metabolite of interest from standard sources such as KEGG, (2) convert generic lipid names (e.g., phosphatidylcholine) on a given biochemical pathway template to systematic subclass names (e.g., 1-acyl-2-acyl-sn-glycero-3-phosphocholine), (3) construct XML schema for a given pathway using a network visualization tool called megNet *(42)*, (4) create an XML document for the instance pathway based on the lipid of interest and gene expression data if available, and (5) visualize the pathway using megNet. Other visualization tools such as VANTED (*see* **Subheading 3.3**) enable the analysis and manipulation of experimental data under multiple conditions such as time series data. These applications will be extremely important for mapping temporal fluctuations in lipid levels and highlight a limitation of the majority of biochemical pathway tools.

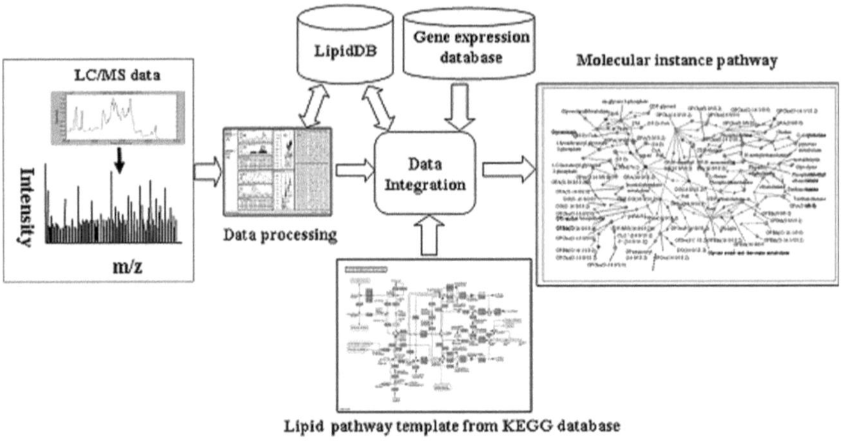

Fig. 14. Workflow of lipid pathway instantiation: data processing software, spectral database, gene expression data, KEGG PATHWAY template and data integration and visualization tools.

5. Conclusions

Recently, a number of different bioinformatics tools for the analysis and interpretation of lipidomics data have become available. However, this field is still in its infancy and further advances are needed. A particular challenge in the pathway reconstruction of lipid molecular species is understanding how individual lipids affect underlying biological phenotypes. The majority of existing pathway tools (e.g., KEGG, Ingenuity, and MetaCore) are not capable of mapping individual lipid species from different lipid classes. For example, KEGG contains a pathway for "Sphingolipid metabolism"; however, this pathway is focused mainly on the biosynthesis of the different sphingolipid components (e.g., ceramide, sphingosine, and sphingomyelin). The pathway cannot account for the different fatty acid species of the individual components. For example, KEGG shows that the enzyme ceramide cholinephosphotransferase (E.C. 2.7.8.3) forms sphingomyelin from ceramide, but the individual fatty acid species is only displayed as "R," making the mapping of "sphingolipidomics" data on the KEGG impossible. The Sphingo MAP consortium (**Table 1**) has identified more than 450 species to date, demonstrating an obvious limitation in current pathway reconstruction efforts. It is necessary to develop bioinformatics techniques capable of analyzing these large and increasingly complex datasets. Efforts such as LIPID MAPS that seek to develop a systematic and universal classification and nomenclature system for individual lipid species will be extremely important for these efforts. However, further developments are clearly needed for context-specific pathway reconstruction. The available tools and databases including

MZmine, LipidDB, LipidBank, LIPID MAPS, and pathway mapping strategies such as VANTED and KEGG PATHWAY will provide the basis for comprehensive pathway reconstruction, but further advances are required. In addition, altered biochemical pathways are not the only cause of lipid concentration changes. The measured lipid concentrations in fact reflect regulation at multiple spatial and dynamic scales, including, for example, systemic lipid metabolism, global changes in cell membrane composition, or lipid oxidation. The inherent difficulty of accounting for such complexity in the analysis of lipidomics data will remain a formidable challenge as well as a research opportunity for some time to come.

Acknowledgments

This research was supported by the Åke Wibergs Stiftelse, the Fredrik and Ingrid Thurings Stiftelse, and The Royal Swedish Academy of Sciences. C.E.W. was supported by a fellowship from the Centre for Allergy Research.

References

1. Fahy E, Subramaniam S, Brown HA, et al. (2005) A comprehensive classification system for lipids. *J Lipid Res* 46, 839–861.
2. van Meer G. (2005) Cellular lipidomics. *Embo J* 24, 3159–3165.
3. Wenk MR. (2006) Lipidomics of host-pathogen interactions. *FEBS Lett* 580, 5541–5551.
4. Adibhatla RM, Hatcher JF. (2007) Role of lipids in brain injury and diseases. *Future Lipidol* 2, 403–422.
5. Wymann MP, Schneiter R. (2008) Lipid signalling in disease. *Nat Rev Mol Cell Biol* 9, 162–176.
6. Wenk MR. (2005) The emerging field of lipidomics. *Nat Rev Drug Discov* 4, 594–610.
7. Mattila I, Seppanen-Laakso T, Suortti T, Oresic M. (2008) Application of lipidomics and metabolomics to the study of adipose tissue. *Methods Mol Biol* 456, 123–130.
8. Jia L, Wang C, Zhao S, Lu X, Xu G. (2007) Metabolomic identification of potential phospholipid biomarkers for chronic glomerulonephritis by using high performance liquid chromatography-mass spectrometry. *J Chromatogr B Analyt Technol Biomed Life Sci* 860, 134–140.
9. Roberts LD, McCombie G, Titman CM, Griffin JL. (2008) A matter of fat: an introduction to lipidomic profiling methods. *J Chromatogr B Analyt Technol Biomed Life Sci* 871, 174–181.
10. Han X, Gross RW. (2005) Shotgun lipidomics: electrospray ionization mass spectrometric analysis and quantitation of cellular lipidomes directly from crude extracts of biological samples. *Mass Spectrom Rev* 24, 367–412.
11. Yetukuri L, Ekroos K, Vidal-Puig A, Oresic M. (2008) Informatics and computational strategies for the study of lipids. *Mol Biosyst* 4, 121–127.
12. Baker CJ, Kanagasabai R, Ang WT, Veeramani A, Low HS, Wenk MR. (2008) Towards ontology-driven navigation of the lipid bibliosphere. *BMC Bioinformatics* 9 Suppl 1, S5.
13. Schmelzer K, Fahy E, Subramaniam S, Dennis EA. (2007) The lipid maps initiative in lipidomics. *Methods Enzymol* 432, 171–183.
14. van Meer G, Leeflang BR, Liebisch G, Schmitz G, Goni FM. (2007) The European lipidomics initiative: enabling technologies. *Methods Enzymol* 432, 213–232.

15. Taguchi R, Nishijima M, Shimizu T. (2007) Basic analytical systems for lipidomics by mass spectrometry in Japan. *Methods Enzymol* 432, 185–211.
16. Watanabe K, Yasugi E, Oshima M. (2000) How to search the glycolipid data in LIPID-BANK for Web: the newly developed lipid database. *Japan Trend Glycosci Glycotechnol* 12, 175–184.
17. Fahy E, Sud M, Cotter D, Subramaniam S. (2007) LIPID MAPS online tools for lipid research. *Nucleic Acids Res* 35, W606–612.
18. Sud M, Fahy E, Cotter D, et al. (2007) LMSD: LIPID MAPS structure database. *Nucleic Acids Res* 35, D527–532.
19. Cotter D, Maer A, Guda C, Saunders B, Subramaniam S. (2006) LMPD: LIPID MAPS proteome database. *Nucleic Acids Res* 34, D507–510.
20. Kanehisa M, Araki M, Goto S, et al. (2008) KEGG for linking genomes to life and the environment. *Nucleic Acids Res* 36, D480–484.
21. Kanehisa M. (1997) A database for post-genome analysis. *Trends Genet* 13, 375–376.
22. Fleischmann RD, Adams MD, White O, et al. (1995) Whole-genome random sequencing and assembly of Haemophilus influenzae Rd. *Science* 269, 496–512.
23. Goto S, Nishioka T, Kanehisa M. (1998) LIGAND: chemical database for enzyme reactions. *Bioinformatics* 14, 591–599.
24. Goto S, Okuno Y, Hattori M, Nishioka T, Kanehisa M. (2002) LIGAND: database of chemical compounds and reactions in biological pathways. *Nucleic Acids Res* 30, 402–404.
25. Hashimoto K, Yoshizawa AC, Okuda S, Kuma K, Goto S, Kanehisa M. (2008) The repertoire of desaturases and elongases reveals fatty acid variations in 56 eukaryotic genomes. *J Lipid Res* 49, 183–191.
26. Smith TF, Waterman MS. (1981) Identification of common molecular subsequences. *J Mol Biol* 147, 195–197.
27. Fujibuchi W, Goto S, Migimatsu H, et al. (1998) DBGET/LinkDB: an integrated database retrieval system. *Pac Symp Biocomput* 683–694.
28. Klukas C, Schreiber F. (2007) Dynamic exploration and editing of KEGG pathway diagrams. *Bioinformatics* 23, 344–350.
29. Suderman M, Hallett M. (2007) Tools for visually exploring biological networks. *Bioinformatics* 23, 2651–2659.
30. Junker BH, Klukas C, Schreiber F. (2006) VANTED: a system for advanced data analysis and visualization in the context of biological networks. *BMC Bioinformatics* 7, 109.
31. Klukas C, Junker BH, Schreiber F. (2006) The VANTED software system for transcriptomics, proteomics and metabolomics analysis. *J Pestic Sci* 31, 289–292.
32. Yoshida K, Kobayashi K, Miwa Y, et al. (2001) Combined transcriptome and proteome analysis as a powerful approach to study genes under glucose repression in *Bacillus subtilis. Nucleic Acids Res* 29, 683–692.
33. Fahy E, Cotter D, Byrnes R, et al. (2007) Bioinformatics for lipidomics. *Methods Enzymol* 432, 247–273.
34. Lu Y, Hong S, Serhan C. (2006) Lipid mediator informatics-lipidomics: novel pathways in mapping resolution. *AAPS Journal* 8, E284–E297.
35. Ejsing CS, Duchoslav E, Sampaio J, et al. (2006) Automated identification and quantification of glycerophospholipid molecular species by multiple precursor ion scanning. *Anal Chem* 78, 6202–6214.
36. Kurvinen J-P, Aaltonan J, Kuksis A, H. Kallio. (2002) Software algorithm for automatic interpretation of mass spectra of glycerolipids. *Rapid Commun Mass Spectrom* 16, 1812–18201.
37. Hermansson M, Uphoff A, Kakela R, Somerharju P. (2005) Automated quantitative analysis of complex lipidomes by liquid chromatography/mass spectrometry. *Anal Chem* 77, 2166–2175.
38. Katajamaa M, Orešič M. (2005) Processing methods for differential analysis of LC/MS profile data. *BMC Bioinformatics* 6, 179–190.
39. Yetukuri L, Katajamaa M, Medina-Gomez G, Seppanen-Laakso T, Vidal-Puig A, Orešič M. (2007) Bioinformatics strategies for lipidomics analysis: characterization of obesity related hepatic steatosis. *BMC Systems Biology* 1, 12.
40. Katajamaa M, Miettinen J, Orešič M. (2006) MZmine: toolbox for processing and visualization of mass spectrometry based molecular profile data. *Bioinformatics* 22, 634–636.
41. Krieger CJ, Zhang P, Mueller LA, et al. (2004) MetaCyc: a multiorganism database of metabolic pathways and enzymes 10.1093/nar/gkh100. *Nucl Acids Res* 32, D438–442.
42. Gopalacharyulu PV, Lindfors E, Bounsaythip C, et al. (2005) Data integration and visualization system for enabling conceptual biology. *Bioinformatics* 21, i177–185.
43. Wheelock CE, Wheelock AM, Kawashima S, Diez D, Kanehisa M, van Erk M, Kleemann R, Haeggström JZ, Goto S. (2009) Systems biology approaches and pathway tools for investigating cardiovascular disease. *Mol Biosyst.* 5(6), 588–602.

Part IV

Biostatistics

Chapter 20

The Effect of Lipid Adjustment on the Analysis of Environmental Contaminants and the Outcome of Human Health Risks

Audrey J. Gaskins and Enrique F. Schisterman

Summary

Past literature on exposure to lipophilic agents such as organochlorines (OCs) is conflicting, posing challenges for the interpretation of their potential human health risks. Since blood is often used as a proxy for adipose tissue, it is necessary to model serum lipids when assessing health risks of OCs. Using a simulation study, we evaluated four statistical models (unadjusted, standardized, adjusted, and two-stage) for the analysis of polychlorinated biphenyls (PCBs) exposure, serum lipids, and health outcome risk. Eight candidate true causal scenarios, depicted by directed acyclic graphs, were used to illustrate the ramifications of misspecification of underlying assumptions when interpreting results. Biased results were produced when statistical models that deviated from the underlying causal assumptions were used with the lipid standardization method found to be particularly prone to bias. We concluded that investigators must consider biology, biological medium, laboratory measurement, and other underlying modeling assumptions when devising a statistical model for assessing health outcomes in relation to environmental exposures.

Key words: Causal modeling, Directed acyclic graphs, Risk estimation, Serum lipids, Organochlorines, Polychlorinated biphenyls

1. Introduction

When assessing potential human health risks, persistent lipophilic xenobiotics pose particular methodological challenges. Current literature on exposure to these lipophilic agents such as organochlorines (OCs) is ambiguous which further impairs the ability to quantify these risks *(1–4)*. Serum OC concentrations are dependent on serum lipid concentrations *(5, 6)* but only

Donald Armstrong (ed.), *Lipidomics*, Methods in Molecular Biology, vol. 580,
DOI 10.1007/978-1-60761-325-1_20, © Humana Press, a part of Springer Science+Business Media, LLC 2009

under certain circumstances, when equilibrium is reached, can information regarding serum OC and serum lipid levels be predictive of the overall OC body burden *(7)*. Higher serum lipid levels should correspond to higher serum OC concentrations but serum OC concentrations and lipids are affected postprandial, so both must be considered in relation to the quantity and timing of food consumption *(8)*. Serum (or plasma) samples are frequently used due to the difficulty of collecting adipose tissue but these samples can introduce methodological challenges to estimating health risks particularly when using nonfasting samples *(9)*. The alternative, fasting samples, has potentially larger drawbacks, hampering the feasibility of epidemiological research and adversely impacting study participation. Thus, to avoid the large drawbacks with fasting samples and methodological challenges posed by nonfasting samples, further attention should be focused on nonfasting serum samples' relation to serum lipids *(5, 7, 10)*.

Our limited understanding of the true relation between serum and adipose tissue concentrations of lipophilic xenobiotics to serum lipids and health outcomes makes model specification difficult *(11, 12)*. To overcome this, investigators typically make assumptions about the relation between serum lipids and serum OCs expressing OC measurements as a wet-weight, lipid-weight, or lipid-standardization value. Lipid standardization (OC concentration per gram of fat) is particularly useful for comparing exposure concentrations across tissue specimens or study populations *(13)*. Lipid weight (OC per unit of serum lipids) is advocated as superior to wet weight (OC per unit serum) in the measurement of persistent lipophilic chemicals *(7)*, especially when assuming body burden equilibrium. Other approaches include the use of a log-linear model with serum lipids included as a separate term in the regression equation *(14)* and a two-stage analysis where serum lipids are regressed on serum OCs with residuals entered as individual risk factors *(2)*. The best way to model the relationship among serum OCs, lipids, and health outcomes remains an understudied area critical for the assessment of health effects. In this analysis, we demonstrate the impact of model (mis)specification and its potential effects on the interpretation of study findings.

2. Materials

2.1. Computer Programs

The SAS software package (SAS Institute, Cary, NC, USA) was used for all simulations and analyses.

3. Methods

3.1. Directed Acyclic Graphs

Optimal modeling of the statistical relations among serum OCs, serum lipids, and health outcomes requires conceiving a causal model that reflects the following considerations: (a) biologic plausibility, (b) laboratory capability for quantifying compounds and lipids, (c) underlying statistical assumptions (e.g., error structure), and (d) other relevant study covariates (e.g., known and potential confounders). A single-headed arrow represents a causal relation, a dashed line represents a noncausal association, and the absence of an arrow signifies no relation between two variables. In our scenarios, the hypothetical "causal truths," are based on the literature and their relation to frequently used models. For bias calculations, we assume perfect laboratory measurements of OCs and the absence of unmeasured confounders. The eight directed acyclic graphs (DAGs) are illustrated in **Fig. 1** with polychlorinated biphenyls (PCBs) chosen to exemplify the role of OCs. These eight DAGS are also described in detail below.

1. PCB and SL are marginally dependent conditional on Y; serum PCB (S-PCB) causes Y, and SL causes Y.
2. PCB is a cause of Y; S-PCB causes Y, independent of SL.
3. PCB and Y are marginally dependent on and blocked by SL; S-PCB causes SL, which causes Y.
4. Y and SL are marginally dependent and blocked by serum PCB; S-PCB causes Y and SL.
5. PCB and SL are marginally dependent conditional on both the shared ancestor variable, A, and Y. An unmeasured variable, A, causes both S-PCB and SL, each of which independently causes the outcome. This is the traditional situation of

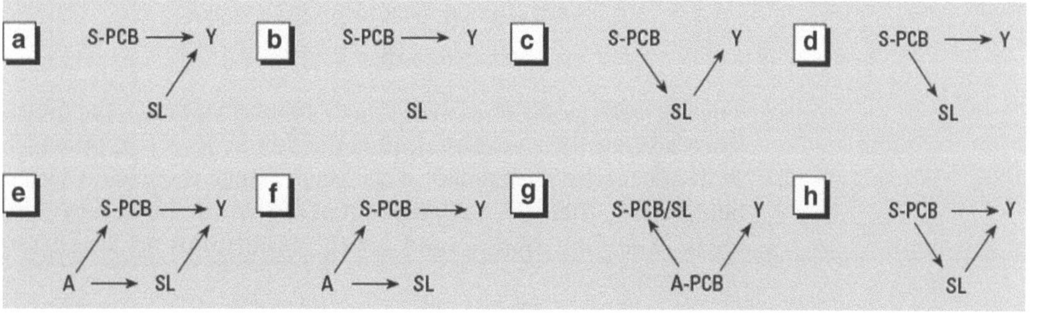

Fig. 1. Causal scenarios for relations among polychlorinated biphenyl (*PCB*), serum lipids (*SL*), and outcome (*Y*)

confounding, with SL acting as a confounder of the relation between serum PCBs, PCBs, and Y.

6. PCB and SL are marginally dependent on the ancestor, A; SL and Y are marginally dependent on A and, thus effectively, on PCB. S-PCB and SL are caused by A, but only PCB is causally related to Y.
7. PCB per unit SL and Y are marginally dependent conditional on adipose tissue PCB. Adipose tissue PCB (A-PCB) causes serum PCB per unit serum lipid and causes Y; PCB and outcome are correlated rather than directly causally related.
8. Blocked and unblocked path. Y is both directly caused by PCB and marginally dependent conditional on SL; S-PCB causes Y, as well as SL, which causes Y.

3.2. Statistical Models

We investigated four statistical models for the analysis of hypothesized PCB exposure, serum lipids, and a health outcome along with eight plausible DAGs for each model. All models assume that there are no unmeasured confounders. The basis of all the models is

$$P = \Pr(Y = 1 \mid X, \mathrm{SL})$$

where Y is the dichotomous-dependent variable representing the presence/absence of the disease, X the PCB, and SL the serum lipids.

3.3. Simulations

A simulation study was conducted to evaluate the utility of the various models for the different scenarios depicted by the DAGs. Using the causal structures, lognormal distributions were assigned for PCB and serum levels and a binomial outcome variable, Y, was assumed with Pr (Y = 1|PCB, serum lipids). Using **Fig. 1h** as an example, the given associations motivate the model

$$\ln(\mathrm{SL}) = \alpha_0 + \gamma[\ln(X)]$$

$$\begin{aligned} \mathrm{logit}(P) &= \boldsymbol{\alpha}_1 + \boldsymbol{\beta}_1 \ln(\mathrm{X}) + \boldsymbol{\beta}_2\{E[\ln(SL) \mid X]\} \\ &= \boldsymbol{\alpha}_0 + \boldsymbol{\alpha}_1 + (\boldsymbol{\beta}_1 + \boldsymbol{\beta}_2\boldsymbol{\gamma})[\ln(X)] \end{aligned} \quad (1)$$

The log odds [logit $P(X, \mathrm{SL})$] equals an intercept (α_0), the prevalence among the unexposed, plus the factor, $\beta_1 + \beta_2\gamma$, by which PCB affects the probability of the event. Since there is no serum lipids term, there is no linear influence from the serum lipid levels. The four models used for the simulations are listed and described below.

1. **Unadjusted Model**. This model is equivalent to the use of wet-weight values when estimating the effect of an exposure

such as PCBs on health outcomes without consideration of serum lipids.

$$\text{logit}(P) = \alpha_1 + \beta_1 \ln(x) \quad (2)$$

This model is only suitable with the assumption that serum lipids are not a confounder regardless of the relation between lipids and the outcome. The inclusion or exclusion of lipids as an adjustor may affect model fit but it will not impact PCB exposure/response estimates.

2. **Standardized model.** This model is one way to account for the effect of serum lipids on serum OC levels by dividing the serum concentration of PCBs by serum lipids. The basis of this model is

$$\text{logit}(P) = \alpha_2 + \beta_2 \ln\left(\frac{X}{\text{SL}^m}\right) = \alpha_2 + \beta_2[\ln(X) - m \times \ln(SL)] \quad (3)$$

where the power, m, is a factor that generalizes the relation of PCBs and serum lipids.

3. **Adjusted Model**. In this model, there is an assumption that PCBs are not standardized for serum lipids. This is reflected in the absence of an association between lipids and the study outcome. The basis of this model is

$$\text{logit}(P) = \alpha_3 + \beta_3 \ln(X) + \beta_4 \ln(\text{SL}) \quad (4)$$

The standardized model is a member of the family of adjusted models and in general, is applicable under the same set of assumptions. Comparing the lipid component in the standardized model $[\ln(X) - m \times \ln(\text{SL})]$ with that in the adjusted $[\beta_4 \ln(\text{SL})]$ demonstrates that equivalent results are produced when β_4 is set equal to $-m$. Because of the added β-coefficient, the adjusted model is generally more flexible than the standardized model.

4. **Two-stage model**. This model includes the effects of both PCBs and serum lipids on the outcome:

$$\ln(\text{SL}) = \alpha + \beta_5 \ln(X) + R$$

$$\text{logit}(P) = \alpha_4 + \beta_6 \ln(X) + \beta_7 \times (R) \quad (5)$$

Both the intercept and the β-coefficient are simple functions of the parameters from the adjusted model and the regression of serum lipids on log PCBs. The coefficient for the residual term, R, is also precisely that of the adjusted model's lipids term:

$$\alpha_4 = \alpha_3 - \beta_4\alpha$$

$$\beta_6 = \beta_3 - \beta_5\beta_4$$

$$\beta_7 = \beta_4$$

To further evaluate the efficiency of the models, the effects that the serum lipid measurement error and the strength of linear relationship have on the outcome were assessed. Measurement error was set to $[\varepsilon \sim N(0,\sigma_e^2)]$ with differing values of σ_e^2. The relation between PCB and serum lipids was analyzed by varying α from the linear regression equation, SL = $\alpha_0 + \alpha X$. In these quantitative representations of DAGs, the magnitude of effects, error, and bias are functions of the values chosen for the parameters. The independent effect of PCB was set as a constant ($\beta_{\ln \text{PCB}}$ = 0.6 in the logistic regression model) with approximate values taken from literature ***(15)***. In the unpublished data, there was a significant linear relation between total serum PCBs and serum lipids with regression coefficient of approximately 0.3. These values represented the strength of the linear relation between PCB while serum lipids values represented a range from very weak association (α = 0.01) to strong association (α = 2.0).

3.4. Results

Table 1 displays the bias and mean square error when using $\sigma_e^2 = 1$ and α = 0.3 as the underlying casual truths for the four statistical models in each DAG scenario.

For **Fig. 1a** that represents PCB and SL as independent causes of the outcome, all the models except the standardized produce minimally biased estimates. The standardized model, instead, results in a largely underestimated bias of the PBC effect on outcome. In **Fig. 1b**, SL is completely extraneous and thus the bias occurs similarly to the previous situation in **Fig. 1a**. **Figure 1c** depicts a scenario where the effect of PCB acts strictly through SL. It was found to be best represented by the two-stage approach. The unadjusted model produced minimal bias but the adjusted and particularly the standardized model resulted in large underestimates of bias (99% and 351%, respectively). This particular scenario was the one instance where the adjusted model produced extremely large bias compared to its performance on the other seven models where bias was kept to almost a fourth of its value.

When SLs are affected by PCBs, but do not directly influence the outcome, as depicted in **Fig. 1d**, standardization is the only modeling approach with substantial bias, underestimating the true effect by nearly 80% compared to the others which are within 1% of the true effect. In **Fig. 1e**, the confounded case, only the adjusted model performed with minimal bias. The lack of adjustment in the other models failed to address the SL confounder and

Table 1
Percent bias of estimates of effect of PCBs on outcome for evaluated statistical models

DAG[b]	Percent bias (MSE)[a] Unadjusted	Standardized	Adjusted	Two-stage
A	1.2 (1.26)	-51.3 (10.3)	1.8 (1.28)	1.8 (1.28)
B	-0.8 (1.34)	-75.9 (21.1)	-.07 (1.35)	-0.7 (1.33)
C	-15.4 (2.78)	-351.3 (161.1)	-99.4 (1.59)	1.1 (2.78)
D	0.4 (1.14)	-79.8 (23.3)	0.8 (1.17)	0.5 (1.14)
E	24.0 (3.37)	-128.8 (60.3)	0.1 (1.39)	27.2 (3.37)
F	-0.4 (1.29)	-85.0 (26.4)	-0.1 (1.41)	-0.3 (1.29)
G	-86.3 (27.0)	-1.0 (1.51)	-1.0 (1.51)	-85.9 (27.0)
H	-11.2 (1.75)	-128.3 (59.7)	-25.4 (3.65)	-8.7 (1.75)

[a]Mean square error multiplied by 100 for illustration (shown in parentheses)
[b]*See* Fig. 1
Serum lipid measurement error distributed normally with mean 0; variance 1; α (strength of linear relation between log PCB and log serum lipids) = 0.3; 500 repetitions; n = 1,000

further indicated that the standardization, when present, was not a sufficient method to account for this confounder. The two-stage model fails because in adjusting for serum lipids via the residuals, the model misattributes the association between PBC and SL as a causal link. This ultimately results in biased estimates of the effect of interest-the total effect of PCB on risk.

Similar to **Fig. 1a**, **b**, and **d**, using the standardized model for **Fig. 1f** produced biased underestimates much larger than those from the other three models. This can be attributed to the noncausal correlation between PCB and SL that is represented in the DAG. **Figure 1g** depicts serum levels of PCB as being dependent on the adipose levels of PCB which are then casually related to the outcome. Given this situation, the standardization model, which up till now had produced substantial bias in all scenarios, functioned optimally. The adjusted model resulted in similar unbiased estimates while neither the unadjusted nor two-stage model worked well. **Figure 1h** represents a direct and indirect causal link of PCB with outcome. PCB not only affects the outcome, it indirectly affects SL which indirectly affects the outcome as well. This relationship was not represented well by any of the models with the least biased estimates resulting from the two-stage (which separates total into estimated direct and indirect) and unadjusted (which estimates total effect) models.

After comparing the standardized and adjusted models over all eight causal scenarios, with the exception of the **Fig. 1g**, the adjusted model consistently produced smaller bias than the standardized model. Even under ideal conditions for the standardized model (as depicted in **Fig. 1g**), the adjusted model produced a nearly identical, unbiased estimate. The two-stage and unadjusted models produced similar results, except in the case of **Fig. 1c** where the two-stage yielded substantially less bias.

1. **Measurement Error**

The potential measurement error accompanying the quantification of serum lipids was addressed through the use of an error term with mean 0 and variance σ_e^2 which was added to the simulated distribution of serum lipids. **Figures 2–4** display bias as a function of error at four values of α for each of the models. The bias as a function of σ_e^2 followed three distinct patterns among the eight DAGs.

The first pattern was displayed by **Fig. 1a, b, d**, and **f** and is shown in **Fig. 2**. In this pattern, bias was stable for the unadjusted, adjusted, and two-stage models consistently staying close to zero. Only for the standardized model was the relation between bias and σ_e^2 more complicated. In this model, bias increased with measurement error when the relation between PCB and lipids was weak (low α) but decreased with measurement error when the relation between the two variables was strong (higher α). The transition between this increase and decrease (the inflection point) occurred at a value of σ_e^2 that ranged from 0.5 for **Fig. 1f** to 3.0 for **Fig. 1a**.

The second pattern was displayed by **Fig. 1c, e**, and **h** and is show in **Fig. 3**. Similar to the first pattern described above, the bias for the standardized model varied in a nonlinear manner, increasing for all values of α except for the highest ($\alpha = 2$). The adjusted and two-stage models were essentially robust to measurement

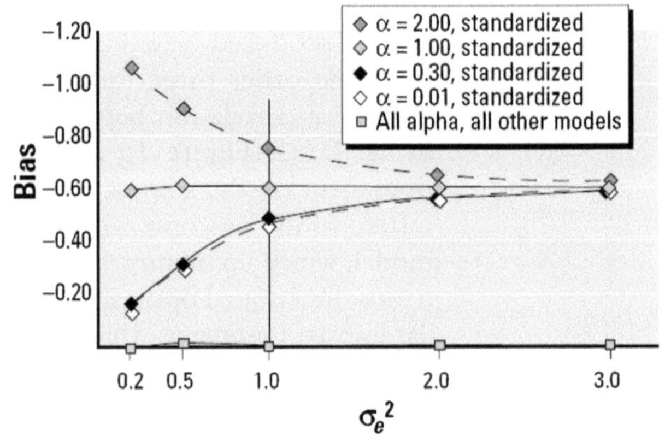

Fig. 2. Comparison of bias for standardization versus all other models as a function of measurement error of serum lipids and strength of linear association of polychlorinated biphenyl (*PCB*) with serum lipids for **Fig. 1a, b, d,** and **f**. Bias for the standardized model was systematically centered on −0.60 (100% underestimation). The *vertical line* at σ_e^2 signifies the level used for **Table 1**.

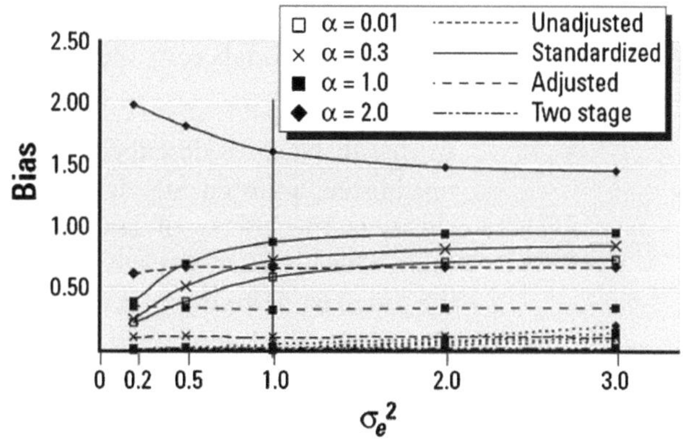

Fig. 3. Bias as a function of measurement error of serum lipids and strength of linear association of polychlorinated biphenyl (*PCB*) with serum lipids for **Fig. 1c, e,** and **h**. The *vertical line* at σ_e^2 signifies the level used for **Table 1**.

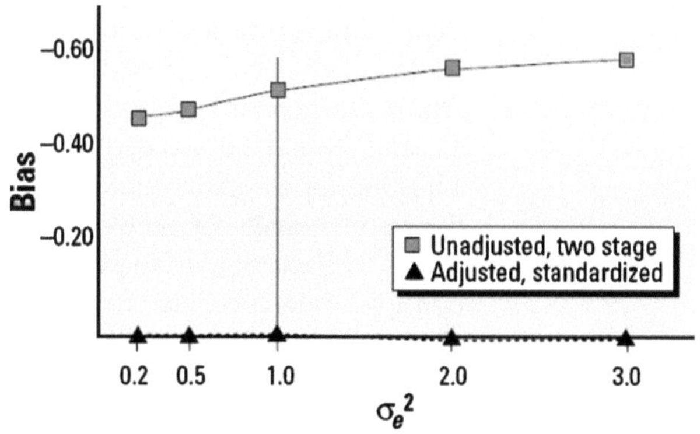

Fig. 4. Bias as a function of measurement error of serum lipids and strength of linear association of polychlorinated biphenyl (*PCB*) with serum lipids for **Fig. 1g**. The *vertical line* at σ_e^2 signifies the level used for **Table 1**.

error; however, they did not always produce unbiased estimates of parameters for all underlying DAGs. This was particularly apparent at different levels of α . In the adjusted model, a stronger relation between PCB and lipids (higher α) resulted in greater bias. The bias of estimates produced by the unadjusted model varied slightly with σ_e^2 depending on the DAG being modeled. For **Fig. 1c** and **h**, the bias increased slightly with increasing measurement error (from 0 to 0.1 for σ_e^2 = 0.8 and from 0 to 0.2 for σ_e^2). For **Fig. 1e**, the bias decreased with increasing measurement error as the strength of the noncausal relation between PCBs and serum lipids was altered by the variance in lipids.

The third pattern was displayed by **Fig. 1g** and is shown in **Fig. 4**. In this pattern, both the standardized and adjusted models produced unbiased estimates robust to measurement error. The unadjusted and two-stage model produced biased estimates that were equally prone to measurement error. Regardless of the strength of the linear relation between PCB and lipids (α), the bias for all four of the models remained unchanged in this scenario.

2. **Application**

In this analysis, we described and evaluated four statistical models – unadjusted, adjusted, standardized, and two-stage – commonly used to assess the effects of lipophilic environmental contaminants on human health. Each statistical model showed minimal bias for at least the causal truth for which it was ideally suited. Every model except for the standardized performed well in all but one scenario. The standardized model, on the other hand, produced large biases for most of the DAGs evaluated and even produced similar biases to the adjusted model for the DAG in which standardization is optimal.

The basic causal scenarios, depicted in the eight DAGs, included only two to four factors which impact levels of both PCB and serum lipids. When additional factors are considered, the evaluation becomes much more complex and the trade-off between efficiency and robustness becomes more important. Even though in our simulation, the adjusted model produced consistently unbiased estimates, there are circumstances where adjustment is inappropriate and should be avoided. Examples of these situations include adjusting for a collider (an effect of two or more other variables in the graph), which has been demonstrated to bias estimators of effect *(16, 17)*. Factors that share a common cause will also give large bias appearing correlated in strata of that common cause. Given an alleged relation between PCB and serum lipids, adjusting for these factors might generate spurious associations if an unmeasured factor is related to both serum lipid levels and the outcome.

Our simulations demonstrated that statistical models that fail to uphold the underlying assumptions about causality lead to biased results. This bias can have negative implications on the interpretation of effects of exposures on human health end points. Equivocal findings may arise in part from the varying laboratory and analytic approaches for specifying serum lipids when using nonfasting blood specimens to estimate risk. Investigators should remember to consider biology, biological medium, and laboratory methodology when specifying a statistical model. They should also take caution to make sure the model's underlying assumptions are appropriate for the study.

4. Notes

1. Bias was stable for the unadjusted, adjusted, and two-stage models consistently staying close to zero. Only for the standardized model was the relation between bias and variance more complicated.

2. The adjusted model consistently produced smaller bias than the standardized model. Even under ideal conditions for the standardized model (DAG G), the adjusted model produced a nearly identical, unbiased estimate.
3. The two-stage and unadjusted models produced similar results except in the case of DAG C where the two-stage yielded substantially less bias.

References

1. Calle EE, Frumkin H, Henley SJ, Savitz DA, Thun MJ. Organochlorines and breast cancer risk. CA Cancer J Clin 2002 Sep;52(5):301–9.
2. Hunter DJ, Hankinson SE, Laden F, Colditz GA, Manson JE, Willett WC, et al. Plasma organochlorine levels and the risk of breast cancer. N Engl J Med 1997 Oct 30;337(18):1253–8.
3. Laden F, Collman G, Iwamoto K, Alberg AJ, Berkowitz GS, Freudenheim JL, et al. 1,1-Dichloro-2,2-bis(p-chlorophenyl)ethylene and polychlorinated biphenyls and breast cancer: combined analysis of five U.S. studies. J Natl Cancer Inst 2001 May 16;93(10):768–76.
4. Laden F, Hankinson SE, Wolff MS, Colditz GA, Willett WC, Speizer FE, et al. Plasma organochlorine levels and the risk of breast cancer: an extended follow-up in the Nurses' Health Study. Int J Cancer 2001 Feb 15;91(4):568–74.
5. Eyster JT, Humphrey HE, Kimbrough RD. Partitioning of polybrominated biphenyls (PBBs) in serum, adipose tissue, breast milk, placenta, cord blood, biliary fluid, and feces. Arch Environ Health 1983 Jan;38(1):47–53.
6. Guo YL, Emmett EA, Pellizzari ED, Rohde CA. Influence of serum cholesterol and albumin on partitioning of PCB congeners between human serum and adipose tissue. Toxicol Appl Pharmacol 1987 Jan;87(1):48–56.
7. Brown JF, Jr., Lawton RW. Polychlorinated biphenyl (PCB) partitioning between adipose tissue and serum. Bull Environ Contam Toxicol 1984 Sep;33(3):277–80.
8. Phillips DL, Smith AB, Burse VW, Steele GK, Needham LL, Hannon WH. Half-life of polychlorinated biphenyls in occupationally exposed workers. Arch Environ Health 1989 Nov;44(6):351–4.
9. Whitcomb BW, Schisterman EF, Buck GM, Weiner JM, Greizerstein H, Kostyniak PJ. Relative concentrations of organochlorides in adipose tissue and serum among reproductive age women. Environ Toxicol Pharmacol 2005;19:203–13.
10. Brown JF, Jr., Lawton RW, Morgan CB. PCB metabolism, persistence, and health effects after occupational exposure: implications for risk assessment. Chemosphere 1994 Nov;29(9–11):2287–94.
11. Calvert GM, Willie KK, Sweeney MH, Fingerhut MA, Halperin WE. Evaluation of serum lipid concentrations among U.S. workers exposed to 2,3,7,8-tetrachlorodibenzo-p-dioxin. Arch Environ Health 1996 Mar;51(2):100–7.
12. Mussalo-Rauhamaa H. Partitioning and levels of neutral organochlorine compounds in human serum, blood cells, and adipose and liver tissue. Sci Total Environ 1991 Apr 15;103(2–3):159–75.
13. Morgan DP, Roan CC. Chlorinated hydrocarbon pesticide residue in human tissues. Arch Environ Health 1970 Apr;20(4):452–7.
14. Moysich KB, Ambrosone CB, Vena JE, Shields PG, Mendola P, Kostyniak P, et al. Environmental organochlorine exposure and postmenopausal breast cancer risk. Cancer Epidemiol Biomarkers Prev 1998 Mar;7(3):181–8.
15. Wolff MS, Toniolo PG. Environmental organochlorine exposure as a potential etiologic factor in breast cancer. Environ Health Perspect 1995 Oct;103(Suppl 7):141–5.
16. Greenland S, Brumback B. An overview of relations among causal modelling methods. Int J Epidemiol 2002 Oct;31(5):1030–7.
17. Hernan MA, Hernandez-Diaz S, Werler MM, Mitchell AA. Causal knowledge as a prerequisite for confounding evaluation: an application to birth defects epidemiology. Am J Epidemiol 2002 Jan 15;155(2):176–84.

INDEX

C

D

E

F

M

N

O

P

R

S